Signaling and Communication in Plants

Series editor
František Baluška
Department of Plant Cell Biology, IZMB, University of Bonn, Kirschallee 1,
D-53115 Bonn, Germany

More information about this series at http://www.springer.com/series/8094

Christine M.F. Vos • Kemal Kazan
Editors

Belowground Defence Strategies in Plants

Springer

Editors
Christine M.F. Vos
CSIRO Agriculture Flagship
St Lucia, Queensland
Australia

Scientia Terrae Research Institute
Sint-Katelijne-Waver, Belgium

Kemal Kazan
CSIRO Agriculture Flagship
St Lucia, Queensland
Australia

ISSN 1867-9048 ISSN 1867-9056 (electronic)
Signaling and Communication in Plants
ISBN 978-3-319-82552-6 ISBN 978-3-319-42319-7 (eBook)
DOI 10.1007/978-3-319-42319-7

Softcover reprint of the hardcover 1st edition 2016

Printed on acid-free paper

This Springer imprint is published by Springer Nature
The registered company is Springer International Publishing AG Switzerland

Preface

Now that scientific focus is increasingly shifting to plant roots, it is a timely occasion to summarize our current knowledge on belowground defence strategies in plants by world-class scientists actively working in the area. The volume includes chapters covering belowground defence to main soil pathogens such as *Fusarium*, *Rhizoctonia*, *Verticillium*, *Phytophthora*, *Pythium*, and *Plasmodiophora*, as well as to migratory and sedentary plant parasitic nematodes. In addition, the role of root exudates in belowground plant defence is highlighted. Finally, accumulating evidence on how plants can differentiate beneficial soil microbes from the pathogenic ones is covered as well. Better understanding of belowground defences can lead to the development of environmentally friendly plant protection strategies effective against soilborne pathogens which cause substantial damage on many crop plants all over the world. The book will be a useful reference material for plant pathologists, agronomists, plant molecular biologists, as well as students working on these and related areas. The editors would like to thank all authors for their valuable contributions to this book.

St Lucia, Australia — Christine M.F. Vos
St Lucia, Australia — Kemal Kazan

Contents

Introduction to Belowground Defence Strategies in Plants

Christine M.F. Vos and Kemal Kazan

Abstract Plant roots have long been literally and figuratively hidden from sight, despite their unmistakable importance in a plant's life. Interactions between plant roots and soil microbes indeed seem to take place in a black box, but science is starting to shed some light into this box. This book aims to bring together our current knowledge on the belowground interactions of plant roots with both detrimental and beneficial microbes. This knowledge can form the basis for more environmentally friendly plant disease management of soil-borne pathogens and pests, and the book will be of interest to both plant scientists and students eager to discover the hidden part of a plant's daily life and survival.

Plants are multicellular photosynthetic organisms that have evolved from unicellular fresh water green algae. During their evolution, plants have acquired diverse capabilities that enabled them not only to survive but also to adapt and successfully colonize diverse land environments. In particular, the acquisition of roots or root-like structures that facilitate extracting water from soil rather than relying on limited amounts of moisture available on the soil surface has no doubt played an important role in plant's adaptation to life on land.

Obviously, roots are also essential for physical attachment of plants to the soil, as well as for nutrient uptake and interaction with soil biota. Plant roots continuously

C.M.F. Vos (✉)
Centre of Microbial and Plant Genetics, KU Leuven, Kasteelpark Arenberg 20, 3001 Leuven, Belgium

Department of Plant Systems Biology, Vlaams Instituut voor Biotechnologie (VIB), Technologie park 927, 9052 Ghent, Belgium

Commonwealth Scientific and Industrial Research Organisation (CSIRO) Agriculture, 306 Carmody Road, St Lucia, Brisbane, QLD 4067, Australia

Scientia Terrae Research Institute, Fortsesteenweg 30A, Sint-Katelijne-Waver, Belgium
e-mail: cvo@scientiaterrae.org

K. Kazan
CSIRO Agriculture St Lucia, 306 Carmody Road, St Lucia, QLD 4067, Australia

The Queensland Alliance for Agriculture & Food Innovation (QAAFI), Queensland Bioscience Precinct, University of Queensland, Brisbane, QLD 4072, Australia

C.M.F. Vos, K. Kazan (eds.), *Belowground Defence Strategies in Plants*, Signaling and Communication in Plants, DOI 10.1007/978-3-319-42319-7_1

explore the soil to sense and transmit diverse belowground signals needed to modify plant architecture. The interaction between plant roots and beneficial microbes (e.g., rhizobia or arbuscular mycorrhiza) can be highly advantageous for both parties and greatly contributes to agriculture. However, the belowground environment can be very hostile as well and plant roots are often threatened by various biotic and abiotic stress factors (e.g., lack of water, oxygen, nutrients; soil acidity, salinity, low temperatures, as well as pathogenic microbes). While the interaction between roots and nonpathogenic microbes can be beneficial, many pathogenic microbes and nematodes can inflict serious damage to roots, restricting plant growth, reducing yield, and even causing plant death. Therefore, plants must differentiate friends from foes to survive in a hostile environment, and the soil and plant roots play essential roles in this process.

Despite the importance of plant roots in the overall well-being of plants, crop breeding efforts aimed at improving biotic and abiotic stress tolerance have so far been mostly focused on the aboveground part of the plant. In fact, the roots are often referred to as "the hidden half," or the "black box," reflecting the neglected nature of plant root research. Similarly, although root pathogens cause enormous losses on our crop plants, root health has always been a difficult issue to deal with. Possible reasons for this are probably numerous but mainly include the complexity of the belowground environment.

Better understanding of the nature of the interaction between plant roots and both beneficial and pathogenic microbes can generate new knowledge leading to the development of novel strategies aimed at boosting plant productivity, while reducing crop losses. As Editors of this Springer book, our objective is to contribute to the ongoing efforts in this area by bringing together contributors who are leading researchers in their respective areas.

The first part of the book focuses on the general plant responses to soil microbes and the role that root exudates play in this process, both highly active research domains. The first chapter of this part (chapter "Belowground Defence Strategies in Plants: Parallels Between Root Responses to Beneficial and Detrimental Microbes") highlights the parallels that are increasingly emerging in plant root responses to beneficial and pathogenic microbes. The next chapter (chapter "Root Exudates as Integral Part of Belowground Plant Defence") details the essential and versatile roles of root exudates in belowground plant defences, impacting both detrimental and beneficial microbes.

The second part of the book then zooms in on the belowground defence strategies against specific root pathogens. Fungal root pathogens are represented by *Fusarium oxysporum* (chapter "Belowground Defence Strategies Against *Fusarium oxysporum*"), *Rhizoctonia* (chapter "Belowground Defence Strategies Against *Rhizoctonia*"), and *Verticillium* (chapter "Belowground Defence Strategies Against *Verticillium* Pathogens"). Next in line are the plant root responses to the oomycete pathogens *Phytophthora* (chapter "Belowground and Aboveground Strategies of Plant Resistance Against *Phytophthora* Species") and *Pythium* (chapter "Belowground Signaling and Defence in Host–*Pythium* Interactions"). Protists are represented by the clubroot pathogen *Plasmodiophora brassicae* (chapter

"Belowground Defence Strategies Against Clubroot (*Plasmodiophora brassicae*)"). Finally, nematodes are another detrimental soil pest with severe consequences for our worldwide food production. Chapter "Belowground Defence Strategies Against Sedentary Nematodes" covers sedentary nematodes, among which the highly damaging cyst and root-knot nematodes, while chapter "Belowground Defence Strategies Against Migratory Nematodes" deals with the migratory nematodes. The chapters in this part mainly focus on pathogen infection strategies and host resistance mechanisms, allowing an overview of the diverse nature of plant belowground defence strategies against pathogens and pests with varying lifestyles and infection strategies.

As already mentioned above, plants also seem to mount an initial defence response against beneficial microbes. Successfully colonizing microbes are able to overcome this and will assist the plant in its further belowground defences. This topic will be covered for the interactions between plant roots and the following beneficial microbes: nonpathogenic *Fusarium oxysporum* (chapter "Root Interactions with Nonpathogenic *Fusarium*"), *Trichoderma* (chapter "Belowground Defence Strategies in Plants: The Plant–*Trichoderma* Dialogue"), *Piriformospora indica* (chapter "Defence Reactions in Roots Elicited by Endofungal Bacteria of the Sebacinalean Symbiosis"), and arbuscular mycorrhizal fungi (chapter "Mitigating Abiotic Stresses in Crop Plants by Arbuscular Mycorrhizal Fungi"). The editors want to thank all authors for their valuable contributions, and wish you enjoyable reading of this book.

Part I
General Principles of Belowground Defence Strategies

Belowground Defence Strategies in Plants: Parallels Between Root Responses to Beneficial and Detrimental Microbes

Ruth Le Fevre and Sebastian Schornack

Abstract Plant roots, as underground structures, are hidden from view, difficult to work with and therefore typically understudied, especially in agricultural research. In addition to providing crucial support for aerial tissues and acquiring nutrients, roots engage with filamentous microorganisms in the soil. These interactions have outcomes ranging from positive to negative and therefore roots must respond appropriately to different microbes to ensure plant survival. While leaf responses to filamentous pathogens have been well researched, we lack comparative information from roots. Moreover, we lack knowledge on the extent of overlap of root responses to microbes that share similarities in morphology, biochemistry and colonisation strategy but that result in different outcomes. In this chapter, we highlight current knowledge on parallels in root responses to beneficial and detrimental filamentous microorganisms. We also emphasise the importance of root studies and advocate the development of new host systems that allow comparative root–microbe interaction research. Ultimately, understanding of this field at the molecular level could inform breeding for pathogen resistance in crops while promoting cooperative root interactions with other microbes.

1 Introduction

Plant roots are in constant contact with microorganisms in the soil. Interactions with specific microbes can lead to beneficial or detrimental outcomes for plants and significantly affect plant growth and development. Therefore, distinguishing between a potential mutualist and pathogen and responding appropriately are paramount to plant survival because pathogenic microorganisms can destroy plant tissue, while beneficial microorganisms can aid nutrient uptake and confer resistance to biotic and abiotic stresses.

R. Le Fevre • S. Schornack (✉)
Sainsbury Laboratory (SLCU), University of Cambridge, Cambridge, UK
e-mail: sebastian.schornack@slcu.cam.ac.uk

C.M.F. Vos, K. Kazan (eds.), *Belowground Defence Strategies in Plants*, Signaling and Communication in Plants, DOI 10.1007/978-3-319-42319-7_2

In leaves, responses to and interactions with pathogens have been well characterised. In roots, pathogen studies are fewer; however, beneficial interactions are well studied. Interestingly, the morphologies and mechanisms of colonisation of plant roots by filamentous microbes that have different effects on plants are similar. Therefore there is likely to be significant overlap in root responses to these different microbes. However, our research into the extent of this overlap is hampered, partly because suitable systems for comparative studies between these different interactions are rare (Rey and Schornack 2013). A greater understanding of microbial interactions with plant roots could enable new ways of protecting crops from those that are detrimental while promoting those that are beneficial. This is especially important considering future agricultural settings where we may rely on beneficial plant–microbe interactions, for enhancing plant nutrition when fertilizers become limited, and simultaneously aim to reduce disease in crops in order to maximise yield.

In this chapter we review recent work that highlights what is known about root responses to beneficial and detrimental filamentous microbes. We highlight the importance of studies in roots and advocate the development of new host systems, both plant and microorganism, which allow comparative root–microbe interaction studies.

2 The Study of Root–Microbe Interactions

The interactions of soil microbes with plant roots are typically understudied, especially in agricultural research, because as underground structures they are hidden from view and difficult to work with (Balmer and Mauch-Mani 2013). However, given the absolute importance of roots for nutrient and water uptake, anchoring and support of aerial tissue and direct interaction with the soil environment and microbiome, it is critical we understand more about these plant tissues and the associations they form with microorganisms. Understanding and engineering root–microbe interactions will help us find possible strategies to improve crop yield, stress resilience and pathogen protection.

Above- and belowground plant tissues are exposed to different microorganisms. The soil environment contains millions of filamentous microbes (fungal and other eukaryotic microorganisms with fungal morphologies, such as oomycetes) that are in constant proximity to or contact with plant roots (van der Heijden et al. 2008). Therefore, it is reasonable to hypothesise that recognition of and downstream responses to microbes in shoots and roots will differ (Balmer and Mauch-Mani 2013). Appropriate and timely responses in roots are especially important so as not to be constitutively activated, as this could impose fitness costs (De Coninck et al. 2015). Schreiber et al. (2011) demonstrated that the roots, but not leaves, of *Arabidopsis thaliana* were susceptible to the pathogenic fungus *Magnaporthe oryzae*, indicating that the defence situation below and above ground to this microbe is indeed different. However, the use of mutants has illustrated that plant defence signalling pathways are generally conserved between above- and below-ground tissues (De Coninck et al. 2015). As most work on plant responses to pathogenic microbes has been done in aboveground tissue, we can use our knowledge from leaves to test root responses to pathogens and highlight common and contrasting principles.

Microbes engage in a range of interactions with plant roots. Beneficial symbioses facilitate plant nutrient uptake and can increase abiotic and biotic stress tolerance. Detrimental pathogenic interactions result in nutrient loss and disease. We know most about the associations at the more extreme ends of the spectrum (Fig. 1b). However, what are less well understood are the intermediate interactions, such as those with endophytes (Jumpponen and Trappe 1998; Franken 2012). Filamentous endophytic fungi (such as the dark septate endophytes, DSE) persist in plant roots seemingly without causing disease, but the outcomes, in terms of effects on the plant, can vary from negative to neutral to positive depending on the specific microbe–host combination (Jumpponen 2001). Given that the microbe and the host environment can influence the outcome of an interaction, comparative studies that keep one interaction partner constant (one microbe in multiple hosts or multiple microbes with similar lifestyles within one host) would allow characterisation of the contribution of each partner. Additionally, appropriate plant host and microbial systems (see Table 2) to study these associations could help to answer many interesting questions arising from the topic of root–microbe interactions:

- Why do some microbes have different lifestyles on different plant tissues? (Sect. 2.2.1)
- How and why do some microbes engage in different interactions with different hosts? (Sect. 3.5)
- Are plant defence responses activated and suppressed in a microbe-specific or lifestyle-specific manner? (Sects. 4.1–4.3)
- Are structures formed by beneficial and detrimental microbes analogous? (Sects. 4.2 and 4.3, Fig. 1)
- Do plant traits similarly or differentially affect filamentous microbes with different lifestyles in roots? (see Table 1)

Understanding how the outcomes of plant root–microbe interactions are controlled would ultimately provide inroads to promote beneficial partnerships while suppressing detrimental ones.

2.1 *Plant Systems*

To better understand root responses to different microbes, a variety of appropriate plant and microbial systems to work with are needed. Studying root responses to different microbes that engage in a range of interactions in the same plant species would be advantageous.

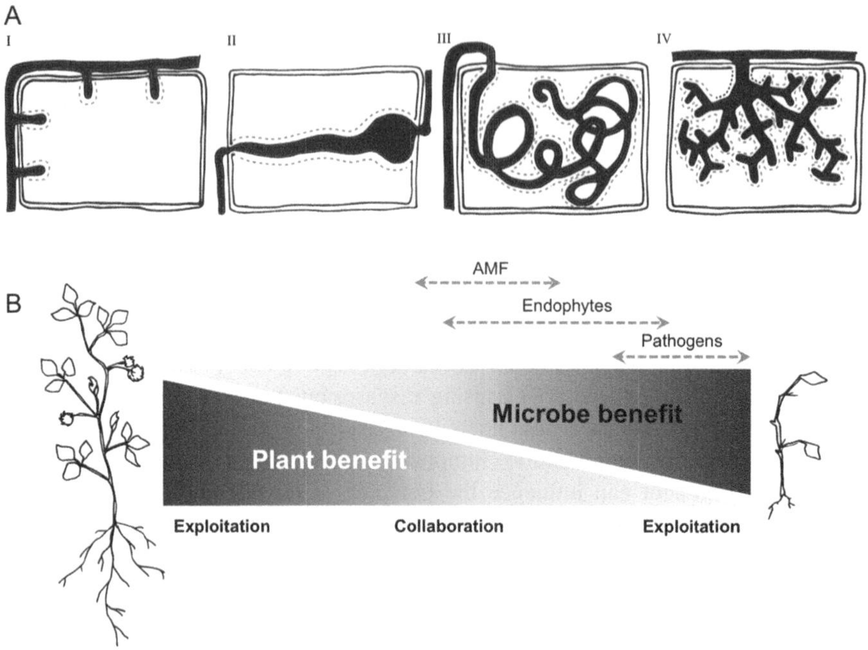

Fig. 1 Microbes engage in a spectrum of interactions with plant roots. (**a**) During root colonisation microbes form a variety of intracellular structures that can facilitate nutrient transfer, effector delivery to modulate host immune responses or simply the progress of growth through root cells. Although the microbe penetrates the cell wall (*outer solid line*), the protoplast remains intact and, at least in the case of **I**, haustoria, and **IV**, arbuscules, a modified membrane (*dashed line*) that contains a distinct protein complement from the rest of the plasma membrane (*inner solid line*) encases the microbial structure. *M. oryzae* transverses root cells as in **II** and *P. indica* forms coils insides cells as in **III**, but nothing is known about the membranes that surround these structures and whether they are also different from the plasma membrane as in **I** and **II**. (**b**) Root–microbe interactions lie on a spectrum and cannot be compartmentalised into beneficial or detrimental without taking into consideration the interaction in context of environmental factors and host/microbe genotype. This spectrum has been described elsewhere as the mutualism–parasitism continuum (Mandyam and Jumpponen 2015). *Dashed arrows* for arbuscular mycorrhizal fungi (AMF), endophytes and pathogens represent perceived extents to which microbe and plant benefit from the interactions they engage in

Medicago truncatula has been used extensively for symbiosis research and has been instrumental for identifying genes affecting interactions with beneficial arbuscular mycorrhizal fungi (AM fungi, Table 1, Ane et al. 2008). With this resource we are now able to determine whether these same genes are important for colonisation of roots by other microbes, including pathogens (Table 1, Wang et al. 2012; Gobbato et al. 2012, 2013; Rey et al. 2013, 2015).

Given that the three most important food crops (maize, wheat and rice) are monocots, with root architectures divergent from dicots, the use of monocot plants is also important for monocot versus dicot root response comparisons. In this regard rice and maize are good candidates as plant systems for root–microbe interactions as they have been used for AM fungi and pathogen research (see Table 2).

Table 1 Examples of plant genes implicated in colonisation of roots by beneficial and detrimental filamentous microbes

Gene	Plant species	Protein	Colonisation phenotype upon gene mutation or knock-down		References
			Beneficial	Detrimental	
HvMLO	*Hordeum vulgare*	Plasma membrane-localised seven trans-membrane domain protein	Reduced colonisation by *Funneliformis mosseae*	Unknown	Ruiz-Lozano et al. (1999)
MtDMI1/ *LjCASTOR* + *LjPOLLUX*	*Medicago truncatula*, *Lotus japonicus*	K^+ ion channel	Myc–	Increased resistance to *Verticillium albo-atrum*	Catoira et al. (2000); Ane et al. (2004); Imaizumi-Anraku et al. (2005); Charpentier et al. (2008); Ben et al. (2013)
MtDMI2/ *MtNORK*/ *LjSymRK*	*M. truncatula*, *Medicago sativa*, *L. japonicus*	MAL-LRR-RLK	Myc–	No alteration in *Phytophthora palmivora* colonisation	Wegel et al. (1998); Catoira et al. (2000); Endre et al. (2002); Rey et al. (2015); Stracke et al. (2002)
MtDMI3/ *LjCCamK*	*M. truncatula*, *L. japonicus*	CCaMK	Myc–, no cytoplasmic aggregations under *Gigaspora gigantea*	No alteration in *Colletotrichum trifolii* colonisation but altered in cytoplasmic aggregation under hyphopodia or from contact with *Phoma medicaginis*	Genre et al. (2009); Levy et al. (2004); Morandi et al. (2005)
MtLIN/ *LjCERBERUS*	*M. truncatula*	E3 ligase	Reduced colonisation by *Rhizophagus irregularis* and *Gigaspora margarita*. Infection structures normal but defective in hyphal elongation	More susceptible to *P. palmivora*	Rey et al. (2015); Takeda et al. (2013)

(continued)

Table 1 (continued)

Gene	Plant species	Protein	Colonisation phenotype upon gene mutation or knock-down		References
			Beneficial	Detrimental	
MtLYK3/ *LjNFR1*	*M. truncatula*, *L. japonicus*	LysM-RLK	Reduced myc colonisation in both *nfr* and *hcl* mutants (phenotype was stronger for *nfr* than *hcl*)	*hcl2* but not *hcl1* is more susceptible to *P. palmivora*	Rey et al. (2015); Zhang et al. (2015)
MtNFP/ *LjNFR5*	*M. truncatula*, *L. japonicus*	LysM-RLK	Myc+ but involved in MYC-signal elicited root branching stimulation	More susceptible to *A. euteiches*, *C. trifolii*, *V. albo-atrum* and *P. palmivora*	Achatz et al. (2010); Ben et al. (2013); Maillet et al. (2011); Rey et al. (2015)
MtNSP1/ *LjNSP1*	*M. truncatula*, *L. japonicus*	GRAS TF	Reduced colonisation and infection frequency by *R. irregularis*	Increased resistance to *V. albo-atrum*	Ben et al. (2013); Liu et al. (2011); Takeda et al. (2013)
MtNSP2	*M. truncatula*	GRAS TF	Reduced colonisation by *R. irregularis*, reduced MYC-signal elicited root branching stimulation	No alteration in *P. palmivora* colonisation	Maillet et al. (2011); Rey et al. (2015)
MtRAM1	*M. truncatula*	GRAS TF	Reduced colonisation by *R. irregularis*, suppressed MYC-signal elicited root branching stimulation	No alteration in *Aphanomyces euteiches* or *P. palmivora* colonisation	Gobbato et al. (2012); Maillet et al. (2011); Rey et al. (2015)
MtRAM2	*M. truncatula*	GPAT	Reduced myc colonisation by *R. irregularis* and *Glomus hoi*	Reduced colonisation by *P. palmivora* and *A. euteiches*	Gobbato et al. (2013); Wang et al. (2012)
MtROP9	*M. truncatula*	G-protein	*R. irregularis* colonisation promoted in *MtROP9* RNAi plants	*A. euteiches* colonisation promoted in *MtROP9* RNAi plants	Kiirika et al. (2012)
MtVPY	*M. truncatula*	Protein with N-terminal major sperm protein domain	Myc+ but *R. irregularis* produces deformed hyphopodia, more intraradical hyphae and no arbuscules	Unknown	Murray et al. (2011); Pumplin et al. (2010)

OsCERK1	*Oryzae sativa*	LysM-RLK	Reduced *R. irregularis* colonisation in *cerk1* rice	Increased *Magnaporthe oryzae* susceptibility (in leaves but roots not yet tested)	Miyata et al. (2014); Kishimoto et al. (2010); Mentlak et al. (2012); Zhang et al. (2015)
OsCEBiP	*O. sativa*	LysM chitin-binding protein	No alteration in myc phenotype in *cebip* rice	Increased *M. oryzae* susceptibility (in leaves but roots not yet tested)	Kaku et al. (2006); Kishimoto et al. (2010); Kouzai et al. (2014); Mentlak et al. (2012); Miyata et al. (2014)

Table 2 Examples of filamentous microorganisms that permit comparative studies of detrimental and beneficial interactions in roots

Microbe species	Taxonomic group	Perceived root interaction	Plant hosts	Penetration of root surface	Intracellular interface	References
Bipolaris sorokiniana	Fungus—Ascomycete	Detrimental	Small grain cereals	Appressoria	Intracellular hyphae	Kumar et al. (2002); Carlson et al. (1991)
Colletotrichum spp. (*Colletotrichum graminicola* and *Colletotrichum trifolii*)	Fungus—Ascomycete	Detrimental after greatly extended biotrophy	Maize, *Medicago truncatula*	Intercellular, melanised appressoria and hyphopodia	Intracellular hyphae	Genre et al. (2009); Sukno et al. (2008); Venard and Vaillancourt (2007)
Dark septate endophytes (e.g. *Phialocephala fortinii* and *Chloridium paucisporum*)	Fungi—Ascomycetes	Detrimental to beneficial	A huge range, including *Arabidopsis thaliana* and leek	Intracellular (root hairs) and intercellular	Intracellular hyphae	Mandyam and Jumpponen (2015)
Ericoid mycorrhiza (e.g. *Rhizoscyphus ericae*)	Fungi—Ascomycetes	Neutral to beneficial	Ericaceous plants and liverworts, e.g. *Pachyschistochila splachnophylla*	Intracellular (through rhizoids in liverworts)	Intracellular coils	Pressel et al. (2008)
Fusarium spp.	Fungus—Ascomycete	Detrimental (hemibiotrophic)	Many, including wheat and barley, *A. thaliana* and *Brachypodium distachyon*	Intercellular and intracellular	Intracellular hyphae	Lyons et al. (2015); Peraldi et al. (2011); Scherm et al. (2013)
Magnaporthe oryzae	Fungus—Ascomycete	Detrimental after greatly extended biotrophy	Rice, barley, *A. thaliana*	Hyphopodia, intracellular	Intracellular hyphae	Marcel et al. (2010); Sesma and Osbourn (2004); Schreiber et al. (2011)

Trichoderma spp.	Fungus—Ascomycete	Neutral to beneficial. Some cell death	Many, including maize, wheat and tomato	Intracellular, through root hairs/rhizodermis	None	Moran-Diez et al. (2015); Shukla et al. (2015); Yedidia et al. (1999)
Verticillium spp. (*Verticillium longisporum* and *Verticillium albo-atrum*)	Fungus—Ascomycete	Detrimental	Many, including *M. truncatula*, *A. thaliana*, alfalfa, tomato and Brassica oilseed crops	Intercellular and intracellular	Intracellular hyphae	Ben et al. (2013); Johansson et al. (2006)
Laccaria bicolor (Ecomycorrhiza)	Fungus—Basidiomycete	Beneficial	Trees, including *Populus* spp. such as black cottonwood (*Populus trichocarpa*)	Intercellular	None	Tschaplinski et al. (2014)
Piriformospora indica	Fungus—Basidiomycete	Neutral to beneficial. Some cell death	Many, including *A. thaliana*, barley and maize	Intercellular	Intracellular hyphae (coils)	Lahrmann et al. (2013); Kumar et al. (2009)
Rhizoctonia solani	Fungus—Basidiomycete	Detrimental	Many, including wheat, *B. distachyon*, rice, potato, maize, sugar beet, bean, lupin, cotton, lettuce, melon, *M. truncatula*	Wounds, intercellular	None	Schneebeli et al. (2015); Anderson et al. (2013); Garcia et al. (2006)
Ustilago maydis	Fungus—Basidiomycete	Neutral	Maize, *M. truncatula*	Intercellular	Intracellular hyphae	Mazaheri-Naeini et al. (2015)
Glomeromycota arbuscular mycorrhizal fungi (*e.g. Rhizophagus irregularis*)	Fungus—Glomeromycete	Beneficial (potentially detrimental on non-mycorrhizal hosts such as *A. thaliana*)	A huge range, including *M. truncatula*, rice, *A. thaliana* and liverworts (although not the model system *Marchantia polymorpha* var. *vulgaris*)	Hyphopodia	Arbuscules	Bonfante and Genre (2008); Ligrone et al. (2007); Parniske (2008); Russell and Bulman (2005); Veiga et al. (2013); Ligrone et al. (2007); Veiga et al. (2013); Wang and Qiu (2006)

(continued)

Table 2 (continued)

Microbe species	Taxonomic group	Perceived root interaction	Plant hosts	Penetration of root surface	Intracellular interface	References
Endogone fungi (Mucoromycotina)	Fungi—Zygomycete	Neutral to beneficial	Basal and higher land plants such as *Marchantia*, tobacco and tomato	Unknown	Intracellular hyphal coils and lumps	Daft and Nicolson (1966, 1969); Field et al. (2015); Russell and Bulman (2005)
Aphanomyces euteiches	Oomycete—Heterokontophyte	Detrimental	*M. truncatula*, pea and other legumes	Intracellular and intercellular	Haustoria reported once	Djebali et al. (2009, 2011); Gaulin et al. (2007); Franken et al. (2007)
Phytophthora spp. (e.g. *P. palmivora*, *P. sojae*)	Oomycete—Heterokontophyte	Detrimental	Many (species dependent) including soybean, lupin, *M. truncatula*	Intracellular, appressoria	Haustoria	Rey et al. (2015); Drenth and Guest (2004); Kroon et al. (2012)
Pythium spp.	Oomycete—Heterokontophyte	Detrimental	Many, including wheat and soybean	Intracellular, swollen hyphae	Intracellular hyphae	Kageyama (2014); Van Buyten and Hofte (2013)
Plasmodiophora brassicae	Rhizaria—Cercozoa	Detrimental	Brassicaceae	Through root hairs	Intracellular plasmodia	Gludovacz et al. (2014); Kageyama and Asano (2009)

Importantly, recent work in rice has shown that there are root type-specific transcriptional responses to colonisation by AM fungi (Gutjahr et al. 2015). This highlights the need for root type-specific microbe interactions to be studied independently.

Barley and wheat are other suitable monocot candidate systems of significant economic relevance. Work in crops is especially advantageous because it negates the need for knowledge transfer from model plant species. Both barley and wheat engage in beneficial symbiotic interactions with AM fungi and are affected by *Fusarium*, *Rhizoctonia* and *Pythium* root pathogens. Additionally the barley–*Piriformospora indica* (a model endophytic fungus) root interaction is already an established research system (Table 2).

Arabidopsis has been used to investigate *P. indica*, *M. oryzae*, *Verticillium* and *Fusarium*–root interactions. While it is a non-mycorrhizal species, it may still undergo interactions with these fungi (Veiga et al. 2013). Other advantages of using *Arabidopsis* as a model include the accessibility of mutants and extent of genome resources and its convenience in size and life cycle.

Ultimately, the use of a range of monocot and dicot model plant species will help to uncover core microbial accommodation programmes and those that are host species specific for microbes with specific lifestyles. The evolutionary conservation of these programmes can also be studied as lower descent plants, such as liverworts and hornworts, are also colonised by AM fungi and other filamentous microbes (see Table 2, Russell and Bulman 2005; Bonfante and Genre 2008).

2.2 Microbial Systems

In the following sections, we introduce additional microbial systems that are particularly suited for comparative studies between root responses to pathogens and mutualists.

2.2.1 Foliar Fungal Pathogens

The study of fungal pathogens and responses to pathogen colonisation in roots has been neglected in comparison to leaves, but this is not for a lack of root pathogens (see, e.g. *Fusarium* in chapter "Belowground Defence Strategies Against *Fusarium oxysporum*", *Rhizoctonia* in chapter "Belowground Defence Strategies Against *Rhizoctonia*", *Verticillium* in chapter "Belowground Defence Strategies Against *Verticillium* Pathogens" and *Pythium* in chapter "Belowground Signalling and Defence in Host–*Pythium* Interactions" in this book). Other non-pathogenic root-infecting fungi have also been introduced elsewhere (*Trichoderma* in chapter "Belowground Defence Strategies in Plants: The Plant–*Trichoderma* Dialogue", *P. indica* in chapter "Defence Reactions in Roots Elicited by Endofungal Bacteria of the Sebacinalean Symbiosis" and AM fungi in chapter "Mitigating Abiotic

Stresses in Crop Plants by Arbuscular Mycorrhizal Fungi"). Interesting, there is accumulating evidence that many foliar pathogens, including the rice blast fungus *M. oryzae*, anthracnose causing hemibiotrophic (i.e. exhibiting both symptomless biotrophic growth and tissue destroying necrotrophic life stages) *Colletotrichum* spp. and smut fungus *Ustilago maydis*, are also able to infect roots—although knowledge on their occurrence as natural root pathogens is often limited (Table 2, Dufresne and Osbourn 2001; Sukno et al. 2008; Mazaheri-Naeini et al. 2015). There is, therefore, the potential to use foliar fungal pathogens to facilitate the study of root–microbe interactions. Their classification as disease-causing pathogens, however, may have to be revisited in the root situation, as their associations with underground plant tissues appear less aggressively parasitic and more endophytic. Interestingly, penetration structures formed by some leaf pathogens on roots appear more similar structurally to those produced by AM fungi (see Sect. 4.2). Additionally, inside root tissue, *M. oryzae*, *Colletotrichum graminicola* and *U. maydis* engage in intercellular and intracellular biotrophic growth, and symptoms of disease are either extremely delayed, as for *M. oryzae* and *C. graminicola*, or do not seem to occur at all, as for *U. maydis* (Sukno et al. 2008; Marcel et al. 2010; Mazaheri-Naeini et al. 2015). In this way, these aggressive foliar pathogens appear to have different programmes for colonisation of different plant tissues and become more endophytic in lifestyle when infecting plant roots. One hypothesis for this is an absence of strong immune response signalling in some root tissues (such as the cortex) compared to leaves, enabling an extended period of biotrophic growth, although this has yet to be tested. As an avenue for future research, it will be especially interesting to discover just how many leaf pathogens also engage in root colonisation.

2.2.2 Oomycete Pathogens

Oomycetes are root- and shoot-infecting fungus-lookalikes which are taxonomically unrelated to fungi and differ from them in some structural and lifestyle features (Fawke et al. 2015). *Aphanomyces euteiches* and *Phytophthora palmivora* are root rot-causing oomycete pathogens. While *A. euteiches* infects legumes, *P. palmivora* has a very broad host range and infects many monocot and dicot species (Drenth and Guest 2004; Agrios 2005). *P. palmivora* is particularly interesting as it forms specialised intracellular lateral hyphal branches, termed haustoria, inside root cells (Rey et al. 2015). *A. euteiches* may also form haustoria, although so far they have only been reported from a single study (Franken et al. 2007). Haustoria have been best studied as structures formed by biotrophic and hemibiotrophic pathogens that cause foliar diseases, and parallels have been drawn between these structures and the intracellular branched hyphal arbuscules formed by AM fungi (chapter "Mitigating Abiotic Stresses in Crop Plants by Arbuscular Mycorrhizal Fungi", Sect. 4.3, Rey and Schornack 2013). Also, specialised plant-derived membranes form around haustoria as they do for AM fungi (see Sect. 4.3.2). Therefore, in comparison with AM fungi, we can use

P. palmivora to increase our understanding of the formation and function of intracellular microbial structures and interfaces.

3 Can I Stay or Must I Go? Parallels in Root Responses to Beneficial and Detrimental Microbes at the Tissue Level

In the interaction of plant roots with filamentous microbes, complex two-way signalling occurs between host and potential invader. Depending on the microbe, root responses can facilitate long-term accommodation and mutualistic associations or act defensively to try and rid plant tissue of the foreign body. Parallels in root responses to microbes with different lifestyles occur at the molecular level (Sect. 4) and also at the tissue level as discussed in the following sections.

3.1 Nutrition Status

The nutrient status of the soil affects root responses to potential microbial interactions. For example, if sufficient, accessible phosphate is present in the soil, it is directly acquired through the roots. As a result, colonisation by AM fungi and the symbiotic-phosphate uptake pathway are suppressed. Additionally, production of strigolactone (SL) phytohormones by plant roots, which stimulate germination of AM fungal spores and hyphal branching, is reduced if phosphate levels are non-limiting (Gu et al. 2011). Conversely, if phosphate and nitrate levels are limiting, roots respond by producing and secreting increased amounts of SL (Yoneyama et al. 2007, 2013). Mutant plants defective in SL production, *nsp1* and *nsp2* (genes that control SL biosynthesis), are compromised in colonisation by AM fungi compared to wild-type plants (Liu et al. 2011; Lauressergues et al. 2012; Takeda et al. 2013; Delaux et al. 2013). Interestingly, SL-deficient *nsp1* mutant *Medicago* plants were more resistant to the pathogenic microbe *Verticillium albo-atrum* than the wild type (Table 1, Ben et al. 2013). Production of SL by roots in response to nutrient status is therefore important for colonisation by beneficial microbes and may also affect colonisation by detrimental microbes, although the effects of SLs on growth and branching of filamentous microbes other than AM fungi are unclear. When the effects of the synthetic strigolactone GR24 were tested on *P. indica* and the root pathogen *Fusarium oxysporum* f. sp. *lycopersici*, no effect in growth or branching was reported (Steinkellner et al. 2007; Steinkellner and Mammerler 2007). However, in another study, GR24 actually inhibited radial growth of *F. oxysporum* and *Fusarium solani* and increased the number of branches in the former, but not the latter microbe (Dor et al. 2011).

3.2 Root System Morphology and Root Branching

Responses to mutualistic and parasitic interactions result in various changes to root system morphology. AM fungi are well noted for their effects on root morphogenesis and can alter the number, length and size of roots, although their modifications to lateral roots seem to be the most frequent effect (Fusconi 2014). Lateral roots in host plants (such as *Medicago*) are induced by recognition of AM fungi lipochitooligosaccharides (LCO) compounds, although both LCO and chitooligosaccharide (CO) compounds can induce them in rice (see Sect. 4.1.4, Maillet et al. 2011; Sun et al. 2015). *Trichoderma* spp. also induces the production of lateral roots and other endophytic fungi cause changes in root diameter and root hair length (Malinowski and Belesky 1999; Contreras-Cornejo et al. 2009). Ectomycorrhizal (EcM) fungi, such as *Laccaria bicolor* (Table 2) that grow intercellularly rather than intracellularly, stimulate lateral root formation and increase root hair length through release of volatile organic compounds and modulation of auxin gradients during the pre-infection stage (Sect. 4.1, Felten et al. 2009; Ditengou et al. 2015). Detrimental microbes can induce similar effects to beneficial microbes on roots, as *A. euteiches* induces lateral root formation in *M. truncatula* during infection (Djebali et al. 2009). *Pythium ultimum* and *Pythium irregulare* infections, however, lead to a smaller root system size and reduced degree of root branching (Larkin et al. 1995).

3.3 Secondary Metabolite Responses

Phytoalexins (PAs) are diverse low molecular weight antimicrobial compounds. Plants produce PAs, most notably after pathogen attack, although beneficial microbes also stimulate their production and this can provide resistance to subsequent infections by pathogenic microbes. Most evidence of these effects is derived from studies on root colonisation effects on aboveground rather than belowground tissues. For example, AM fungi, especially *Funneliformis mosseae*, stimulate capsidiol PA production in pepper stems (Ozgonen and Erkilic 2007). Supporting a role for AM fungi-based protection of belowground tissues, *F. mosseae* colonisation also provides a bioprotector effect against *Phytophthora parasitica* infection in tomato roots (Pozo et al. 2002). Endophytes also induce PA production. A type II hydrophobin protein produced by *Trichoderma longibrachiatum* induces the production of the PA rishitin in tomato leaves (Ruocco et al. 2015). Interestingly the induction of secondary metabolite compounds may be host and/or microbe specific as a different species of *Trichoderma* was shown to suppress expression of genes involved in the production of the PA vestitol in *Lotus japonicus* (Masunaka et al. 2011).

Microbes have evolved to utilise the production of secondary metabolites to their benefit. For example, *Phytophthora sojae* is attracted to soybean roots that exude isoflavone compounds and *Aphanomyces cochlioides* zoospores display a homing response to host-specific signals (Morris and Ward 1992; Islam and Tahara 2001). Chemicals released by plant roots also help orient the spores of fungi and oomycetes so they do not germinate in the wrong direction away from the host (Deacon 1996). Other compounds, such as flavonoids, may regulate initial stages of AM fungal colonisation and influence hyphal growth and branching, while in pathogenic interactions they are implicated in inhibition of growth (see chapter "Mitigating Abiotic Stresses in Crop Plants by Arbuscular Mycorrhizal Fungi", Hassan and Mathesius 2012 and references therein).

3.4 Systemic Responses to Microbial Colonisation

Colonisation of roots by detrimental microbes can inhibit growth and development of shoots. Conversely, colonisation of roots by beneficial microbes can induce systemic responses such as increases in shoot biomass and greater abiotic and biotic stress resistance in aerial plant tissue. This indicates that root responses to local microbial interactions induce signalling to influence the shoot. AM fungi, *Trichoderma* spp., *P. indica* and DSE interactions (which can all aid nutrient uptake) confer increases in shoot biomass in some plant species (Ozgonen and Erkilic 2007; Fakhro et al. 2010; Andrade-Linares et al. 2011b; Maag et al. 2014). While such growth increases are probably due to the improved nutrient situation of the plant, other systemic responses, such as increased stress tolerance, are conferred by microbe-induced increases in antioxidative capacity through regulation of genes involved in oxidative stress (Brotman et al. 2013). Interestingly, the AM fungus *Rhizophagus irregularis* confers a growth reduction in the non-mycorrhizal plant *A. thaliana*, again highlighting that root–microbe interactions are dependent on the specific organisms involved (see as well Sect. 3.5, Veiga et al. 2013).

As could be expected, signalling between above- and belowground plant tissues during microbial interactions also works in the other direction—microbial colonisation of leaves influences plant roots. For example, colonisation of bean roots with AM fungi was reduced if plant leaves were infected with the pathogen *Colletotrichum gloeosporioides* (Ballhorn et al. 2014).

3.5 Host-Dependent Responses

The outcome of root–microbe interactions can depend on the plant host. Whereas the majority of plants that form interactions with AM fungi form a beneficial symbiotic relationship, in the case of non-mycorrhizal species, the fungi may actually exert a detrimental effect. This indicates that the response of roots to a particular microbe and the outcome of an interaction are case-specific depending on the host and microbe involved. For example, the interaction of AM fungi with

A. thaliana results in root colonisation without arbuscule formation and plant growth is reduced (Veiga et al. 2013). Additionally, the interaction with *Trichoderma* spp. can be swung from neutral endophytic to detrimental depending on the host genetic background (Tucci et al. 2011). Encouragingly, these results suggest that the interaction with these microbes, and the benefits they induce, could be improved through breeding. Finally, the colonisation strategy and lifestyle of *P. indica* also varies in a host-dependent manner, specifically depending on the availability of nitrogen in colonised tissue (Lahrmann et al. 2013). The root responses of these specific individual interactions are likely very different and therefore need to be studied on a case-by-case basis.

4 Parallels in Molecular and Cellular Responses to Beneficial and Detrimental Microbes

To assess parallels in root responses to beneficial and detrimental filamentous microbes, it is pertinent to consider the similarities and differences in their infection strategies and colonisation of root tissue. In order to facilitate effective growth in the plant host, different filamentous microorganisms must perceive chemical and physical signals from the host and modify their growth accordingly. There are different microbial colonisation stages at which root responses can be considered. These are pre-infection (Sect. 4.1), the targeting of microbes to roots and microbial recognition by the root; penetration (Sect. 4.2), root responses to microbial attachment and surface invasion of the host; accommodation (Sect. 4.3), the housing of specialised microbial structures in plant cells; and collaboration or eviction (Sect. 3), the overall response to the interaction, which can be for better or for worse for the plant host.

4.1 Pre-infection

Regardless of whether the outcome of the interaction is beneficial or detrimental, both host plant and invading filamentous microbes release signals signifying their presence in the soil. There is substantial overlap in root responses to these signals, which involve activation of plant defences, but beneficial microbes also produce additional signals to induce symbiosis-related responses in the plant.

4.1.1 Transcriptional Responses Preceding Microbial Contact

In *M. truncatula*, expression of the GRAS transcription factor encoding gene, *RAM1*, is induced before physical contact is made with the AM fungus

R. irregularis and *RAM1* is required for mycorrhizal colonisation and arbuscule formation. However, it is not required for colonisation by the pathogenic oomycetes *P. palmivora* or *A. euteiches* (Gobbato et al. 2013). RAM1 regulates the expression of *RAM2*, a gene encoding a glycerol-3-phosphate acyl transferase, involved in cutin biosynthesis. Later in the mycorrhizal interaction, both *RAM1* and *RAM2* expressions are induced (Gobbato et al. 2012). RAM2 function is important for colonisation of *M. truncatula* roots by *R. irregularis*, *P. palmivora* and *A. euteiches* (Wang et al. 2012; Gobbato et al. 2013). The AM fungi *R. irregularis* and the oomycete pathogen *P. palmivora* both recognise cutin monomers from plant roots as a signal to promote formation of their respective penetrations structures (Table 2). Consequently, colonisation of *ram2-1* plants by *R. irregularis*, *P. palmivora* and also by *A. euteiches* was reduced (Wang et al. 2012; Gobbato et al. 2013).

4.1.2 Responses to the Microbe-Associated Molecular Pattern Chitin

Filamentous microbes display their presence to plants by the release of microbe-associated molecular patterns (MAMPs) (Newman et al. 2013 and references therein). Typically, the presence of true fungi is announced when chitin polymers are released from fungal cell walls by the activities of plant chitinases (Kaku et al. 2006; Silipo et al. 2010). While oomycete cell walls are mainly cellulosic, evidence indicates that chitin is also integral to the cell wall structure of at least some groups of root-infecting oomycetes—*A. euteiches*, for example (Badreddine et al. 2008; Nars et al. 2013a). In *M. truncatula*, chitinase expression in roots was induced by interaction with microbes with different lifestyles. Interestingly, the AM fungi tested induced some different chitinases compared to the pathogens, indicating there may be microbe–lifestyle-specific effects for these enzymes (Salzer et al. 2000).

Most work on chitin perception has been conducted in suspension-cultured rice cells (Kaku et al. 2006; Kishimoto et al. 2010; Shimizu et al. 2010; Kouzai et al. 2014). Preferential recognition of octameric chitooligosaccharide polymers (CO8, chitin) at the plant cell surface triggers a cascade of downstream signalling leading to the activation of plant defence responses (Hamel and Beaudoin 2010; Shimizu et al. 2010). The lysin motif (LysM)-containing proteins OsCERK1 and OsCEBiP are required for pathogen chitin recognition in rice, where they function as a heterodimer (Miya et al. 2007; Liu et al. 2012; Shimizu et al. 2010). On binding CO8 from filamentous microbes, OsCEBiP recruits OsCERK1 that then phosphorylates OsRacGEF1, enabling the activation of signalling pathways that lead to activation of MAPK cascades and the production of reactive oxygen species, PAs (Sect. 3.3), lignins and pathogenesis-related proteins in rice (see Sanchez-Vallet et al. 2015). Similarly in *M. truncatula* roots, chitin fractions induced the production of extracellular reactive oxygen species and the transient expression of defence-associated genes (Nars et al. 2013b).

$[Ca^{2+}]_{cyt}$ increases are also observed in response to MAMP recognition. The use of $[Ca^{2+}]_{cyt}$ elevation mutants has demonstrated the importance of this response for

P. indica-mediated growth promotion in *A. thaliana* (Vadassery and Oelmuller 2009; Vadassery et al. 2009). *P. indica* induces different $[Ca^{2+}]_{cyt}$ responses in tobacco, suggesting there are host species-specific responses to the same microbe (Vadassery and Oelmuller 2009). *T. atroviride* and AM fungi culture exudates were also found to increase $[Ca^{2+}]_{cyt}$ levels (Navazio et al. 2007). Therefore, Ca^{2+} responses in roots are a common feature of interactions with both detrimental and beneficial microbes (see also Sect. 4.1.4).

Recently, *OsCERK1* was shown to be required for colonisation by AM fungi in rice roots, as well as for pathogenic *M. oryzae* colonisation in leaves (Zhang et al. 2015). OsCEBiP, the interacting partner of OsCERK1 in chitin perception, does not appear to play a role in mycorrhization, as the colonisation phenotype of mutant *cebip* plants was normal (Miyata et al. 2014). However, OsCEBiP is important for resistance to the fungal pathogen *M. oryzae* in leaves (Kishimoto et al. 2010; Mentlak et al. 2012; Kouzai et al. 2014). This implies, therefore, that there are different *OsCERK1*-dependent signalling complexes responsible for the detection of different microbes (Table 1). Both *OsCERK1* and *OsCEBiP* are expressed in rice roots; however, crucial information is still missing about the role of these genes in pathogen infection in this plant tissue (Shimizu et al. 2010).

4.1.3 Oomycete Elicitins

Phytophthora and *Pythium* oomycete pathogens also produce elicitin MAMPs (structurally conserved extracellular proteins with lipid binding roles) that trigger plant immunity. Plant recognition of elicitin proteins has only recently been described. The elicitin response (ELR) receptor-like protein was identified in a wild potato species and mediates extracellular recognition of a conserved pathogen elicitin domain in leaves (Du et al. 2015). Again it remains to be shown whether ELR is important for defence responses upon recognition of elicitins in roots.

4.1.4 Responses to Short (Lipo)chitooligosaccharides

In addition to the release of MAMPs, AM fungi also produce MYC factors which are diffusible lipochitooligosaccharide (LCO) and short-chain chitooligosaccharide (CO) signals that promote symbiosis-related responses in host–plant roots (Maillet et al. 2011; Genre et al. 2013). LCOs are mostly tetrameric or pentameric, β-1-4 linked *N*-acetylglucosamine chitooligosaccharide backbones decorated with various chemical groups, including sulphates, while short-chain COs are undecorated (Gough and Cullimore 2011; Genre et al. 2013; Maillet et al. 2011; Oldroyd 2013). AM fungal LCOs promote lateral root development (see Sect. 3.2) and enhance the formation of mycorrhizal symbiosis in *Medicago* but stimulate symbiosis-related nuclear Ca^{2+} spiking (an early event in the development of symbiosis) less efficiently than short-chain COs (Genre et al. 2013).

Exudates from the pathogenic fungus *Colletotrichum trifolii* also contain short-chain COs, but these do not elicit the symbiosis-related nuclear Ca^{2+} spiking in *M. truncatula* root epidermal cells seen with exudates from AM fungi (Genre et al. 2013). A specific cell wall fraction from the oomycete root pathogen *A. euteiches*, however, can induce some form of nuclear Ca^{2+} spiking in *M. truncatula* root cells, suggesting this response may depend on the microbe (Nars et al. 2013a). The requirement for functional symbiosis pathway genes *DMI1* and *DMI2* for AM fungi-induced nuclear Ca^{2+} spiking, but not for *A. euteiches* induced spiking, suggests this response occurs via different pathways for detrimental and beneficial microbes (Table 1, Genre et al. 2013; Nars et al. 2013a).

The hypothesis that nuclear Ca^{2+} responses may be microbe specific is further supported by evidence from two studies (using beneficial endophytic fungi) that show that *P. indica* does and *Trichoderma atroviride* does not induce nuclear Ca^{2+} responses, respectively (Vadassery et al. 2009; Lace et al. 2015). The host plant species, cell type and position of cells along the roots are also important factors in determining the response to different microbial signals (Chabaud et al. 2011; Sun et al. 2015). These important findings should influence future research in this area.

4.1.5 Responses to Diffusible Molecules from Other Filamentous Microbes

A recent report provides evidence for the release of diffusible chemical compounds from endophytic fungus *P. indica* in the early stages of an interaction with plant roots, before contact has been made. These signals induce a number of responses in the plant including transcriptomic changes in stress and defence-related genes, accumulation of phytohormones and stomatal closure; however, the compounds responsible have not yet been identified (Vahabi et al. 2015). Another report speculates on the production of MYC factor-like compounds by the endophyte *Trichoderma koningii*, which may be responsible for mediating its mutualistic lifestyle in *Lotus* (Masunaka et al. 2011).

4.1.6 Microbial Effector-Mediated Suppression of MAMP Recognition

The activation of defence responses from recognition of filamentous microbial MAMPs is unfavourable because it hampers development of both parasitic and mutualistic interactions. Therefore, filamentous microbes evolved solutions to suppress host defence and facilitate colonisation—they secrete effector proteins to manipulate the interaction with the host and suppress host defences. Both beneficial and detrimental microbes produce chitin-binding LysM domain containing proteins that interfere with chitin-triggered immunity to protect themselves from host recognition (see Sanchez-Vallet et al. 2015). Also, a small secreted protein, homologous to the leaf pathogen *Cladosporium fulvum* effector Avr4, found in *Trichoderma harzianum* and *T. atroviride* may bind chitin and protect the fungi from plant hydrolytic enzymes (Stergiopoulos and de Wit 2009). Slp1 (which competes with OsCEBiP for CO sequestration) and ECP6 apoplastic LysM proteins from *M. oryzae* and *C. fulvum*, respectively, also suppress chitin-triggered immunity (de Jonge et al. 2010; Mentlak et al. 2012).

4.2 Microbial Penetration Structures

Once contact is made between microbe and root, the next stage of colonisation requires penetration of the root surface. There are clear structural similarities between penetration strategies of beneficial and detrimental microbes. For example, AM fungi and some root pathogens produce specialised differentiated structures at the tips of their hyphae termed hyphopodia and appressoria, respectively. Some endophytes produce appressoria-like structures or swollen cells (Andrade-Linares et al. 2011a). These structures fulfil similar roles for both detrimental and beneficial microbes for mediating attachment to the plant surface and penetration of the root surface/epidermis. While on leaves *M. oryzae* and *C. graminicola* produce appressoria, on roots *M. oryzae* forms swollen hyphal tips and *C. graminicola* produces hyphopodia (Sukno et al. 2008; Marcel et al. 2010). Therefore, the structures are more reminiscent of those formed by beneficial AM fungi (see chapter "Mitigating Abiotic Stresses in Crop Plants by Arbuscular Mycorrhizal Fungi"). *P. palmivora*, however, produces appressoria on roots (Rey et al. 2015). The endophyte *P. indica* and *Verticillium* spp. pathogens do not produce appressoria or hyphopodia but penetrate the root directly or in the anticlinal space between rhizodermal cells (Deshmukh et al. 2006; Eynck et al. 2007). Similarly, *U. maydis* penetrates at the junction of root epidermal cells (Freitag et al. 2011; Mazaheri-Naeini et al. 2015).

Hyphopodia of AM fungi anchor to the root surface using many protrusions that penetrate the plant cell wall (Bonfante and Genre 2010). In appressoria-forming microbes, this may be achieved with extracellular matrix-derived glycoproteins, at least on leaves (Bircher and Hohl 1997). Beneficial microbes such as AM fungi lack cell wall-degrading enzymes, perhaps to avoid the release of fragments which may induce immune responses in the host, and thus their mechanism of cell wall penetration remains elusive (Tisserant et al. 2012, 2013). Both AM fungi and *P. palmivora* require cutin monomers produced by RAM2 for surface penetration-structure development (Sect. 4.1.1, Table 2, Wang et al. 2012). *A. euteiches* colonisation is reduced in *ram2* mutant plants suggesting surface penetration may also be impaired for this microbe (Gobbato et al. 2013).

Development of microbial penetration structures on the cell surface triggers cellular rearrangements (see Takemoto and Hardham 2004). So far, nearly all work on this subject in roots has been elucidated using AM fungi; therefore, we can only speculate as to the cellular responses to detrimental and endophytic surface penetration in this tissue. However, we can draw on knowledge from studies with pathogenic microbes in leaves to find parallels between detrimental and beneficial effects.

4.2.1 Nuclear Repositioning

Plant nucleus repositioning to the point of microbial contact on the cell surface is a well-characterised cellular response. This could well be a mechanical stimulus, rather than related to the recognition of MAMPs, as microneedle pinching caused a similar response in root cells (Genre et al. 2009). Evidence from leaf pathogen studies suggests that, for oomycete interactions, nuclear movement depends on whether the interaction is compatible or incompatible (Freytag et al. 1994; Caillaud et al. 2012). For fungi, however, the nucleus repositions in both types of interaction (see references in Griffis et al. 2014). Nuclear movement may also be cell type dependent, as well as microbe dependent, as no organelle movement was detected for intercellular hyphae of *Hyaloperonospora arabidopsidis* in *A. thaliana* mesophyll cells (Hermanns et al. 2008). Alternatively, perhaps different cells types in different tissues have varying mechanical thresholds to stimulate nuclear repositioning.

4.2.2 Cytoplasmic Aggregations

Cytoplasmic aggregations, the actin filament-driven accumulation of cellular organelles, are an important response to different types of microbes. These aggregates are associated with defence against fungal and oomycete filamentous pathogens as they occur before the development of cell wall apposition, or papillae, barriers against microbial ingress (see Takemoto and Hardham 2004). Cytoplasmic aggregations occur under the hyphopodia of the AM fungus *Gigaspora margarita*, the pathogens *C. trifolii* and *Phoma medicaginis* in *M. truncatula* roots (Genre et al. 2009). No aggregations were observed, however, for the ericoid endomycorrhizal fungus *Oidiodendron maius—perhaps because Medicago is a non-host for this microbe* (Genre et al. 2009). While this suggests aggregation of cytoplasm is a common, general process in compatible microbe–root interactions, none occurred under contact points of *Medicago* roots in the compatible interaction with *T. atroviride* (Lace et al. 2015).

In root–AM fungi interactions, but not in root–pathogen interaction, a pre-penetration apparatus (PPA) forms, after initial cytoplasmic aggregation of organelles at the plant–fungus contact point, under where the fungus will penetrate the epidermis (Genre et al. 2005, 2009). The PPA is a transient structure of microtubule bundles and ER patches that guides the growth of the penetrating hyphae through the plant cell (Genre et al. 2005). In the interaction of pathogenic oomycetes on *A. thaliana* leaves, actin filaments form bundles focused on the microbial penetration sites and ER and Golgi stacks also accumulate at these positions (Takemoto et al. 2003). In barley leaf–powdery mildew interactions, the actin cytoskeleton is differently organised depending on whether the host is susceptible or resistant. In a resistant host, actin filaments become strongly focused on the penetration site and are associated with penetration resistance, whereas in susceptible hosts actin is only weakly focused. If the epidermal surface penetration event is successful, the resulting powdery mildew haustorium becomes surrounded by a ring of host actin filaments (Opalski et al. 2005). Similarly to actin, the pattern of microtubules accumulating at the entry point of powdery mildew in barley leaves also depends on whether the penetration event is successful or not (Hoefle et al. 2011).

Genre et al. (2009) found that the symbiosis pathway gene *Does not make infections 3* (*DMI3*, Table 2) is required for cytoplasmic aggregation in the *C. trifolii* and *P. medicaginis* interaction and PPA development in AM–fungi–root interactions. This suggests the existence of a general genetic pathway in roots that mediates interactions with filamentous microbes.

4.3 Microbial Accommodation Structures

Intracellular infection structures, listed in Table 2, act as nutrient exchange sites as well as sites for effector delivery (Fig. 1a). Arbuscules are the site of nutrient exchange for AM fungi where phosphate, nitrogen and sulphur are transferred to the plant in exchange for a plant-derived carbohydrate source (Kiers et al. 2011). In DSE interactions, there is evidence to support both two-way nutrient transfer, as well as a role in increasing nutrient availability in the rhizosphere by mineralisation (Usuki and Narisawa 2007; Upson et al. 2009). Detrimental microbes also acquire nutrients from plant tissue via intracellular structures, although transfer occurs in one direction only. Evidence for this has been elucidated from work on leaf pathogens, which acquire phytoassimilates from the host via haustoria, as is the case for powdery mildews, although in other leaf pathogens some nutrients can be transferred even before haustoria form (see Harrison 1999).

During AM fungal interactions, evidence suggests neither plant nor microbe is exploiting the other (or there is mutual exploitation) during this symbiosis and both are able to enforce an established interaction by enhancing nutrient transfer to cooperative partners (Kiers et al. 2011). In the case of the endophyte *P. indica*, however, an interaction can occur regardless of nutrient availability (Achatz et al. 2010). This suggests this microbe exerts more control in the interaction.

4.3.1 Arbuscules and Haustoria

After surface penetration, later in the colonisation of plant root tissue, AM fungi and some hemibiotrophic root pathogens produce specialised intracellular accommodation structures termed arbuscules and haustoria, respectively (Table 2, Fig. 1a). Endophytes such as *P. indica* may also produce differentiated intracellular structures and also produce coils inside root cells (Fig. 1a, Rafiqi et al. 2013). Other fungi, such as *M. oryzae*, produce intracellular hyphae that pass from cell to cell (Fig. 1a, Marcel et al. 2010). Arbuscules, which senesce 2–3 days after maturity, and haustoria, which become encased in plant cell wall deposits (at least in leaves), are both transient structures (Wang et al. 2009; Kobae and Hata 2010). While active, these structures act as intimate communication points between microbe and plant cell and mediate the receipt and delivery of nutrients as well as delivery of effector proteins that suppress plant defence responses. Although the plant cell wall is breached during the formation of both arbuscules and haustoria, the plasma membrane remains intact. During microbial ingress, a new membrane, continuous with the plasma membrane but different in composition, develops to surround the invading microbial structure—termed periarbuscular membrane or PAM and extrahaustorial membrane or EHM as appropriate (Yi and Valent 2013). The complement of proteins included in this membrane help to determine how the plant can respond to the invading microbial structure.

Again, nearly all work concerning accommodation of intracellular fungal structures is derived from AM fungi interactions. After surface penetration, AM fungi grow either intercellularly or intracellularly through root tissue to reach the cortex where they form arbuscules inside cortical cells. Arbuscules occupy a significant volume of root cells and induce a substantial reorganisation of host cellular components (Harrison 2012). Arbuscule formation and intracellular growth displays similarities with PPA formation in that nuclear-headed cytoplasmic bridges form to guide the growth of the fungal structure into the cell. Interestingly, in carrot roots, multiple adjacent cells undergo simultaneous cellular rearrangements to prepare for the passage of intracellular fungal hyphae en route to the cortex. During arbuscule formation, localised aggregations of ER around the penetrating hyphae predict the emergence of lateral arbuscule branches (Genre et al. 2008). Additionally, as the arbuscule develops, microtubules, that are normally helically oriented in uncolonised cells, orient to outline the hyphal trunk and branches. Microtubule reorganisation also occurs in adjacent cortical cells, preempting arbuscules formation (Blancaflor et al. 2001).

Nothing is known about how roots respond to haustoria formation. In leaves, ER cisternae and Golgi stacks were found to accumulate around the neck of *Peronospora parasitica* haustoria (Takemoto et al. 2003). Actin rings form around developing powdery mildew haustoria in barley (Opalski et al. 2005).

4.3.2 Formation of Specialised Membranes Around Microbial Structures

Intracellular microbial structures become enclosed in specialised membranes (Fig. 1a). Work has shown that plant secretory pathways are involved in formation of these membranes and both endo- and exocytosis are crucial for the accommodation of beneficial and detrimental microbial structures in plant cells (Yamazaki and Hayashi 2015). VAMP proteins, that mediate exocytosis in plants, are important for both interactions with beneficial and detrimental microbes. VAMP721d and VAMP721e are required for AM fungal symbiosome formation in *M. truncatula*, while in *A. thaliana* VAMP721 and 722 function in defence against powdery mildew (Ivanov et al. 2012; Wang et al. 2013; Kim et al. 2014; Dormann et al. 2014). So far evidence supports the hypothesis for de novo EHM and PAM biosynthesis, rather than selective sorting of proteins from pre-existing membrane (Koh et al. 2005).

The AM fungal PAM is composed of at least two specific domains determined by plant proteins that specifically localise to the branches, such as phosphate transporter PT4 (Pumplin and Harrison 2009; Pumplin et al. 2012). Furthermore, the apoplastic compartment surrounding the trunk and branch domains seems different, as evidenced through differential GFP-/RFP-labelled blue copper protein 1 (Ivanov and Harrison 2014). Construction of the PAM has been studied with the use of fluorescently labelled components of the plant secretory pathway, which have a fundamental role in PAM biogenesis. PAM formation begins inside the PPA with an accumulation of Golgi stacks and components of the exocytotic pathway just ahead of the growing hyphae (Ivanov et al. 2012; Genre et al. 2012).

Similar, comparative work with accommodation structures of detrimental root microbes is lacking. However, we can again compare membranes around symbiotic accommodation structures from AM fungi with those around structures formed by biotrophic plant pathogens in leaves. Most information concerning the EHM around haustoria in detrimental microbe–plant interactions has been elucidated from leaf studies using *P. infestans*, downy and powdery mildew pathogens.

The EHM around haustoria in leaves, like the PAM in AM fungi–root interactions, has a protein composition distinct from that of the plant cell plasma membrane (Koh et al. 2005; Micali et al. 2011; Lu et al. 2012; Pumplin et al. 2012). Additionally, some membrane-localised proteins appear to be restricted to specific locations of the EHM—corresponding to the neck or the rim of the haustoria (Micali et al. 2011; Pumplin and Harrison 2009). Studies have also reported the exclusion of plasma membrane-localised proteins specifically from the EHM, such as the *A. thaliana* aquaporin PIP1;4 and calcium ATPase ACA8 in the interaction with *Phytophthora infestans* and *H. arabidopsidis* (Koh et al. 2005; Micali et al. 2011; Lu et al. 2012). Other proteins such as the immunity-related FLS2 and EFR appear to be differentially targeted depending on the microbe. FLS2 accumulates in the EHM around *H. arabidopsidis* haustoria but neither FLS2 nor EFR accumulate around *P. infestans* haustoria (Lu et al. 2012). Accumulation patterns of the immune protein RPW8.2 was also different depending on whether the pathogen was an oomycete or a fungus indicating distinct pathogen-specific roles for this protein (Wang et al. 2009; Lu et al. 2012). Whether the accumulation of immunity-related plant proteins is similar around haustoria of pathogens in roots remains to be seen.

4.3.3 Cytoplasmic Microbial Effectors

Some plants are capable of perceiving effectors through cognate disease resistance proteins and mount effector-triggered immunity (ETI) responses. In leaves, ETI is often, but not always, concomitant with a hypersensitive response resulting in cell death and resistance (Lo Presti et al. 2015). The activity and role of R genes in resistance to root pathogens is not well understood, but some evidence suggests R genes active in leaves are also active in roots. For example, the R gene *Pi-CO39(t)* is active against *M. oryzae* in roots and leaves of rice (Sesma and Osbourn 2004). The authors speculate that the maintenance of this activity in root tissue implies root infection by this foliar pathogen may be of biological significance. It remains to be studied whether hypersensitive cell death also widely occurs during root ETI. Future research will also reveal whether the effector complement of leaf pathogens that infect roots (e.g. *M. oryzae*) is the same during above- and belowground infection.

Effector proteins have been better characterised from pathogenic microbes that usually infect leaves while only a couple have so far been characterised from filamentous microbes that engage in beneficial interactions with plant roots. The SP7 effector protein from *R. irregularis* suppresses expression of a pathogenesis-related transcription factor, ERF19, which is highly induced on root infection with *C. trifolii*. Expression of SP7 by *M. oryzae* resulted in reduced induction of defence-related genes and delayed root decay, indicating that this protein is involved in the maintenance of biotrophic growth in plant tissue (Kloppholz et al. 2011). Effector candidates have also been predicted in silico in the endophyte *P. indica* (Rafiqi et al. 2013). A recent report showed that at 6 days after infection with this microbe, responses that were induced at 2 days were suppressed, suggesting the dampening of the plant defence response (Vahabi et al. 2015). The identity and function of cytoplasmic *P. indica* effectors however is still unknown.

5 Outlook and Conclusions

Filamentous microbes engage with plant roots in a spectrum of interactions and they share many morphological and biochemical traits that plants must accurately distinguish between and respond to in order to survive (Fig. 1b). Some of these responses appear to be more general (i.e. microbe non-specific), such as the elicitation of defence responses through MAMP perception. Others, such as the recognition of specific signals (e.g. from AM fungi), induce a cascade of specific responses that facilitate mutualistic symbiosis and the long-term accommodation of the microbe in host root tissue—promoting reciprocal exchange of nutrients. We try to categorise microbes as beneficial or detrimental, but it is clear that plant responses and interaction outcomes can depend on host genotype, environmental factors and tissue type. In order to understand what determines the outcome of a specific root–microbe interaction, we need to utilise a range of plant and microbial systems in our research. Conducting comparative experiments between root and shoot interactions with the same microbe will also be instrumental in elucidating how defence responses are similar or different in these tissues. Microbes including *M. oryzae*, which alters its lifestyle between roots and shoots, and *P. palmivora*, which maintains a hemibiotrophic lifestyle in both tissues, will therefore be important for these studies.

Due to the overlap in colonisation strategy of root-infecting microbes with different lifestyles, and the parallels in root responses to them, we may not be able to develop a molecular handle to promote specific interactions while suppressing others. One possible solution might be to understand more about the function of R genes in roots. If R gene-mediated resistance is as effective in roots as in shoots, we may be able to tailor resistance to certain root diseases while maintaining symbiotic interactions. However, our current knowledge of R gene-mediated resistance in roots is lacking behind and requires further attention.

There are many other avenues for further research into root–microbe interactions. In particular, we should focus on elucidating the mechanisms behind mycorrhizal and endophyte-mediated suppression of pathogen infection and the roles of effectors from these microbes in both suppression of host defence responses and maintenance of biotrophic lifestyles. It also remains to be discovered how the mutualistic nature of interactions with filamentous microbes such as *Trichoderma* spp. and *P. indica* arise and whether additional symbiotic signals, such as the MYC-factor LCOs for AM fungi, are required. A focus on underground interactions and continued collaboration between the fields of immunity and symbiosis will uncover how roots respond to and balance beneficial and detrimental interactions.

Acknowledgements We thank The Gatsby Charitable Foundation (RG62472 to R.L.F and S.S.) and Royal Society (RG69135 to S. S.) for funding, Dario Fisher for assistance with the production of Fig. 1 and Dr. Féi Mão for motivational support.

References

Achatz B, Kogel KH, Franken P, Waller F (2010) Piriformospora indica mycorrhization increases grain yield by accelerating early development of barley plants. Plant Signal Behav 5 (12):1685–1687

Agrios GN (2005) Plant pathology, 5th edn. Elsevier Academic, Amsterdam

Anderson JP, Lichtenzveig J, Oliver RP, Singh KB (2013) Medicago truncatula as a model host for studying legume infecting Rhizoctonia solani and identification of a locus affecting resistance to root canker. Plant Pathol 62(4):908–921. doi:10.1111/J.1365-3059.2012.02694.X

Andrade-Linares DR, Grosch R, Franken P, Rexer KH, Kost G, Restrepo S, de Garcia MCC, Maximova E (2011a) Colonization of roots of cultivated Solanum lycopersicum by dark septate and other ascomycetous endophytes. Mycologia 103(4):710–721. doi:10.3852/10-329

Andrade-Linares DR, Grosch R, Restrepo S, Krumbein A, Franken P (2011b) Effects of dark septate endophytes on tomato plant performance. Mycorrhiza 21(5):413–422. doi:10.1007/s00572-010-0351-1

Ane JM, Kiss GB, Riely BK, Penmetsa RV, Oldroyd GE, Ayax C, Levy J, Debelle F, Baek JM, Kalo P, Rosenberg C, Roe BA, Long SR, Denarie J, Cook DR (2004) Medicago truncatula DMI1 required for bacterial and fungal symbioses in legumes. Science 303(5662):1364–1367. doi:10.1126/science.1092986

Ane JM, Zhu H, Frugoli J (2008) Recent advances in Medicago truncatula genomics. Int J Plant Genome 2008:256597. doi:10.1155/2008/256597

Badreddine I, Lafitte C, Heux L, Skandalis N, Spanou Z, Martinez Y, Esquerre-Tugaye MT, Bulone V, Dumas B, Bottin A (2008) Cell wall chitosaccharides are essential components and exposed patterns of the phytopathogenic oomycete Aphanomyces euteiches. Eukaryot Cell 7 (11):1980–1993. doi:10.1128/EC.00091-08

Ballhorn DJ, Younginger BS, Kautz S (2014) An aboveground pathogen inhibits belowground rhizobia and arbuscular mycorrhizal fungi in Phaseolus vulgaris. BMC Plant Biol 14:321. doi:10.1186/s12870-014-0321-4

Balmer D, Mauch-Mani B (2013) More beneath the surface? Root versus shoot antifungal plant defenses. Front Plant Sci 4:256. doi:10.3389/fpls.2013.00256

Ben C, Toueni M, Montanari S, Tardin MC, Fervel M, Negahi A, Saint-Pierre L, Mathieu G, Gras MC, Noel D, Prosperi JM, Pilet-Nayel ML, Baranger A, Huguet T, Julier B, Rickauer M, Gentzbittel L (2013) Natural diversity in the model legume Medicago truncatula allows identifying distinct genetic mechanisms conferring partial resistance to Verticillium wilt. J Exp Bot 64(1):317–332. doi:10.1093/jxb/ers337

Bircher U, Hohl HR (1997) Surface glycoproteins associated with appressorium formation and adhesion in Phytophthora palmivora. Mycol Res 101:769–775. doi:10.1017/S0953756296003279

Blancaflor EB, Zhao L, Harrison MJ (2001) Microtubule organization in root cells of Medicago truncatula during development of an arbuscular mycorrhizal symbiosis with Glomus versiforme. Protoplasma 217(4):154–165

Bonfante P, Genre A (2008) Plants and arbuscular mycorrhizal fungi: an evolutionary-developmental perspective. Trends Plant Sci 13(9):492–498. doi:10.1016/j.tplants.2008.07.001

Bonfante P, Genre A (2010) Mechanisms underlying beneficial plant-fungus interactions in mycorrhizal symbiosis. Nat Commun 1:48. doi:10.1038/ncomms1046

Brotman Y, Landau U, Cuadros-Inostroza A, Tohge T, Fernie AR, Chet I, Viterbo A, Willmitzer L (2013) Trichoderma-plant root colonization: escaping early plant defense responses and activation of the antioxidant machinery for saline stress tolerance. PLoS Pathog 9(3): e1003221. doi:10.1371/journal.ppat.1003221

Caillaud MC, Piquerez SJ, Fabro G, Steinbrenner J, Ishaque N, Beynon J, Jones JD (2012) Subcellular localization of the Hpa RxLR effector repertoire identifies a tonoplast-associated protein HaRxL17 that confers enhanced plant susceptibility. Plant J 69(2):252–265. doi:10.1111/j.1365-313X.2011.04787.x

Carlson H, Stenram U, Gustafsson M, Jansson HB (1991) Electron-microscopy of barley root infection by the fungal pathogen Bipolaris-Sorokiniana. Can J Bot 69(12):2724–2731

Catoira R, Galera C, de Billy F, Penmetsa RV, Journet EP, Maillet F, Rosenberg C, Cook D, Gough C, Denarie J (2000) Four genes of Medicago truncatula controlling components of a nod factor transduction pathway. Plant Cell 12(9):1647–1666

Chabaud M, Genre A, Sieberer BJ, Faccio A, Fournier J, Novero M, Barker DG, Bonfante P (2011) Arbuscular mycorrhizal hyphopodia and germinated spore exudates trigger Ca2+ spiking in the legume and nonlegume root epidermis. New Phytol 189(1):347–355. doi:10.1111/j.1469-8137.2010.03464.x

Charpentier M, Bredemeier R, Wanner G, Takeda N, Schleiff E, Parniske M (2008) Lotus japonicus CASTOR and POLLUX are ion channels essential for perinuclear calcium spiking in legume root endosymbiosis. Plant Cell 20(12):3467–3479. doi:10.1105/tpc.108.063255

Contreras-Cornejo HA, Macias-Rodriguez L, Cortes-Penagos C, Lopez-Bucio J (2009) Trichoderma virens, a plant beneficial fungus, enhances biomass production and promotes

lateral root growth through an auxin-dependent mechanism in Arabidopsis. Plant Physiol 149 (3):1579–1592. doi:10.1104/pp.108.130369
Daft MJ, Nicolson TH (1966) Effect of Endogone mycorrhiza on plant growth. I. New Phytol 65:343–350
Daft MJ, Nicolson TH (1969) Effect of endogone mycorrhiza on plant growth. II. New Phytol 68:953–963
Deacon JW (1996) Ecological implications of recognition events in the pre-infection stages of root pathogens. New Phytol 133(1):135–145. doi:10.1111/J.1469-8137.1996.Tb04349.X
De Coninck B, Timmermans P, Vos C, Cammue BP, Kazan K (2015) What lies beneath: belowground defense strategies in plants. Trends Plant Sci 20(2):91–101. doi:10.1016/j.tplants.2014.09.007
de Jonge R, van Esse HP, Kombrink A, Shinya T, Desaki Y, Bours R, van der Krol S, Shibuya N, Joosten MHAJ, Thomma BPHJ (2010) Conserved fungal LysM effector Ecp6 prevents chitin-triggered immunity in plants. Science 329(5994):953–955. doi:10.1126/Science.1190859
Delaux PM, Becard G, Combier JP (2013) NSP1 is a component of the Myc signaling pathway. New Phytol 199(1):59–65. doi:10.1111/nph.12340
Deshmukh S, Huckelhoven R, Schafer P, Imani J, Sharma M, Weiss M, Waller F, Kogel KH (2006) The root endophytic fungus Piriformospora indica requires host cell death for proliferation during mutualistic symbiosis with barley. Proc Natl Acad Sci U S A 103 (49):18450–18457. doi:10.1073/pnas.0605697103
Ditengou FA, Muller A, Rosenkranz M, Felten J, Lasok H, van Doorn MM, Legue V, Palme K, Schnitzler JP, Polle A (2015) Volatile signalling by sesquiterpenes from ectomycorrhizal fungi reprogrammes root architecture. Nat Commun 6:Artn 6279. doi:10.1038/Ncomms7279
Djebali N, Jauneau A, Ameline-Torregrosa C, Chardon F, Jaulneau V, Mathe C, Bottin A, Cazaux M, Pilet-Nayel ML, Baranger A, Aouani ME, Esquerre-Tugaye MT, Dumas B, Huguet T, Jacquet C (2009) Partial resistance of Medicago truncatula to Aphanomyces euteiches is associated with protection of the root stele and is controlled by a major QTL rich in proteasome-related genes. Mol Plant Microbe Interact 22(9):1043–1055. doi:10.1094/MPMI-22-9-1043
Djebali N, Mhadhbi H, Lafitte C, Dumas B, Esquerre-Tugaye MT, Aouani ME, Jacquet C (2011) Hydrogen peroxide scavenging mechanisms are components of Medicago truncatula partial resistance to Aphanomyces euteiches. Eur J Plant Pathol 131(4):559–571. doi:10.1007/S10658-011-9831-1
Dor E, Joel DM, Kapulnik Y, Koltai H, Hershenhorn J (2011) The synthetic strigolactone GR24 influences the growth pattern of phytopathogenic fungi. Planta 234(2):419–427. doi:10.1007/s00425-011-1452-6
Dormann P, Kim H, Ott T, Schulze-Lefert P, Trujillo M, Wewer V, Huckelhoven R (2014) Cell-autonomous defense, re-organization and trafficking of membranes in plant-microbe interactions. New Phytol 204(4):815–822. doi:10.1111/nph.12978
Drenth A, Guest DI (2004) Diversity and management of Phytophthora in Southeast Asia, vol 114, ACIAR monograph series. Australian Centre for International Agricultural Research, Canberra
Du J, Verzaux E, Chaparro-Garcia A, Bijsterbosch G, Keizer LCP, Zhou J, Liebrand TWH, Xie C, Govers F, Robatzek S, van der Vossen EAG, Jacobsen E, Visser RGF, Kamoun S, Vleeshouwers VGAA (2015) Elicitin recognition confers enhanced resistance to Phytophthora infestans in potato. Nat Plants 1(4):15034. doi:10.1038/nplants.2015.34
Dufresne M, Osbourn AE (2001) Definition of tissue-specific and general requirements for plant infection in a phytopathogenic fungus. Mol Plant Microbe Interact 14(3):300–307. doi:10.1094/MPMI.2001.14.3.300
Endre G, Kereszt A, Kevei Z, Mihacea S, Kalo P, Kiss GB (2002) A receptor kinase gene regulating symbiotic nodule development. Nature 417(6892):962–966. doi:10.1038/nature00842

Eynck C, Koopmann B, Grunewaldt-Stoecker G, Karlovsky P, von Tiedemann A (2007) Differential interactions of Verticillium longisporum and V-dahliae with Brassica napus detected with molecular and histological techniques. Eur J Plant Pathol 118(3):259–274. doi:10.1007/S10658-007-9144-6

Fakhro A, Andrade-Linares DR, von Bargen S, Bandte M, Buttner C, Grosch R, Schwarz D, Franken P (2010) Impact of Piriformospora indica on tomato growth and on interaction with fungal and viral pathogens. Mycorrhiza 20(3):191–200. doi:10.1007/S00572-009-0279-5

Fawke S, Doumane M, Schornack S (2015) Oomycete interactions with plants: infection strategies and resistance principles. Microbiol Mol Biol Rev 79(3):263–280. doi:10.1128/MMBR.00010-15

Felten J, Kohler A, Morin E, Bhalerao RP, Palme K, Martin F, Ditengou FA, Legue V (2009) The ectomycorrhizal fungus Laccaria bicolor stimulates lateral root formation in poplar and Arabidopsis through auxin transport and signaling. Plant Physiol 151(4):1991–2005. doi:10.1104/pp.109.147231

Field KJ, Rimington WR, Bidartondo MI, Allinson KE, Beerling DJ, Cameron DD, Duckett JG, Leake JR, Pressel S (2015) First evidence of mutualism between ancient plant lineages (Haplomitriopsida liverworts) and Mucoromycotina fungi and its response to simulated Palaeozoic changes in atmospheric CO2. New Phytol 205(2):743–756. doi:10.1111/Nph.13024

Franken P (2012) The plant strengthening root endophyte Piriformospora indica: potential application and the biology behind. Appl Microbiol Biotechnol 96(6):1455–1464. doi:10.1007/s00253-012-4506-1

Franken P, Donges K, Grunwald U, Kost G, Rexer KH, Tamasloukht M, Waschke A, Zeuske D (2007) Gene expression analysis of arbuscule development and functioning. Phytochemistry 68(1):68–74. doi:10.1016/j.phytochem.2006.09.027

Freitag J, Lanver D, Bohmer C, Schink KO, Bolker M, Sandrock B (2011) Septation of infectious hyphae is critical for appressoria formation and virulence in the smut fungus Ustilago maydis. PLoS Pathog 7(5):e1002044. doi:10.1371/journal.ppat.1002044

Freytag S, Arabatzis N, Hahlbrock K, Schmelzer E (1994) Reversible cytoplasmic rearrangements precede wall apposition, hypersensitive cell-death and defense-related gene activation in potato Phytophthora-Infestans interactions. Planta 194(1):123–135

Fusconi A (2014) Regulation of root morphogenesis in arbuscular mycorrhizae: what role do fungal exudates, phosphate, sugars and hormones play in lateral root formation? Ann Bot 113 (1):19–33. doi:10.1093/aob/mct258

Garcia VG, Onco MAP, Susan VR (2006) Review. Biology and systematics of the form genus Rhizoctonia. Span J Agric Res 4(1):55–79

Gaulin E, Jacquet C, Bottin A, Dumas B (2007) Root rot disease of legumes caused by Aphanomyces euteiches. Mol Plant Pathol 8(5):539–548. doi:10.1111/J.1364-3703.2007.00413.X

Genre A, Chabaud M, Timmers T, Bonfante P, Barker DG (2005) Arbuscular mycorrhizal fungi elicit a novel intracellular apparatus in Medicago truncatula root epidermal cells before infection. Plant Cell 17(12):3489–3499. doi:10.1105/Tpc.105.035410

Genre A, Chabaud M, Faccio A, Barker DG, Bonfante P (2008) Prepenetration apparatus assembly precedes and predicts the colonization patterns of arbuscular mycorrhizal fungi within the root cortex of both Medicago truncatula and Daucus carota. Plant Cell 20(5):1407–1420. doi:10.1105/tpc.108.059014

Genre A, Ortu G, Bertoldo C, Martino E, Bonfante P (2009) Biotic and abiotic stimulation of root epidermal cells reveals common and specific responses to Arbuscular mycorrhizal fungi. Plant Physiol 149(3):1424–1434. doi:10.1104/Pp.108.132225

Genre A, Ivanov S, Fendrych M, Faccio A, Zarsky V, Bisseling T, Bonfante P (2012) Multiple exocytotic markers accumulate at the sites of perifungal membrane biogenesis in arbuscular mycorrhizas. Plant Cell Physiol 53(1):244–255. doi:10.1093/pcp/pcr170

Genre A, Chabaud M, Balzergue C, Puech-Pages V, Novero M, Rey T, Fournier J, Rochange S, Becard G, Bonfante P, Barker DG (2013) Short-chain chitin oligomers from arbuscular mycorrhizal fungi trigger nuclear Ca2+ spiking in Medicago truncatula roots and their production is enhanced by strigolactone. New Phytol 198(1):190–202. doi:10.1111/nph.12146

Gludovacz TV, Deora A, McDonald MR, Gossen BD (2014) Cortical colonization by Plasmodiophora brassicae in susceptible and resistant cabbage cultivars. Eur J Plant Pathol 140(4):859–862. doi:10.1007/S10658-014-0492-8

Gobbato E, Marsh JF, Vernie T, Wang E, Maillet F, Kim J, Miller JB, Sun J, Bano SA, Ratet P, Mysore KS, Denarie J, Schultze M, Oldroyd GE (2012) A GRAS-type transcription factor with a specific function in mycorrhizal signaling. Curr Biol 22(23):2236–2241. doi:10.1016/j.cub.2012.09.044

Gobbato E, Wang E, Higgins G, Bano SA, Henry C, Schultze M, Oldroyd GE (2013) RAM1 and RAM2 function and expression during arbuscular mycorrhizal symbiosis and Aphanomyces euteiches colonization. Plant Signal Behav 8(10):pii: e26049. doi:10.4161/psb.26049

Gough C, Cullimore J (2011) Lipo-chitooligosaccharide signaling in endosymbiotic plant-microbe interactions. Mol Plant Microbe Interact 24(8):867–878. doi:10.1094/MPMI-01-11-0019

Griffis AH, Groves NR, Zhou X, Meier I (2014) Nuclei in motion: movement and positioning of plant nuclei in development, signaling, symbiosis, and disease. Front Plant Sci 5:129. doi:10.3389/fpls.2014.00129

Gu M, Chen A, Dai X, Liu W, Xu G (2011) How does phosphate status influence the development of the arbuscular mycorrhizal symbiosis? Plant Signal Behav 6(9):1300–1304. doi:10.4161/psb.6.9.16365

Gutjahr C, Siegler H, Haga K, Iino M, Paszkowski U (2015) Full establishment of arbuscular mycorrhizal symbiosis in rice occurs independently of enzymatic jasmonate biosynthesis. PLoS One 10(4):e0123422. doi:10.1371/journal.pone.0123422

Hamel LP, Beaudoin N (2010) Chitooligosaccharide sensing and downstream signaling: contrasted outcomes in pathogenic and beneficial plant-microbe interactions. Planta 232 (4):787–806. doi:10.1007/s00425-010-1215-9

Harrison MJ (1999) Biotrophic interfaces and nutrient transport in plant fungal symbioses. J Exp Bot 50:1013–1022. doi:10.1093/Jexbot/50.Suppl_1.1013

Harrison MJ (2012) Cellular programs for arbuscular mycorrhizal symbiosis. Curr Opin Plant Biol 15(6):691–698. doi:10.1016/j.pbi.2012.08.010

Hassan S, Mathesius U (2012) The role of flavonoids in root-rhizosphere signalling: opportunities and challenges for improving plant-microbe interactions. J Exp Bot 63(9):3429–3444. doi:10.1093/jxb/err430

Hermanns M, Slusarenko AJ, Schlaich NL (2008) The early organelle migration response of Arabidopsis to Hyaloperonospora arabidopsidis is independent of RAR1, SGT1b, PAD4 and NPR1. Physiol Mol Plant Pathol 72(1–3):96–101. doi:10.1016/J.Pmpp.2008.06.002

Hoefle C, Huesmann C, Schultheiss H, Bornke F, Hensel G, Kumlehn J, Huckelhoven R (2011) A barley ROP GTPase ACTIVATING PROTEIN associates with microtubules and regulates entry of the barley powdery mildew fungus into leaf epidermal cells. Plant Cell 23 (6):2422–2439. doi:10.1105/tpc.110.082131

Imaizumi-Anraku H, Takeda N, Charpentier M, Perry J, Miwa H, Umehara Y, Kouchi H, Murakami Y, Mulder L, Vickers K, Pike J, Downie JA, Wang T, Sato S, Asamizu E, Tabata S, Yoshikawa M, Murooka Y, Wu GJ, Kawaguchi M, Kawasaki S, Parniske M, Hayashi M (2005) Plastid proteins crucial for symbiotic fungal and bacterial entry into plant roots. Nature 433(7025):527–531. doi:10.1038/nature03237

Islam MT, Tahara S (2001) Chemotaxis of fungal zoospores, with special reference to Aphanomyces cochlioides. Biosci Biotechnol Biochem 65(9):1933–1948. doi:10.1271/bbb.65.1933

Ivanov S, Harrison MJ (2014) A set of fluorescent protein-based markers expressed from constitutive and arbuscular mycorrhiza-inducible promoters to label organelles, membranes and cytoskeletal elements in Medicago truncatula. Plant J 80(6):1151–1163. doi:10.1111/tpj.12706

Ivanov S, Fedorova EE, Limpens E, De Mita S, Genre A, Bonfante P, Bisseling T (2012) Rhizobium-legume symbiosis shares an exocytotic pathway required for arbuscule formation. Proc Natl Acad Sci U S A 109(21):8316–8321. doi:10.1073/pnas.1200407109

Johansson A, Goud JKC, Dixelius C (2006) Plant host range of Verticillium longisporum and microsclerotia density in Swedish soils. Eur J Plant Pathol 114(2):139–149. doi:10.1007/S10658-005-2333-2

Jumpponen A (2001) Dark septate endophytes—are they mycorrhizal? Mycorrhiza 11 (4):207–211. doi:10.1007/S005720100112

Jumpponen A, Trappe JM (1998) Dark septate endophytes: a review of facultative biotrophic root-colonizing fungi. New Phytol 140(2):295–310. doi:10.1046/J.1469-8137.1998.00265.X

Kageyama K (2014) Molecular taxonomy and its application to ecological studies of Pythium species. J Gen Plant Pathol 80(4):314–326. doi:10.1007/S10327-014-0526-2

Kageyama K, Asano T (2009) Life cycle of Plasmodiophora brassicae. J Plant Growth Regul 28 (3):203–211. doi:10.1007/S00344-009-9101-Z

Kaku H, Nishizawa Y, Ishii-Minami N, Akimoto-Tomiyama C, Dohmae N, Takio K, Minami E, Shibuya N (2006) Plant cells recognize chitin fragments for defense signaling through a plasma membrane receptor. Proc Natl Acad Sci U S A 103(29):11086–11091. doi:10.1073/pnas.0508882103

Kiers ET, Duhamel M, Beesetty Y, Mensah JA, Franken O, Verbruggen E, Fellbaum CR, Kowalchuk GA, Hart MM, Bago A, Palmer TM, West SA, Vandenkoornhuyse P, Jansa J, Bucking H (2011) Reciprocal rewards stabilize cooperation in the mycorrhizal symbiosis. Science 333(6044):880–882. doi:10.1126/science.1208473

Kiirika LM, Bergmann HF, Schikowsky C, Wimmer D, Korte J, Schmitz U, Niehaus K, Colditz F (2012) Silencing of the Rac1 GTPase MtROP9 in Medicago truncatula stimulates early mycorrhizal and oomycete root colonizations but negatively affects rhizobial infection. Plant Physiol 159(1):501–516. doi:10.1104/pp.112.193706

Kim H, O'Connell R, Maekawa-Yoshikawa M, Uemura T, Neumann U, Schulze-Lefert P (2014) The powdery mildew resistance protein RPW8.2 is carried on VAMP721/722 vesicles to the extrahaustorial membrane of haustorial complexes. Plant J 79(5):835–847. doi:10.1111/tpj.12591

Kishimoto K, Kouzai Y, Kaku H, Shibuya N, Minami E, Nishizawa Y (2010) Perception of the chitin oligosaccharides contributes to disease resistance to blast fungus Magnaporthe oryzae in rice. Plant J 64(2):343–354. doi:10.1111/j.1365-313X.2010.04328.x

Kloppholz S, Kuhn H, Requena N (2011) A secreted fungal effector of Glomus intraradices promotes symbiotic biotrophy. Curr Biol 21(14):1204–1209. doi:10.1016/j.cub.2011.06.044

Kobae Y, Hata S (2010) Dynamics of periarbuscular membranes visualized with a fluorescent phosphate transporter in arbuscular mycorrhizal roots of rice. Plant Cell Physiol 51 (3):341–353. doi:10.1093/pcp/pcq013

Koh S, Andre A, Edwards H, Ehrhardt D, Somerville S (2005) Arabidopsis thaliana subcellular responses to compatible Erysiphe cichoracearum infections. Plant J 44(3):516–529. doi:10.1111/j.1365-313X.2005.02545.x

Kouzai Y, Nakajima K, Hayafune M, Ozawa K, Kaku H, Shibuya N, Minami E, Nishizawa Y (2014) CEBiP is the major chitin oligomer-binding protein in rice and plays a main role in the perception of chitin oligomers. Plant Mol Biol 84(4-5):519–528. doi:10.1007/s11103-013-0149-6

Kroon LPNM, Brouwer H, de Cock AWAM, Govers F (2012) The genus Phytophthora anno 2012. Phytopathology 102(4):348–364. doi:10.1094/Phyto-01-11-0025

Kumar J, Schafer P, Huckelhoven R, Langen G, Baltruschat H, Stein E, Nagarajan S, Kogel KH (2002) Bipolaris sorokiniana, a cereal pathogen of global concern: cytological and molecular approaches towards better control. Mol Plant Pathol 3(4):185–195. doi:10.1046/J.1364-3703.2002.00120.X

Kumar M, Yadav V, Tuteja N, Johri AK (2009) Antioxidant enzyme activities in maize plants colonized with Piriformospora indica. Microbiology 155:780–790. doi:10.1099/Mic.0.019869-0

Lace B, Genre A, Woo S, Faccio A, Lorito M, Bonfante P (2015) Gate crashing arbuscular mycorrhizas: in vivo imaging shows the extensive colonization of both symbionts by Trichoderma atroviride. Environ Microbiol Rep 7(1):64–77. doi:10.1111/1758-2229.12221

Lahrmann U, Ding Y, Banhara A, Rath M, Hajirezaei MR, Dohlemann S, von Wiren N, Parniske M, Zuccaro A (2013) Host-related metabolic cues affect colonization strategies of a root endophyte. Proc Natl Acad Sci USA 110(34):13965–13970. doi:10.1073/Pnas.1301653110

Larkin RP, English JT, Mihail JD (1995) Effects of infection by Pythium spp on root-system morphology of alfalfa seedlings. Phytopathology 85(4):430–435. doi:10.1094/Phyto-85-430

Lauressergues D, Delaux PM, Formey D, Lelandais-Briere C, Fort S, Cottaz S, Becard G, Niebel A, Roux C, Combier JP (2012) The microRNA miR171h modulates arbuscular mycorrhizal colonization of Medicago truncatula by targeting NSP2. Plant J 72(3):512–522. doi:10.1111/j.1365-313X.2012.05099.x

Levy J, Bres C, Geurts R, Chalhoub B, Kulikova O, Duc G, Journet EP, Ane JM, Lauber E, Bisseling T, Denarie J, Rosenberg C, Debelle F (2004) A putative Ca2+ and calmodulin-dependent protein kinase required for bacterial and fungal symbioses. Science 303 (5662):1361–1364. doi:10.1126/science.1093038

Ligrone R, Carafa A, Lumini E, Bianciotto V, Bonfante P, Duckett JG (2007) Glomeromycotean associations in liverworts: a molecular, cellular, and taxonomic analysis. Am J Bot 94 (11):1756–1777. doi:10.3732/ajb.94.11.1756

Liu W, Kohlen W, Lillo A, Op den Camp R, Ivanov S, Hartog M, Limpens E, Jamil M, Smaczniak C, Kaufmann K, Yang WC, Hooiveld GJ, Charnikhova T, Bouwmeester HJ, Bisseling T, Geurts R (2011) Strigolactone biosynthesis in Medicago truncatula and rice requires the symbiotic GRAS-type transcription factors NSP1 and NSP2. Plant Cell 23 (10):3853–3865. doi:10.1105/tpc.111.089771

Liu T, Liu Z, Song C, Hu Y, Han Z, She J, Fan F, Wang J, Jin C, Chang J, Zhou JM, Chai J (2012) Chitin-induced dimerization activates a plant immune receptor. Science 336 (6085):1160–1164. doi:10.1126/science.1218867

Lo Presti L, Lanver D, Schweizer G, Tanaka S, Liang L, Tollot M, Zuccaro A, Reissmann S, Kahmann R (2015) Fungal effectors and plant susceptibility. Annu Rev Plant Biol 66:513–545. doi:10.1146/annurev-arplant-043014-114623

Lu YJ, Schornack S, Spallek T, Geldner N, Chory J, Schellmann S, Schumacher K, Kamoun S, Robatzek S (2012) Patterns of plant subcellular responses to successful oomycete infections reveal differences in host cell reprogramming and endocytic trafficking. Cell Microbiol 14 (5):682–697. doi:10.1111/j.1462-5822.2012.01751.x

Lyons R, Stiller J, Powell J, Rusu A, Manners JM, Kazan K (2015) Fusarium oxysporum triggers tissue-specific transcriptional reprogramming in Arabidopsis thaliana. PLoS One 10(4):Artn E0121902. doi:10.1371/Journal.Pone.0121902

Maag D, Kandula DRW, Muller C, Mendoza-Mendoza A, Wratten SD, Stewart A, Rostas M (2014) Trichoderma atroviride LU132 promotes plant growth but not induced systemic resistance to Plutella xylostella in oilseed rape. Biocontrol 59(2):241–252. doi:10.1007/S10526-013-9554-7

Maillet F, Poinsot V, Andre O, Puech-Pages V, Haouy A, Gueunier M, Cromer L, Giraudet D, Formey D, Niebel A, Martinez EA, Driguez H, Becard G, Denarie J (2011) Fungal lipochitooligosaccharide symbiotic signals in arbuscular mycorrhiza. Nature 469 (7328):58–63. doi:10.1038/nature09622

Malinowski DP, Belesky DP (1999) Neotyphodium coenophialum-endophyte infection affects the ability of tall fescue to use sparingly available phosphorus. J Plant Nutr 22(4-5):835–853. doi:10.1080/01904169909365675

Mandyam KG, Jumpponen A (2015) Mutualism-parasitism paradigm synthesized from results of root-endophyte models. Front Microbiol 5:Artn 776. doi:10.3389/Fmicb.2014.00776
Marcel S, Sawers R, Oakeley E, Angliker H, Paszkowski U (2010) Tissue-adapted invasion strategies of the rice blast fungus Magnaporthe oryzae. Plant Cell 22(9):3177–3187. doi:10.1105/tpc.110.078048
Masunaka A, Hyakumachi M, Takenaka S (2011) Plant growth-promoting fungus, Trichoderma koningii suppresses isoflavonoid phytoalexin vestitol production for colonization on/in the roots of Lotus japonicus. Microbes Environ 26(2):128–134. doi:10.1264/Jsme2.Me10176
Mazaheri-Naeini M, Sabbagh SK, Martinez Y, Sejalon-Delmas N, Roux C (2015) Assessment of Ustilago maydis as a fungal model for root infection studies. Fungal Biol 119(2–3):145–153. doi:10.1016/J.Funbio.2014.12.002
Mentlak TA, Kombrink A, Shinya T, Ryder LS, Otomo I, Saitoh H, Terauchi R, Nishizawa Y, Shibuya N, Thomma BP, Talbot NJ (2012) Effector-mediated suppression of chitin-triggered immunity by Magnaporthe oryzae is necessary for rice blast disease. Plant Cell 24(1):322–335. doi:10.1105/tpc.111.092957
Micali CO, Neumann U, Grunewald D, Panstruga R, O'Connell R (2011) Biogenesis of a specialized plant-fungal interface during host cell internalization of Golovinomyces orontii haustoria. Cell Microbiol 13(2):210–226. doi:10.1111/j.1462-5822.2010.01530.x
Miya A, Albert P, Shinya T, Desaki Y, Ichimura K, Shirasu K, Narusaka Y, Kawakami N, Kaku H, Shibuya N (2007) CERK1, a LysM receptor kinase, is essential for chitin elicitor signaling in Arabidopsis. Proc Natl Acad Sci U S A 104(49):19613–19618. doi:10.1073/pnas.0705147104
Miyata K, Kozaki T, Kouzai Y, Ozawa K, Ishii K, Asamizu E, Okabe Y, Umehara Y, Miyamoto A, Kobae Y, Akiyama K, Kaku H, Nishizawa Y, Shibuya N, Nakagawa T (2014) The bifunctional plant receptor, OsCERK1, regulates both chitin-triggered immunity and arbuscular mycorrhizal symbiosis in rice. Plant Cell Physiol 55(11):1864–1872. doi:10.1093/pcp/pcu129
Morandi D, Prado E, Sagan M, Duc G (2005) Characterisation of new symbiotic Medicago truncatula (Gaertn.) mutants, and phenotypic or genotypic complementary information on previously described mutants. Mycorrhiza 15(4):283–289. doi:10.1007/s00572-004-0331-4
Moran-Diez ME, Trushina N, Lamdan NL, Rosenfelder L, Mukherjee PK, Kenerley CM, Horwitz BA (2015) Host-specific transcriptomic pattern of Trichoderma virens during interaction with maize or tomato roots. BMC Genomics 16:Artn 8. doi:10.1186/S12864-014-1208-3
Morris PF, Ward EWB (1992) Chemoattraction of zoospores of the soybean pathogen, Phytophthora sojae, by isoflavones. Physiol Mol Plant Pathol 40(1):17–22. doi:10.1016/0885-5765(92)90067-6
Murray JD, Muni RR, Torres-Jerez I, Tang Y, Allen S, Andriankaja M, Li G, Laxmi A, Cheng X, Wen J, Vaughan D, Schultze M, Sun J, Charpentier M, Oldroyd G, Tadege M, Ratet P, Mysore KS, Chen R, Udvardi MK (2011) Vapyrin, a gene essential for intracellular progression of arbuscular mycorrhizal symbiosis, is also essential for infection by rhizobia in the nodule symbiosis of Medicago truncatula. Plant J 65(2):244–252. doi:10.1111/j.1365-313X.2010.04415.x
Nars A, Lafitte C, Chabaud M, Drouillard S, Melida H, Danoun S, Le Costaouec T, Rey T, Benedetti J, Bulone V, Barker DG, Bono JJ, Dumas B, Jacquet C, Heux L, Fliegmann J, Bottin A (2013a) Aphanomyces euteiches cell wall fractions containing novel glucan-chitosaccharides induce defense genes and nuclear calcium oscillations in the plant host Medicago truncatula. PLoS One 8(9):e75039. doi:10.1371/journal.pone.0075039
Nars A, Rey T, Lafitte C, Vergnes S, Amatya S, Jacquet C, Dumas B, Thibaudeau C, Heux L, Bottin A, Fliegmann J (2013b) An experimental system to study responses of Medicago truncatula roots to chitin oligomers of high degree of polymerization and other microbial elicitors. Plant Cell Rep 32(4):489–502. doi:10.1007/s00299-012-1380-3
Navazio L, Moscatiello R, Genre A, Novero M, Baldan B, Bonfante P, Mariani P (2007) A diffusible signal from arbuscular mycorrhizal fungi elicits a transient cytosolic calcium elevation in host plant cells. Plant Physiol 144(2):673–681. doi:10.1104/pp.106.086959

Newman MA, Sundelin T, Nielsen JT, Erbs G (2013) MAMP (microbe-associated molecular pattern) triggered immunity in plants. Front Plant Sci 4:139. doi:10.3389/fpls.2013.00139

Oldroyd GE (2013) Speak, friend, and enter: signalling systems that promote beneficial symbiotic associations in plants. Nat Rev Microbiol 11(4):252–263. doi:10.1038/nrmicro2990

Opalski KS, Schultheiss H, Kogel KH, Huckelhoven R (2005) The receptor-like MLO protein and the RAC/ROP family G-protein RACB modulate actin reorganization in barley attacked by the biotrophic powdery mildew fungus Blumeria graminis f.sp. hordei. Plant J 41(2):291–303. doi:10.1111/j.1365-313X.2004.02292.x

Ozgonen H, Erkilic A (2007) Growth enhancement and Phytophthora blight (Phytophthora capsici Leonian) control by arbuscular mycorrhizal fungal inoculation in pepper. Crop Prot 26 (11):1682–1688. doi:10.1016/J.Cropro.2007.02.010

Parniske M (2008) Arbuscular mycorrhiza: the mother of plant root endosymbioses. Nat Rev Microbiol 6(10):763–775. doi:10.1038/Nrmicro1987

Peraldi A, Beccari G, Steed A, Nicholson P (2011) Brachypodium distachyon: a new pathosystem to study Fusarium head blight and other Fusarium diseases of wheat. BMC Plant Biol 11:Artn 100. doi:10.1186/1471-2229-11-100

Pozo MJ, Cordier C, Dumas-Gaudot E, Gianinazzi S, Barea JM, Azcon-Aguilar C (2002) Localized versus systemic effect of arbuscular mycorrhizal fungi on defence responses to Phytophthora infection in tomato plants. J Exp Bot 53(368):525–534

Pressel S, Ligrone R, Duckett JG, Davis EC (2008) A novel ascomycetous endophytic association in the rhizoids of the leafy liverwort family, Schistochilaceae (Jungermanniidae, Hepaticopsida). Am J Bot 95(5):531–541. doi:10.3732/ajb.2007171

Pumplin N, Harrison MJ (2009) Live-cell imaging reveals periarbuscular membrane domains and organelle location in Medicago truncatula roots during arbuscular mycorrhizal symbiosis. Plant Physiol 151(2):809–819. doi:10.1104/pp.109.141879

Pumplin N, Mondo SJ, Topp S, Starker CG, Gantt JS, Harrison MJ (2010) Medicago truncatula Vapyrin is a novel protein required for arbuscular mycorrhizal symbiosis. Plant J 61 (3):482–494. doi:10.1111/j.1365-313X.2009.04072.x

Pumplin N, Zhang X, Noar RD, Harrison MJ (2012) Polar localization of a symbiosis-specific phosphate transporter is mediated by a transient reorientation of secretion. Proc Natl Acad Sci U S A 109(11):E665–E672. doi:10.1073/pnas.1110215109

Rafiqi M, Jelonek L, Akum NF, Zhang F, Kogel KH (2013) Effector candidates in the secretome of Piriformospora indica, a ubiquitous plant-associated fungus. Front Plant Sci 4:228. doi:10.3389/fpls.2013.00228

Rey T, Schornack S (2013) Interactions of beneficial and detrimental root-colonizing filamentous microbes with plant hosts. Genome Biol 14(6):121. doi:10.1186/gb-2013-14-6-121

Rey T, Nars A, Bonhomme M, Bottin A, Huguet S, Balzergue S, Jardinaud MF, Bono JJ, Cullimore J, Dumas B, Gough C, Jacquet C (2013) NFP, a LysM protein controlling Nod factor perception, also intervenes in Medicago truncatula resistance to pathogens. New Phytol 198(3):875–886. doi:10.1111/nph.12198

Rey T, Chatterjee A, Buttay M, Toulotte J, Schornack S (2015) Medicago truncatula symbiosis mutants affected in the interaction with a biotrophic root pathogen. New Phytol 206 (2):497–500. doi:10.1111/Nph.13233

Ruiz-Lozano JM, Gianinazzi S, Gianinazzi-Pearson V (1999) Genes involved in resistance to powdery mildew in barley differentially modulate root colonization by the mycorrhizal fungus Glomus mosseae. Mycorrhiza 9(4):237–240. doi:10.1007/S005720050273

Ruocco M, Lanzuise S, Lombardi N, Woo SL, Vinale F, Marra R, Varlese R, Manganiello G, Pascale A, Scala V, Turra D, Scala F, Lorito M (2015) Multiple roles and effects of a novel Trichoderma hydrophobin. Mol Plant Microbe Interact 28(2):167–179. doi:10.1094/MPMI-07-14-0194-R

Russell J, Bulman S (2005) The liverwort Marchantia foliacea forms a specialized symbiosis with arbuscular mycorrhizal fungi in the genus Glomus. New Phytol 165(2):567–579. doi:10.1111/j.1469-8137.2004.01251.x

Salzer P, Bonanomi A, Beyer K, Vogeli-Lange R, Aeschbacher RA, Lange J, Wiemken A, Kim D, Cook DR, Boller T (2000) Differential expression of eight chitinase genes in Medicago truncatula roots during mycorrhiza formation, nodulation, and pathogen infection. Mol Plant Microbe Interact 13(7):763–777. doi:10.1094/MPMI.2000.13.7.763

Sanchez-Vallet A, Mesters JR, Thomma BP (2015) The battle for chitin recognition in plant-microbe interactions. FEMS Microbiol Rev 39(2):171–183. doi:10.1093/femsre/fuu003

Scherm B, Balmas V, Spanu F, Pani G, Delogu G, Pasquali M, Migheli Q (2013) Fusarium culmorum: causal agent of foot and root rot and head blight on wheat. Mol Plant Pathol 14 (4):323–341. doi:10.1111/Mpp.12011

Schneebeli K, Mathesius U, Watt M (2015) Brachypodium distachyon is a pathosystem model for the study of the wheat disease rhizoctonia root rot. Plant Pathol 64(1):91–100. doi:10.1111/ Ppa.12227

Schreiber C, Slusarenko AJ, Schaffrath U (2011) Organ identity and environmental conditions determine the effectiveness of nonhost resistance in the interaction between Arabidopsis thaliana and Magnaporthe oryzae. Mol Plant Pathol 12(4):397–402. doi:10.1111/j.1364-3703.2010.00682.x

Sesma A, Osbourn AE (2004) The rice leaf blast pathogen undergoes developmental processes typical of root-infecting fungi. Nature 431(7008):582–586. doi:10.1038/nature02880

Shimizu T, Nakano T, Takamizawa D, Desaki Y, Ishii-Minami N, Nishizawa Y, Minami E, Okada K, Yamane H, Kaku H, Shibuya N (2010) Two LysM receptor molecules, CEBiP and OsCERK1, cooperatively regulate chitin elicitor signaling in rice. Plant J 64(2):204–214. doi:10.1111/j.1365-313X.2010.04324.x

Shukla N, Awasthi RP, Rawat L, Kumar J (2015) Seed biopriming with drought tolerant isolates of Trichoderma harzianum promote growth and drought tolerance in Triticum aestivum. Ann Appl Biol 166(2):171–182. doi:10.1111/Aab.12160

Silipo A, Erbs G, Shinya T, Dow JM, Parrilli M, Lanzetta R, Shibuya N, Newman MA, Molinaro A (2010) Glyco-conjugates as elicitors or suppressors of plant innate immunity. Glycobiology 20 (4):406–419. doi:10.1093/Glycob/Cwp201

Steinkellner S, Mammerler R (2007) Effect of flavonoids on the development of Fusarium oxysporum f. sp lycopersici. J Plant Interact 2(1):17–23. doi:10.1080/17429140701409352

Steinkellner S, Lendzemo V, Langer I, Schweiger P, Khaosaad T, Toussaint JP, Vierheilig H (2007) Flavonoids and strigolactones in root exudates as signals in symbiotic and pathogenic plant-fungus interactions. Molecules 12(7):1290–1306

Stergiopoulos I, de Wit PJ (2009) Fungal effector proteins. Ann Rev Phytopathol 47:233–263. doi:10.1146/annurev.phyto.112408.132637

Stracke S, Kistner C, Yoshida S, Mulder L, Sato S, Kaneko T, Tabata S, Sandal N, Stougaard J, Szczyglowski K, Parniske M (2002) A plant receptor-like kinase required for both bacterial and fungal symbiosis. Nature 417(6892):959–962. doi:10.1038/nature00841

Sukno SA, Garcia VM, Shaw BD, Thon MR (2008) Root infection and systemic colonization of maize by Colletotrichum graminicola. Appl Environ Microb 74(3):823–832. doi:10.1128/ Aem.01165-07

Sun J, Miller JB, Granqvist E, Wiley-Kalil A, Gobbato E, Maillet F, Cottaz S, Samain E, Venkateshwaran M, Fort S, Morris RJ, Ane JM, Denarie J, Oldroyd GE (2015) Activation of symbiosis signaling by arbuscular mycorrhizal fungi in legumes and rice. Plant Cell 27 (3):823–838. doi:10.1105/tpc.114.131326

Takeda N, Tsuzuki S, Suzaki T, Parniske M, Kawaguchi M (2013) CERBERUS and NSP1 of Lotus japonicus are common symbiosis genes that modulate arbuscular mycorrhiza development. Plant Cell Physiol 54(10):1711–1723. doi:10.1093/pcp/pct114

Takemoto D, Hardham AR (2004) The cytoskeleton as a regulator and target of biotic interactions in plants. Plant Physiol 136(4):3864–3876. doi:10.1104/pp.104.052159

Takemoto D, Jones DA, Hardham AR (2003) GFP-tagging of cell components reveals the dynamics of subcellular re-organization in response to infection of Arabidopsis by oomycete pathogens. Plant J 33(4):775–792. doi:10.1046/J.1365-313x.2003.01673.X

Tisserant E, Kohler A, Dozolme-Seddas P, Balestrini R, Benabdellah K, Colard A, Croll D, Da Silva C, Gomez SK, Koul R, Ferrol N, Fiorilli V, Formey D, Franken P, Helber N, Hijri M, Lanfranco L, Lindquist E, Liu Y, Malbreil M, Morin E, Poulain J, Shapiro H, van Tuinen D, Waschke A, Azcon-Aguilar C, Becard G, Bonfante P, Harrison MJ, Kuster H, Lammers P, Paszkowski U, Requena N, Rensing SA, Roux C, Sanders IR, Shachar-Hill Y, Tuskan G, Young JPW, Gianinazzi-Pearson V, Martin F (2012) The transcriptome of the arbuscular mycorrhizal fungus Glomus intraradices (DAOM 197198) reveals functional tradeoffs in an obligate symbiont. New Phytol 193(3):755–769. doi:10.1111/J.1469-8137.2011.03948.X

Tisserant E, Malbreil M, Kuo A, Kohler A, Symeonidi A, Balestrini R, Charron P, Duensing N, Frey NFD, Gianinazzi-Pearson V, Gilbert LB, Handa Y, Herr JR, Hijri M, Koul R, Kawaguchi M, Krajinski F, Lammers PJ, Masclauxm FG, Murat C, Morin E, Ndikumana S, Pagni M, Petitpierre D, Requena N, Rosikiewicz P, Riley R, Saito K, Clemente HS, Shapiro H, Van Tuinen D, Becard G, Bonfante P, Paszkowski U, Shachar-Hill YY, Tuskan GA, Young PW, Sanders IR, Henrissat B, Rensing SA, Grigoriev IV, Corradi N, Roux C, Martin F (2013) Genome of an arbuscular mycorrhizal fungus provides insight into the oldest plant symbiosis. Proc Natl Acad Sci USA 110(50):20117–20122. doi:10.1073/Pnas.1313452110

Tschaplinski TJ, Plett JM, Engle NL, Deveau A, Cushman KC, Martin MZ, Doktycz MJ, Tuskan GA, Brun A, Kohler A, Martin F (2014) Populus trichocarpa and Populus deltoides exhibit different metabolomic responses to colonization by the symbiotic fungus Laccaria bicolor. Mol Plant Microbe Interact 27(6):546–556. doi:10.1094/Mpmi-09-13-0286-R

Tucci M, Ruocco M, De Masi L, De Palma M, Lorito M (2011) The beneficial effect of Trichoderma spp. on tomato is modulated by the plant genotype. Mol Plant Pathol 12 (4):341–354. doi:10.1111/j.1364-3703.2010.00674.x

Upson R, Read DJ, Newsham KK (2009) Nitrogen form influences the response of Deschampsia antarctica to dark septate root endophytes. Mycorrhiza 20(1):1–11. doi:10.1007/s00572-009-0260-3

Usuki F, Narisawa K (2007) A mutualistic symbiosis between a dark septate endophytic fungus, Heteroconium chaetospira, and a nonmycorrhizal plant, Chinese cabbage. Mycologia 99 (2):175–184

Vadassery J, Oelmuller R (2009) Calcium signaling in pathogenic and beneficial plant microbe interactions: what can we learn from the interaction between Piriformospora indica and Arabidopsis thaliana. Plant Signal Behav 4(11):1024–1027

Vadassery J, Ranf S, Drzewiecki C, Mithofer A, Mazars C, Scheel D, Lee J, Oelmuller R (2009) A cell wall extract from the endophytic fungus Piriformospora indica promotes growth of Arabidopsis seedlings and induces intracellular calcium elevation in roots. Plant J 59 (2):193–206. doi:10.1111/j.1365-313X.2009.03867.x

Vahabi K, Sherameti I, Bakshi M, Mrozinska A, Ludwig A, Reichelt M, Oelmuller R (2015) The interaction of Arabidopsis with Piriformospora indica shifts from initial transient stress induced by fungus-released chemical mediators to a mutualistic interaction after physical contact of the two symbionts. BMC Plant Biol 15:58. doi:10.1186/s12870-015-0419-3

Van Buyten E, Hofte M (2013) Pythium species from rice roots differ in virulence, host colonization and nutritional profile. BMC Plant Biol 13:203. doi:10.1186/1471-2229-13-203

van der Heijden MG, Bardgett RD, van Straalen NM (2008) The unseen majority: soil microbes as drivers of plant diversity and productivity in terrestrial ecosystems. Ecol Lett 11(3):296–310. doi:10.1111/j.1461-0248.2007.01139.x

Veiga RSL, Faccio A, Genre A, Pieterse CMJ, Bonfante P, van der Heijden MGA (2013) Arbuscular mycorrhizal fungi reduce growth and infect roots of the non-host plant Arabidopsis thaliana. Plant Cell Environ 36(11):1926–1937. doi:10.1111/Pce.12102

Venard C, Vaillancourt L (2007) Penetration and colonization of unwounded maize tissues by the maize anthracnose pathogen Colletotrichum graminicola and the related nonpathogen C. sublineolum. Mycologia 99(3):368–377

Wang B, Qiu YL (2006) Phylogenetic distribution and evolution of mycorrhizas in land plants. Mycorrhiza 16(5):299–363. doi:10.1007/s00572-005-0033-6

Wang W, Wen Y, Berkey R, Xiao S (2009) Specific targeting of the Arabidopsis resistance protein RPW8.2 to the interfacial membrane encasing the fungal Haustorium renders broad-spectrum resistance to powdery mildew. Plant Cell 21(9):2898–2913. doi:10.1105/tpc.109.067587

Wang E, Schornack S, Marsh JF, Gobbato E, Schwessinger B, Eastmond P, Schultze M, Kamoun S, Oldroyd GE (2012) A common signaling process that promotes mycorrhizal and oomycete colonization of plants. Curr Biol 22(23):2242–2246. doi:10.1016/j.cub.2012.09.043

Wang W, Zhang Y, Wen Y, Berkey R, Ma X, Pan Z, Bendigeri D, King H, Zhang Q, Xiao S (2013) A comprehensive mutational analysis of the Arabidopsis resistance protein RPW8.2 reveals key amino acids for defense activation and protein targeting. Plant Cell 25(10):4242–4261. doi:10.1105/tpc.113.117226

Wegel E, Schauser L, Sandal N, Stougaard J, Parniske M (1998) Mycorrhiza mutants of Lotus japonicus define genetically independent steps during symbiotic infection. Mol Plant Microbe Interact 11(9):933–936. doi:10.1094/Mpmi.1998.11.9.933

Yamazaki A, Hayashi M (2015) Building the interaction interfaces: host responses upon infection with microorganisms. Curr Opin Plant Biol 23:132–139. doi:10.1016/j.pbi.2014.12.003

Yedidia II, Benhamou N, Chet II (1999) Induction of defense responses in cucumber plants (Cucumis sativus L.) by the biocontrol agent trichoderma harzianum. Appl Environ Microbiol 65(3):1061–1070

Yi M, Valent B (2013) Communication between filamentous pathogens and plants at the biotrophic interface. Ann Rev Phytopathol 51:587–611. doi:10.1146/annurev-phyto-081211-172916

Yoneyama K, Xie X, Kusumoto D, Sekimoto H, Sugimoto Y, Takeuchi Y, Yoneyama K (2007) Nitrogen deficiency as well as phosphorus deficiency in sorghum promotes the production and exudation of 5-deoxystrigol, the host recognition signal for arbuscular mycorrhizal fungi and root parasites. Planta 227(1):125–132. doi:10.1007/s00425-007-0600-5

Yoneyama K, Xie X, Kisugi T, Nomura T, Yoneyama K (2013) Nitrogen and phosphorus fertilization negatively affects strigolactone production and exudation in sorghum. Planta 238(5):885–894. doi:10.1007/s00425-013-1943-8

Zhang X, Dong W, Sun J, Feng F, Deng Y, He Z, Oldroyd GE, Wang E (2015) The receptor kinase CERK1 has dual functions in symbiosis and immunity signalling. Plant J 81(2):258–267. doi:10.1111/tpj.12723

Root Exudates as Integral Part of Belowground Plant Defence

Ulrike Baetz

Abstract Root exudates comprise a heterogeneous group of compounds that display various effects on soilborne organisms, including stimulation, attraction, but also repellence and inhibition. Therefore, root-secreted chemicals can assist belowground plant defence through direct and/or indirect mechanisms. Direct defence strategies exploited by roots include the secretion of phytochemicals with antimicrobial, insecticide, or nematicide properties. In contrast, other root exudates recruit or influence beneficial organisms to serve as biological weapons against plant aggressors, a mechanism termed indirect plant defence. Since rhizosecretion fundamentally shapes the composition of soil-inhabiting organisms and contributes to plant survival, the quality and quantity of defence root exudates are tightly controlled. Various environmental and endogenous factors can stimulate the release of phytochemicals that exhibit precisely targeted bioactivities. On the molecular level, several primary active transport proteins have been demonstrated to affect the composition of defence root exudates in the rhizosphere. In this chapter, we will focus our attention on direct and indirect defence strategies mediated by root exudates. In addition, we will shed light on regulatory mechanisms of defence-related root exudation that prevent belowground disease and ensure optimal plant performance.

1 Introduction

Plants interact with a multitude of soilborne organisms in complex biological and ecological processes in the narrow zone surrounding the root system, termed the rhizosphere. These beneficial, antagonistic, or neutral interactions have a profound effect on plant health and survival and shape the soil microbiome.

Within the rhizosphere, roots are constantly exposed to biotic stressors, ranging from plant disease-causing pathogens such as bacteria, fungi, and oomycetes to nematodes and insects. Although being sessile organisms anchored to the soil, plants are not just passive victims of these antagonistic microbes and invertebrates

U. Baetz (✉)
Department of Plant and Microbial Biology, University of Zurich, Zollikerstrasse 107, 8008 Zurich, Switzerland
e-mail: baetzu@botinst.uzh.ch

C.M.F. Vos, K. Kazan (eds.), *Belowground Defence Strategies in Plants*, Signaling and Communication in Plants, DOI 10.1007/978-3-319-42319-7_3

that occur in the vicinity of roots. In fact, roots are equipped with an arsenal of defence compounds that can be released into the rhizosphere to counteract plant attackers (Baetz and Martinoia 2014). However, the significance of root exudates as a direct or indirect belowground protection has long been underestimated, presumably due to literally being out of sight.

Secreted substances can be of low or high molecular weight. Low-molecular-weight root exudates include a variety of defence secondary metabolites such as flavonoids, glucosinolates, and terpenoids. Protective high-molecular-weight compounds such as antimicrobial proteins and secreted extracellular DNA also contribute to the local belowground resistance. The tremendous metabolic diversity of root exudates has been progressively elucidated in the past decade through the identification and characterization of numerous novel constitutively secreted and inducible compounds and previously undescribed classes of defence molecules. Equally, genes and biosynthetic pathways involved in the production of these phytochemicals have been gradually deciphered. A deepened knowledge of phytochemical properties, their composition in the rhizosphere, and their impact on soil-inhabiting organisms is crucial to understand the diverse nature of root-exudate-mediated defence mechanisms that protect plants against pathogens and invaders. It has been demonstrated that some root exudates exhibit antibacterial, antifungal, nematicide, or insecticide properties that directly assist the plant in coping with antagonistic organisms. Other root exudates are released from damaged roots to attract natural enemies of the attackers (such as carnivorous nematodes) to indirectly protect plants. Another highly sophisticated indirect defence strategy of plants is to outsource defence compound production. On that purpose, root exudates attract beneficial microorganisms that release secondary metabolites such as antibiotics with an antagonistic effect on the root-attacking pathogen.

In this chapter, we will compile the roles of root exudates in various direct and indirect, targeted belowground defence processes that protect plants against soilborne diseases. In addition, we will discuss regulation mechanisms of root exudation, e.g., inducible substance production and controlled secretion, that collectively make root-exudate-mediated belowground plant defence a highly efficient process.

2 Root Exudation as a Direct Defence Strategy Against Detrimental Soilborne Organisms

In the rhizosphere, roots face relentless harmful attack through the presence of plant disease-causing pathogens (e.g., bacteria, fungi, and oomycetes), as well as root-damaging animals (in particular nematodes and insects). In the following, we will illustrate with selected examples how aggressors are being repelled, inhibited, or killed by certain root-secreted phytochemicals in order to confer direct defence against belowground plant diseases.

2.1 Bacteria

The bacterial community in the soil is diverse in its composition, ranging from beneficial plant growth-promoting bacteria to bacteria that infect roots and exhibit harmful effects. Plant-derived molecules can act as chemical signals that stimulate or repress microbes. Thereby, root exudates fundamentally drive the selection of bacteria inhabiting the rhizosphere. Shifts in root-exudate blends, as observed in an *Arabidopsis* (*Arabidopsis thaliana*) mutant impaired in root exudation, elicited significant compositional alterations in bacteria that colonize the rhizosphere (Badri et al. 2009). Furthermore, it has been recently reported that merely the application of root exudates collected from *Arabidopsis* modulated the overall native bacterial community in the soil, even in the absence of the plant (Badri et al. 2013). Conversely, the chemical profile of root-secreted molecules is largely dependent on distinct bacterial members present in the vicinity of roots. For instance, the formation and release of the antimicrobial monoterpene 1,8-cineole were induced upon compatible interactions between *Arabidopsis* roots and the bacterial pathogen *Pseudomonas syringae* DC3000 (Steeghs et al. 2004; Kalemba et al. 2002). In another study, *Arabidopsis* roots that were exposed to *P. syringae* secreted significantly higher amounts of defence-related proteins, whereas the incompatible interaction with a bacterial symbiont did not induce the secretion of these protective proteins (De-la-Peña et al. 2008).

A phytochemical known to feature direct antibacterial activity particularly against *Pseudomonas aeruginosa* is rosmarinic acid (RA) (Bais et al. 2002). This multifunctional caffeic acid ester is produced in hairy root cultures of sweet basil (*Ocimum basilicum* L.) and exuded in response to pathogen attack. However, the compound is absent from exudates of unchallenged root cultures (Bais et al. 2002). *Arabidopsis* root exudates that were supplemented with exogenous RA prior infection with pathogenic *P. aeruginosa* strains highly reduced pathogenicity under in vitro and in vivo conditions (Walker et al. 2004). Without supplementation, *Arabidopsis* roots displayed a high level of susceptibility to *P. aeruginosa* resulting in mortality. Similarly, the induction of RA secretion by sweet basil roots before infection conferred resistance to *P. aeruginosa* (Walker et al. 2004). Hence, host plants can deliberately release antibacterial molecules into the rhizosphere that directly counteract root colonization of pathogenic bacteria and plant mortality.

2.2 Fungi and Oomycetes

Tremendous yield losses result from fungal root invasion every year, emphasizing the necessity to study the cross talk between plants and fungi and to elucidate root exudates that confer direct disease resistance. In fact, oomycetes are phylogenetically distinct organisms but show high physiological and morphological similarities to fungi. Therefore, fungi and oomycetes will be both covered in this section.

A potent root-secreted antimicrobial compound that is implemented into defence mechanisms against oomycete pathogens is the pea (*Pisum sativum*) isoflavonoid pisatin (Cannesan et al. 2011). Once pea roots were challenged with the oomycete *Aphanomyces euteiches*, the biosynthesis and release of pisatin into the rhizosphere were induced (Cannesan et al. 2011). Interestingly, the inoculation also had a stimulatory effect on border cell production of pea. Border cells are metabolically active cells at the root periphery that originate and detach from the root cap meristem (Stubbs et al. 2004; Vicré et al. 2005; Driouich et al. 2007). They assist the growing root tip during the mechanical penetration of the soil by decreasing frictional resistance at the root–soil interface (Driouich et al. 2007). In addition, antimicrobial molecules in the rhizosphere largely derive from cap and border cells (Hawes et al. 2012; Griffin et al. 1976; Odell et al. 2008), revealing a link between the *A. euteiches* induced formation of border cells and the increased pisatin exudation (Cannesan et al. 2011). The exposure of pea root tips encompassing border cells to exogenous pisatin, in turn, led to the upregulation of border cell production in vitro (Curlango-Rivera et al. 2010). Hence, border cells and their exudates account for a local protective shield that is strengthened in response to pathogen invasion (Cannesan et al. 2011; Hawes et al. 2012; Curlango-Rivera et al. 2010). Because a correlation was observed between border cell separation and the induction of protein secretion, Wen et al. (2007) proteolytically degraded the root cap secretome during inoculation with the pea-pathogenic fungus *N. haematococca*. The researcher demonstrated that protease treatment increased the percentage of infected root tips significantly, providing evidence that root-secreted defence proteins from border cells contribute fundamentally to the resistance of pea roots to fungal infection (Wen et al. 2007). Detailed proteome analysis of root exudates of several plant species confirmed the secretion of antimicrobial enzymes and demonstrated dynamic compositional changes during development and upon pathogenic interactions (De-la-Peña et al. 2010; Shinano et al. 2011; Liao et al. 2012; Ma et al. 2010; De-la-Peña et al. 2008; Wen et al. 2007). Unexpectedly, besides defence-related proteins, also the DNA-binding protein histone H4 was detected in border cell exudates of pea (Wen et al. 2007). Histone-linked extracellular DNA (exDNA) is thought to have a critical role in defence against microbial pathogens in mammals (von Köckritz-Blickwede and Nizet 2009; Brinkmann et al. 2004; Medina 2009). In plants, exDNA linked to histone proteins has been found to be exuded from root border cells and suggested to be a component of direct belowground defence against fungal invasion (Wen et al. 2009). Similar to proteolytic solubilization of exuded proteins, nuclease treatment of pea root tips resulted in enhanced susceptibility to fungal infection by *N. haematococca* (Wen et al. 2009). However, the distinct mechanism of how exDNA inhibits pathogen infection awaits elucidation (Hawes et al. 2011, 2012). In addition to protective proteins and exDNA, also low-molecular-weight antimicrobial root exudates are proved direct chemical weapons against soilborne diseases of fungal origin. For instance, the phenolic compound *t*-cinnamic acid potently protects barley (*Hordeum vulgare*) against the soilborne fungus *Fusarium graminearum* (Lanoue et al. 2010a, b).

2.3 Nematodes

Nematodes are wormlike eukaryotic invertebrates that consume bacteria, fungi, or other nematodes, and some can parasitize plants. Intense research on root-secreted compounds uncovered attractants that influence the chemotaxis response of beneficial nematodes or assist pathogenic nematodes in host recognition. Other phytochemicals have been found to exhibit nematode-antagonistic properties (Reynolds et al. 2011; Curtis 2008; Hiltpold and Turlings 2012). Lilley et al. (2011) investigated the potency of a root-exuded direct defence compound against nematodes. The researchers showed that the root cap targeted expression and release of a nematode-repellent chemodisruptive peptide in *Arabidopsis thaliana* reduced the establishment of the beet cyst nematode *Heterodera schachtii* (Lilley et al. 2011). In line with this, it was found that transgenic *Solanum tuberosum* (potato) that secreted this repellent peptide from their roots suppressed parasitism by the potato cyst nematode *Globodera pallida* (Lilley et al. 2011; Liu et al. 2005). In another study, a genetic approach was used to broaden the resistance of soybean (*Glycine max*) against nematodes. An *Arabidopsis* gene that modulates synthesis of the antimicrobial camalexin and other defence-related responses was ectopically overexpressed in roots of soybean (Youssef et al. 2013), resulting in enhanced resistance to the parasitic soybean cyst nematode (*Heterodera glycines*) and the root-knot nematode (*Meloidogyne incognita*). Lauric acid, a naturally occurring, highly abundant root exudate from crown daisy (*Chrysanthemum coronarium*) also limited parasitic damage by decreasing the number of *M. incognita* and suppressing nematode infection (Dong et al. 2014). Likewise, total root-cap exudates from various legumes showed the ability to repel root-knot nematodes in sand assays (Zhao et al. 2000). In summary, root exudates can have direct nematotoxic or repelling effects to ensure protection of the roots. However, in contrast to compounds with antimicrobial activity, examples for nematicide root exudates remain limited.

2.4 Insects

As plants cannot escape belowground insects and root feeding causes tremendous tissue damage, roots employ elegant defence strategies to counteract herbivory. For instance, the semi-volatile diterpene hydrocarbon, rhizathalene A, is constitutively produced and released by noninfected *Arabidopsis* roots (Vaughan et al. 2013). Plants that are deficient in rhizathalene A production were found to be less resistant to herbivory by the fungus gnat (*Bradysia* spp.) and suffered considerable removal of peripheral tissue at larval feeding sites. In this study it was comprehensively shown that rhizathalene A is a local antiherbivore metabolite that is implicated in the direct belowground defence against insect herbivory (Vaughan et al. 2013). The monoterpene 1,8-cineole is another volatile compound that exhibits defence activity. It is released from *Arabidopsis thaliana* roots upon compatible interaction with the herbivore *Diuraphis noxia* (Steeghs et al. 2004). However, little is known about root-released volatiles and other root exudates with insecticidal properties that directly defend plants against root-feeding arthropods. Nevertheless, as discussed in the following, belowground volatile compounds and their protective role were extensively studied as an indirect defence trait.

3 Root Exudates Are a Tool to Establish Indirect Plant Defence

Direct defence via root exudation is an effective mean of plants to deal with the constant exposure to pathogenic microbes and invertebrates in the rhizosphere. Besides, by root exudation plants can influence the behavior of phytobeneficial soil organisms to serve defensive roles during belowground diseases. For instance, the orientation of rhizospheric nematodes that are predators of insect aggressors can be altered by root-released signals, thereby indirectly conferring resistance to the roots against herbivory (Rasmann et al. 2005). Furthermore, some rhizobacteria species are known for their production of toxic compounds targeting plant pathogens, a process that has been hypothesized to be regulated by root exudates upon infection (Jousset et al. 2011; Haas and Défago 2005). A scenario in which plants recruit defence-assisting organisms to counteract pathogen attack is considered indirect belowground plant defence. This tripartite interaction is mediated by root exudates.

3.1 Recruitment of "Natural Soldiers" by Root Exudates

The concept of indirect defence and the corresponding plant-released signaling compounds has been examined thoroughly in the aboveground terrestrial environment. Leaves emit a complex battery of volatile organic compounds to communicate with their environment and attract predators. Intriguingly, when attacked by belowground herbivores, plants can also attract soilborne mobile predators such as entomopathogenic nematodes (EPNs). In fact, EPNs are plant protagonists but obligate parasites that kill insect hosts. The pivotal role of root-emitted volatile compounds that act as efficient cues to direct natural enemies such as EPNs specifically to the sites where potential hosts are damaging roots has become increasingly evident in the last years (Hiltpold and Turlings 2008; Hiltpold et al. 2011). The best studied example of a volatile signal that mediates belowground indirect plant defence is the maize (*Zea mays* L.) sesquiterpene olefin (*E*)-β-caryophyllene (EβC) (Rasmann et al. 2005). EβC is completely absent in healthy maize roots but emitted upon feeding by voracious larvae of the Western corn rootworm (WCR), *Diabrotica virgifera virgifera*. Herbivore attack induces the expression of the *terpene synthase 23* (*tps23*) gene, which is involved in the biosynthesis of EβC (Capra et al. 2014; Köllner et al. 2008). The released volatile signal strongly attracts the EPN *Heterorhabditis megidis*, a natural enemy of root-feeding herbivores that assists maize defence by killing WCR larvae (Rasmann et al. 2005).

WCR is a severe pest causing tremendous yield losses particularly on maize roots (Miller et al. 2005). Exploiting naturally produced indirect defence compounds against WCR could provide an effective biological control strategy for crop protection. Degenhardt et al. (2009) aimed at promoting plant attractiveness to natural enemies of WCR larvae by genetically introducing EβC emission in maize varieties that are not capable of synthesizing the sesquiterpene due to a lack of *tps23* transcript. On that purpose, a non-emitting maize line was transformed with an (*E*)-β-caryophyllene synthase from oregano (*Origanum vulgare*), resulting in a constitutive emission of EβC (Degenhardt et al. 2009). In field experiments, transformed plants attracted EPNs more efficiently and consequently suffered less root feeding by WCR larvae compared to non-emitting maize plants. In a subsequent study, it has been demonstrated that a constitutive emission of the volatile signal generated also physiological costs such as compromised seed germination, plant growth, and yield (Robert et al. 2013). This negative effect on plant fitness was possibly due to an increased attraction of herbivores, including aboveground pests. Ali et al. (2010, 2012) similarly exercised caution when investigating the complex effects of belowground volatiles on indirect plant defence. Citrus roots release volatile compounds such as pregeijerene (1,5-dimethylcyclodeca-1,5,7-triene) in response to feeding by the larvae of the root weevil, *Diaprepes abbreviatus* (Ali et al. 2010, 2012). The herbivore-induced volatile emission recruited a naturally occurring EPN (*Steinernema diaprepesi*), resulting in an increase of root weevil mortality and, hence, the control of herbivore infestation (Ali et al. 2010, 2012). Yet, further research uncovered that besides the recruitment of beneficial nematodes, herbivore-induced volatiles also allowed more efficient host localization by phytopathogenic nematodes (Ali et al. 2011). Collectively these studies illustrate clearly that consequences evoked by the manipulation of belowground volatile emission should be carefully assessed on multitrophic levels and under field conditions in order to understand their specificity and minimize detrimental physiological or ecological effects for plants or nontarget organisms.

Besides targeting volatile emission, another elegant approach to enhance the effectiveness of indirect plant defence is selective breeding of natural enemies for increased responsiveness to a volatile host signal in order to obtain a more efficient natural finding and killing of pests. Hiltpold et al. (2010a) aimed at improving the attraction of *Heterorhabditis bacteriophora*, one of the most virulent nematodes against WCR larvae, toward EβC (Hiltpold et al. 2010a). After few generations of selection, the researchers isolated an *H. bacteriophora* strain that was significantly more attracted to the EβC source than the original strain. Consistently, in field experiments WCR populations that attacked EβC-emitting maize roots were more effectively reduced by the selected strain compared to the original strain. Importantly, control experiments showed that this artificial selection for the responsiveness trait of *H. bacteriophora* toward the volatile signal has not considerably altered other essential properties for controlling WCR populations such as the infectiveness of *H. bacteriophora* (Hiltpold et al. 2010a, b).

Taken together, the research shows that plants can recruit natural enemies of their soilborne aggressors through root-released volatiles to indirectly defend the root system. Thoroughly exploited manipulation of indirect plant defence has a great potential as an alternative method to traditional broad-spectrum pesticides in controlling root pests in agroecosystems.

3.2 Root Exudates Can Stimulate the Antimicrobial Potency of Phytobeneficial Microbes

Besides attracting natural predators of their enemies, plants have established dialogues with beneficial root-colonizing bacteria to protect roots against the attack of deleterious rhizosphere microorganisms. Defence-assisting microbes belong to so-called plant growth-promoting rhizobacteria (PGPR) (Compant et al. 2005). PGPR primarily stimulate plant growth by, e.g., the production of phytohormones or the enhancement of plant nutrition (Vacheron et al. 2013). In contrast, defence-assisting PGPR can improve plant health either directly by repelling plant aggressors with the production of antibiotics or indirectly by eliciting induced systemic resistance in host plants (Compant et al. 2005; Haas and Défago 2005; Doornbos et al. 2012). However, to date only few studies addressed the role and the chemical nature of plant-derived exudates in the suppression of soilborne diseases via direct bacterial antagonism (Neal et al. 2012; Neal and Ton 2013; Santos et al. 2014; Jousset et al. 2011; Haas and Défago 2005; Notz et al. 2001; Baehler et al. 2005; de Werra et al. 2008, 2011). Jousset et al. (2011) made an elaborate experiment providing compelling evidence that plants are able to influence the metabolism of beneficial rhizosphere-colonizing bacteria through root exudates as part of the indirect belowground plant defence against pathogens. In order to prevent physical contact between the microorganisms, barley plants were grown in a split-root system in which one part of the roots was challenged by the pathogenic oomycete *Pythium ultimum*. The other side was inoculated with the biocontrol bacterium *Pseudomonas fluorescens* CHA0, a PGPR known to assist crop plant defence by producing antifungal chemicals against pathogenic fungi (Haas and Défago 2005). This separation system allowed the investigation of alterations of bacterial gene expression patterns that are induced by pathogens but mediated by systemic signaling of plants and root exudation (Fig. 1). The researchers found that the expression of the bacterial *phlA* gene was considerably stimulated following pathogen infection at the other side of the root (Jousset et al. 2011). The expression of this gene reflects the production of the antifungal metabolite 2,4-diacetylphloroglucinol (DAPG), a key component of the biocontrol activity of root-associated bacteria acting in disease suppression (Notz et al. 2001; Bangera

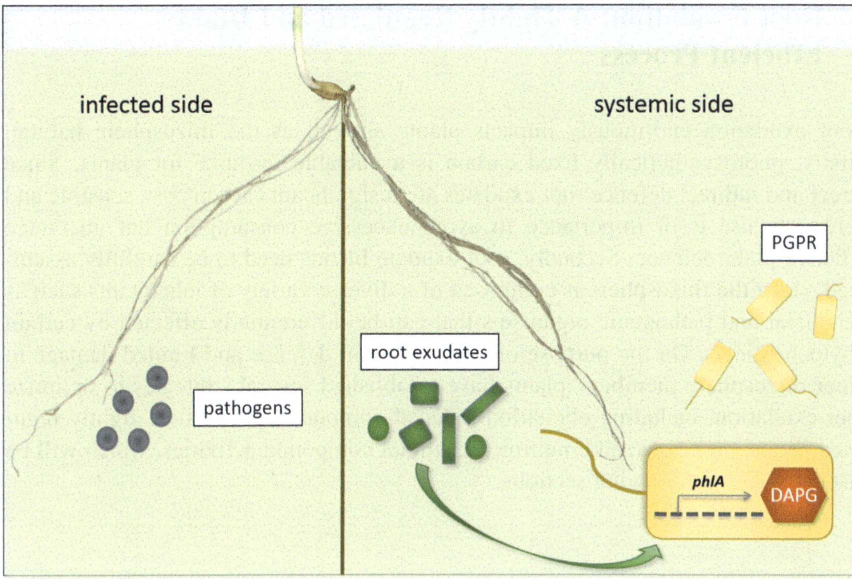

Fig. 1 Relevance of systemic plant signaling and root exudation in a tripartite interaction that confers indirect plant defence. To investigate pathogen-induced but plant-mediated modulation of bacterial gene expression and antifungal activity, Jousset et al. (2011) grew barley in a split-root system (Jousset et al. 2011). One part of the root (infected side) was challenged with the pathogen *Pythium ultimum*, whereas the other part of the root (systemic side) was exposed to the beneficial plant growth-promoting rhizobacterium (PGPR) *Pseudomonas fluorescens* CHA0. Without physical contact but through systemic plant signaling, pathogen attack induced compositional changes in root exudates on the systemic side. These changes, in turn, stimulated bacterial *phlA* expression. The transcript levels of this gene directly correlate with the production of the antifungal compound 2,4-diacetylphloroglucinol (DAPG)

and Thomashow 1999; de Souza et al. 2003). Interestingly, also the composition of exudates from the systemic side at which roots were inoculated with *P. fluorescens* changed in response to the presence of the pathogen *Pythium ultimum* at the other side of the root system (Fig. 1), uncovering candidates of signaling root exudates that provoke changes in antifungal gene expression of beneficial bacteria (Jousset et al. 2011). In summary, first insights have been gained on how antifungal activities of rhizobacteria can be adjusted by root exudates to provide service of indirect defence against plant pathogens. It will be of interest to further explore this tripartite interaction and investigate how and which plant-derived compounds are released under pathogen pressure and subsequently modulate rapidly the activity of plant growth-promoting rhizobacteria.

4 Root Exudation: A Tightly Regulated and Highly Efficient Process

Root exudation enormously impacts plants as well as the rhizosphere habitat. Firstly, photosynthetically fixed carbon is a valuable resource for plants. Since direct and indirect defence root exudates are a significant carbon cost, sensible and deliberate use is of importance to avoid excessive consumption but guarantee efficient plant defence. Secondly, root-exudate blends need to be carefully assembled, since the rhizosphere is composed of a diverse variety of inhabitants such as beneficial and pathogenic organisms that can be differentially affected by certain phytochemicals. On the purpose of accurate plant defence and limited damage to other rhizosphere members, plants have established several strategies to optimize root exudation, including elicitation-induced compound production, tightly regulated export processes, and multiple beneficial compound activities, which will be discussed in the following sections.

4.1 Constitutive Versus Induced Exudation of Phytochemicals

Plants are constantly exposed to soilborne antagonists. To form a protective buffer zone around roots, certain defence root exudates are constitutively released into the rhizosphere. For instance, rhizathalene A, an antifeedant involved in direct plant defence, is synthesized and secreted from *Arabidopsis* roots even in the absence of root-feeding insects (Vaughan et al. 2013). Plants secrete a wide array of other defence molecules before pathogen elicitation (Kato-Noguchi et al. 2008; Toyomasu et al. 2008; Wen et al. 2009; De-la-Peña et al. 2010; Shinano et al. 2011; Chaparro et al. 2013; Badri et al. 2010; Liao et al. 2012; Ma et al. 2010; McCully et al. 2008; Dong et al. 2014). Besides a constitutive root exudation, the biosynthesis, accumulation, and secretion of certain defence molecules can be induced in the presence of aggressors in the rhizosphere. The phenolic compound *t*-cinnamic acid is an antifungal exudate of barley roots (Lanoue et al. 2010a, b). Upon attack of a fungal pathogen, labeling experiments demonstrated the de novo biosynthesis and secretion of this aromatic defence metabolite into the rhizosphere. Another example is rosmarinic acid, which is constitutively produced in root tissue but exclusively released into the rhizosphere in response to root infection (Bais et al. 2002). These studies illustrate that the profile of root exudates is not just diverse in its composition but also strikingly dynamic, to adjust the identity and amount of defence compounds toward necessity in heterogeneous environments.

4.2 Stimuli That Control Defence Root Exudation

As discussed above, the belowground attack by antagonistic organisms can induce the release of a multitude of defence compounds into the rhizosphere. Astonishingly, upon aboveground attack, intraplant chemical signals can be relayed to influence root exudation (Bezemer and van Dam 2005; Robert et al. 2012; Pangesti et al. 2013). Secretion of L-malic acid from *Arabidopsis* roots is stimulated by infection with the bacterial foliar pathogen *Pseudomonas syringae* pv. *tomato* Pst DC3000 (Rudrappa et al. 2008; Lakshmanan et al. 2012). Elevated levels of malic acid in the rhizosphere in turn recruit the beneficial *Bacillus subtilis* FB17 and promote rhizobacterial colonization to enhance plant defence (Rudrappa et al. 2008; Lakshmanan et al. 2012).

Under laboratory conditions, the rhizosecretion process can be elicited also by exogenous application of biotic stress-related signaling molecules such as salicylic acid, nitric oxide, or methyl jasmonate (Badri et al. 2008b; Badri and Vivanco 2009; Ruiz-May et al. 2009; Schreiner et al. 2011). Likewise, an ectopic expression of the oomycetal elicitor β-cryptogein in hairy roots of *Coleus blumei* mimics pathogen attack resulting in an enhanced level of secreted antimicrobial rosmarinic acid in the external culture medium (Vuković et al. 2013). Recently it has been reported that the presence of phytobeneficial bacteria can enhance root volatile emission required for indirect plant defence (Santos et al. 2014). Root colonization with *Azospirillum brasilense* induced higher release of (*E*)-β-caryophyllene from maize roots. Furthermore, larvae of the South American corn rootworm, *Diabrotica speciosa*, gained less weight when feeding on rhizobacterium-inoculated roots (Santos et al. 2014).

Besides exogenous stimuli that influence the release of compounds implemented in direct and indirect plant defence, root exudation is also under the control of endogenous genetic programs such as the developmental stage of the plant. In maize benzoxazinoids form a class of defence molecules (Ahmad et al. 2011) that are released during the emergence of lateral and crown roots when the plant is locally and temporally more susceptible (Park et al. 2004). Hence, benzoxazinoid secretion presents a genetically regulated, protective process that alleviates damage at local sites or during discrete developmental stages when infection is more deleterious for the plant. In accordance, the peak of defence-related protein exudation into the rhizosphere can be observed just before flowering (De-la-Peña et al. 2010). Toward later stages of the *Arabidopsis* life cycle, also the level of putatively antimicrobial phenolic compounds increases in the root-exudate profile (Chaparro et al. 2013). Again, the recruitment of phytobeneficial microbes that indirectly prevent root infection through the production of antibacterial compounds is dependent on the growth stage of the plant (Picard et al. 2000, 2004).

Taken together, these studies exemplify that the secretion of defence compounds into the rhizosphere is a tightly controlled, spatiotemporal dynamic process that is regulated by various endogenous and exogenous factors.

4.3 The Role of Transport Proteins in Root Exudation

Root exudation is in part mediated by diffusion, channels, and vesicle transport. However, a substantial proportion of root exudates is also secreted actively by transport proteins. First indirect evidence of a primary and secondary active secretion process of plant-derived molecules across the root plasma membrane came from comprehensive pharmacological studies. The use of various inhibitors revealed that the secretion of some root-derived phytochemicals was dependent on ATP hydrolysis (Loyola-Vargas et al. 2007), indicating that active transport systems such as ATP-binding cassette (ABC) transporters might be involved in the release of constituents of the root phytochemical cocktail into the rhizosphere. ABC-type proteins constitute a large family of transporters that are involved in mediating the transport of a wide array of organic substances (Yazaki et al. 2008, 2009; Kang et al. 2011). More than 120 genes in the *Arabidopsis thaliana* genome encode for ABC transporter proteins, and some of these genes exhibit strikingly high expression in root cells, raising the potential for their involvement in rhizosecretion processes (Badri et al. 2008a). Subsequent studies in which root-exudate (Badri et al. 2008a, 2009) and microbial (Badri et al. 2009) compositions of ABC transporter mutants differed significantly from those of corresponding wild-type plants confirmed the essential role of ABC proteins in root exudation. In addition, these studies revealed that multiple ABC transporters can release the same substrate and that a discrete ABC transporter can have low substrate specificity and export multiple structurally and functionally unrelated substances (Fig. 2a). The role of *At*ABCG37/*At*PDR9 in mediating the rhizosecretion of not only auxinic compounds (Ito and Gray 2006; Ruzicka et al. 2010) but also of phenolics as an iron acquisition strategy (Rodríguez-Celma et al. 2013; Fourcroy et al. 2014) supports this observation. Likewise, *At*ABCG36/*At*PDR8 is suggested to export cadmium (Kim et al. 2007) as well as indole-3-butyric acid (Strader and Bartel 2009) into the rhizosphere.

To date, few ABC transport proteins were proposed to be implemented in the export and accumulation of phytochemicals that confer resistance against soilborne diseases. For example, silencing *Nt*ABCG5/*Nt*PDR5 from tobacco (*Nicotiana tabacum*) improved larval performance of the herbivore *Manduca sexta* but also increased slightly the susceptibility to the soilborne fungus *Fusarium oxysporum*, suggesting a role of this transport protein in defence inter alia through root exudation (Bienert et al. 2012). More evidently, the transporter *Np*PDR1 of *Nicotiana plumbaginifolia* was shown to be involved in belowground plant defence against pathogen invasion (Bultreys et al. 2009; Stukkens et al. 2005). Silencing the ABC transporter accounted for enhanced sensitivity of roots and petals toward several fungal and oomycetal pathogens, possibly due to diminished secretion of antimicrobial compounds such as the diterpene sclareol (Bultreys et al. 2009; Stukkens et al. 2005; Jasiński et al. 2001). Besides these obvious connections between a transporter, its substrate, and a direct effect on the rhizosphere microbiome, further research on ABC proteins implemented in root exudation

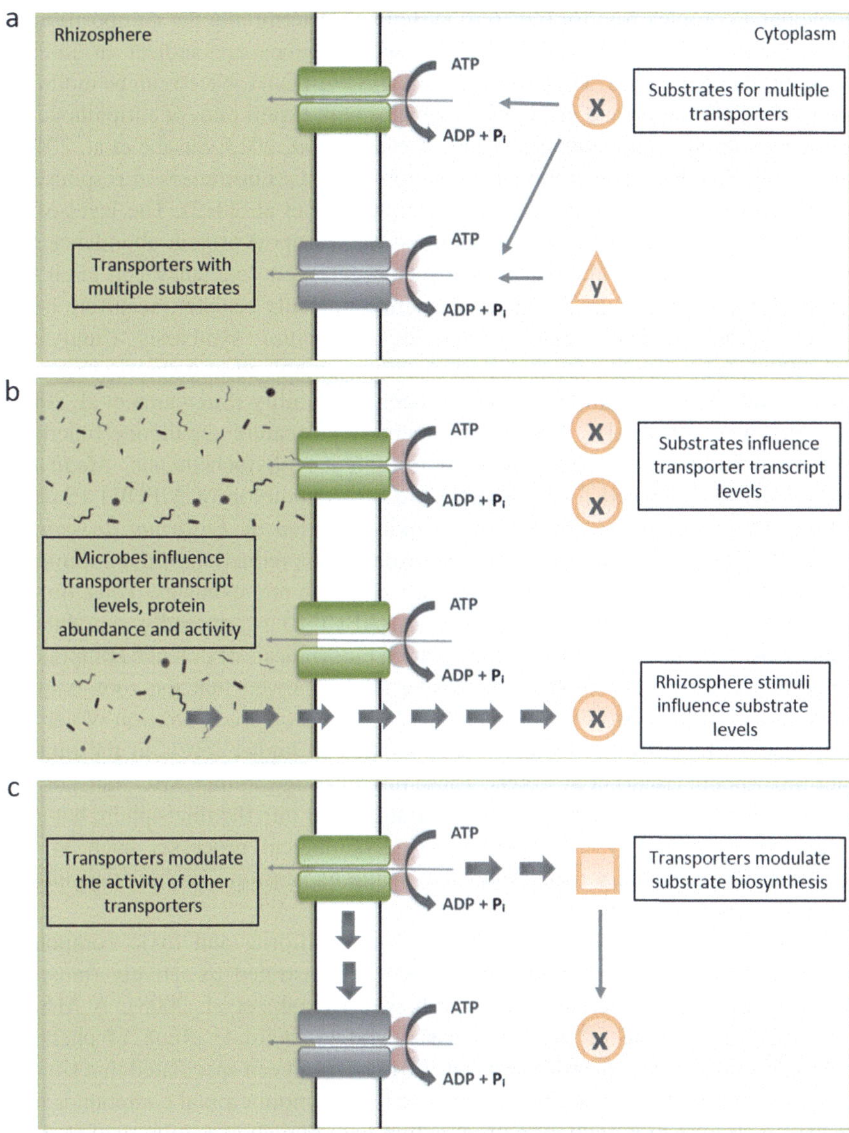

Fig. 2 ABC proteins are complex transport systems that modulate root exudation. (**a**) Some ABC proteins transport multiple substrates. Equally, some compounds can be a substrate of several transporters. (**b**) Transporter transcript levels, protein abundance, and activity can be dependent on substrate availability, elicitors, and microbial presence. In addition, rhizosphere stimuli can influence substrate production. (**c**) ABC transporters can pleiotropically modulate cell physiology, e.g., by influencing substrate biosynthesis or the activity of other transporters

uncovered a complex role for transport systems in determining the composition of root exudates (Fig. 2). Certain ABC transporter genes are subject of intense transcriptional regulation. The expression of *Nt*PDR1 from tobacco can be modified by microbial elicitation and positively correlates with export rates of antipathogenic diterpenes into the extracellular medium (Crouzet et al. 2013; Sasabe et al. 2002). In line with this, the transcriptional regulation of ABC transporters in response to their substrates has been reported (e.g., Kretzschmar et al. 2012). The level of an external phytochemical can be dependent on the transport protein abundance but also on the substrate availability. For instance, nitrogen deficiency can elicit the increased production of the flavonoid signaling molecule genistein resulting in its secretion from soybean roots to initiate rhizobium symbiosis (Sugiyama et al. 2008). Interestingly, the transport machinery involved in genistein export is constitutively active, regardless of the nitrogen availability (Sugiyama et al. 2007) (Fig. 2b). Yet, other ABC transporters themselves feature regulatory functions influencing biosynthesis and exudation of defence phytochemicals. *Medicago truncatula* roots silenced for *Mt*ABCG10, a close homolog of *Nt*PDR1 (Sasabe et al. 2002; Crouzet et al. 2013), were rapidly infected by *Fusarium oxysporum* (Banasiak et al. 2013). The silencing resulted also in a reduction of the antimicrobial medicarpin as well as its precursors in root tissue and exudates. Thus, during belowground biotic stress response, *Mt*ABCG10 supposedly modulates isoflavonoid levels associated with the de novo biosynthesis of defence compounds (Banasiak et al. 2013). Another persuasive study showed that the root-exudate profile of the *Arabidopsis* mutant *abcg30* exhibits a decreased secretion of certain compounds, whereas other exudates accumulated to higher levels in the mutant plant rhizosphere (Badri et al. 2009). These findings suggest that ABC transporters have a sophisticated role in mediating substrate export into the rhizosphere but also in directly or indirectly modifying other physiological processes such as the biosynthesis of secondary metabolites and/or the expression of other transporters involved in root exudation (Fig. 2c).

Besides ABC transporters, members of the multidrug and toxic compound extrusion (MATE) protein family have been demonstrated to actively transport secondary metabolites across plant membranes (Yazaki et al. 2008). A MATE transporter in the stele of rice roots was found to facilitate efflux of phenolic compounds into the xylem (Ishimaru et al. 2011). It has been speculated that similar transporters might be responsible for the secretion of antimicrobial compounds into the soil. A crucial root exudation process that has been shown to be mediated by MATE proteins is the release of citrate into the rhizosphere (Furukawa et al. 2007; Fujii et al. 2012; Magalhaes et al. 2007; Liu et al. 2009; Maron et al. 2010). Since citrate is a carbon source for many microorganisms, this exudation may have a vital impact on microbial soil communities. However, to our knowledge, no evidence has been provided for an implementation of MATE transport proteins in direct or indirect belowground plant defence.

Taken together, active transport systems largely influence the composition of root exudates and can dynamically adjust the quality and quantity of certain phytochemicals in response to changes in microbial rhizosphere communities.

Identification and investigation of transporter proteins implemented in regulated rhizosecretion processes are fundamental to understand belowground direct and indirect plant defence.

4.4 One Phytochemical- Additive Defence Functions

In the previous sections, we demonstrated that the release of defence-related root exudates is inducible, how this induction can be elicited, and that regulated secretion is mediated on the molecular level by transport proteins. In this section, we will highlight that single root exudates can target multiple rhizosphere organisms and may elicit dissimilar responses. Belowground plant defence becomes highly efficient if different exudate bioactivities are appropriately fine-tuned to allow an opposite effect on plant mutualists and antagonists.

Some root-secreted defence compounds affect a highly specific spectrum of rhizosphere organisms. For instance, the legume root-exudate canavanine exhibits cytotoxic properties against many soil bacteria but initiates the detoxification machinery of rhizobia, accounting for their resistance to canavanine (Cai et al. 2009). In *Arabidopsis*, resistance to *Phytophthora capsici* relies on the production of the antimicrobial camalexin (Wang et al. 2013); however, this defence compound does not confer resistance to the oomycetes pathogen *Phytophthora cinnamomi* (Rookes et al. 2008). Notably, this high target specificity of root exudates can be partially explained by variations in the tolerance to specific defence molecules based on the efficiency of active detoxification and efflux processes between different microbes (Cai et al. 2009; Bouarab et al. 2002).

Other root exudates have a broader recipient spectrum and affect various rhizosphere organisms, including beneficial and pathogenic members (Badri et al. 2013). This can be exemplified by the different effects of green pea (*Pisum sativa*) root exudates on the behavior of beneficial and plant-parasitic nematodes (Hiltpold et al. 2015). Low concentrations of root exudates induced the loss of mobility and a state of reversible quiescence in antagonistic nematodes, protecting the roots against infection. In sharp contrast, the activity and infectiousness of beneficial entomopathogenic nematodes (EPNs) enhanced markedly under low root-exudate concentrations. Dual bioactivity in the rhizosphere was also observed for benzoxazinoids, a class of phytochemicals detected in maize root exudates. Plant-beneficial *Pseudomonas putida* was found to be recruited in response to exudation of a benzoxazinoid metabolite from maize roots during relatively young and vulnerable growth stages (Neal et al. 2012). The root colonization stimulated jasmonic acid-dependent defence pathways in maize entailing a beneficial systemic defence priming in the plant (Neal and Ton 2013). Conversely, benzoxazinoids were previously shown to exert antimicrobial and insecticidal activities and function in direct above- and belowground plant defence against pests and diseases (Niemeyer 2009; Park et al. 2004; Ahmad et al. 2011). Hence, released benzoxazinoids provide coupled profitable service for the plant by attracting beneficial microbes (indirect plant defence) and repelling pathogenic organisms in the maize rhizosphere (direct plant defence). Similarly, dimethyl disulfide emitted from cabbage (*Brassica napus*) roots invested by the cabbage root fly *Delia radicum* showed multiple defence bioactivities, the inhibition of oviposition by cabbage root fly females and the attraction of natural enemies of *D. radicum* (Ferry et al. 2007, 2009). In summary, root exudates with directed dual functions that complement each other enhance the efficiency of belowground plant protection by broadening the spectrum of defence modes and lowering carbon costs for the plant.

5 Summary

Interactions between plants and other organisms are as fascinating as they are complex. Plants can, for instance, communicate with arbuscular mycorrhizal fungi to initiate a mutually beneficial symbiosis. However, not all organisms that plants are exposed to have neutral or even advantageous impacts. Negative interactions and defence strategies against antagonistic organisms are an intensively investigated field of biology. Previously, researchers focused on interactions and processes that appear in the visible, more easily accessible half of the plant, the aerial part. However, since tremendous yield losses are caused by root feeding and infection, it is equally crucial to study plant defence mechanisms belowground.

Root exudates in the rhizosphere serve as chemical mediators of positive interactions between plants and soilborne organisms and as defence compounds in negative interactions. During plant attack root exudates are engaged in two types of defence traits, the direct and the indirect defence. Root exudates with direct defence properties act repelling, inhibiting, or killing on plant aggressors such as pathogens and feeders. In contrast, root exudates incorporated in indirect plant defence initiate the interaction with beneficial organisms that counteract aggressors. The chemical nature and mode of action of various compounds involved in direct and indirect defence have been progressively elucidated in the past years. Interestingly, several compounds were found to exhibit multiple bioactivities in the rhizosphere and influence organisms differently. In other words, a single phytochemical might act synergistically in direct and indirect plant defence. Nevertheless, another compound might recruit beneficial and detrimental organisms. Therefore, it is of importance to carefully assess the targets and effects of root exudates on multitrophic levels. In addition to the discovery of various root-secreted defence compounds and their role in the rhizosphere, the understanding of the stimulation and regulation of root exudation has advanced dramatically. Root exudation is a dynamic and bidirectional process: root exudates shape the soil inhabitants and rhizosphere members modulate the root-exudate ensemble. Besides the presence of soilborne organisms, several other exogenous as well as endogenous factors can rapidly and precisely adjust the nature of root-secreted phytochemicals. On the molecular level, transporter proteins have been shown to modulate rhizosecretion processes in a complex manner that goes beyond a role as pure substrate carriers.

Consequently, also the stimuli and regulatory mechanisms that modify the quality and quantity of the root-exudate cocktail require thorough investigation.

Taken together, root exudates impact the rhizosphere inhabitants markedly. Accordingly, they are a powerful tool that can be exploited to enhance natural defence properties of plants. Deepening our knowledge of the targets and effects of root exudates, as well as the regulation of root secretion processes, will unravel the path for more efficient disease management in the rhizosphere.

References

Ahmad S, Veyrat N, Gordon-Weeks R, Zhang Y, Martin J, Smart L, Glauser G, Erb M, Flors V, Frey M, Ton J (2011) Benzoxazinoid metabolites regulate innate immunity against aphids, fungi in maize. Plant Physiol 157:317–327

Ali JG, Alborn HT, Stelinski LL (2010) Subterranean herbivore-induced volatiles released by citrus roots upon feeding by Diaprepes abbreviatus recruit entomopathogenic nematodes. J Chem Ecol 36(4):361–368

Ali JG, Alborn HT, Stelinski LL (2011) Constitutive, induced subterranean plant volatiles attract both entomopathogenic, plant parasitic nematodes. J Ecol 99(1):26–35

Ali JG, Alborn HT, Campos-Herrera R, Kaplan F, Duncan LW, Rodriguez-Saona C, Koppenhöfer AM, Stelinski LL (2012) Subterranean, herbivore-induced plant volatile increases biological control activity of multiple beneficial nematode species in distinct habitats. PLoS One 7(6): e38146

Badri DV, Vivanco JM (2009) Regulation, function of root exudates. Plant Cell Environ 32 (6):666–681

Badri DV, Loyola-Vargas VM, Broeckling CD, De-la-Peña C, Jasinski M, Santelia D, Martinoia E, Sumner LW, Banta LM, Stermitz F, Vivanco JM (2008a) Altered profile of secondary metabolites in the root exudates of Arabidopsis ATP-binding cassette transporter mutants. Plant Physiol 146(2):762–771

Badri DV, Loyola-Vargas VM, Du J, Stermitz FR, Broeckling CD, Iglesias-Andreu L, Vivanco JM (2008b) Transcriptome analysis of Arabidopsis roots treated with signaling compounds: a focus on signal transduction, metabolic regulation, secretion. New Phytol 179(1):209–223

Badri DV, Quintana N, El Kassis EG, Kim HK, Choi YH, Sugiyama A, Verpoorte R, Martinoia E, Manter DK, Vivanco JM (2009) An ABC transporter mutation alters root exudation of phytochemicals that provoke an overhaul of natural soil microbiota. Plant Physiol 151 (4):2006–2017

Badri DV, Loyola-Vargas VM, Broeckling CD, Vivanco JM (2010) Root secretion of phytochemicals in Arabidopsis is predominantly not influenced by diurnal rhythms. Mol Plant 3 (3):491–498

Badri DV, Chaparro JM, Zhang R, Shen Q, Vivanco JM (2013) Application of natural blends of phytochemicals derived from the root exudates of Arabidopsis to the soil reveal that phenolic-related compounds predominantly modulate the soil microbiome. J Biol Chem 288 (7):4502–4512

Baehler E, Bottiglieri M, Péchy-Tarr M, Maurhofer M, Keel C (2005) Use of green fluorescent protein-based reporters to monitor balanced production of antifungal compounds in the biocontrol agent *Pseudomonas fluorescens* CHA0. J Appl Microbiol 99(1):24–38

Baetz U, Martinoia E (2014) Root exudates: the hidden part of plant defense. Trends Plant Sci 19 (2):90–98

Bais HP, Walker TS, Schweizer HP, Vivanco JM (2002) Root specific elicitation, antimicrobial activity of rosmarinic acid in hairy root cultures of *Ocimum basilicum*. Plant Physiol Biochem 40:983–995

Banasiak J, Biala W, Staszków A, Swarcewicz B, Kepczynska E, Figlerowicz M, Jasinski M (2013) A *Medicago truncatula* ABC transporter belonging to subfamily G modulates the level of isoflavonoids. J Exp Bot 64(4):1005–1015

Bangera MG, Thomashow LS (1999) Identification, characterization of a gene cluster for synthesis of the polyketide antibiotic 2,4-diacetylphloroglucinol from *Pseudomonas fluorescens* Q2-87. J Bacteriol 181(10):3155–3163

Bezemer TM, van Dam NM (2005) Linking aboveground, belowground interactions via induced plant defenses. Trends Ecol Evol 20(11):617–624

Bienert MD, Siegmund SE, Drozak A, Trombik T, Bultreys A, Baldwin IT, Boutry M (2012) A pleiotropic drug resistance transporter in *Nicotiana tabacum* is involved in defense against the herbivore *Manduca sexta*. Plant J 72(5):745–757

Bouarab K, Melton R, Peart J, Baulcombe D, Osbourn A (2002) A saponin-detoxifying enzyme mediates suppression of plant defences. Nature 418(6900):889–892

Brinkmann V, Reichard U, Goosmann C, Fauler B, Uhlemann Y, Weiss DS, Weinrauch Y, Zychlinsky A (2004) Neutrophil extracellular traps kill bacteria. Science 303 (5663):1532–1535

Bultreys A, Trombik T, Drozak A, Boutry M (2009) *Nicotiana plumbaginifolia* plants silenced for the ATP-binding cassette transporter gene NpPDR1 show increased susceptibility to a group of fungal and oomycete pathogens. Mol Plant Pathol 10(5):651–663

Cai T, Cai W, Zhang J, Zheng H, Tsou AM, Xiao L, Zhong Z, Zhu J (2009) Host legume-exuded antimetabolites optimize the symbiotic rhizosphere. Mol Microbiol 73(3):507–517

Cannesan MA, Gangneux C, Lanoue A, Giron D, Laval K, Hawes M, Driouich A, Vicré-Gibouin M (2011) Association between border cell responses, localized root infection by pathogenic *Aphanomyces euteiches*. Ann Bot 108(3):459–469

Capra E, Colombi C, De Poli P, Nocito FF, Cocucci M, Vecchietti A, Marocco A, Stile MR, Rossini L (2014) Protein profiling, tps23 induction in different maize lines in response to methyl jasmonate treatment and *Diabrotica virgifera* infestation. J Plant Physiol 175C:68–77

Chaparro JM, Badri DV, Bakker MG, Sugiyama A, Manter DK, Vivanco JM (2013) Root exudation of phytochemicals in Arabidopsis follows specific patterns that are developmentally programmed and correlate with soil microbial functions. PLoS One 8(2):e55731

Compant S, Duffy B, Nowak J, Clément C, Barka EA (2005) Use of plant growth-promoting bacteria for biocontrol of plant diseases: principles, mechanisms of action, future prospects. Appl Environ Microbiol 71(9):4951–4959

Crouzet J, Roland J, Peeters E, Trombik T, Ducos E, Nader J, Boutry M (2013) NtPDR1, a plasma membrane ABC transporter from *Nicotiana tabacum*, is involved in diterpene transport. Plant Mol Biol 82(1–2):181–192

Curlango-Rivera G, Duclos DV, Ebolo JJ, Hawes MC (2010) Transient exposure of root tips to Primary and secondary metabolites: impact on root growth and production of border cells. Plant Soil 332:267–275

Curtis RH (2008) Plant-nematode interactions: environmental signals detected by the nematodes chemosensory organs control changes in the surface cuticle and behaviour. Parasite 15 (3):310–316

Degenhardt J, Hiltpold I, Köllner TG, Frey M, Gierl A, Gershenzon J, Hibbard BE, Ellersieck MR, Turlings TC (2009) Restoring a maize root signal that attracts insect-killing nematodes to control a major pest. Proc Natl Acad Sci USA 106(32):13213–13218

De-la-Peña C, Lei Z, Watson BS, Sumner LW, Vivanco JM (2008) Root-microbe communication through protein secretion. J Biol Chem 283(37):25247–25255

De-la-Peña C, Badri DV, Lei Z, Watson BS, Brandão MM, Silva-Filho MC, Sumner LW, Vivanco JM (2010) Root secretion of defense-related proteins is development-dependent and correlated with flowering time. J Biol Chem 285(40):30654–30665

de Souza JT, Arnould C, Deulvot C, Lemanceau P, Gianinazzi-Pearson V, Raaijmakers JM (2003) Effect of 2,4-diacetylphloroglucinol on pythium: cellular responses, variation in sensitivity among propagules and species. Phytopathology 93(8):966–975

de Werra P, Baehler E, Huser A, Keel C, Maurhofer M (2008) Detection of plant-modulated alterations in antifungal gene expression in *Pseudomonas fluorescens* CHA0 on roots by flow cytometry. Appl Environ Microbiol 74(5):1339–1349

de Werra P, Huser A, Tabacchi R, Keel C, Maurhofer M (2011) Plant-, microbe-derived compounds affect the expression of genes encoding antifungal compounds in a pseudomonad with biocontrol activity. Appl Environ Microbiol 77(8):2807–2812

Dong L, Li X, Huang L, Gao Y, Zhong L, Zheng Y, Zuo Y (2014) Lauric acid in crown daisy root exudate potently regulates root-knot nematode chemotaxis, disrupts Mi-flp-18 expression to block infection. J Exp Bot 65(1):131–141

Doornbos RF, van Loon LC, Bakker PAHM (2012) Impact of root exudates and plant defense signaling on bacterial communities in the rhizosphere A review. Agron Sustain Dev 32:227–243

Driouich A, Durand C, Vicré-Gibouin M (2007) Formation, separation of root border cells. Trends Plant Sci 12(1):14–19

Ferry A, Dugravot S, Delattre T, Christides JP, Auger J, Bagnères AG, Poinsot D, Cortesero AM (2007) Identification of a widespread monomolecular odor differentially attractive to several *Delia radicum* ground-dwelling predators in the field. J Chem Ecol 33(11):2064–2077

Ferry A, Le Tron S, Dugravot S, Cortesero AM (2009) Field evaluation of the combined deterrent, attractive effects of dimethyl disulfide on *Delia radicum* and its natural enemies. Biol Control 49(3):219–226

Fourcroy P, Sisó-Terraza P, Sudre D, Savirón M, Reyt G, Gaymard F, Abadía A, Abadia J, Alvarez-Fernández A, Briat JF (2014) Involvement of the ABCG37 transporter in secretion of scopoletin and derivatives by Arabidopsis roots in response to iron deficiency. New Phytol 201 (1):155–167

Fujii M, Yokosho K, Yamaji N, Saisho D, Yamane M, Takahashi H, Sato K, Nakazono M, Ma JF (2012) Acquisition of aluminium tolerance by modification of a single gene in barley. Nat Commun 3:713

Furukawa J, Yamaji N, Wang H, Mitani N, Murata Y, Sato K, Katsuhara M, Takeda K, Ma JF (2007) An aluminum-activated citrate transporter in barley. Plant Cell Physiol 48 (8):1081–1091

Griffin G, Hale M, Shay FJ (1976) Nature and quantity of sloughed organic matter produced by roots of axenic peanut plants. Soil Biol Biochem 8:29–32

Haas D, Défago G (2005) Biological control of soil-borne pathogens by fluorescent pseudomonads. Nat Rev Microbiol 3(4):307–319

Hawes MC, Curlango-Rivera G, Wen F, White GJ, Vanetten HD, Xiong Z (2011) Extracellular DNA: the tip of root defenses? Plant Sci 180(6):741–745

Hawes MC, Curlango-Rivera G, Xiong Z, Kessler JO (2012) Roles of root border cells in plant defense, regulation of rhizosphere microbial populations by extracellular DNA trapping. Plant Soil 355(1):1–16

Hiltpold I, Turlings TC (2008) Belowground chemical signaling in maize: when simplicity rhymes with efficiency. J Chem Ecol 34(5):628–635

Hiltpold I, Turlings TC (2012) Manipulation of chemically mediated interactions in agricultural soils to enhance the control of crop pests, to improve crop yield. J Chem Ecol 38(6):641–650

Hiltpold I, Baroni M, Toepfer S, Kuhlmann U, Turlings TC (2010a) Selection of entomopathogenic nematodes for enhanced responsiveness to a volatile root signal helps to control a major root pest. J Exp Biol 213(Pt 14):2417–2423

Hiltpold I, Baroni M, Toepfer S, Kuhlmann U, Turlings TC (2010b) Selective breeding of entomopathogenic nematodes for enhanced attraction to a root signal did not reduce their establishment or persistence after field release. Plant Signal Behav 5(11):1450–1452

Hiltpold I, Erb M, Robert CA, Turlings TC (2011) Systemic root signalling in a belowground and volatile-mediated tritrophic interaction. Plant Cell Environ 34(8):1267–1275

Hiltpold I, Jaffuel G, Turlings TC (2015) The dual effects of root-cap exudates on nematodes: from quiescence in plant-parasitic nematodes to frenzy in entomopathogenic nematodes. J Exp Bot 66(2):603–611

Ishimaru Y, Kakei Y, Shimo H, Bashir K, Sato Y, Uozumi N, Nakanishi H, Nishizawa NK (2011) A rice phenolic efflux transporter is essential for solubilizing precipitated apoplasmic iron in the plant stele. J Biol Chem 286(28):24649–24655

Ito H, Gray WM (2006) A gain-of-function mutation in the Arabidopsis pleiotropic drug resistance transporter PDR9 confers resistance to auxinic herbicides. Plant Physiol 142(1):63–74

Jasiński M, Stukkens Y, Degand H, Purnelle B, Marchand-Brynaert J, Boutry M (2001) A plant plasma membrane ATP binding cassette-type transporter is involved in antifungal terpenoid secretion. Plant Cell 13(5):1095–1107

Jousset A, Rochat L, Lanoue A, Bonkowski M, Keel C, Scheu S (2011) Plants respond to pathogen infection by enhancing the antifungal gene expression of root-associated bacteria. Mol Plant Microbe Interact 24(3):352–358

Kalemba D, Kusewicz D, Swiader K (2002) Antimicrobial properties of the essential oil of *Artemisia asiatica* Nakai. Phytother Res 16(3):288–291

Kang J, Park J, Choi H, Burla B, Kretzschmar T, Lee Y, Martinoia E (2011) Plant ABC transporters. Arabidopsis Book 9:e0153

Kato-Noguchi H, Ino T, Ota K (2008) Secretion of momilactone A from rice roots to the rhizosphere. J Plant Physiol 165(7):691–696

Kim DY, Bovet L, Maeshima M, Martinoia E, Lee Y (2007) The ABC transporter AtPDR8 is a cadmium extrusion pump conferring heavy metal resistance. Plant J 50(2):207–218

Köllner TG, Held M, Lenk C, Hiltpold I, Turlings TC, Gershenzon J, Degenhardt J (2008) A maize (E)-beta-caryophyllene synthase implicated in indirect defense responses against herbivores is not expressed in most American maize varieties. Plant Cell 20(2):482–494

Kretzschmar T, Kohlen W, Sasse J, Borghi L, Schlegel M, Bachelier JB, Reinhardt D, Bours R, Bouwmeester HJ, Martinoia E (2012) A petunia ABC protein controls strigolactone-dependent symbiotic signalling and branching. Nature 483(7389):341–344

Lakshmanan V, Kitto SL, Caplan JL, Hsueh YH, Kearns DB, Wu YS, Bais HP (2012) Microbe-associated molecular patterns-triggered root responses mediate beneficial rhizobacterial recruitment in Arabidopsis. Plant Physiol 160(3):1642–1661

Lanoue A, Burlat V, Henkes GJ, Koch I, Schurr U, Röse US (2010a) De novo biosynthesis of defense root exudates in response to Fusarium attack in barley. New Phytol 185(2):577–588

Lanoue A, Burlat V, Schurr U, Röse US (2010b) Induced root-secreted phenolic compounds as a belowground plant defense. Plant Signal Behav 5(8):1037–1038

Liao C, Hochholdinger F, Li C (2012) Comparative analyses of three legume species reveals conserved, unique root extracellular proteins. Proteomics 12(21):3219–3228

Lilley CJ, Wang D, Atkinson HJ, Urwin PE (2011) Effective delivery of a nematode-repellent peptide using a root-cap-specific promoter. Plant Biotechnol J 9(2):151–161

Liu B, Hibbard JK, Urwin PE, Atkinson HJ (2005) The production of synthetic chemodisruptive peptides in planta disrupts the establishment of cyst nematodes. Plant Biotechnol J 3 (5):487–496

Liu J, Magalhaes JV, Shaff J, Kochian LV (2009) Aluminum-activated citrate, malate transporters from the MATE and ALMT families function independently to confer Arabidopsis aluminum tolerance. Plant J 57(3):389–399

Loyola-Vargas VM, Broeckling CD, Badri D, Vivanco JM (2007) Effect of transporters on the secretion of phytochemicals by the roots of *Arabidopsis thaliana*. Planta 225(2):301–310

Ma W, Muthreich N, Liao C, Franz-Wachtel M, Schütz W, Zhang F, Hochholdinger F, Li C (2010) The mucilage proteome of maize (*Zea mays* L) primary roots. J Proteome Res 9(6):2968–2976

Magalhaes JV, Liu J, Guimarães CT, Lana UG, Alves VM, Wang YH, Schaffert RE, Hoekenga OA, Piñeros MA, Shaff JE, Klein PE, Carneiro NP, Coelho CM, Trick HN, Kochian LV (2007)

A gene in the multidrug, toxic compound extrusion (MATE) family confers aluminum tolerance in sorghum. Nat Genet 39(9):1156–1161
Maron LG, Piñeros MA, Guimarães CT, Magalhaes JV, Pleiman JK, Mao C, Shaff J, Belicuas SN, Kochian LV (2010) Two functionally distinct members of the MATE (multi-drug, toxic compound extrusion) family of transporters potentially underlie two major aluminum tolerance QTLs in maize. Plant J 61(5):728–740
McCully ME, Miller C, Sprague SJ, Huang CX, Kirkegaard JA (2008) Distribution of glucosinolates, sulphur-rich cells in roots of field-grown canola (*Brassica napus*). New Phytol 180(1):193–205
Medina E (2009) Neutrophil extracellular traps: a strategic tactic to defeat pathogens with potential consequences for the host. J Innate Immun 1(3):176–180
Miller N, Estoup A, Toepfer S, Bourguet D, Lapchin L, Derridj S, Kim KS, Reynaud P, Furlan L, Guillemaud T (2005) Multiple transatlantic introductions of the western corn rootworm. Science 310(5750):992
Neal AL, Ton J (2013) Systemic defense priming by *Pseudomonas putida* KT2440 in maize depends on benzoxazinoid exudation from the roots. Plant Signal Behav 8(1):e22655
Neal AL, Ahmad S, Gordon-Weeks R, Ton J (2012) Benzoxazinoids in root exudates of maize attract *Pseudomonas putida* to the rhizosphere. PLoS One 7(4):e35498
Niemeyer HM (2009) Hydroxamic acids derived from 2-hydroxy-2H-1,4-benzoxazin-3(4H)-one: key defense chemicals of cereals. J Agric Food Chem 57(5):1677–1696
Notz R, Maurhofer M, Schnider-Keel U, Duffy B, Haas D, Défago G (2001) Biotic factors affecting expression of the 2,4-diacetylphloroglucinol biosynthesis gene phlA in *Pseudomonas fluorescens* biocontrol strain CHA0 in the rhizosphere. Phytopathology 91(9):873–881
Odell RE, Dumlao MR, Samar D, Silk WK (2008) Stage-dependent border cell, carbon flow from roots to rhizosphere. Am J Bot 95(4):441–446
Pangesti N, Pineda A, Pieterse CM, Dicke M, van Loon JJ (2013) Two-way plant mediated interactions between root-associated microbes and insects: from ecology to mechanisms. Front Plant Sci 4:414
Park WJ, Hochholdinger F, Gierl A (2004) Release of the benzoxazinoids defense molecules during lateral-, and crown root emergence in *Zea mays*. J Plant Physiol 161(8):981–985
Picard C, Di Cello F, Ventura M, Fani R, Guckert A (2000) Frequency and biodiversity of 2,4-diacetylphloroglucinol-producing bacteria isolated from the maize rhizosphere at different stages of plant growth. Appl Environ Microbiol 66(3):948–955
Picard C, Frascaroli E, Bosco M (2004) Frequency and biodiversity of 2,4-diacetylphloroglucinol-producing rhizobacteria are differentially affected by the genotype of two maize inbred lines and their hybrid. FEMS Microbiol Ecol 49(2):207–215
Rasmann S, Köllner TG, Degenhardt J, Hiltpold I, Toepfer S, Kuhlmann U, Gershenzon J, Turlings TC (2005) Recruitment of entomopathogenic nematodes by insect-damaged maize roots. Nature 434(7034):732–737
Reynolds AM, Dutta TK, Curtis RH, Powers SJ, Gaur HS, Kerry BR (2011) Chemotaxis can take plant-parasitic nematodes to the source of a chemo-attractant via the shortest possible routes. J R Soc Interface 8(57):568–577
Robert CA, Erb M, Duployer M, Zwahlen C, Doyen GR, Turlings TC (2012) Herbivore-induced plant volatiles mediate host selection by a root herbivore. New Phytol 194(4):1061–1069
Robert CA, Erb M, Hiltpold I, Hibbard BE, Gaillard MD, Bilat J, Degenhardt J, Cambet-Petit-Jean X, Turlings TC, Zwahlen C (2013) Genetically engineered maize plants reveal distinct costs and benefits of constitutive volatile emissions in the field. Plant Biotechnol J 11 (5):628–639
Rodríguez-Celma J, Lin WD, Fu GM, Abadia J, López-Míllán AF, Schmidt W (2013) Mutually exclusive alterations in secondary metabolism are critical for the uptake of insoluble iron compounds by Arabidopsis and *Medicago truncatula*. Plant Physiol 162(3):1473–1485

Rookes JE, Wright ML, Cahill DM (2008) Elucidation of defence responses and signalling pathways induced in *Arabidopsis thaliana* following challenge with *Phytophthora cinnamomi*. Physiol Mol Plant Pathol 72:151–161

Rudrappa T, Czymmek KJ, Paré PW, Bais HP (2008) Root-secreted malic acid recruits beneficial soil bacteria. Plant Physiol 148(3):1547–1556

Ruiz-May E, Galaz-Avalos RM, Loyola-Vargas VM (2009) Differential secretion, accumulation of terpene indole alkaloids in hairy roots of *Catharanthus roseus* treated with methyl jasmonate. Mol Biotechnol 41(3):278–285

Ruzicka K, Strader LC, Bailly A, Yang H, Blakeslee J, Langowski L, Nejedlá E, Fujita H, Itoh H, Syono K, Hejátko J, Gray WM, Martinoia E, Geisler M, Bartel B, Murphy AS, Friml J (2010) Arabidopsis PIS1 encodes the ABCG37 transporter of auxinic compounds including the auxin precursor indole-3-butyric acid. Proc Natl Acad Sci U S A 107(23):10749–10753

Santos F, Peñaflor MF, Paré PW, Sanches PA, Kamiya AC, Tonelli M, Nardi C, Bento JM (2014) A novel interaction between plant-beneficial rhizobacteria and roots: colonization induces corn resistance against the root herbivore *Diabrotica speciosa*. PLoS One 9(11):e113280

Sasabe M, Toyoda K, Shiraishi T, Inagaki Y, Ichinose Y (2002) cDNA cloning, characterization of tobacco ABC transporter: NtPDR1 is a novel elicitor-responsive gene. FEBS Lett 518 (1–3):164–168

Schreiner M, Krumbein A, Knorr D, Smetanska I (2011) Enhanced glucosinolates in root exudates of *Brassica rapa ssp rapa* mediated by salicylic acid and methyl jasmonate. J Agric Food Chem 59(4):1400–1405

Shinano T, Komatsu S, Yoshimura T, Tokutake S, Kong FJ, Watanabe T, Wasaki J, Osaki M (2011) Proteomic analysis of secreted proteins from aseptically grown rice. Phytochemistry 72 (4-5):312–320

Steeghs M, Bais HP, de Gouw J, Goldan P, Kuster W, Northway M, Fall R, Vivanco JM (2004) Proton-transfer-reaction mass spectrometry as a new tool for real time analysis of root-secreted volatile organic compounds in Arabidopsis. Plant Physiol 135(1):47–58

Strader LC, Bartel B (2009) The Arabidopsis PLEIOTROPIC DRUG RESISTANCE8/ABCG36 ATP binding cassette transporter modulates sensitivity to the auxin precursor indole-3-butyric acid. Plant Cell 21(7):1992–2007

Stubbs VE, Standing D, Knox OG, Killham K, Bengough AG, Griffiths B (2004) Root border cells take up and release glucose-C. Ann Bot 93(2):221–224

Stukkens Y, Bultreys A, Grec S, Trombik T, Vanham D, Boutry M (2005) NpPDR1, a pleiotropic drug resistance-type ATP-binding cassette transporter from *Nicotiana plumbaginifolia*, plays a major role in plant pathogen defense. Plant Physiol 139(1):341–352

Sugiyama A, Shitan N, Yazaki K (2007) Involvement of a soybean ATP-binding cassette-type transporter in the secretion of genistein, a signal flavonoid in legume-Rhizobium symbiosis. Plant Physiol 144(4):2000–2008

Sugiyama A, Shitan N, Yazaki K (2008) Signaling from soybean roots to rhizobium: an ATP-binding cassette-type transporter mediates genistein secretion. Plant Signal Behav 3 (1):38–40

Toyomasu T, Kagahara T, Okada K, Koga J, Hasegawa M, Mitsuhashi W, Sassa T, Yamane H (2008) Diterpene phytoalexins are biosynthesized in and exuded from the roots of rice seedlings. Biosci Biotechnol Biochem 72(2):562–567

Vacheron J, Desbrosses G, Bouffaud ML, Touraine B, Moënne-Loccoz Y, Muller D, Legendre L, Wisniewski-Dyé F, Prigent-Combaret C (2013) Plant growth-promoting rhizobacteria and root system functioning. Front Plant Sci 4:356

Vaughan MM, Wang Q, Webster FX, Kiemle D, Hong YJ, Tantillo DJ, Coates RM, Wray AT, Askew W, ODonnell C, Tokuhisa JG, Tholl D (2013) Formation of the unusual semivolatile diterpene rhizathalene by the Arabidopsis class I terpene synthase TPS08 in the root stele is involved in defense against belowground herbivory. Plant Cell 25(3):1108–1125

Vicré M, Santaella C, Blanchet S, Gateau A, Driouich A (2005) Root border-like cells of Arabidopsis microscopical characterization and role in the interaction with rhizobacteria. Plant Physiol 138(2):998–1008

von Köckritz-Blickwede M, Nizet V (2009) Innate immunity turned inside-out: antimicrobial defense by phagocyte extracellular traps. J Mol Med 87(8):775–783

Vuković R, Bauer N, Curković-Perica M (2013) Genetic elicitation by inducible expression of β-cryptogein stimulates secretion of phenolics from Coleus blumei hairy roots. Plant Sci 199–200:18–28

Walker TS, Bais HP, Déziel E, Schweizer HP, Rahme LG, Fall R, Vivanco JM (2004) *Pseudomonas aeruginosa*-plant root interactions Pathogenicity, biofilm formation, and root exudation. Plant Physiol 134(1):320–331

Wang Y, Bouwmeester K, van de Mortel JE, Shan W, Govers F (2013) A novel Arabidopsis-oomycete pathosystem: differential interactions with *Phytophthora capsici* reveal a role for camalexin, indole glucosinolates and salicylic acid in defence. Plant Cell Environ 36 (6):1192–1203

Wen F, VanEtten HD, Tsaprailis G, Hawes MC (2007) Extracellular proteins in pea root tip and border cell exudates. Plant Physiol 143(2):773–783

Wen F, White GJ, VanEtten HD, Xiong Z, Hawes MC (2009) Extracellular DNA is required for root tip resistance to fungal infection. Plant Physiol 151(2):820–829

Yazaki K, Sugiyama A, Morita M, Shitan N (2008) Secondary transport as an efficient membrane transport mechanism for plant secondary metabolites. Phytochem Rev 7(3):513–524

Yazaki K, Shitan N, Sugiyama A, Takanashi K (2009) Cell, molecular biology of ATP-binding cassette proteins in plants. Int Rev Cell Mol Biol 276:263–299

Youssef RM, Macdonald MH, Brewer EP, Bauchan GR, Kim KH, Matthews BF (2013) Ectopic expression of AtPAD4 broadens resistance of soybean to soybean cyst and root-knot nematodes. BMC Plant Biol 13(1):67

Zhao X, Schmitt M, Hawes MC (2000) Species-dependent effects of border cell, root tip exudates on nematode behavior. Phytopathology 90(11):1239–1245

Part II
Belowground Defence Strategies to Root Pathogens

Belowground Defence Strategies Against *Fusarium oxysporum*

Louise F. Thatcher, Brendan N. Kidd, and Kemal Kazan

Abstract The root-infecting pathogen *Fusarium oxysporum* (causative agent of the *Fusarium* wilt disease) causes widespread losses in many plant species, including important crop plants such as cotton, melons, bananas and tomatoes; many legume species such as chickpeas, peas, lentils and *Medicago*; and various tree species such as palms. The spores of this pathogen survive in soil for long periods; thus, it is notoriously difficult to eradicate following soil contamination. The pathogen enters into the compatible plants through root tips and lateral root initials, initially invading the cortex tissue. It then gradually moves through the xylem tissue to the upper part of the plant. In addition to the secretion of effectors (e.g. toxins) into the plant cell, the infection by this pathogen can lead to the deposition of plant defence substances such as gums and tyloses in the xylem, which then blocks the water and solute transport to the upper parts of the plant. This leads to wilting, discolouration of xylem, followed by senescence and infection-associated necrotic symptom development in the leaves of infected plants. A number of other developmental changes can also be observed in pathogen-infected plants. Here we describe *F. oxysporum*–host interactions, highlighting recent updates on pathogen infection strategies and host resistance mechanisms.

1 Introduction

Fusarium oxysporum strains that are specialised on specific host plants are classified into *formae speciales* (ff. spp.) (singular *forma specialis*, abbr. f. sp.), such as *Fusarium oxysporum* f. sp. *asparagi* (asparagus); f. sp. *cubense* (banana); f. sp.

L.F. Thatcher (✉) • B.N. Kidd
Centre for Environment and Life Sciences, Commonwealth Scientific and Industrial Research Organization Agriculture, Wembley, WA 6913, Australia
e-mail: Louise.Thatcher@csiro.au

K. Kazan
Queensland Bioscience Precinct, CSIRO Agriculture, St. Lucia, QLD 4067, Australia

The Queensland Alliance for Agriculture & Food Innovation (QAAFI), Queensland Bioscience Precinct, The University of Queensland, Brisbane, QLD 4072, Australia

C.M.F. Vos, K. Kazan (eds.), *Belowground Defence Strategies in Plants*, Signaling and Communication in Plants, DOI 10.1007/978-3-319-42319-7_4

dianthi (carnation); f. sp. *lycopersici* (tomato); f. sp. *melonis* (melon); f. sp. *niveum* (watermelon); f. sp. *pisi* (pea); f. sp. *zingiberi* (ginger); f. sp. *vasinfectum* (cotton); f. sp. *medicaginis* (*Medicago*); f. sp. *ciceris* (chickpea); f. sp. *citri* (orange); f. sp. *cucumerinum* (cucumber) and f. sp. *conglutinans* (canola and *Brassica* crops). While most of the above cause vascular wilts, not all *formae speciales* are primarily vascular pathogens, but cause foot, root rot, crown or bulb rots such as *F. oxysporum* f. sp. *radicis-lycopersici* (Agrios 2005).

Fusarium wilts are most destructive under warm conditions and thus particularly to horticultural production in greenhouses or in tropical climates. For example, *Fusarium oxysporum* f. sp. *cubense* (*Foc*) causes Panama disease on banana. Bananas are the world's most popular fruit (FAO: www.fao.org) and have an estimated value of $44 billion globally (Ploetz 2015). In the 1950s the race 1 strain of *Foc* wiped out almost all banana production in South America and subsequently spread to other banana-growing regions of the world. Due to their susceptibility to *Foc*, the commercial Gros Michel banana cultivars were replaced by race 1-resistant Cavendish cultivars. However, the Cavendish variety is now under threat by *Foc* TR4 (tropical race 4) (reviewed by Ploetz 2015). Also of major concern is *F. oxysporum* f. sp. *ciceris*, which is a major pathogen of chickpea, the second most important legume crop worldwide with countries of tropical/subtropical South Asia by far the largest producers (FAO: www.fao.org). Typically this chickpea pathogen causes yield losses of 10–15 %, but complete loss can occur under conducive conditions (Trapero-Casas and Jiménez-Díaz 1985; Abera et al. 2011; Sharma et al. 2014).

2 Disease Symptoms and Pathogen Movement

F. oxysporum causes a number of symptoms depending on plant species, but common symptoms include leaf vein clearing, epinasty, wilting, stunting, yellowing of older leaves, browning of vascular tissue, necrosis and plant death (Agrios 2005). Its saprophytic ability enables it to survive in the soil between crop cycles in infected plant debris. The fungus can survive either as mycelium or as asexual spores: microconidia, macroconidia and chlamydospores (Agrios 2005). To initiate its life cycle (Fig. 1), the pathogen often directly infects the plants by entering through root tips, wounds or natural openings at lateral root initials. The pathogen then invades the root cortex first and then the xylem tissue, potentially blocking water movement leading to the appearance of wilting. The fungus will stay in xylem vessels (and some surrounding cells) as long as the plant is alive and move to other cells when the plant is dead so it can sporulate at or near the plant surface (Agrios 2005). The fungus sporulates on the dead tissue where these spores can initiate new infection cycles. The pathogen often spreads within short distances through irrigation water and through the use of contaminated equipment. It is also possible for the fungus to spread over long distances through infected plant material or contaminated soil. Therefore, hygiene (disinfection of planting materials/

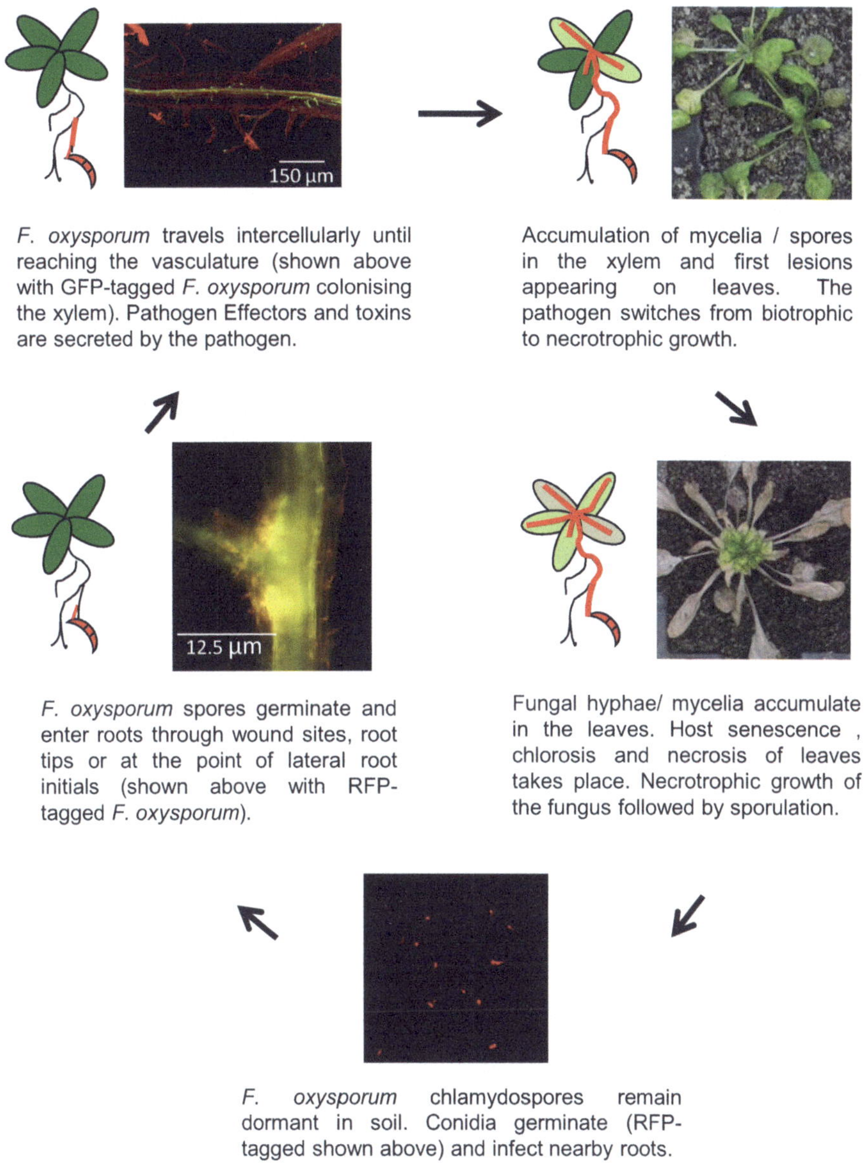

Fig. 1 *F. oxysporum* life cycle. Shown is a schematic of *F. oxysporum* life cycle as represented by *F. oxysporum* strain Fo5176 infecting wild-type *Arabidopsis* (Col-0)

equipment) and quarantine measures (e.g. inhibiting the transfer of infected plant and soil material from one region to another) can be effective to stop the disease spreading although it is often quite difficult to eradicate the fungus from the soil as its chlamydospores can survive there for decades. To manage this disease, the use of resistant cultivar crop rotation with non-host plants is often recommended (Agrios 2005).

3 Pathogen Infection Strategies

Pathogenic and non-pathogenic strains of *F. oxysporum* exist, both of which colonise host roots albeit to different degrees depending on the host but with initial root penetration favoured through wounds or at natural openings at the base of lateral root initials (Beckman 1987; Gordon and Martyn 1997; Recorbet et al. 2003; Michielse and Rep 2009; Kidd et al. 2011; Ma 2014; Perez-Nadales et al. 2014). Pathogenic strains have evolved to overcome host defence and cause disease. In such infected plants, wilting and eventual death occur largely as a result of water stress caused by proliferating spore and hyphae clogging the xylem vessels of roots and the stem and the action of secreted fungal proteins and toxins potentially blocking water movement and enhancing the appearance of wilting. The secreted molecules can differentially affect leaf and root tissues. For example, in roots toxins can initiate excessive division of parenchyma cells that encompass the xylem resulting in the collapse of xylem vessels or restricting their water flow, while the movement of toxins to leaves can affect chlorophyll synthesis (Di Pietro et al. 2003; Agrios 2005; Czymmek et al. 2007; Ramírez-Suero et al. 2010; Perez-Nadales et al. 2014; Li et al. 2015; Wang et al. 2015).

3.1 Pathogen Versus Non-pathogen

The ability of both pathogenic and non-pathogenic isolates to colonialise and penetrate the roots of hosts and non-hosts (Olivain et al. 2006; Ma 2014) suggests following colonisation plants adequately defend themselves against most *F. oxysporum* isolates, likely due to their recognition of conserved fungal molecules called microbe-associated molecular patterns (MAMPs) (also known as pathogen-associated molecular patterns (PAMPs) as they are present in pathogens). These include molecules such as chitin and β-glucan. PAMPs are typically recognised at the plant cell surface by membrane-bound receptor kinases and receptor-like proteins called pattern recognition receptors (PRRs) and induce PAMP-triggered immunity (PTI). PTI can also be triggered by host-derived products of infection called damage-associated molecular patterns (DAMPs) (e.g. plant cell wall fragments). Non-pathogenic *F. oxysporum* isolates would be recognised by these receptors; however, some isolates have become pathogenic by producing host-specific effectors that suppress or overcome PTI resulting in effector-triggered susceptibility (ETS). These effectors may mask MAMPs, manipulate host cell physiology or modify, inhibit or remove host immune response targets. Although an increasing list of candidate *F. oxysporum* effectors have been identified, relatively few *F. oxysporum* effectors have been functionally characterised. These are discussed in detail in further sections. Under selective pressure, plants have evolved receptors (resistance (R) proteins) to recognise specific effectors (avirulence (Avr) gene products) and mount resistance in a process termed effector-triggered immunity (ETI). ETI only occurs when specific *F. oxysporum* f. sp. isolates, known as races, express Avr products recognised by the corresponding host receptor, and unlike a classical ETI response of hypersensitive cell death to biotrophic pathogens, ETI in known *F. oxysporum Avr–R*-gene responses results in callose deposition, the vascular accumulation of phenolics, tyloses and gels (Takken and Rep 2010; De Coninck et al. 2015). See recent reviews for overviews of PTI and ETI triggered against plant–fungal pathogens (Win et al. 2012; van Schie and Takken 2014; Lo Presti et al. 2015).

3.2 Origins of Pathogenicity

3.2.1 Evolution of Pathogenicity

As stated above, pathogenic strains of *F. oxysporum* are classified into *formae speciales* (ff. spp.) based on the host species they cause disease on. For example, *F. oxysporum* f. sp. *lycopersici* (*Fol*) causes disease on tomato (*Solanum lycopersicum*) but no other plant species. While it was assumed isolates of a f. sp. arose through descent from a monophyletic origin, it has been demonstrated for some that this is not the case and that their genetic heterogeneity is polyphyletic in origin (Gordon and Martyn 1997; O'Donnell et al. 1998; Michielse and Rep 2009). That is, pathogenicity on a specific host may have arisen independently several times.

The polyphyletic origins of host specificity observed in some f. sp. can be explained by the recent demonstration of whole chromosome horizontal transfer. Experimentally it was shown a so-called pathogenicity chromosome containing most known effectors from a virulent *Fol* isolate was transferred to a non-pathogenic isolate, conferring its virulence on tomato (Ma et al. 2010). While horizontal gene transfer (HGT) has been demonstrated amongst many fungi, this was one the first demonstrations of whole chromosome transfer conferring host-specific pathogenicity. This pathogenicity chromosome could also transfer to another f. sp. (*melonis*); however, virulence of this isolate on tomato was not conferred suggesting other genetic content defines disease-causing host specificity.

3.2.2 Genomic Organisation of Pathogenicity Components

The sequencing of *F. oxysporum* genomes and their comparative analysis amongst ff. spp. and other fusaria has allowed identification of chromosomes and gene content geared towards pathogenicity. For example, the 15 chromosomes of the reference *F. oxysporum* genome (*Fol* race 2 isolate 4287) can be divided into "core" and "lineage specific" (Ma et al. 2010). Core chromosomes are conserved across fusaria and contain genes required for normal growth and metabolism, while lineage-specific chromosomes are absent or poorly conserved across fusaria or other fungi and lack house-keeping genes. For this reason, the latter chromosomes are also often referred to as "conditionally dispensable" or "accessory".

The lineage-specific chromosomes of *Fol* refer to chromosomes 3, 6, 14 and 15 and telomere-proximal parts of chromosomes 1 and 2. These chromosomes are enriched in rapidly evolving genes and in transposable elements (TEs), remarkably accounting for nearly 75 % of all TEs in the *Fol* genome with Chr 14 comprised of 87 % TEs (Ma et al. 2010; Schmidt et al. 2013; Sperschneider et al. 2015). Further, only 20 % of genes on these chromosomes can be functionally classified and are enriched for genes related to pathogenicity such as known and putative effectors, fungal transcription factors and genes with roles in signal transduction and secondary metabolism.

The smaller lineage-specific chromosome 14 is referred to as the "pathogenicity" chromosome as it contains the majority of known *Fol in planta* expressed effectors and its horizontal transfer of pathogenicity to a non-pathogenic isolate (Michielse et al. 2009a; Ma et al. 2010; de Sain and Rep 2015). Interestingly, the most virulent of the newly created pathogenic isolates following HGT also contained additional parts of the lineage-specific chromosomes 3 and 6 (Ma et al. 2010). Loss of pathogenicity or virulence is also associated with the spontaneous loss of all or parts of *Fol* Chr 14 (Rep et al. 2004, 2015). This gain and loss of genetic material are likely associated with the enrichment of transposable and/or repetitive elements on the lineage-specific chromosomes surrounding effectors and other pathogenicity-related genes (Ma et al. 2010; Schmidt et al. 2013). The impact of transposable element activity combined with horizontal gene/chromosome transfer may facilitate the rapid modification of genetic material and ability for *F. oxysporum* to cause disease on so many diverse hosts.

With the advent of short-read sequencing technology, the list of available *F. oxysporum* genomes is increasing at a solid rate and covers ff. spp. causing disease over a range of economically important crops such as banana, brassicas, melons, cotton and legumes (Table 1). This not only facilitates the prediction of effectors and other pathogenicity components but also enables genome-wide analyses and comparative studies. For example, it was suggested the *Fol* (4287) effector *Avr3* and its homologous pseudogene may undergo accelerated evolution (Rep 2005). Unbiased whole-genome comparative analysis of diversifying selection between *Fol* 4287 and another f. sp., *conglutinans* Fo5176, indeed identified *Avr3*, as well as other candidate effectors, as undergoing diversifying selection (Sperschneider et al. 2015). Even small modifications in avirulence proteins can affect their recognition by host receptors (e.g. a single amino acid change in *Fol* SIX3 (Avr2) confers a loss of recognition by the host receptor I-2 (Immunity-2), but interestingly does not affect its virulence phenotype (Houterman et al. 2009)). Comparative genomic analysis of ff. spp. pathogenic to three different legume species enabled the discovery of several effector candidates and a previously

Table 1 *F. oxysporum* genomes deposited at NCBI. Accessed 11-06-2015

Strain	Forma specialis (f. sp.)	Race/VCG	Host	Disease	NCBI or BioProject accession	Reference
FOL 4287	*lycopersici*	Race 2/VCG 0030	*Lycopersicum* (tomato)	Tomato wilt	AAXH01000000	Ma et al. (2010)
MN25	*lycopersici*	Race 3/VCG 0033	*Lycopersicum* (tomato)	Tomato wilt	AGBH01000000	FCD, Broad Institute
CL57/ 26381	*radicis-lycopersici*	VCG 0094	*Lycopersicum* (tomato)	Tomato crown rot	AGNB01000000	FCD, Broad Institute
PHW808/ 54008	*conglutinans*	Race 2/VCG 0101	*Brassica*/*Arabidopsis*	Cabbage yellows disease, crucifer wilt	AGNF01000000	FCD, Broad Institute
PHW815/ 54005	*raphani*	VCG 0102	*Raphanus*/*Arabidopsis*	Crucifer wilt	AGNG01000000	FCD, Broad Institute
Fo5176	*conglutinans*		*Brassica*/*Arabidopsis*	Crucifer wilt	AFQF01000000	Thatcher et al. (2012a)
II5/54006	*cubense*	Tropical race 4/VCG01213	*Musa* (banana)	Banana wilt	AGND01000000	FCD, Broad Institute
N2	*cubense*	Race 1	*Musa* (banana)	Banana wilt	AMGP01000000	Guo et al. (2014)
B2	*cubense*	Race 4	*Musa* (banana)	Banana wilt	AMGQ01000000	Guo et al. (2014)
26406	*melonis*		*Cucurbita* (melon)	Wilt	AGNE01000000	FCD, Broad Institute
Various	*melonis*		*Cucurbita* (melon)	Wilt	PRJNA251724	Broad Institute
25433	*vasinfectum*		*Gossypium* (cotton)	Wilt	AGNC01000000	FCD, Broad Institute
HDV247	*pisi*		*Pisum* (pea)	Wilt	AGBI01000000	FCD, Broad Institute
Foc-38-1	*ciceris*		*Cicer arietinum* (chickpea)	Wilt	PRJNA188291	Williams et al. (2016) International Crops Research Institute for the Semi-Arid Tropics

(continued)

Table 1 (continued)

Strain	Forma specialis (f. sp.)	Race/VCG	Host	Disease	NCBI or BioProject accession	Reference
Fom-5190a	*medicaginis*		*Medicago* species (includes *M. sativa* and *M. truncatula*)	Wilt	PRJNA294248	Williams et al. (2016)
Fo47			Soil	Biocontrol	AFMM01000000	FCD, Broad Institute

FCD Fusarium comparative database

unrecognised gene region specifically conserved amongst legume-infecting isolates (Williams et al. 2016). These types of analyses expedite the identification of effectors responsible for inciting disease on specific hosts, an area of research that will hopefully identify the genetic determinants for classifying an isolate into a f. sp.

3.3 Pathogenicity Machinery

To invade and initiate disease on a host, pathogenic *F. oxysporum* secrete an arsenal of enzymes, toxins, secondary metabolites and effectors. Effectors suppress or overcome PTI to induce host susceptibility, and while typically classified as host specific, a broader definition of effectors includes many molecules such as toxins (e.g. fusaric acid), degradative enzymes and even PAMPs/MAMPs (Hogenhout et al. 2009; Stergiopoulos and de Wit 2009; Dong et al. 2014; Pusztahelyi et al. 2015). This is supported by the finding that genes encoding some of the latter molecules are induced upon plant contact. Large-scale fungal mutagenesis and xylem sap proteomics facilitated the initial discovery of *F. oxysporum* effectors and pathogenicity-related proteins, but more recently comparative genomics and high-coverage *in planta* transcriptome sequencing (RNA-seq) have increased the rate of candidate effector identification across ff. spp. The rate-limiting step here is still functional characterisation which is best studied in knockout and mutant lines.

3.3.1 General Pathogenicity Machinery

Like other pathogenic plant–fungal pathogens, the genomes of *F. oxysporum* ff. spp. are enriched in genes encoding plant cell wall-degrading enzymes (CWDEs) (Ma et al. 2010; Zhao et al. 2013; Williams et al. 2016) and are known to secrete these enzymes during host colonisation (Beckman 1987; Roncero et al. 2003). These include polygalacturonases, pectate lyases, xylanases and proteases and act by degrading cell walls and membranes, releasing nutrient sources such as sugars (Yadeta and Thomma 2013). While these enzymes play key roles in pathogenicity, are expressed during infection and likely contribute to virulence, individual gene knockouts have failed to produce altered disease phenotypes, which is expected in multi-gene families like these where functional redundancy may exist (Di Pietro et al. 2003; Recorbet et al. 2003; McFadden et al. 2006; Guo et al. 2014; Kubicek et al. 2014). Functional analysis therefore requires the generation of at least double deletions, for example, as shown in a *Fol* polygalacturonase and endopolygalacturonase double mutant ($\Delta pg1\Delta pgx6$) which exhibited reduced virulence on tomato (Ruiz et al. 2015).

Two other classes of secreted effector proteins found in *F. oxysporum* are the necrosis and ethylene-inducing-like proteins (NLPs) and lysine motifs (LysMs). Nep1 was first identified in *F. oxysporum* culture filtrates, but NLPs are present in other fungi as well as oomycetes and even bacteria (Bailey 1995; Pemberton and Salmond 2004; Bae et al. 2006; Böhm et al. 2014; Oome et al. 2014). LysM effectors contain the LysM carbohydrate-binding domain that mediates recognition of fungal chitin, an essential component of the fungal cell wall, and is found in some membrane-localised plant receptors (Gust et al. 2012; Kombrink and Thomma 2013). It is proposed that LysM effectors (most well characterised in *Cladosporium fulvum* and *Verticillium* pathogens) contribute to virulence through mechanisms such as suppression of chitin-triggered PTI. For example, by protecting fungal hyphae from hydrolytic plant enzymes or to scavenge hydrolytically derived chitin oligomers produced during invasion and subsequently avoid or delay host detection (Kombrink and Thomma 2013). Further, knockouts in several *Fol* chitin synthase genes are associated with a loss of pathogenicity phenotype or reduced virulence (reviewed in Michielse and Rep 2009). *Fol* also produces enzymes that neutralise host-produced chitinases that bind chitin. A recent study identified a secreted metalloprotease and a serine protease that were responsible for the cleavage of chitinases. When the genes encoding these enzymes were deleted, the mutant showed reduced virulence against tomato, suggesting that these enzymes are important for fungal virulence (Karimi Jashni et al. 2015). Although not functionally characterised in *F. oxysporum*, LysM domain-containing proteins are present in most if not all ff. spp. (Thatcher unpublished) with some expressed *in planta* (Williams et al. 2016). As effectors are often defined by the absence of detectable orthologous proteins outside the genus, the wide distribution of NLPs and LysMs suggests these are best designated as PAMPs (Thomma et al. 2011).

Other *F. oxysporum* proteins found to be secreted during infection include a catalase-peroxidase, a serine protease and the oxidoreductase Orx1 which is a homologue of the Ave1 avirulence protein from *Verticillium dahliae*. These proteins were detected in the xylem sap of *Fol*-infected tomato plants, suggesting they are important for infection (Houterman et al. 2007; Schmidt et al. 2013). Some enzymes such as catalase-peroxidase, galactosidase and chitinase might also contribute to the strong virulence of *Foc* TR4 (Sun et al. 2014).

3.3.2 *F. oxysporum* Signal Transduction Machinery Involved in Pathogen Virulence

Signalling processes and the coordinated control of *F. oxysporum* pathogenicity machinery have been shown in several cases to be critical for host colonisation, penetration or virulence. Components of signal transduction such as kinases and transcription factors are expressed during host infection, and in several cases, their targeted gene knockouts show reduced pathogenicity (Guo et al. 2014; Michielse et al. 2009a, b). For example, mutants of G-protein-coupled receptor subunits α (FGA1, FGA2) and β (FGB1) are impaired in or have lost pathogenicity in *Fol* and *F. oxysporum* f. sp. *cucumerinum* (Jain et al. 2002, 2003, 2005). Mutants of the *Fol* mitogen-activated protein kinase (MAPK) genes *FMK1* and *SNF1* (Di Pietro et al. 2001; Michielse et al. 2009b) are impaired in root penetration and pathogenicity (see reviews by Di Pietro et al. 2003; Michielse and Rep 2009). The constitutively expressed *Fol* F-box gene *FRP1* may function in SCF-mediated ubiquitination processes and is required for pathogenicity as knockouts are non-pathogenic and unable to colonise roots (Duyvesteijn et al. 2005; Jonkers and Rep 2009).

Several transcription factors with roles in pathogenicity have been functionally characterised. For example, a knockout of the zinc finger XlnR is severely impaired in extracellular xylanase activity (Calero-Nieto et al. 2007). The transcription factor gene *FOW2* encoding a Zn(II)2Cys6 family transcriptional regulator appears conserved amongst *F. oxysporum* ff. spp. and in *Fol*, and *F. oxysporum* f. sp. *melonis* is required for colonisation and pathogenicity (Imazaki et al. 2007; Michielse et al. 2009b). And another transcription factor (SGE1, SIX gene expression 1) is not required for root colonisation or penetration, but is essential for pathogenicity in *Fol* where its expression is upregulated during infection of tomato roots and is required for expression of most secreted *Fol* effectors as discussed in the following section (Michielse et al. 2009a).

3.3.3 Effectors

While general machinery necessary for host colonisation tends to be expressed constitutively, genes necessary for pathogenicity and virulence are typically only expressed upon plant contact (lowly or not expressed under axenic conditions) (Rep 2005). The most well-characterised effectors from *F. oxysporum* belong to a class termed the secreted in xylem or SIX effectors, first identified in the xylem sap proteome of tomato plants infected with *Fol*, with roles in virulence and/or avirulence determined for some depending on the host genotype (Rep et al. 2004, 2005; Houterman et al. 2007; de Sain and Rep 2015). So far, 14 families of SIX proteins have been identified (Rep et al. 2004; Houterman et al. 2007; van der Does and Rep 2007; Lievens et al. 2009; Ma et al. 2010; Rep and Kistler 2010; Schmidt et al. 2013), and these are typically only found in *F. oxysporum* isolates, although some, such as SIX6, are present in other fungi such as *Colletotrichum* species (Gawehns et al. 2014). The SIX effectors were originally thought to be unique to *Fol* but have since been identified in several *F. oxysporum* ff. spp. with some sharing high levels of sequence identity (Lievens et al. 2009; Meldrum et al. 2012; Thatcher et al. 2012a; Laurence et al. 2015; Schmidt et al. 2016). For example, the *Arabidopsis* infecting isolate Fo5176 contains a highly conserved SIX4 homologue, only differing from the *Fol* SIX4 by two amino acids (Thatcher et al. 2012a). Interestingly, in the tomato pathosystem, *Fol* SIX4 (Avr1) is not required for general virulence but acts by suppressing ETI mediated by two resistance genes (*immunity-2* (*I-2*) and *immunity-3* (*I-3*)), whereas in *Arabidopsis* lacking immunity resistance genes, Fo5176 SIX4 is required for full virulence (Rep et al. 2005; Houterman et al. 2008; Thatcher et al. 2012a). *Fol* SIX4 (Avr1), as well as *Fol* SIX6, can also suppress cell death triggered by I-2 (Gawehns et al. 2014).

Similar to most known fungal effectors, SIX proteins are small and generally cysteine rich and most contain a signal peptide for secretion (Houterman et al. 2007; Schmidt et al. 2013), but apart from these characteristics, they share little similarity with each other and other known fungal proteins (Rep 2005). Secreted into apoplast or xylem, the cysteine-rich nature of these extracellular proteins creates disulphide bridges that stabilises the protein against protease degradation (Takken and Rep 2010). The majority of *Fol SIX* genes reside on pathogenicity Chr 14 or in some cases on other dispensable chromosomes and are located within transposon-rich regions often associated with miniature transposable elements (MITE) present in their promoters (Ma et al. 2010; Schmidt et al. 2013). Some are even co-located at the same loci and share common promoters (e.g. SIX3 (Avr2) and SIX5) and may also physically interact with each other at the protein level (Schmidt et al. 2013; Ma et al. 2015).

For most SIX effectors, their expression requires the core-chromosome-encoded transcription factor Sge1 (SIX gene expression 1) (Michielse et al. 2009a). The expression profiles of *SIX* genes from other *F. oxysporum* ff. spp. confirm that most are either highly *in planta* inducible or only expressed *in planta* (McFadden et al. 2006; van der Does et al. 2008; Thatcher et al. 2012a; Gawehns et al. 2014; Guo et al. 2014; Williams et al. 2016). *In planta* gene expression has also been used in other *F. oxysporum* ff. spp. to identify putative effectors (e.g. f. sp. *cubense*, f. sp. *vasinfectum*, f. sp. *medicaginis* (McFadden et al. 2006; Guo et al. 2014; Williams et al. 2016)), and the associated presence of MITEs helped identify the *F. oxysporum* f. sp. *melonis* avirulence protein AvrFOM2 that is recognised by the melon resistance gene *Fom-2* (Schmidt et al. 2016).

4 Host Resistance

The genetic and molecular *F. oxysporum*–plant interaction is best understood in the tomato pathosystem where R-gene resistance is available (Takken and Rep 2010), with other model pathosystems in *Arabidopsis thaliana* and *Medicago truncatula* also studied (Diener and Ausubel 2005; Lichtenzveig et al. 2006; Berrocal-Lobo and Molina 2008; Ramírez-Suero et al. 2010; Lyons et al. 2015a; Rispail et al. 2015). The following sections will discuss the findings from studying host resistance to *F. oxysporum*.

4.1 Transcriptome Studies

Plant responses to *F. oxysporum* infection have been studied using genome-wide expression profiling using microarray and RNA-seq analyses (see Table 2 for examples). Most of the earlier efforts investigated defence responses occurring in the leaves. A recent study that comparatively analysed defence responses triggered

Table 2 Recent transcriptome, metabolome and proteome studies analysing *F. oxysporum* infection

Fo–plant interaction	Tissue type	Technique	Reference
Fo5176—*Arabidopsis*	Roots and shoots	Transcriptome (RNA-seq)	Lyons et al. (2015a)
Fo5176—*Arabidopsis*	Roots and shoots	Transcriptome (RNA-seq)	Lyons et al. (2015b)
Fo5176—*Arabidopsis*	Root	Transcriptome (Agilent GeneChip)	Chen et al. (2014a)
Fo5176—*Arabidopsis*	Seedlings	Transcriptome (RNA-seq)	Zhu et al. (2013)
Fo5176 *SIX4* overexpression—*Arabidopsis*	Root	Transcriptome (Affy array)	Thatcher et al. (2012a) and this publication, microarray data deposited at NCBI under accession number GSE75928
F. oxysporum f. sp. *phaseoli*—bean	Seedlings	Transcriptome (cDNA-AFLP)	Xue et al. (2015a)
F. oxysporum f. sp. *ciceris*—chickpea	Root	Metabolome	Kumar et al. (2015)
F. oxysporum f. sp. *ciceris*—chickpea	Root	Proteome	Chatterjee et al. (2014)
F. oxysporum f. sp. *pisi*—pea	Root	Proteome	Castillejo et al. (2015)
F. oxysporum f. sp. *cubense* TR4—banana	Root	Transcriptome	Li et al. (2012)
F. oxysporum f. sp. *cubense TR1* and *TR4—banana*	Root	Transcriptome	Li et al. (2013a)
F. oxysporum f. sp. *cubense* TR4—banana	Root	Transcriptome	Wang et al. (2012)
F. oxysporum f. sp. *cubense* TR4—banana	Root	Transcriptome	Bai et al. (2013)
F. oxysporum f. sp. *cubense*—banana	Root	Proteome	Sun et al. (2014)
F. oxysporum f. sp. *cubense* TR4—banana	Root	Proteome	Li et al. (2013b)
F. oxysporum f. sp. *radicis-lycopersici*—tomato	Root	Proteome	Mazzeo et al. (2014)

by *Fusarium* infection revealed that the infection triggers expression from separate classes of defence-associated genes in the roots and shoots (leaves or rosettes), suggesting that different physiological and defence-associated processes might be operational in these tissues (Lyons et al. 2015a). Plant development and flowering time seem to have a major effect on *F. oxysporum* disease symptom expression. It was shown recently that diverse *Arabidopsis* ecotypes and various mutants affected in flowering time also show altered disease development (Lyons et al. 2015b). In particular, late flowering time is associated with increased disease resistance. It was speculated that delayed senescence as a result of late flowering could be a reason explaining this delay in disease progression.

Other studies (Table 2) have compared differentially expressed genes between resistant and susceptible genotypes to determine what makes the plant resistant or susceptible to infection. For instance, Xue et al. (2015a) recently compared resistant and susceptible bean plants, while Bai et al. (2013) looked at resistant and susceptible banana cultivars. As a result, large numbers of genes corresponding to certain defence categories have been identified. These studies have certainly provided useful candidates that can be further studied functionally, and if their association with disease resistance is confirmed, they may be useful targets for marker-assisted selection studies. However, it should be remembered that some of the host genes induced by the pathogen may also be associated with susceptibility.

Interestingly, a recent study comparing transcriptomes of banana roots inoculated with either race 1 or tropical race 4 shows that both *Foc* race 1 and *Foc* TR4 triggered similar gene expression profiles in banana roots, despite their differing pathogenicity/virulence (Li et al. 2013a). Following *F. oxysporum* Fo5176 infection, we have also analysed the root transcriptomes of wild-type *Arabidopsis* plants and *Arabidopsis* overexpressing the Fo5176 *SIX4* effector (arrays conducted on root tissue from Col-0 or 35sSIX4 plants (Thatcher et al. 2012a) 4 days postinoculation, pathogen infection and microarray analysis conducted as described previously (Kidd et al. 2009), microarray data deposited at NCBI under accession number GSE75928). This process identified genes downregulated >1.5-fold in the effector overexpression plants to be enriched in genes associated with oxidative stress and wound/defence responses suggesting virulence function of the SIX4 effector is associated with modifying host-signalling processes.

4.2 Genetics of Host Resistance in Arabidopsis

Analysis of mutants affected in disease resistance against *F. oxysporum* has identified a number of genes that regulate resistance or susceptibility in *Arabidopsis*. So far a number of transcription factors altering disease resistance to *F. oxysporum* have been identified. This has also helped in the development of a model that explains host susceptibility or resistance. In particular, the SA signalling pathway seems to be required for increased resistance, while *F. oxysporum* seems to exploit the JA signalling pathway to cause disease. The evidence for this comes from the observation that *Arabidopsis* JA signalling mutants such as *coi1*, *myc2* and *pft1* but not JA biosynthesis mutants show increased resistance to *F. oxysporum* (Anderson et al. 2004; Thatcher et al. 2009; Kidd et al. 2009). The *esr1-1* (enhanced stress response 1) mutant defective in a KH domain containing RNA-binding protein (At5g53060) also confers increased resistance to *F. oxysporum*. Similar to other JA signalling genes that make *Arabidopsis* susceptible to *F. oxysporum* infection, ESR1 seems to modulate JA responses as well (Thatcher et al. 2015). It is possible that pathogen-produced JA-like compounds secreted by the pathogen activate the host's JA signalling pathway, which then promotes senescence (Thatcher et al. 2009; Cole et al. 2014). In the banana–*Foc* interaction, fusaric acid secreted by *Foc* also seems to play a role in promoting senescence (Dong et al. 2014). Transgenic expression of JA-responsive transcription factors such as ethylene response factors (ERFs) can also positively contribute to disease inhibition by modulating defence gene expression without promoting senescence. For instance, overexpression of ERF1 in *Arabidopsis* increases *F. oxysporum* resistance by altering the expression of defence-related genes (Berrocal-Lobo and Molina 2004). Similarly, another *Arabidopsis* ERF transcription factor, ERF14, is required for wild-type resistance to *F. oxysporum* in *Arabidopsis* as *erf14* loss-of-function mutants show reduced defence gene expression and increased susceptibility to this pathogen (Onate-Sanchez et al. 2007).

In addition, it was reported that auxin signalling and biosynthesis mutants show increased susceptibility to *F. oxysporum* as a number of auxin mutants show altered *F. oxysporum* resistance (Kidd et al. 2011). A *F. oxysporum* strain genetically modified to produce increased levels of auxin shows hypervirulence (Cohen et al. 2002), further suggesting that auxin is associated with increased disease. However, how auxin promotes disease susceptibility is currently unknown. One possibility is that auxin signalling and transport are required for lateral root formation and increased lateral root formation may provide a higher number of infection sites. *F. oxysporum* is known to infect the plant lateral root initials and root tips that are also auxin-rich regions. Interestingly a recent study showed that volatiles produced by *F. oxysporum* improve plant growth and were dependent on a functional auxin signalling pathway in *Arabidopsis* (Bitas et al. 2015) (Table 3).

4.3 Deployment of Resistance Genes and Marker-Based Selection Approaches

In several crops resistance against specific pathogenic f. sp. or races of *F. oxysporum* have been identified enabling researchers to develop molecular markers that can be used for germplasm-screening purposes (Jimenez-Gasco et al. 2004; reviewed Michielse and Rep 2009; Sharma et al. 2014; Schmidt et al. 2016). However, only a handful of the underlying R-genes have been cloned

Table 3 *Arabidopsis* genes that regulate resistance or susceptibility to the *F. oxysporum* strain Fo5176

Gene	Signalling pathway	Reference
COI1	Jasmonate	Thatcher et al. (2009); Trusov et al. (2009)
LBD20	Jasmonate	Thatcher et al. (2012b)
ERF2	Jasmonate	McGrath et al. (2005)
ERF14	Jasmonate/ethylene	Onate-Sanchez et al. (2007)
ERF72	Ethylene/ROS	Chen et al. (2014a, b)
MYC2	Jasmonate	Anderson et al. (2004)
G proteins, AGB1-1, AGB1-2, AGG1-1 and AGG1-2	G-protein signalling	Trusov et al. (2009)
ABA2–1	ABA	Anderson et al. (2004)
AXR1, AXR2, AXR3, AXR4, SGT1B, AUX1, PIN2 and BIG	Auxin signalling and transport	Kidd et al. (2011)
PFT1	Jasmonate	Kidd et al. (2009)
MED8	Defence and development	Kidd et al. (2009)
ESR1	Jasmonate	Thatcher et al. (2015)
Gigantea	Circadian	Lyons et al. (2015b)
ARF2	Auxin signalling	Lyons et al. (2015a)
PRX33	ROS production	Lyons et al. (2015a)
ATAF2	Negative regulator of defence gene expression	Delessert et al. (2005)
RBOHD and RBOHF	ROS production	Zhu et al. (2013)

See Swarupa et al. (2014) for additional genes that regulate resistance to other *Arabidopsis*-infecting *F. oxysporum* strains

(Table 4), the majority of which isolated from tomato are based on monogenetic resistance conferring classical gene-for-gene-mediated interactions. Plant resistance genes can be divided into two main categories, the leucine-rich repeat (LRR) and intracellular nucleotide-binding site (NBS)-LRR-containing R proteins, with the latter mediating recognition of intracellular pathogen-derived signals (Martin et al. 2003). Some transmembrane LRR proteins also have an intracellular protein kinase (PK) domain and belong to the larger class of receptor-like protein kinases (RLKs). The extracellular LRR domain of LRR-TM and LRR-TM-PK proteins is thought to function as receptors for extracellular pathogen-derived signals such as conserved pathogen molecules (MAMPs) and damage-associated molecules.

For some R proteins, their cellular localisation has been determined. For example, the cytosolic R-protein I-2 from tomato mainly localises to xylem tissues of roots, stems and leaves, where it intracellularly perceives the *Fol* effector SIX3 (Avr2) (Mes et al. 2000; Houterman et al. 2009; Gawehns et al. 2014; Ma et al. 2015). The tomato I-3 protein is a plasma membrane-bound receptor with a cytoplasmic kinase domain and an extracellular S-domain (Catanzariti et al. 2015).

Table 4 Summary of cloned or well-characterised *F. oxysporum* resistance genes

Plant	Gene/ loci	Resistance against	Protein description	Effector recognition	Reference
Tomato	*I*	*Fol* race 1	–	SIX4 (Avr1)	Bohn and Tucker (1939); Houterman et al. (2008)
	I-1		–	SIX4 (Avr1)	Sarfatti et al. (1991); Houterman et al. (2008)
	I-2	*Fol* race 2	CC-NBS-LRR	SIX3 (Avr2)	Simons et al. (1998); Houterman et al. (2009)
	I-3	*Fol* race 3	S-receptor-like kinase (SRLK)	SIX1 (Avr3)	Rep et al. (2004); Catanzariti et al. (2015)
	I-7	*Fol* race 3	Membrane-anchored LRR-receptor-like protein (RLP)	–	Gonzalez-Cendales et al. (2015)
Melon	*Fom-1*		Toll/interleukin-1 receptor (TIR)–NB-LRR		Brotman et al. (2013)
	Fom-2		NB-LRR	AvrFOM2	Joobeur et al. (2004); Schmidt et al. (2016)
Arabidopsis	*RFO1*	f. sp. *conglutinans*, *raphani* and *matthioli*	At1g79670 WALL-ASSOCI-ATED KINASE-LIKE KINASE 22 (WAKL22)		Diener and Ausubel (2005)
	RFO2	f. sp. *matthioli*	Extracellular RLP, At1g17250		Shen and Diener (2013)
	RFO3	f. sp. *matthioli*	S-receptor-like kinase (SRLK)		Cole and Diener (2013)
	RFO7	f. sp. *conglutinans* race 1			Diener (2013)

Interestingly, *I-3* gene expression is higher in leaf tissues compared to root or stem tissues where initial stages of *Fol* infection take place. Like I-3, I-7 also contains an extracellular recognition domain suggesting SIX1 (Avr3) and Avr7 are recognised at the cell surface and may not be taken up by host plant cells (Catanzariti et al. 2015; Gonzalez-Cendales et al. 2015).

In the *Arabidopsis* pathosystem, several resistance loci have been identified using various f. sp. such as f. sp. c*onglutinans*, f. sp. *raphani* and f. sp. *matthioli*

(Diener and Ausubel 2005). Crosses between the *F. oxysporum* f. sp. *matthioli*-resistant Col-0 ecotype and the susceptible Ty-0 ecotype identified six resistance loci (RFO1–6) with RFO1 the largest contributor to resistance encoding a WALL-ASSOCIATED KINASE-LIKE KINASE (WAKL22) that provides resistance to three isolates of *F. oxysporum* (Diener and Ausubel 2005), while RFO2 and RFO3 encode a receptor-like protein and a receptor-like kinase, respectively, which have undergone duplication in the parent ecotypes (Cole and Diener 2013; Shen and Diener 2013). Identification of RFO2 also leads to a role for tyrosine-sulphated peptide signalling in the *F. oxysporum* interaction (Shen and Diener 2013). Therefore the identification and characterisation of R-genes effective against *F. oxysporum* provide an opportunity to understand effective resistance strategies against this pathogen. Transcriptome analysis of *F. oxysporum* infected *Arabidopsis* also identified significant upregulation of several receptor-associated genes including a wall-associated kinase-like gene, lectin receptor kinases, receptor-like protein kinase 1 and TIR-NBS-LRR genes suggesting roles in resistance (Zhu et al. 2013). Using a comparative transcriptome approach between resistant and susceptible Chinese cabbage (*Brassica rapa* var. *pekinensis*), Shimizu et al. (2014) were also able to narrow a single dominant *R*-gene down to two possible candidates encoding TIR-NBS-LRRs.

4.4 Resistance Through the Application of Biological and Chemical Agents

Given the long-term survival of *F. oxysporum* in the soil, attention has been given to treatments that can suppress disease. Silicon addition has been observed to provide increased tolerance to *Foc* in banana (Fortunato et al. 2012a, b). While the role that silicon plays in protecting plants against plant pathogens is debated, a recent study found that silicon may act by stimulating lignin and products of the phenylpropanoid pathway in infected banana plants (Fortunato et al. 2014).

Non-pathogenic isolates of *F. oxysporum* may also be employed to manage pathogenic isolates of *F. oxysporum* (Forsyth et al. 2006). For instance incompatible *Foc* race 1 was used to induce systemic resistance against *Foc* TR 4 (Wu et al. 2013). This increased resistance state was accompanied by systemic upregulation of defence-related genes such as *MaNPR1A*, *MaNPR1B*, *PR1* and *PR3* as well as upregulation of SA and JA pathways (Wu et al. 2013). Similar findings were found with *Fo47*, a protective strain of *Fusarium* wilt in tomato (Olivain et al. 2006). *Fo47* reduces the growth of pathogenic *F. oxysporum* f. sp. *lycopersici* isolate *Fol8* and induces the expression of defence genes CHI3, GLUA and *PR1a* in tomato (Aimé et al. 2013). Understanding the microbiome may also provide protection against *F. oxysporum*. Studying the microbial components of disease-suppressive soils has been a popular area of research (see Ajilogba and Babalola 2013 for research in tomato), and recent reports have focussed on the banana rhizosphere given the current global outbreak of TR4 (Huang et al. 2015; Xue et al. 2015b). A recent analysis of soils suppressive to *Fusarium* wilt of strawberry identified members of the Actinobacteria and the identification of a novel antifungal thiopeptide from one of these bacteria which targeted fungal cell wall biosynthesis (Cha et al. 2015). While many microbial isolates appear beneficial in suppressing the disease in particular soil types, so far those identified haven't been sufficient to prevent disease occurrence globally, but this is a promising area of research.

4.5 Engineering of Resistance

Given that *F. oxysporum* infection leads to widespread cell death and necrosis on the above-ground tissues, genes that play roles in inhibiting apoptosis or cell death (namely, Bcl-xL, Ced-9) can play a role in disease resistance. Indeed, transgenic expression of apoptosis-related genes enhanced banana resistance to *Foc* and is undergoing field testing (Paul et al. 2011). Transgenic plants expressing a defensive gene from *Nicotiana alata* were recently shown to provide a quantitative resistance to *Fusarium oxysporum* and *Verticillium dahliae* in cotton (Gaspar et al. 2014). The expression of defensin chitinase and/or thaumatin-like genes from other plant species also shows promise as candidates for increasing *Fusarium* wilt resistance in tomato and banana (Abdallah et al. 2010; Ghag et al. 2012; Mahdavi et al. 2012; Jabeen et al. 2015).

4.6 Host-Induced Gene Silencing

Inhibiting the expression of genes involved in fungal growth and development and pathogenicity through host-delivered (host-induced) gene silencing seems to be a promising way to engineer disease resistance against *F. oxysporum*. In a recent study, transgenic banana plants expressing hairpin RNA against *Velvet* and *FTF1* genes (*Fusarium transcription factor 1*) showed complete resistance to *Foc* in greenhouse bioassays (Ghag et al. 2014). In *Arabidopsis*, survival rates of transgenic lines expressing dsRNA against three *F. oxysporum* genes (*FOW2*, *FRP1* and *OPR*) were found to be higher than wild-type plants (Hu et al. 2015). *FOW2* encodes a Zn(II)2Cys6 TF that is required for the pathogenicity of *F. oxysporum* f. sp. *melonis* (Imazaki et al. 2007), *FRP1* encodes an F-box protein involved in protein ubiquitination, which was also required for *F. oxysporum* f. sp. *lycopersici* pathogenicity, and *OPR* encodes a 12-oxo-phytodienoate-10-11-reductase-like protein potentially involved in JA biosynthesis in *F. oxysporum* (Hu et al. 2015). These studies show promising results; however, commercialisation of transgenic plants is dependent on a number of factors including regulatory (no adverse health and environmental effects), legal (e.g. patenting and licensing issues) as well as economic and social consideration (consumer acceptance). Therefore, genetic modification approaches can be difficult to commercialise under the current climate but provide potential solutions for combatting *F. oxysporum*.

5 Conclusion

The ubiquitous and persistent nature of *F. oxysporum* as well as its ability to evolve new pathogenic strains makes *F. oxysporum* a particularly difficult pathogen to control. Despite this, significant progress has been made in recent years in understanding the factors responsible for both virulence in the pathogen and resistance or susceptibility in the host. Building upon these studies will hopefully lead to the identification of additional resistance genes that can be implemented in crops where resistance is lacking. Hopefully, continual research may lead to protection against the current forms of *F. oxysporum* but ideally lead to strategies that may protect against future evolving strains.

Acknowledgements This work is supported by funds from the Commonwealth Scientific and Industrial Research Organization (CSIRO). BNK is supported by a CSIRO Office of the Chief Executive (OCE) postdoctoral fellowship. The 35sSIX4 microarray results presented herein were undertaken within an OCE postdoctoral fellowship awarded to LFT and with the assistance of resources from the Australian Genome Research Facility (AGRF) which is supported by the Australian Government. We thank past members of our labs for their valuable contributions to some of the work reviewed here.

References

Abdallah NA, Shah D, Abbas D, Madkour M (2010) Stable integration and expression of a plant defensin in tomato confers resistance to fusarium wilt. GM Crops 1:344–350

Abera M, Sakhuja PK, Fininsa C, Ahmed S (2011) Status of chickpea fusarium wilt (*Fusarium oxysporum* f. sp. *ciceris*) in northwestern Ethiopia. Arch Phytopathol Plant Prot 44:1261–1272

Agrios GN (2005) Chapter 11—Plant diseases caused by fungi. In: Agrios GN (ed) Plant pathology, 5th edn. Academic, San Diego, CA, pp 385–614

Aimé S, Alabouvette C, Steinberg C, Olivain C (2013) The endophytic strain Fusarium oxysporum Fo47: a good candidate for priming the defense responses in tomato roots. Mol Plant Microbe Interact 26:918–926

Ajilogba CF, Babalola OO (2013) Integrated management strategies for tomato Fusarium wilt. Biocontrol Sci 18:117–127

Anderson JP, Badruzsaufari E, Schenk PM, Manners JM, Desmond OJ, Ehlert C, Maclean DJ, Ebert PR, Kazan K (2004) Antagonistic interaction between abscisic acid and jasmonate-ethylene signaling pathways modulates defense gene expression and disease resistance in Arabidopsis. Plant Cell 16:3460–3479

Bae H, Kim MS, Sicher RC, Bae HJ, Bailey BA (2006) Necrosis- and ethylene-inducing peptide from Fusarium oxysporum induces a complex cascade of transcripts associated with signal transduction and cell death in Arabidopsis. Plant Physiol 141:1056–1067

Bai TT, Xie WB, Zhou PP, Wu ZL, Xiao WC, Zhou L, Sun J, Ruan XL, Li HP (2013) Transcriptome and expression profile analysis of highly resistant and susceptible banana roots challenged with Fusarium oxysporum f. sp. cubense tropical race 4. PLoS One 8(9): e73945

Bailey BA (1995) Purification of a protein from culture filtrates of Fusarium oxysporum that induces ethylene and necrosis in leaves of Erythroxylum coca. Phytopathology 85 (10):1250–1255

Beckman CH (1987) The nature of wilt diseases of plants. APS Press, St. Paul, MN, accessed from http://nla.gov.au/nla.cat-vn281454

Berrocal-Lobo M, Molina A (2004) Ethylene response factor 1 mediates Arabidopsis resistance to the soilborne fungus Fusarium oxysporum. Mol Plant Microbe Interact 17:763–770

Berrocal-Lobo M, Molina A (2008) Arabidopsis defense response against Fusarium oxysporum. Trends Plant Sci 13:145–150

Bitas V, McCartney N, Li N, Demers J, Kim J, Kim H, Brown K, Kang S (2015) Fusarium oxysporum volatiles enhance plant growth via affecting auxin transport and signalling. Front Microbiol 6:1248. doi:10.3389/fmicb.2015.01248

Böhm H, Albert I, Oome S, Raaymakers T, den Ackerveken G, Nürnberger T (2014) A conserved peptide pattern from a widespread microbial virulence factor triggers pattern-induced immunity in Arabidopsis. PLoS Pathog 10(11):e1004491. doi:10.1371/journal.ppat.1004491

Bohn GW, Tucker CM (1939) Immunity to Fusarium wilt in the tomato. Science 89:603–604

Brotman Y, Normantovich M, Goldenberg Z, Zvirin Z, Kovalski I, Stovbun N, Doniger T, Bolger AM, Troadec C, Bendahmane A et al (2013) Dual resistance of melon to *Fusarium oxysporum* races 0 and 2 and to Papaya ring-spot virus is controlled by a pair of head-to-head-oriented NB-LRR genes of unusual architecture. Mol Plant 6:235–238

Calero-Nieto F, Di Pietro A, Roncero MI, Hera C (2007) Role of the transcriptional activator xlnR of Fusarium oxysporum in regulation of xylanase genes and virulence. Mol Plant Microbe Interact 20:977–985

Castillejo MÁ, Bani M, Rubiales D (2015) Understanding pea resistance mechanisms in response to Fusarium oxysporum through proteomic analysis. Phytochemistry 115:44–58

Catanzariti A-M, Lim GTT, Jones DA (2015) The tomato I-3 gene: a novel gene for resistance to Fusarium wilt disease. New Phytol 207:106–118. doi:10.1111/nph.13348

Cha JY, Han S, Hong HJ, Cho H, Kim D, Kwon Y, Kwon SK, Crüsemann M, Bok Lee Y, Kim JF, Giaever G, Nislow C, Moore BS, Thomashow LS, Weller DM, Kwak YS (2015) Microbial and biochemical basis of a Fusarium wilt-suppressive soil. ISME J. doi:10.1038/ismej.2015.95

Chatterjee M, Gupta S, Bhar A, Chakraborti D, Basu D, Das S (2014) Analysis of root proteome unravels differential molecular responses during compatible and incompatible interaction between chickpea (Cicer arietinum L.) and Fusarium oxysporum f. sp. ciceri Race1 (Foc1). BMC Genomics 15:949. doi:10.1186/1471-2164-15-949

Chen YC, Wong CL, Muzzi F, Vlaardingerbroek I, Kidd BN, Schenk PM (2014a) Root defense analysis against Fusarium oxysporum reveals new regulators to confer resistance. Sci Rep 4:5584

Chen YC, Kidd BN, Carvalhais LC, Schenk PM (2014b) Molecular defense responses in roots and the rhizosphere against Fusarium oxysporum. Plant Signal Behav 9:e977710

Cohen BA, Amsellem Z, Maor R, Sharon A, Gressel J (2002) Transgenically enhanced expression of indole-3-acetic Acid confers hypervirulence to plant pathogens. Phytopathology 92:590–596

Cole SJ, Diener AC (2013) Diversity in receptor-like kinase genes is a major determinant of quantitative resistance to Fusarium oxysporum f.sp. matthioli. New Phytol 200:172–184. doi:10.1111/nph.12368

Cole SJ, Yoon AJ, Faull KF, Diener AC (2014) Host perception of jasmonates promotes infection by Fusarium oxysporum formae speciales that produce isoleucine- and leucine-conjugated jasmonates. Mol Plant Pathol 15:589–600

Czymmek KJ, Fogg M, Powell DH, Sweigard J, Park S-Y, Kang S (2007) In vivo time-lapse documentation using confocal and multi-photon microscopy reveals the mechanisms of

invasion into the Arabidopsis root vascular system by Fusarium oxysporum. Fungal Genet Biol 44:1011–1023
De Coninck B, Timmermans P, Vos C, Cammue BP, Kazan K (2015) What lies beneath: belowground defense strategies in plants. Trends Plant Sci 20:91–101
Delessert C, Kazan K, Wilson IW, Van Der Straeten D, Manners J, Dennis ES, Dolferus R (2005) The transcription factor ATAF2 represses the expression of pathogenesis-related genes in Arabidopsis. Plant J 43:745–757
de Sain M, Rep M (2015) The role of pathogen-secreted proteins in fungal vascular wilt diseases. Int J Mol Sci 16:23970–23993
Diener AC (2013) Routine mapping of Fusarium wilt resistance in BC(1) populations of Arabidopsis thaliana. BMC Plant Biol 13:171
Diener AC, Ausubel FM (2005) RESISTANCE TO FUSARIUM OXYSPORUM 1, a dominant Arabidopsis disease-resistance gene, is not race specific. Genetics 171:305–321
Di Pietro A, García-Maceira FI, Méglecz E, Roncero MIG (2001) A MAP kinase of the vascular wilt fungus Fusarium oxysporum is essential for root penetration and pathogenesis. Mol Microbiol 39:1140–1152
Di Pietro A, Madrid MP, Caracuel Z, Delgado-Jarana J, Roncero MIG (2003) Fusarium oxysporum: exploring the molecular arsenal of a vascular wilt fungus. Mol Plant Pathol 4:315–325
Dong X, Xiong Y, Ling N, Shen Q, Guo S (2014) Fusaric acid accelerates the senescence of leaf in banana when infected by Fusarium. World J Microbiol Biotechnol 30:1399–1408
Duyvesteijn RG, van Wijk R, Boer Y, Rep M, Cornelissen BJ, Haring MA (2005) Frp1 is a Fusarium oxysporum F-box protein required for pathogenicity on tomato. Mol Microbiol 57:1051–1063
Forsyth LM, Smith LJ, Aitken EA (2006) Identification and characterization of non-pathogenic Fusarium oxysporum capable of increasing and decreasing Fusarium wilt severity. Mycol Res 110:929–935
Fortunato AA, Rodrigues FÁ, do Nascimento KJ (2012a) Physiological and biochemical aspects of the resistance of banana plants to Fusarium wilt potentiated by silicon. Phytopathology 102:957–966
Fortunato AA, Rodrigues FA, Baroni JCP, Soares GCB, Rodiguez MAD, Pereira OL (2012b) Silicon suppresses fusarium wilt development in banana plants. J Phytopathol 160:674–679
Fortunato AA, da Silva WL, Rodrigues FÁ (2014) Phenylpropanoid pathway is potentiated by silicon in the roots of banana plants during the infection process of Fusarium oxysporum f. sp. cubense. Phytopathology 104:597–603
Gaspar YM, McKenna JA, McGinness BS, Hinch J, Poon S, Connelly AA, Anderson MA, Heath RL (2014) Field resistance to Fusarium oxysporum and Verticillium dahliae in transgenic cotton expressing the plant defensin NaD1. J Exp Bot 65:1541–1550
Gawehns F, Houterman PM, Ichou FA, Michielse CB, Hijdra M, Cornelissen BJC, Rep M, Takken FLW (2014) The Fusarium oxysporum effector Six6 contributes to virulence and suppresses I-2-mediated cell death. Mol Plant Microbe Interact 27:336–348
Ghag SB, Shekhawat UK, Ganapathi TR (2012) Petunia floral defensins with unique prodomains as novel candidates for development of fusarium wilt resistance in transgenic banana plants. PLoS One 7:e39557
Ghag SB, Shekhawat UK, Ganapathi TR (2014) Host-induced post-transcriptional hairpin RNA-mediated gene silencing of vital fungal genes confers efficient resistance against Fusarium wilt in banana. Plant Biotechnol J 12:541–553
Gonzalez-Cendales Y, Catanzariti AM, Baker B, McGrath DJ, Jones DA (2015) Identification of I-7 expands the repertoire of genes for resistance to Fusarium wilt in tomato to three resistance gene classes. Mol Plant Pathol. doi:10.1111/mpp.12294
Gordon TR, Martyn RD (1997) The evolutionary biology of Fusarium oxysporum. Ann Rev Phytopathol 35:111–128

Guo L, Han L, Yang L, Zeng H, Fan D, Zhu Y, Feng Y, Wang G, Peng C, Jiang X, Zhou D, Ni P, Liang C, Liu L, Wang J, Mao C, Fang X, Peng M, Huang J (2014) Genome and transcriptome analysis of the fungal pathogen Fusarium oxysporum f. sp. cubense causing banana vascular wilt disease. PLoS One 9(4):e95543

Gust AA, Willmann R, Desaki Y, Grabherr HM, Nurnberger T (2012) Plant LysM proteins: modules mediating symbiosis and immunity. Trends Plant Sci 17:495–502

Hogenhout SA, Van der Hoorn RAL, Terauchi R, Kamoun S (2009) Emerging concepts in effector biology of plant-associated organisms. Mol Plant Microbe Interact 22:115–122

Houterman PM, Speijer D, Dekker HL, De Koster CG, Cornelissen BJ, Rep M (2007) The mixed xylem sap proteome of Fusarium oxysporum-infected tomato plants. Mol Plant Pathol 8:215–221

Houterman PM, Cornelissen BJ, Rep M (2008) Suppression of plant resistance gene-based immunity by a fungal effector. PLoS Pathog 4:e1000061

Houterman PM, Ma L, van Ooijen G, de Vroomen MJ, Cornelissen BJ, Takken FL, Rep M (2009) The effector protein Avr2 of the xylem-colonizing fungus Fusarium oxysporum activates the tomato resistance protein I-2 intracellularly. Plant J 58:970–978

Hu Z, Parekh U, Maruta N, Trusov Y, Botella JR (2015) Down-regulation of Fusarium oxysporum endogenous genes by host-delivered RNA interference enhances disease resistance. Front Chem 3:1

Huang X, Liu L, Wen T, Zhu R, Zhang J, Cai Z (2015) Illumina MiSeq investigations on the changes of microbial community in the Fusarium oxysporum f.sp. cubense infected soil during and after reductive soil disinfestation. Microbiol Res 181:33–42

Imazaki I, Kurahashi M, Iida Y, Tsuge T (2007) Fow2, a Zn(II)2Cys6-type transcription regulator, controls plant infection of the vascular wilt fungus Fusarium oxysporum. Mol Microbiol 63:737–753

Jabeen N, Chaudhary Z, Gulfraz M, Rashid H, Mirza B (2015) Expression of rice chitinase gene in genetically engineered tomato confers enhanced resistance to Fusarium wilt and early blight. Plant Pathol J 31:252–258

Jain S, Akiyama K, Mae K, Ohguchi T, Takata R (2002) Targeted disruption of a G protein alpha subunit gene results in reduced pathogenicity in Fusarium oxysporum. Curr Genet 41:407–413

Jain S, Akiyama K, Kan T, Ohguchi T, Takata R (2003) The G protein beta subunit FGB1 regulates development and pathogenicity in Fusarium oxysporum. Curr Genet 43:79–86

Jain S, Akiyama K, Takata R, Ohguchi T (2005) Signaling via the G protein α subunit FGA2 is necessary for pathogenesis in Fusarium oxysporum. FEMS Microbiol Lett 243:165–172

Jimenez-Gasco MM, Navas-Cortes JA, Jimenez-Diaz RM (2004) The Fusarium oxysporum f. sp. ciceris/Cicer arietinum pathosystem: a case study of the evolution of plant-pathogenic fungi into races and pathotypes. Int Microbiol 7:95–104

Jonkers W, Rep M (2009) Mutation of CRE1 in Fusarium oxysporum reverts the pathogenicity defects of the FRP1 deletion mutant. Mol Microbiol 74:1100–1113

Joobeur T, King JJ, Nolin SJ, Thomas CE, Dean RA (2004) The Fusarium wilt resistance locus *FOM-2* of melon contains a single resistance gene with complex features. Plant J 39:283–297

Karimi Jashni M, Dols IH, Iida Y, Boeren S, Beenen HG, Mehabi R, Collemare J, De Wit PJ (2015) Synergistic action of serine- and metallo-proteases from Fusarium oxysporum f. sp. lycopersici cleaves chitin-binding tomato chitinases, reduces their antifungal activity and enhances fungal virulence. Mol Plant Microbe Interact 28(9):996–1008

Kidd BN, Edgar CI, Kumar KK, Aitken EA, Schenk PM, Manners JM, Kazan K (2009) The mediator complex subunit PFT1 is a key regulator of jasmonate-dependent defense in Arabidopsis. Plant Cell 21:2237–2252

Kidd BN, Kadoo NY, Dombrecht B, Tekeoglu M, Gardiner DM, Thatcher LF, Aitken EA, Schenk PM, Manners JM, Kazan K (2011) Auxin signaling and transport promote susceptibility to the root-infecting fungal pathogen Fusarium oxysporum in Arabidopsis. Mol Plant Microbe Interact 24:733–748

Kombrink A, Thomma BPHJ (2013) LysM effectors: secreted proteins supporting fungal life. PLoS Pathog 9:e1003769

Kubicek CP, Starr TL, Glass NL (2014) Plant cell wall-degrading enzymes and their secretion in plant-pathogenic fungi. Annu Rev Phytopathol 52:427–451

Kumar Y, Dholakia BB, Panigrahi P, Kadoo NY, Giri AP, Gupta VS (2015) Metabolic profiling of chickpea-Fusarium interaction identifies differential modulation of disease resistance pathways. Phytochemistry 116:120–129

Laurence MH, Summerell BA, Liew ECY (2015) Fusarium oxysporum f. sp. canariensis: evidence for horizontal gene transfer of putative pathogenicity genes. Plant Pathol 5:1068–1075

Li CY, Deng GM, Yang J, Viljoen A, Jin Y, Kuang RB, Zuo CW, Lv ZC, Yang QS, Sheng O, Wei YR, Hu CH, Dong T, Yi GJ (2012) Transcriptome profiling of resistant and susceptible Cavendish banana roots following inoculation with Fusarium oxysporum f. sp. cubense tropical race 4. BMC Genomics 13:374

Li C, Shao J, Wang Y, Li W, Guo D, Yan B, Xia Y, Peng M (2013a) Analysis of banana transcriptome and global gene expression profiles in banana roots in response to infection by race 1 and tropical race 4 of Fusarium oxysporum f. sp. cubense. BMC Genomics 14:851

Li X, Bai T, Li Y, Ruan X, Li H (2013b) Proteomic analysis of Fusarium oxysporum f. sp. cubense tropical race 4-inoculated response to Fusarium wilts in the banana root cells. Proteome Sci 11:41

Li E, Wang G, Yang Y, Xiao J, Mao Z, Xie B (2015) Microscopic analysis of the compatible and incompatible interactions between Fusarium oxysporum f. sp. conglutinans and cabbage. Eur J Plant Pathol 141:597–609

Lichtenzveig J, Anderson J, Thomas G, Oliver R, Singh K (2006) Inoculation and growth with soil borne pathogenic fungi. In: Mathesius U, Journet EP, Sumner LW (eds) The Medicago truncatula handbook. Samuel Roberts Noble Foundation, Ardmore, OK

Lievens B, Houterman PM, Rep M (2009) Effector gene screening allows unambiguous identification of *Fusarium oxysporum* f. sp. *lycopersici* races and discrimination from other formae speciales. FEMS Microbiol Lett 300(2):201–215

Lo Presti L, Lanver D, Schweizer G, Tanaka S, Liang L, Tollot M, Zuccaro A, Reissmann S, Kahmann R (2015) Fungal effectors and plant susceptibility. Ann Rev Plant Biol 66:513–545

Lyons R, Stiller J, Powell J, Rusu A, Manners JM, Kazan K (2015a) *Fusarium oxysporum* triggers tissue-specific transcriptional reprogramming in *Arabidopsis thaliana*. PLoS One 10(4): e0121902

Lyons R, Rusu A, Stiller J, Powell J, Manners JM, Kazan K (2015b) Investigating the association between flowering time and defense in the Arabidopsis thaliana-Fusarium oxysporum interaction. PLoS One 10:e0127699

Ma L-J (2014) Horizontal chromosome transfer and rational strategies to manage Fusarium vascular wilt diseases. Mol Plant Pathol 15:763–766

Ma L-J, van der Does HC, Borkovich KA, Coleman JJ, Daboussi M-J, Di Pietro A, Dufresne M, Freitag M, Grabherr M, Henrissat B, Houterman PM, Kang S, Shim W-B, Woloshuk C, Xie X, Xu J-R, Antoniw J, Baker SE, Bluhm BH, Breakspear A, Brown DW, Butchko RAE, Chapman S, Coulson R, Coutinho PM, Danchin EGJ, Diener A, Gale LR, Gardiner DM, Goff S, Hammond-Kosack KE, Hilburn K, Hua-Van A, Jonkers W, Kazan K, Kodira CD, Koehrsen M, Kumar L, Lee Y-H, Li L, Manners JM, Miranda-Saavedra D, Mukherjee M, Park G, Park J, Park S-Y, Proctor RH, Regev A, Ruiz-Roldan MC, Sain D, Sakthikumar S, Sykes S, Schwartz DC, Turgeon BG, Wapinski I, Yoder O, Young S, Zeng Q, Zhou S, Galagan J, Cuomo CA, Kistler HC, Rep M (2010) Comparative genomics reveals mobile pathogenicity chromosomes in Fusarium. Nature 464:367–373

Ma L, Houterman PM, Gawehns F, Cao L, Sillo F, Richter H, Clavijo-Ortiz MJ, Schmidt SM, Boeren S, Vervoort J, Cornelissen BJC, Rep M, Takken FLW (2015) The AVR2–SIX5 gene pair is required to activate I-2-mediated immunity in tomato. New Phytol 208(2):507–518. doi:10.1111/nph.13455

Mahdavi F, Sariah M, Maziah M (2012) Expression of rice thaumatin-like protein gene in transgenic banana plants enhances resistance to fusarium wilt. Appl Biochem Biotechnol 166:1008–1019

Martin GB, Bogdanove AJ, Sessa G (2003) Understanding the functions of plant disease resistance proteins. Annu Rev Plant Biol 54:23–61

Mazzeo MF, Cacace G, Ferriello F, Puopolo G, Zoina A, Ercolano MR, Siciliano RA (2014) Proteomic investigation of response to FORL infection in tomato roots. Plant Physiol Biochem 74:42–49

McFadden HG, Wilson IW, Chapple RM, Dowd C (2006) Fusarium wilt (Fusarium oxysporum f. sp. vasinfectum) genes expressed during infection of cotton (Gossypium hirsutum) dagger. Mol Plant Pathol 7:87–101

McGrath KC, Dombrecht B, Manners JM, Schenk PM, Edgar CI, Maclean DJ, Scheible WR, Udvardi MK, Kazan K (2005) Repressor- and activator-type ethylene response factors functioning in jasmonate signaling and disease resistance identified via a genome-wide screen of Arabidopsis transcription factor gene expression. Plant Physiol 139:949–959

Meldrum RA, Fraser-Smith S, Tran-Nguyen LTT, Daly AM, Aitken EAB (2012) Presence of putative pathogenicity genes in isolates of Fusarium oxysporum f. sp. cubense from Australia. Australas Plant Pathol 41:551–557

Mes JJ, van Doorn AA, Wijbrandi J, Simons G, Cornelissen BJ, Haring MA (2000) Expression of the Fusarium resistance gene I-2 colocalizes with the site of fungal containment. Plant J 23:183–193

Michielse CB, Rep M (2009) Pathogen profile update: Fusarium oxysporum. Mol Plant Pathol 10:311–324

Michielse CB, van Wijk R, Reijnen L, Manders EM, Boas S, Olivain C, Alabouvette C, Rep M (2009a) The nuclear protein Sge1 of Fusarium oxysporum is required for parasitic growth. PLoS Pathog 5(10):e1000637. doi:10.1371/journal.ppat.1000637

Michielse CB, van Wijk R, Reijnen L, Cornelissen BJ, Rep M (2009b) Insight into the molecular requirements for pathogenicity of Fusarium oxysporum f. sp. lycopersici through large-scale insertional mutagenesis. Genome Biol 10(1):R4. doi:10.1186/gb-2009-10-1-r4

O'Donnell K, Kistler HC, Cigelnik E, Ploetz RC (1998) Multiple evolutionary origins of the fungus causing Panama disease of banana: concordant evidence from nuclear and mitochondrial gene genealogies. Proc Natl Acad Sci U S A 95:2044–2049

Olivain C, Humbert C, Nahalkova J, Fatehi J, L'Haridon F, Alabouvette C (2006) Colonization of tomato root by pathogenic and nonpathogenic *Fusarium oxysporum* strains inoculated together and separately into the soil. App Environ Microbiol 72:1523–1531

Onate-Sanchez L, Anderson JP, Young J, Singh KB (2007) AtERF14, a member of the ERF family of transcription factors, plays a nonredundant role in plant defense. Plant Physiol 143:400–409

Oome S et al (2014) Nep1-like proteins from three kingdoms of life act as a microbe-associated molecular pattern in Arabidopsis. Proc Natl Acad Sci U S A 111(47):16955–16960. doi:10.1073/pnas.1410031111

Paul JY, Becker DK, Dickman MB, Harding RM, Khanna HK, Dale JL (2011) Apoptosis-related genes confer resistance to Fusarium wilt in transgenic 'Lady Finger' bananas. Plant Biotechnol J 9:1141–1148

Pemberton CL, Salmond GP (2004) The Nep1-like proteins-a growing family of microbial elicitors of plant necrosis. Mol Plant Pathol 5:353–359

Perez-Nadales E, Almeida Nogueira MF, Baldin C, Castanheira S, El Ghalid M, Grund E, Lengeler K, Marchegiani E, Mehrotra PV, Moretti M, Naik V, Oses-Ruiz M, Oskarsson T, Schäfer K, Wasserstrom L, Brakhage AA, Gow NAR, Kahmann R, Lebrun M-H, Perez-Martin J, Di Pietro A, Talbot NJ, Toquin V, Walther A, Wendland J (2014) Fungal model systems and the elucidation of pathogenicity determinants. Fungal Genet Biol 70:42–67

Ploetz R (2015) Fusarium wilt of banana. Phytopathology 105(12):1512–1521

Pusztahelyi T, Holb IJ, Pócsi I (2015) Secondary metabolites in fungus-plant interactions. Front Plant Sci 6:573. doi:10.3389/fpls.2015.00573

Ramírez-Suero M, Khanshour A, Martinez Y, Rickauer M (2010) A study on the susceptibility of the model legume plant Medicago truncatula to the soil-borne pathogen Fusarium oxysporum. Eur J Plant Pathol 126:517–530
Recorbet G, Steinberg C, Olivain C, Edel V, Trouvelot S, Dumas-Gaudot E, Gianinazzi S, Alabouvette C (2003) Wanted: pathogenesis-related marker molecules for Fusarium oxysporum. New Phytol 159:73–92
Rep M (2005) Small proteins of plant-pathogenic fungi secreted during host colonization. FEMS Microbiol Lett 253:19–27
Rep M, Kistler HC (2010) The genomic organization of plant pathogenicity in *Fusarium* species. Curr Opin Plant Biol 13:420–426
Rep M, van der Does HC, Meijer M, van Wijk R, Houterman PM, Dekker HL, de Koster CG, Cornelissen BJ (2004) A small, cysteine-rich protein secreted by Fusarium oxysporum during colonization of xylem vessels is required for I-3-mediated resistance in tomato. Mol Microbiol 53:1373–1383
Rep M, Meijer M, Houterman PM, van der Does HC, Cornelissen BJ (2005) Fusarium oxysporum evades I-3-mediated resistance without altering the matching avirulence gene. Mol Plant Microbe Interact 18:15–23
Rep M, Schmidt S, van Dam P, de Sain M, Vlaardingerbroek I, Shahi S, Widinugraheni S, Fokkens L, Tintor N, Beerens B, Houterman P, van der Does C (2015) Effectors of *Fusarium oxysporum*: identification, function, evolution and regulation of gene expression. Fungal Genetics Reports 61S. http://www.fgsc.net/FungalGeneticsReports.htm
Rispail N, Bani M, Rubiales D (2015) Resistance reaction of Medicago truncatula genotypes to Fusarium oxysporum: effect of plant age, substrate and inoculation method. Crop Pasture Sci 66:506–515
Roncero MIG, Hera C, Ruiz-Rubio M, Garcıa Maceira FI, Madrid MP, Caracuel Z, Calero F, Delgado-Jarana J, Roldán R, Guez R, Martınez-Rocha AL, Velasco C, Roa J, Martın-Urdiroz M, Córdoba D, Di Pietro A (2003) Fusarium as a model for studying virulence in soilborne plant pathogens. Physiol Mol Plant Pathol 62:87–98
Ruiz GB, Di Pietro A, Roncero MIG (2015) Combined action of the major secreted exo- and endopolygalacturonase is required for full virulence of Fusarium oxysporum. Mol Plant Pathol. doi:10.1111/mpp.12283
Sarfatti M, Abu-Abied M, Katan J, Zamir D (1991) RFLP mapping of *I-1*, a new locus in tomato conferring resistance against *Fusarium oxysporum* f. sp. *lycopersici* race 1. Theor Appl Genet 82:22–26
Schmidt SM, Houterman PM, Schreiver I, Ma L, Amyotte S, Chellappan B, Boeren S, Takken FL, Rep M (2013) MITEs in the promoters of effector genes allow prediction of novel virulence genes in *Fusarium oxysporum*. BMC Genomics 14(1):119
Schmidt SM, Lukasiewicz J, Farrer R, van Dam P, Bertoldo C, Rep M (2016) Comparative genomics of Fusarium oxysporum f. sp. melonis reveals the secreted protein recognized by the Fom-2 resistance gene in melon. New Phytol 209:307–318
Sharma M, Nagavardhini A, Thudi M, Ghosh R, Pande S, Varshney RK (2014) Development of DArT markers and assessment of diversity in Fusarium oxysporum f. sp. ciceris, wilt pathogen of chickpea (Cicer arietinum L.). BMC Genomics 15:454. doi:10.1186/1471-2164-15-454
Shen Y, Diener AC (2013) *Arabidopsis thaliana RESISTANCE TO FUSARIUM OXYSPORUM 2* implicates tyrosine-sulfated peptide signaling in susceptibility and resistance to root infection. PLoS Genet 9(5):e1003525. doi:10.1371/journal.pgen.1003525
Shimizu M, Fujimoto R, Ying H, Z-j P, Ebe Y, Kawanabe T, Saeki N, Taylor J, Kaji M, Dennis E, Okazaki K (2014) Identification of candidate genes for fusarium yellows resistance in Chinese cabbage by differential expression analysis. Plant Mol Biol 85:247–257
Simons G, Groenendijk J, Wijbrandi J, Reijans M, Groenen J, Diergaarde P, van der Lee T, Bleeker M, Onstenk J, de Both M et al (1998) Dissection of the *Fusarium I-2* gene cluster in tomato reveals six homologs and one active gene copy. Plant Cell 10:1055–1068

Sperschneider J, Gardiner DM, Thatcher LF, Lyons R, Singh KB, Manners JM, Taylor JM (2015) Genome-wide analysis in three Fusarium pathogens identifies rapidly evolving chromosomes and genes associated with pathogenicity. Genome Biol Evol 7:1613–1627

Stergiopoulos I, de Wit PJ (2009) Fungal effector proteins. Annu Rev Phytopathol 47:233–263

Sun Y, Yi X, Peng M, Zeng H, Wang D, Li B, Tong Z, Chang L, Jin X, Wang X (2014) Proteomics of Fusarium oxysporum race 1 and race 4 reveals enzymes involved in carbohydrate metabolism and ion transport that might play important roles in banana Fusarium wilt. PLoS One 9: e113818

Swarupa V, Ravishankar KV, Rekha A (2014) Plant defense response against Fusarium oxysporum and strategies to develop tolerant genotypes in banana. Planta 239:735–751

Takken F, Rep M (2010) The arms race between tomato and Fusarium oxysporum. Mol Plant Pathol 11:309–314

Thatcher LF, Manners JM, Kazan K (2009) Fusarium oxysporum hijacks COI1-mediated jasmonate signaling to promote disease development in Arabidopsis. Plant J 58:927–939

Thatcher LF, Gardiner DM, Kazan K, Manners JM (2012a) A highly conserved effector in Fusarium oxysporum is required for full virulence on Arabidopsis. Mol Plant Microbe Interact 25:180–190

Thatcher LF, Powell JJ, Aitken EA, Kazan K, Manners JM (2012b) The lateral organ boundaries domain transcription factor LBD20 functions in Fusarium wilt Susceptibility and jasmonate signaling in Arabidopsis. Plant Physiol 160:407–418

Thatcher LF, Kamphuis LG, Hane JK, Oñate-Sánchez L, Singh KB (2015) The Arabidopsis KH-domain RNA-binding protein ESR1 functions in components of jasmonate signalling, unlinking growth restraint and resistance to stress. PLoS One 10:e0126978

Thomma BPHJ, Nürnberger T, Joosten MHAJ (2011) Of PAMPs and effectors: the blurred PTI-ETI dichotomy. Plant Cell 23:4–15

Trapero-Casas A, Jiménez-Díaz RM (1985) Fungal wilt and root rot diseases of chickpea in southern Spain. Phytopathology 75:146–1151

Trusov Y, Sewelam N, Rookes JE, Kunkel M, Nowak E, Schenk PM, Botella JR (2009) Heterotrimeric G proteins-mediated resistance to necrotrophic pathogens includes mechanisms independent of salicylic acid-, jasmonic acid/ethylene- and abscisic acid-mediated defense signaling. Plant J 58:69–81

van der Does HC, Rep M (2007) Virulence genes and the evolution of host specificity in plant-pathogenic fungi. Mol Plant Microbe Interact 20:1175–1182

van der Does HC, Duyvesteijn RG, Goltstein PM, van Schie CC, Manders EM, Cornelissen BJ, Rep M (2008) Expression of effector gene SIX1 of Fusarium oxysporum requires living plant cells. Fungal Genet Biol 45:1257–1264

van Schie CCN, Takken FLW (2014) Susceptibility genes 101: how to be a good host. Annu Rev Phytopathol 52:551–581

Wang Z, Zhang J, Jia C, Liu J, Li Y, Yin X, Xu B, Jin Z (2012) De novo characterization of the banana root transcriptome and analysis of gene expression under Fusarium oxysporum f. sp. Cubense tropical race 4 infection. BMC Genomics 13:650

Wang M, Sun Y, Sun G, Liu X, Zhai L, Shen Q, Guo S (2015) Water balance altered in cucumber plants infected with Fusarium oxysporum f. sp. cucumerinum. Sci Rep 5:7722

Williams AH, Sharma M, Thatcher LF, Azam S, Hane JK, Sperschneider J, Kidd BN, Anderson JP, Ghosh R, Garg G, Lichtenzveig J, Kistler HC, Shea T, Young S, Buck SA, Kamphuis LG, Saxena R, Pande S, Ma LJ, Varshney RK, Singh KB (2016) Comparative genomics and prediction of conditionally dispensable sequences in legume-infecting Fusarium oxysporum formae speciales facilitates identification of candidate effectors. BMC Genomics 17(1):191

Win J, Chaparro-Garcia A, Belhaj K, Saunders DGO, Yoshida K, Dong S, Schornack S, Zipfel C, Robatzek S, Hogenhout SA, Kamoun S (2012) Effector biology of plant-associated organisms: concepts and perspectives. Cold Spring Harbor Symp Quant Biol 77:235–247

Wu Y, Yi G, Peng X, Huang B, Liu E, Zhang J (2013) Systemic acquired resistance in Cavendish banana induced by infection with an incompatible strain of Fusarium oxysporum f. sp. cubense. J Plant Physiol 170:1039–1046
Xue R, Wu J, Zhu Z, Wang L, Wang X, Wang S, Blair MW (2015a) Differentially expressed genes in resistant and susceptible common bean (Phaseolus vulgaris L.) genotypes in response to Fusarium oxysporum f. sp. phaseoli. PLoS One 10:e0127698
Xue C, Penton CR, Shen Z, Zhang R, Huang Q, Li R, Ruan Y, Shen Q (2015b) Manipulating the banana rhizosphere microbiome for biological control of Panama disease. Sci Rep 5:11124. doi:10.1038/srep11124
Yadeta K, Thomma B (2013) The xylem as battleground for plant hosts and vascular wilt pathogens. Front Plant Sci 4:97. doi:10.3389/fpls.2013.00097
Zhao Z, Liu H, Wang C, Xu JR (2013) Comparative analysis of fungal genomes reveals different plant cell wall degrading capacity in fungi. BMC Genomics 14:274
Zhu QH, Stephen S, Kazan K, Jin G, Fan L, Taylor J, Dennis ES, Helliwell CA, Wang MB (2013) Characterization of the defense transcriptome responsive to Fusarium oxysporum-infection in Arabidopsis using RNA-seq. Gene 512:259–266

Belowground Defence Strategies Against *Rhizoctonia*

Brendan N. Kidd, Kathleen D. DeBoer, Karam B. Singh, and Jonathan P. Anderson

Abstract *Rhizoctonia solani* is a species complex of soilborne fungi that are known for their ability to infect a broad range of plant species. Notoriously, isolates of *R. solani* cause bare-patch and sheath blight diseases on wheat and rice, respectively, and therefore jeopardise global production of these two major cereals. One of the pressing problems in combating *R. solani* is the lack of strong genetic resistance despite broad germplasm screening programmes. In order to determine future approaches for improving resistance, this chapter summarises the current research into *R. solani* pathosystems and the types of control strategies that have been employed to protect plants against this disease. Opportunities and challenges for improving resistance to this pathogen will also be discussed.

1 Introduction

The genus *Rhizoctonia* is home to a broad collection of fungi with diverse lifestyles, ranging from pathogenic, saprophytic to symbiotic organisms. The plant pathogenic isolates of *Rhizoctonia* are predominantly classified into the species complex *Rhizoctonia solani* Kühn (teleomorph, *Thanatephorus cucumeris* (Frank) Donk) and are the focus of this chapter. *R. solani* infects over 188 plant species including a range of economically important crops such as rice, wheat, potato, canola, maize as well as legumes and ornamentals (Anderson 1982). Most *R. solani* host–pathogen interactions are associated with root rot or hypocotyl rot which leads to plant collapse or severe stunting. However, in some plant interactions, *R. solani* can also infect leaves, for example, in rice where it causes rice sheath blight and in tobacco and soybean where it causes target spot or aerial blight in addition to causing root and stem rots (Gonzalez et al. 2011; Okubara et al. 2014).

Globally the largest losses due to *R. solani* infection occur in rice, with rice sheath blight, the second most devastating disease after rice blast, and under favourable conditions *R. solani* can cause up to 50 % decrease in rice yields in

---

B.N. Kidd (✉) • K.D. DeBoer • K.B. Singh • J.P. Anderson (✉)
Centre for Environment and Life Sciences, Commonwealth Scientific and Industrial Research Organization Agriculture Flagship, Perth, WA 6014, Australia
e-mail: Brendan.kidd@csiro.au; Jonathan.anderson@csiro.au

C.M.F. Vos, K. Kazan (eds.), *Belowground Defence Strategies in Plants*, Signaling and Communication in Plants, DOI 10.1007/978-3-319-42319-7_5

Asia (Lee and Rush 1983; Wang et al. 2013). In potato, yields can be reduced around 30 % by *R. solani* (Banville 1989; Carling et al. 1989). Whilst in wheat and barley, *R. solani* has been known to cause up to $100 million in losses in the state of Washington in the United States alone, and worldwide losses are considerably larger (Okubara et al. 2014). In Australia, *R. solani* average losses of wheat and barley are approximately Australian $78 million per year, with a potential loss of Australian $165 million in wheat and Australian $64 million in barley during heavy disease years (Murray and Brennan 2009, 2010). *R. solani* can also infect canola, legume and tobacco crops with losses in canola up to 36 % observed in some Canadian growing regions (Gugel et al. 1987).

The ability of *R. solani* to infect rice, wheat, potato and even maize, makes it a potential threat to the production of the world's major staple crops. Investigation of *Rhizoctonia* pathosystems can help minimise the losses to these important crop species.

2 Fungal Biology and Taxonomy of *Rhizoctonia* spp.

R. solani belongs to the class Agaricomycetes in the phylum *Basidiomycota* and is therefore phylogenetically quite distant from the more well-known ascomycete fungal pathogens. Moreover, it is also genetically distant from other notable Basidiomycete pathogens that cause rust and smut diseases. *R. solani* is predominantly found in the asexual form, and the sexual stages are rarely seen on plant hosts. Also, depending on the isolate, the production of basidiospores can be difficult to induce in vitro (Stretton et al. 1964). As such the identification of *R. solani* is primarily based on vegetative characteristics, such as brown colouration of hyphae, constriction of hyphae at septa, branching of hyphae near the distal septum of cells in young hyphae, multinucleate cells and dolipore septa (Parmeter and Whitney 1970). Within the species complex, isolates differ broadly in their genetics as well as their ability to cause disease on different hosts. Whilst some *R. solani* isolates cause disease on a very broad range of host plants (Fig. 1), others have a very narrow host range with some even forming symbiotic mycorrhizal associations with orchids (Sneh et al. 1991). To gain a better understanding of the diversity within this group of fungi, techniques such as hyphal fusion (Parmeter et al. 1969), DNA sequence comparison (Kuninaga and Yokosawa 1980; Pannecoucque and Hofte 2009; Broders et al. 2014), host range analysis and biochemical methods such as pectin zymograms (Sweetingham et al. 1986) have been used to further characterise the isolates into subgroups.

Fig. 1 The *R. solani* AG8 isolate WAC10335 is able to cause root or hypocotyl diseases on a broad range of host plants. (**a**) Healthy seedling of the model legume, *Medicago truncatula*, accession A17. (**b**) *M. truncatula* infected with *R. solani* WAC10335. (**c**) Healthy narrow leaf lupin, *Lupinus angustifolius*, c.v. tanjil. (**d**) Narrow leaf lupin infected with *R. solani* WAC10335. (**e**) Healthy wheat c.v. Chinese Spring. (**f**) Chinese Spring infected with *R. solani* WAC10335

2.1 The Classification of R. solani *into Anastomosis Groups*

One of the most successful methods for classifying *R. solani* is based on anastomosis groupings (AGs) (Sneh et al. 1991). There are currently 13 *R. solani* AGs as well as a bridging group AG-B1 that can anastomose with more than one group; however, Carling et al. (2002) have suggested that the bridging group AG-B1 may potentially be grouped into AG2. In addition AGs 1–4, 6, 8 and 9 have been further classified into subgroupings (Cubeta and Vilgalys 1997). For example, AG1 can be divided into AG1-IA, AG1-IB, AG1-IC, AG1-ID, AG1-IE and AG1-IF and AG2 into AG2-1, AG2-2-2 IIIB, AG2-2-2 IV, AG2-2-2 LP, AG2-2-3 and AG2-2-4 (Carling et al. 2002). Although there are exceptions to the rule, DNA sequencing of ribosomal ITS sequences, as well as host range analysis, has generally confirmed AG groupings demonstrating their usefulness as an inexpensive and simple classification system (Salazar et al. 2000; Broders et al. 2014).

2.2 Biochemical Classification of R. solani

An alternate method for classifying *R. solani*, which can assist in differentiating members of the same anastomosis group, is through the use of pectin zymograms. Classifying *R. solani* isolates through pectin zymograms involves running the soluble fraction of induced pectinases from liquid culture grown *R. solani* through an acrylamide gel and analysing the enzyme separation patterns (Sweetingham et al. 1986). The technique provides an additional characteristic to divide isolates within anastomosis groupings and can be useful for matching isolates within an AG to their different host preferences. However, for a thorough identification of new isolates, a combination of anastomosis reactions, ITS sequencing and host range verification is preferable.

3 The Infection Process of *R. solani*

Being a soilborne pathogen, *R. solani* has the ability to survive for long periods as sclerotia in soil, and the presence of suitable hosts or plant debris allows *R. solani* to extract nutrients in order to maintain its survival. After obtaining nutrients, *R. solani* mycelia grows outwards in a circular pattern, and it is these regions of increased fungal biomass that lead to the characteristic "bare-patch" phenotype. However, in rice-infecting isolates, a different infection strategy is employed. Lesions caused by *R. solani* are often formed at above the water level where hyphae derived from floating sclerotia form infection structures on the leaf sheath (Banniza and Holderness 2001). Successful penetration and colonisation of the host tissue lead to nutrient acquisition which allows continued hyphal growth to infect aerial leaves (Sivalingam et al. 2006).

Regardless of the tissue type that the isolate prefers, the infection process broadly follows the following steps: superficial growth to surround the plant surface, adhesion and the transition to directed growth along cell walls, formation of infection structures, penetration which leads to degradation of the host tissue and increasing proliferation which leads to the formation of sclerotia to complete the cycle (Keijer 1996).

4 Biocontrol, Chemical and Management Practices to Control *Rhizoctonia* Diseases

The ability of *R. solani* to persist in the soil, as well as the ability of some isolates to infect a broad range of plants, makes *R. solani* a difficult pathogen to control. In addition, its aggressiveness on young seedlings provides an impossibly short window for chemical control once an outbreak is detected. Despite this, positive effects have been reported for certain chemical controls applied in furrow or as soil treatments at the time of sowing (McKay and Huberli 2014). For diseases of turfgrass, iprodione and propiconazole are reported to assist in preventative and curative control (Tisserat et al. 1994).

However given the ability of *R. solani* to survive in the soil, chemical treatment, in addition to being costly, often leads to reductions in pathogen levels in the field rather than eradicating the fungus completely. In addition to traditional chemical fungicides, a novel fungicide in the form of carbon nanohorn particles has recently been reported (Dharni et al. 2016). The graphene-derived carbon nanohorn inhibited *R. solani* growth in vitro and is predicted to bind to *R. solani* endochitinase; however, the cost of such a treatment and its impact on beneficial fungi such as mycorrhizae have not yet been fully assessed.

The use of fungal or bacterial biocontrol agents such as those belonging to the genera *Bacillus* (Elkahoui et al. 2014; Luo et al. 2015), *Paenibacillus* (Xu et al. 2014), *Pseudomonas* (Jung et al. 2011), *Streptomyces* (Boukaew and Prasertsan 2014; Harikrishnan et al. 2014) and *Trichoderma* (Asad et al. 2014) has shown some success in reducing disease symptoms during *R. solani* infection assays or can reduce the growth of *R. solani* when cocultured in vitro. In addition, non-pathogenic isolates of *Rhizoctonia* have also been used as biocontrol agents with significant reductions in disease severity observed in pot trials (Sneh and Ichielevich-Auster 1998). Unfortunately, despite small-scale successes, biocontrol strains have proven difficult to deploy on a broad scale against *R. solani* and do not appear to work in all environment and soil types. To improve the utility of these treatments, Boukaew et al. (2013) assessed a combination of three bacterial biocontrol agents with four chemical fungicides to achieve better control. The authors found a reduction in rice sheath blight symptoms between 47 and 74 % with the greatest success obtained from a combined treatment with Carbendazim® and *Streptomyces philanthi* strain RM-1-138. Further experimentation in field studies is required to ensure the results are applicable to individual farming systems. Rather than screening commercial bacterial preparations for efficacy against *R. solani*, two recent studies have examined the bacterial populations present in *R. solani* suppressive fields (Yin et al. 2013; Donn et al. 2014). Yin et al. (2013) looked at identifying bacterial isolates from soils where *R. solani* AG8 had declined over an 11-year period. The authors used 16S rRNA sequencing to profile microbial communities that were enriched in bulk and rhizosphere soils obtained from *R. solani* patches as well as recovered patches, to identify candidate bacteria responsible for *R. solani* suppression. Six isolates were identified that suppressed *R. solani* in vitro, three of which were identified as *Chryseobacterium soldanellicola* isolates. Subsequent greenhouse tests showed that the *C. soldanellicola* isolates also reduced root rot in wheat seedlings (Yin et al. 2013).

For root-infecting isolates, mechanical disruption and solarisation of mycelia in bare patches are also possible through tilling. However, conservation cropping and no-till systems often prevent the use of this form of mechanical control. Therefore, in-built genetic resistance is the desired form of protection against *R. solani* in these systems. Unfortunately for most crop species, a strong source of resistance to *R. solani* is not available in commercial varieties.

5 Identifying Resistance to *R. solani* Through Germplasm Screening

Despite the absence of strong resistance in commercial populations of wheat, a slightly improved level of resistance to *R. solani* has been identified in wild relatives compared to the commercial cultivars. Smith et al. (2003) used two isolates of *R. solani* AG8 to screen a commercial and synthetic wheat gene pool as well as secondary and tertiary gene pools consisting of germplasm from wheat relatives, *Aegilops cylindrica* and *Dasypyrum villosum* as well as barley (*Hordeum vulgare*). Amongst the different genetic sources, *D. villosum* showed some improved resistance against one isolate of *R. solani* (Smith et al. 2003). In another study, the addition of chromosome 4E from wheatgrass, *Thinopyrum elongatum* into Chinese Spring wheat was found to provide enhanced resistance to *R. solani* AG8 (Okubara and Jones 2011). However as the *Thinopyrum* chromosome does not recombine with wheat chromosomes, introgression of these genes for the propagation of commercial wheat may be difficult. To improve genetic resistance in the existing commercial wheat population, a mutant wheat line, Scarlet-RZ1, with increased resistance to *R. solani*, was generated through ethyl methanesulfonate (EMS) mutagenesis (Okubara et al. 2009). The Scarlet-RZ1 mutant displayed substantial root and shoot growth after *R. solani* AG8 and *R. oryzae* inoculation in greenhouse assays. Efforts to replicate this resistance in other wheat varieties are ongoing (Okubara et al. 2014). Recently, an initial study to assess the resistance of different accessions of the model grass *Brachypodium distachyon* to *R. solani* AG8 was performed (Schneebeli et al. 2015). Variation in resistance was found within the accessions screened, and given the tools available within the *B. distachyon* community for molecular biology as well as its high level of synteny with wheat, this pathosystem could prove to be an interesting resource to study the response of wheat to *R. solani* AG8 infection. Meanwhile, continued screening of synthetic wheat lines is also continuing in the hope of finding a resistance source that can be integrated into commercial wheat varieties (Okubara et al. 2014).

Lastly, commercial rice cultivars have also been screened for *R. solani* resistance (Srinivasachary et al. 2011; Jia et al. 2012). One cultivar, Yangdao 4, has shown some resistance to *R. solani* (Pan et al. 2001), and when crossed with a susceptible cultivar, Lemont, several resistance-associated quantitative trait loci (QTL) were recently found in F2 mapping populations (Wen et al. 2015). The ability of these QTL from the Yangdao 4/Lemont cross to provide stable resistance in subsequent generations as well as in other rice cultivars will need to be explored further. Continued investigation of rice germplasm using association mapping (Jia et al. 2012) as well as further targeted dissection of the many rice QTL that have been identified may one day lead to a resistant variety for better sheath blight protection.

6 Transgenic Strategies for Enhancing Resistance to *R. solani*

Over the last 20 years, molecular research in plant pathology has focused on studying the transcriptional responses of the pathogen and host to design novel strategies to boost plant immunity. Whilst comparatively less studied relative to leaf fungal pathosystems, the knowledge gained from plant defence research on a whole has provided a platform for studying *R. solani*–host interactions. The next few paragraphs of the chapter focus on the ways in which knowledge from defence signalling may be applied to improve *R. solani* resistance in crop species and also what has been learned from studying the molecular response to *R. solani* infection.

6.1 Overexpression of PATHOGENESIS-RELATED *Genes*

One of the earliest findings from studying plant–pathogen interactions was an observed increase in the expression of *PATHOGENESIS-RELATED* (*PR*) genes (van Loon 1985). Given their involvement in the plant defence response and often direct antifungal effect in vitro, several attempts have been made to overexpress these genes in the hope of achieving increased resistance to *R. solani*. For example, transgenic tobacco and canola plants overexpressing a bean endochitinase were found to be more resistant to *R. solani* (Brogue et al. 1991). In a subsequent study, it was shown that fungal penetration in these plants was restricted, and hyphae showed evidence of degradation by the host-expressed chitinases (Benhamou et al. 1993). In *Arabidopsis thaliana*, expression of a sugar beet *GERMIN-LIKE PROTEIN* (*BtGLP-1*) led to increased resistance to *R. solani* as well as *Verticillium longisporum* (Knecht et al. 2010). The authors found increased reactive oxygen species (ROS) levels as well as higher expression of *PR1* to *PR4* and the *PLANT DEFENSIN1.2* (*PDF1.2*) gene in the transgenic *Arabidopsis* plants. This suggests that overexpression of the *BtGLP-1* gene using the constitutive cauliflower mosaic virus (CaMV) 35S promoter leads to increased activation of broader defence pathways and may contribute to the increased resistance observed to the two fungal root pathogens.

Given the importance of *R. solani* to rice production, several attempts have been made to overexpress *PR* genes in rice, with the use of rice chitinase genes either singularly or together with an additional *PR* protein being a popular approach (Lin et al. 1995; Datta et al. 2002; Kalpana et al. 2006; Maruthasalam et al. 2007; Sridevi et al. 2008; Shah et al. 2009; Mao et al. 2014). For example, Kalpana et al. (2006) used a rice *THAUMATIN-LIKE* protein (*OsTLP*) together with the rice *CHITINASE11* (*OsCHI11*) gene and found increased resistance in T2-transformed lines. Sridevi et al. (2008) co-transformed *OsCHI11* and a tobacco *B-1,3-GLUCANASE* gene into rice and observed decreased disease symptoms in T3 transgenic, whilst Maruthasalam et al. (2007) transformed basmati rice with

OsCHI11, *OsTLP* and a serine–threonine kinase from wild rice *Oryza longistaminata* (*XA21*) involved in bacterial resistance against *Xanthomonas oryzae* pv. *oryzae* (*Xoo*) (Song et al. 1995). The authors found resistance to both sheath blight and bacterial blight in transgenic plants; however, yield penalties or the effect of other agriculturally important traits due to the transgenes were not investigated. More recently, co-expression of a *RICE BASIC CHITINASE10* gene (*OsRCH10*) and an *ALFALFA B-1,3-GLUCANASE1* gene (*AGLU1*) was found to provide increased resistance to *R. solani* and *Magnaporthe grisea* in field disease studies (Mao et al. 2014). Whilst the mature transgenic plants appeared morphologically normal, the transgenic lines had lower germination and seed vigour compared to untransformed lines, suggesting that the transgenic lines are not without side effects.

Recently, an additional study using a rice *POLYGALACTURONASE-INHIBITING PROTEIN* (*OsPGIP1*) overexpressed in Zhonghua 11 rice, a japonica variety, has had success in field trials (Wang et al. 2015). PGI proteins act by inhibiting the polygalacturonase enzymes expressed by pathogens. Wang et al. (2015) found that rice *OsPGIP1* possessed polygalacturonase inhibition activity in vitro and showed that two independent transgenic lines expressing the *OsPGIP1* gene had reduced disease symptoms in field trials. Whilst the symptom suppression was not dramatic, in areas where *R. solani* causes yield decline, transgenic lines such as those mentioned above could be an option, if given regulatory and public acceptance.

Whilst seemingly a good choice for improving pathogen defence, constitutive expression of *PR* genes often comes at a cost to yield as *PR* proteins can be damaging to the cell homeostasis or activate additional plant defence responses. Previously identified *Arabidopsis* mutant lines with increased *PR* gene expression show either spontaneous lesions or dwarf phenotypes, e.g. constitutive *PR* (Bowling et al. 1994, 1997) and accelerated cell death mutants (Greenberg and Ausubel 1993; Greenberg et al. 1994). Therefore, linking defence proteins such as polygalacturonase proteins or chitinases to a temporal or spatially specific promoter might be a more successful approach for improving resistance in crops. Continued examination of the transcriptome of *R. solani*-infected plants may provide better candidates for such a task. One recent study went someway to addressing this problem by using a green tissue-specific promoter to express the rice *OXALATE OXIDASE 4* (*OsOXO4*) gene (Molla et al. 2013). The authors used GUS staining to delineate the regions that the *OsOXO4* gene would be expressed. Oxalic acid is a nonhost-selective toxin that is secreted by some isolates of *R. solani* as well as other necrotrophic pathogens to manipulate the host redox environment to suppress plant defences and promote cell death (Dutton et al. 1993; Cessna et al. 2000; Nagarajkumar et al. 2005; Williams et al. 2011). Oxalate oxidases such as germin-like proteins can degrade oxalic acid and subsequently initiate an efficient immune response (Lane 1994; Dunwell et al. 2000; Livingstone et al. 2005). Molla et al. (2013) found that expression of rice *OsOXO4* in leaves using the *D540* promoter led to reduced disease symptoms in detached leaf experiments as well as whole plant experiments. The transgenic plants also had no significant difference in agronomic traits such as panicle number or 100 seed weight compared to a non-transgenic control.

6.2 Manipulation of the Ethylene Pathway for Improved Resistance

An alternative approach to using one or two *PR* genes for overexpression is to use plant hormone modulation or an upstream transcription factor (TF) to activate plant defence. This would have the advantage that resistance comes from a pathway with multiple endpoint genes and may potentially be more difficult for the pathogen to break resistance. For instance, expressing the rice *OsACS2* gene under a pathogen-inducible promoter leads to enhanced resistance to both *R. solani* and *M. grisea* (Helliwell et al. 2013). *OsACS2* encodes one of six 1-aminocyclopropane-1-carboxylic acid (ACC) synthase enzymes in rice that converts *S*-adenosyl-L-methionine to ACC as part of the first steps in ethylene (ET) synthesis (Chae and Kieber 2005). Despite being controlled by a pathogen-inducible promoter, transgenic *OsACS2* lines had increased basal expression of *OsPR1b* and *OsPR5* genes and showed a 35–45 % reduction in lesion size using a mycelial ball inoculation method (Park et al. 2008; Helliwell et al. 2013). Interestingly, despite having increased basal levels of PR gene expression, the transgenic lines showed no difference in yield characteristics such as the number of panicles per plant, the number of seeds per panicle and the weight of 100 seeds under glasshouse conditions.

A role for ethylene in *R. solani* resistance had previously been shown in the *Medicago truncatula* (Penmetsa et al. 2008; Anderson et al. 2010) and soybean pathosystems (Hoffman et al. 1999). The *Medicago* ethylene-insensitive mutant *sickle*, which is an EMS mutant in *MtSkl*, the homolog of the *Arabidopsis* ethylene-signalling gene *EIN2* (Guzman and Ecker 1990; Ju et al. 2012) shows increased susceptibility to *R. solani* AG8 with a tenfold decrease in survival recorded when infected with *R. solani* AG8 (Penmetsa et al. 2008; Anderson et al. 2013). The *sickle* mutant also shows susceptibility to other legume-infecting isolates of *R. solani*, as well as the root rot pathogen *Phytophthora medicaginis*, suggesting that the ET pathway plays an important role in *Medicago* defence against necrotrophic root-infecting fungi. In addition, overexpression of *MtERF1-1*, an ethylene response transcription factor (ERF) in *M. truncatula* composite roots, led to increased resistance to both *R. solani* and *P. medicaginis* (Anderson et al. 2010). *MtERF1-1* belongs to the B3 clade of ERFs which in *Arabidopsis* are associated with plant defence (Onate-Sanchez and Singh 2002; McGrath et al. 2005; Nakano et al. 2006), and other *M. truncatula* B3 *ERFs* were found to be inducible by *R. solani* (Anderson et al. 2010). These results suggest that overexpression of ERF TFs is sufficient to boost the defence response of *Medicago* against root pathogens. However, loss of function mutations in *AtERF14*, a master regulator of ERFs and homolog of *MtERF1-1* in *Arabidopsis*, did not increase susceptibility to the same *R. solani* isolate even though susceptibility to another root-infecting fungus, *Fusarium oxysporum*, was increased (Onate-Sanchez et al. 2007; Anderson and Singh 2011). These findings suggest that different plant species may employ different defence strategies against the same pathogen.

Importantly, whilst the overexpression of some *ERF* TFs in *Arabidopsis* has shown growth penalties due to the overexpression of defence genes (Solano et al. 1998; Onate-Sanchez et al. 2007), the composite plants with *MtERF1-1* expressed in the roots did not show any phenotypic differences in growth and development. Also of interest was that nodulation of the *MtERF1-1* roots occurred at frequencies similar to a GFP-expressing root, and overexpression of *MtERF1-1* in the *sickle* mutant could restore the hypernodulation phenotype to clearly defined nodules (Anderson et al. 2010).

7 The Role of *Arabidopsis* Defence Pathways in *R. solani* Infection

The model plant *Arabidopsis* has provided substantial advances in the field of plant pathology and has helped to identify the genes involved in defence responses against a wide range of plant pathogens (Thatcher et al. 2005; Piquerez et al. 2014). Recently, the genetic resources of *Arabidopsis* were utilised to try to identify key components for resistance and susceptibility to *R. solani*. To identify novel sources of resistance, 36 *Arabidopsis* ecotypes and 14 mutants associated with plant defence and hormone signalling were assessed for resistance or susceptibility to two isolates of *R. solani*: the wheat-infecting AG8 isolate which is non-pathogenic on *Arabidopsis* and an AG2-1 isolate which infects *Arabidopsis* and crucifers (Perl-Treves et al. 2004); however, none of the mutants or ecotypes tested showed a pathogen phenotype that differed from the wild type Columbia-0 phenotype (Foley et al. 2013). The results suggested that resistance and susceptibility against *R. solani* in *Arabidopsis* are not affected by the major defence pathways such as the jasmonate (JA), salicylate (SA) and ET pathways or by defence-associated phytoalexins (camalexin) or the auxin and abscisic acid pathways. As mutation in the ethylene regulatory gene *ein2* did not affect resistance or susceptibility to either of the two *R. solani* pathogens in *Arabidopsis* but it did in *Medicago*, this suggests that *R. solani* adopts different infection strategies on different hosts (Anderson and Singh 2011).

To delve further into what might be occurring during *R. solani* infection in *Arabidopsis*, Foley et al. (2013) examined gene expression profiles of *Arabidopsis* seedlings infected with AG8 or AG2-1 using Affymetrix microarray experiments. Cell wall-associated proteins were one of the largest responses to *R. solani* infection, but significant changes were also observed in genes involved in stress responses, such as heat shock proteins and oxidative stress such as the *ALTERNATIVE OXIDASE 1D* (*AOX1D*) gene and the *RESPIRATORY BURST OXIDASE*

HOMOLOG D (*RBOHD*) gene. Screening loss of function mutants for four heat shock proteins as well as an *rbohd* mutant failed to identify a change in disease phenotypes against AG8 and AG2-1. However, inoculation of an *rbohd rbohf* double mutant was found to have increased susceptibility to AG8 (Foley et al. 2013). *RBOHD* and *RBOHF* are thought to be the main respiratory burst oxidases involved in pathogen-responsive reactive oxygen species (ROS) production (Torres et al. 2002). The breakdown in resistance of the *rbohd rbohf* double mutant to the AG8 isolate of *R. solani* suggests a role for ROS production in the maintenance of nonhost resistance in *Arabidopsis* against the wheat-infecting isolate. Additional support for this hypothesis came from the *Arabidopsis dsr1* mutant (Gleason et al. 2011). The *dsr1* mutant possesses a point mutation in the mitochondrial *SUCCINATE DEHYDROGENASE 1* gene and displays diminished mitochondrial ROS production. The *dsr1* line was identified from a genetic screen, involving an *Arabidopsis* line expressing luciferase from a stress-responsive *GLUTATHIONE S-TRANSFERASE 8* (*GSTF8*) promoter (Perl-Treves et al. 2004; Gleason et al. 2011). The *GSTF8* promoter is known to be inducible by auxin, SA and ROS treatments (Chen et al. 1996; Chen and Singh 1999) but also by *R. solani* AG8 (Perl-Treves et al. 2004). Interestingly, compatible isolates of *R. solani* did not induce *GSTF8:LUC*, expression suggesting that this gene may act as a marker of an effective defence response against *R. solani* infection, and compatible isolates of *R. solani* may have a way of preventing this host response. Together these studies suggest an important role for ROS as a signalling component in resistance to *R. solani* AG8.

7.1 Assessment of Resistance Pathways Induced in **Arabidopsis thaliana** *by Hypovirulent* **Rhizoctonia** *spp. Isolates*

As compatible isolates of *R. solani* may potentially avoid or suppress defence responses in their respective hosts to cause disease, a key challenge is to be able to activate these defence pathways to provide better protection against *R. solani*. As mentioned previously, the use of biocontrol organisms can provide an enhanced level of protection against *R. solani*, but the mechanism behind this enhanced resistance responses is relatively unknown. To investigate the underlying mechanism of biocontrol-enhanced resistance further, a study was performed to analyse the effectiveness of enhanced protection by non-pathogenic binucleate isolates of *Rhizoctonia* in known *Arabidopsis* defence mutants (Sharon et al. 2011). The authors showed that defence genes belonging to both SA- and JA-associated defence pathways were induced by the protective isolates. In addition, using an agar plate assay, reduced protection from the binucleate *Rhizoctonia* strains was observed in almost all of the *Arabidopsis* defence mutants that were screened compared to the protection provided to the WT *Arabidopsis* plants from *R. solani* infection. These results suggest that non-pathogenic *Rhizoctonia* isolates can activate *Arabidopsis* defence pathways, and this may be one of the factors contributing to the enhanced protection phenotype provided by these isolates.

7.2 Gene Expression Responses to R. solani *in Other Plant Species*

RNA transcript profiling has not been limited to *Arabidopsis*, and efforts have been made to identify cDNAs that are induced in response to *R. solani* infection in bean, rice and potato. Guerrero-Gonzalez et al. (2011) identified 136 cDNA transcripts using a suppressive subtraction library from a moderately resistant variety of common bean infected with *R. solani*. Interestingly, the authors identified pathogenesis-associated proteins such as PR1, a PGIP protein and an ethylene response factor, confirming the role of these genes in *R. solani* defence. Induction of genes associated with the phenylpropanoid pathway was also identified such as phenylalanine ammonia lyase (*PAL*), 4-coumarate-COA-ligase and chalcone synthase (Guerrero-Gonzalez et al. 2011). Additional studies in bean (Guillon et al. 2002), soybean (Chen et al. 2009) and rice (Deborah et al. 2001; Venu et al. 2007) also show upregulation of genes involved in the phenylpropanoid pathway in response to *R. solani* infection. The PAL enzyme catalyses the first step in the phenylpropanoid pathway, a pathway that produces a number of secondary metabolites with roles in plant defence as well as being a biosynthetic pathway for the production of SA (Mauch-Mani and Slusarenko 1996).

The expression of *PAL* was also induced systemically in the apical tip of potato sprouts inoculated with *R. solani* AG3 at 48 h; however, the expression declined by the 120 h time point (Lehtonen et al. 2008). Using a potato cDNA microarray, Lehtonen et al. (2008) identified 122 and 779 genes differentially expressed in systemic tissue of infected potato sprouts, with a number of pathogenesis-related proteins induced at both time points. The systemic defence response provided some protection against *R. solani*, as challenging the non-inoculated portions of the potato sprout at 120 h after the initial infection at the base of the sprout resulted in reduced infection structures on the apical sprout surface. Therefore, upregulation of defence pathways by *R. solani* can provide protection to adjacent surfaces; the question remains how to enhance this resistance at the initial infection site, to prevent root and stem rots from impacting yield and ultimately plant survival.

8 Conclusions and Future Research Priorities

It has been 200 years since the conception of the *Rhizoctonia* genus by De Candolle (1815); however, there is still much to uncover regarding the interaction of *Rhizoctonia* with host plants and its role within the rhizosphere. Despite significant research efforts to find durable genetic resistance to *R. solani*, an effective source of resistance has so far been elusive in the major crop species that *R. solani* infects. Given the extent of germplasm screening that has already occurred, finding enhanced natural genetic resistance to *R. solani* is becoming increasingly unlikely. New sources of resistance may need to be sourced from distant relatives that possess nonhost-type resistance to the major crop-infecting isolates. However, given the genetic distance between wild relatives and elite crop varieties, identifying and then transferring these resistance loci are a significant challenge. Research into understanding nonhost resistance mechanisms in model organisms may help to narrow down the genes or QTL responsible to be able to transfer the resistant phenotype through genetic engineering approaches.

Whilst a limited amount of transcriptional profiling has been performed in moderately resistant and susceptible crop plants, the studies performed have primarily used cDNA-based subtractive libraries or custom microarrays and therefore do not capture the full dynamic range of the transcriptional response to infection. The advancements in gene expression profiling such as second- and third-generation sequencing technologies have not yet been fully exploited to studying *R. solani* interactions and therefore present an opportunity to uncover new strategies for improving resistance. In addition, whilst outside of the scope of this review, recent genome sequencing as well as subsequent comparative genomics of different *R. solani* AG groups will also provide valuable insight into the virulence strategies that *R. solani* employs to cause disease (Wibberg et al. 2013; Zheng et al. 2013; Cubeta et al. 2014; Hane et al. 2014; Wibberg et al. 2015). Again, the reduced cost and increased depth of sequencing technologies will enable an unprecedented window into the molecular processes that occur during *R. solani* infection.

To improve current levels of resistance, management practices such as crop rotations and chemical applications have been utilised, and whilst they continue to be useful strategies to manage the disease, these practices often fail to truly control or eradicate the pathogen. One research area that has gained significant attention in recent years is the investigation of both the composition and relationships between soil microbiota in the rhizosphere. A strategy to exploit the soil microbial community to suppress soilborne diseases such as *R. solani* levels is a potential outcome for research in this area. Regardless of the approach taken, a sustainable and durable solution to combat *R. solani* would be a valuable discovery for improving crop yields necessary to sustain a growing population.

Acknowledgements We wish to acknowledge the Grains Research & Development Corporation for financial support and apologise to the authors of the work we could not cite in this chapter.

References

Anderson NA (1982) The genetics and pathology of *Rhizoctonia solani*. Annu Rev Phytopathol 20:329–347

Anderson JP, Singh KB (2011) Interactions of *Arabidopsis* and *M. truncatula* with the same pathogens differ in dependence on ethylene and ethylene response factors. Plant Signal Behav 6:551–552

Anderson JP, Lichtenzveig J, Gleason C, Oliver RP, Singh KB (2010) The B-3 ethylene response factor MtERF1-1 mediates resistance to a subset of root pathogens in *Medicago truncatula* without adversely affecting symbiosis with rhizobia. Plant Physiol 154:861–873

Anderson JP, Lichtenzveig J, Oliver RP, Singh KB (2013) *Medicago truncatula* as a model host for studying legume infecting *Rhizoctonia solani* and identification of a locus affecting resistance to root canker. Plant Pathol 62:908–921

Asad SA, Ali N, Hameed A, Khan SA, Ahmad R, Bilal M, Shahzad M, Tabassum A (2014) Biocontrol efficacy of different isolates of *Trichoderma* against soil borne pathogen *Rhizoctonia solani*. Pol J Microbiol 63:95–103

Banniza S, Holderness M (2001) Rice sheath blight—pathogen biology and diversity. In: Sreenivasaprasad S, Johnson R (eds) Major fungal diseases of rice: recent advances. Kluwer, Dordrecht, pp 201–211

Banville GJ (1989) Yield losses and damage to potato plants caused by *Rhizoctonia solani* Kuhn. Am Potato J 66:821–834

Benhamou N, Broglie K, Broglie R, Chet I (1993) Antifungal effect of bean endochitinase on *Rhizoctonia solani*: ultrastructural changes and cytochemical aspects of chitin breakdown. Can J Microbiol 39:318–328

Boukaew S, Prasertsan P (2014) Factors affecting antifungal activity of *Streptomyces philanthi* RM-1-138 against *Rhizoctonia solani*. World J Microbiol Biotechnol 30:323–329

Boukaew S, Klinmanee C, Prasertsan P (2013) Potential for the integration of biological and chemical control of sheath blight disease caused by *Rhizoctonia solani* on rice. World J Microbiol Biotechnol 29:1885–1893

Bowling SA, Guo A, Cao H, Gordon AS, Klessig DF, Dong X (1994) A mutation in *Arabidopsis* that leads to constitutive expression of systemic acquired resistance. Plant Cell 6:1845–1857

Bowling SA, Clarke JD, Liu Y, Klessig DF, Dong X (1997) The *cpr5* mutant of *Arabidopsis* expresses both NPR1-dependent and NPR1-independent resistance. Plant Cell 9:1573–1584

Broders KD, Parker ML, Melzer MS, Boland GJ (2014) Phylogenetic diversity of *Rhizoctonia solani* associated with Canola and Wheat in Alberta, Manitoba, and Saskatchewan. Plant Dis 98:1695–1701

Brogue K, Chet I, Holliday M, Cressman R, Biddle P, Knowlton S, Mauvais CJ, Broglie R (1991) Transgenic plants with enhanced resistance to the fungal pathogen *Rhizoctonia solani*. Science 254:1194–1197

Carling DE, Leiner RH, Westphale PC (1989) Symptoms, signs and yield reduction associated with *Rhizoctonia* disease of potato induced by tuberborne inoculum of *Rhizoctonia solani* AG-3. Am Potato J 66:693–701

Carling DE, Kuninaga S, Brainard KA (2002) Hyphal anastomosis reactions, rDNA-internal transcribed spacer sequences, and virulence levels among subsets of Rhizoctonia solani anastomosis group-2 (AG-2) and AG-BI. Phytopathology 92:43–50

Cessna SG, Sears VE, Dickman MB, Low PS (2000) Oxalic acid, a pathogenicity factor for *Sclerotinia sclerotiorum*, suppresses the oxidative burst of the host plant. Plant Cell 12: 2191–2200

Chae HS, Kieber JJ (2005) Eto Brute? Role of ACS turnover in regulating ethylene biosynthesis. Trends Plant Sci 10:291–296

Chen W, Singh KB (1999) The auxin, hydrogen peroxide and salicylic acid induced expression of the *Arabidopsis* GST6 promoter is mediated in part by an ocs element. Plant J 19:667–677

Chen W, Chao G, Singh KB (1996) The promoter of a H2O2-inducible, *Arabidopsis* glutathione S-transferase gene contains closely linked OBF- and OBP1-binding sites. Plant J 10:955–966
Chen H, Seguin P, Jabaji SH (2009) Differential expression of genes encoding the phenylpropanoid pathway upon infection of soybean seedlings by *Rhizoctonia solani*. Can J Plant Pathol 31:356–367
Cubeta MA, Vilgalys R (1997) Population biology of the *Rhizoctonia solani* complex. Phytopathology 87:480–484
Cubeta MA, Thomas E, Dean RA, Jabaji S, Neate SM, Tavantzis S, Toda T, Vilgalys R, Bharathan N, Fedorova-Abrams N, Pakala SB, Pakala SM, Zafar N, Joardar V, Losada L, Nierman WC (2014) Draft genome sequence of the plant-pathogenic soil fungus *Rhizoctonia solani* anastomosis group 3 strain Rhs1AP. Genome Announc 2(5):e01072-14
Datta K, Baisakh N, Thet KM, Tu J, Datta SK (2002) Pyramiding transgenes for multiple resistance in rice against bacterial blight, yellow stem borer and sheath blight. Theor Appl Genet 106:1–8
Deborah SD, Palaniswami A, Vidhyasekaran P, Velazhahan R (2001) Time-course study of the induction of defense enzymes, phenolics and lignin in rice in response to infection by pathogen and non-pathogen. Z Pflanzenk Pflanzen 108:204–216
De Candolle AP (1815) Memoire sur les rhizoctones, nouveau genre de champignons qui attaque les racines, des plantes et en particulier celle de la luzerne cultivee. Mem Mus d'Hist Nat 2:209–216
Dharni S, Sanchita, Unni SM, Kurungot S, Samad A, Sharma A, Patra DD (2016) In vitro and in silico antifungal efficacy of nitrogen-doped carbon nanohorn (NCNH) against *Rhizoctonia solani*. J Biomol Struct Dyn 34(1):152–162
Donn S, Almario J, Mullerc D, Moenne-Loccoz Y, Gupta VVSR, Kirkegaard JA, Richardson AE (2014) Rhizosphere microbial communities associated with *Rhizoctonia* damage at the field and disease patch scale. Appl Soil Ecol 78:37–47
Dunwell JM, Khuri S, Gane PJ (2000) Microbial relatives of the seed storage proteins of higher plants: conservation of structure and diversification of function during evolution of the cupin superfamily. Microbiol Mol Biol Rev 64:153–179
Dutton MV, Evans CS, Atkey PT, Wood DA (1993) Oxalate production by basidiomycetes, including the white-rot species *Coriolus versicolor* and *Phanerochaete chrysosporium*. Appl Microbiol Biot 39:5–10
Elkahoui S, Djebali N, Karkouch I, Ibrahim AH, Kalai L, Bachkovel S, Tabbene O, Limam F (2014) Mass spectrometry identification of antifungal lipopeptides from *Bacillus* sp. BCLRB2 against *Rhizoctonia solani* and *Sclerotinia sclerotiorum*. Appl Biochem Micro 50:184–188
Foley RC, Gleason CA, Anderson JP, Hamann T, Singh KB (2013) Genetic and genomic analysis of *Rhizoctonia solani* interactions with *Arabidopsis*; evidence of resistance mediated through NADPH oxidases. PLoS One 8:e56814
Gleason C, Huang S, Thatcher LF, Foley RC, Anderson CR, Carroll AJ, Millar AH, Singh KB (2011) Mitochondrial complex II has a key role in mitochondrial-derived reactive oxygen species influence on plant stress gene regulation and defense. Proc Natl Acad Sci U S A 108: 10768–10773
Gonzalez M, Pujol M, Metraux JP, Gonzalez-Garcia V, Bolton MD, Borras-Hidalgo O (2011) Tobacco leaf spot and root rot caused by *Rhizoctonia solani* Kuhn. Mol Plant Pathol 12: 209–216
Greenberg JT, Ausubel FM (1993) Arabidopsis mutants compromised for the control of cellular damage during pathogenesis and aging. Plant J 4:327–341
Greenberg JT, Guo A, Klessig DF, Ausubel FM (1994) Programmed cell death in plants: a pathogen-triggered response activated coordinately with multiple defense functions. Cell 77: 551–563
Guerrero-Gonzalez ML, Rodriguez-Kessler M, Rodriguez-Guerra R, Gonzalez-Chavira M, Simpson J, Sanchez F, Jimenez-Bremont JF (2011) Differential expression of *Phaseolus*

vulgaris genes induced during the interaction with *Rhizoctonia solani*. Plant Cell Rep 30: 1465–1473
Gugel RK, Yitbarek SM, Verma PR, Morrall RAA, Sadasivaiah RS (1987) Etiology of the *Rhizoctonia* root-rot complex of canola in the Peace River region of Alberta. Can J Plant Pathol 9:119–128
Guillon C, St-Arnaud M, Hamel C, Jabaji-Hare SH (2002) Differential and systemic alteration of defence-related gene transcript levels in mycorrhizal bean plants infected with *Rhizoctonia solani*. Can J Bot 80:305–315
Guzman P, Ecker JR (1990) Exploiting the triple response of *Arabidopsis* to identify ethylene-related mutants. Plant Cell 2:513–523
Hane JK, Anderson JP, Williams AH, Sperschneider J, Singh KB (2014) Genome sequencing and comparative genomics of the broad host-range pathogen *Rhizoctonia solani* AG8. PLoS Genet 10:e1004281
Harikrishnan H, Shanmugaiah V, Balasubramanian N, Sharma MP, Kotchoni SO (2014) Antagonistic potential of native strain *Streptomyces aurantiogriseus* VSMGT1014 against sheath blight of rice disease. World J Microbiol Biotechnol 30:3149–3161
Helliwell EE, Wang Q, Yang Y (2013) Transgenic rice with inducible ethylene production exhibits broad-spectrum disease resistance to the fungal pathogens *Magnaporthe oryzae* and *Rhizoctonia solani*. Plant Biotechnol J 11:33–42
Hoffman T, Schmidt JS, Zheng X, Bent AF (1999) Isolation of ethylene-insensitive soybean mutants that are altered in pathogen susceptibility and gene-for-gene disease resistance. Plant Physiol 119:935–950
Jia L, Yan W, Zhu C, Agrama HA, Jackson A, Yeater K, Li X, Huang B, Hu B, McClung A, Wu D (2012) Allelic analysis of sheath blight resistance with association mapping in rice. PLoS One 7:e32703
Ju C, Yoon GM, Shemansky JM, Lin DY, Ying ZI, Chang J, Garrett WM, Kessenbrock M, Groth G, Tucker ML, Cooper B, Kieber JJ, Chang C (2012) CTR1 phosphorylates the central regulator EIN2 to control ethylene hormone signaling from the ER membrane to the nucleus in *Arabidopsis*. Proc Natl Acad Sci U S A 109:19486–19491
Jung WJ, Park RD, Mabood F, Souleimanov A, Smith DL (2011) Effects of *Pseudomonas aureofaciens* 63-28 on defense responses in soybean plants infected by *Rhizoctonia solani*. J Microbiol Biotechnol 21:379–386
Kalpana K, Maruthasalam S, Rajesh T, Poovannan K, Kumar KK, Kokiladevi E, Raja JAJ, Sudhakar D, Velazhahan R, Samiyappan R, Balasubramanian P (2006) Engineering sheath blight resistance in elite indica rice cultivars using genes encoding defense proteins. Plant Sci 170:203–215
Keijer J (1996) The initial steps of the infection process in *Rhizoctonia solani*. In: Sneh B, Jabaji-Hare S, Neate S, Dijst G (eds) *Rhizoctonia* species: taxonomy, molecular biology, ecology, pathology and disease control. Kluwer Academic, Dordrecht, pp 149–162
Knecht K, Seyffarth M, Desel C, Thurau T, Sherameti I, Lou B, Oelmuller R, Cai D (2010) Expression of *BvGLP-1* encoding a germin-like protein from sugar beet in *Arabidopsis thaliana* leads to resistance against phytopathogenic fungi. Mol Plant Microbe Interact 23: 446–457
Kuninaga S, Yokosawa R (1980) A comparison of DNA base compositions among anastomosis groups in *Rhizoctonia solani*. Ann Phytopathol Soc Jpn 46:150–158
Lane BG (1994) Oxalate, germin, and the extracellular matrix of higher plants. FASEB J 8: 294–301
Lee FN, Rush MC (1983) Rice sheath blight—a major rice disease. Plant Dis 67:829–832
Lehtonen MJ, Somervuo P, Valkonen JP (2008) Infection with *Rhizoctonia solani* induces defense genes and systemic resistance in potato sprouts grown without light. Phytopathology 98: 1190–1198
Lin W, Anuratha CS, Datta K, Potrykus I, Muthukrishnan S, Datta SK (1995) Genetic engineering of rice for resistance to sheath blight. Bio-Technology 13:686–691

Livingstone DM, Hampton JL, Phipps PM, Grabau EA (2005) Enhancing resistance to *Sclerotinia* minor in peanut by expressing a barley oxalate oxidase gene. Plant Physiol 137:1354–1362

Luo C, Zhou H, Zou J, Wang X, Zhang R, Xiang Y, Chen Z (2015) Bacillomycin L and surfactin contribute synergistically to the phenotypic features of *Bacillus subtilis* 916 and the biocontrol of rice sheath blight induced by *Rhizoctonia solani*. Appl Microbiol Biotechnol 99:1897–1910

Mao B, Liu X, Hu D, Li D (2014) Co-expression of *RCH10* and *AGLU1* confers rice resistance to fungal sheath blight *Rhizoctonia solani* and blast *Magnorpathe oryzae* and reveals impact on seed germination. World J Microbiol Biotechnol 30:1229–1238

Maruthasalam S, Kalpana K, Kumar KK, Loganathan M, Poovannan K, Raja JA, Kokiladevi E, Samiyappan R, Sudhakar D, Balasubramanian P (2007) Pyramiding transgenic resistance in elite indica rice cultivars against the sheath blight and bacterial blight. Plant Cell Rep 26: 791–804

Mauch-Mani B, Slusarenko AJ (1996) Production of salicylic acid precursors is a major function of phenylalanine ammonia lyase in the resistance of *Arabidopsis* to *Peronospora parasitica*. Plant Cell 8:203–212

McGrath KC, Dombrecht B, Manners JM, Schenk PM, Edgar CI, Maclean DJ, Scheible WR, Udvardi MK, Kazan K (2005) Repressor- and activator-type ethylene response factors functioning in jasmonate signaling and disease resistance identified via a genome-wide screen of *Arabidopsis* transcription factor gene expression. Plant Physiol 139:949–959

McKay A, Huberli D (2014) New fungicide options on way for *Rhizoctonia*. In: Ground Cover. Grains Research and Development Corporation. https://www.grdc.com.au/Media-Centre/Ground-Cover-Supplements/GCS111/New-fungicide-options-on-way-for-rhizoctonia. Cited 2 May 2015

Molla KA, Karmakar S, Chanda PK, Ghosh S, Sarkar SN, Datta SK, Datta K (2013) Rice oxalate oxidase gene driven by green tissue-specific promoter increases tolerance to sheath blight pathogen (*Rhizoctonia solani*) in transgenic rice. Mol Plant Pathol 14:910–922

Murray GM, Brennan JP (2009) Estimating disease losses to the Australian wheat industry. Australas Plant Pathol 38:558–570

Murray GM, Brennan JP (2010) Estimating disease losses to the Australian barley industry. Australas Plant Pathol 39:85–96

Nagarajkumar M, Jayaraj J, Muthukrishnan S, Bhaskaran R, Velazhahan R (2005) Detoxification of oxalic acid by pseudomonas fluorescens strain pfMDU2: implications for the biological control of rice sheath blight caused by *Rhizoctonia solani*. Microbiol Res 160:291–298

Nakano T, Suzuki K, Fujimura T, Shinshi H (2006) Genome-wide analysis of the *ERF* gene family in *Arabidopsis* and rice. Plant Physiol 140:411–432

Okubara PA, Jones SS (2011) Seedling resistance to *Rhizoctonia* and *Pythium* spp. in wheat chromosome group 4 addition lines from *Thinopyrum* spp. Can J Plant Pathol 33:416–423

Okubara PA, Steber CM, Demacon VL, Walter NL, Paulitz TC, Kidwell KK (2009) Scarlet-Rz1, an EMS-generated hexaploid wheat with tolerance to the soilborne necrotrophic pathogens *Rhizoctonia solani* AG-8 and *R. oryzae*. Theor Appl Genet 119:293–303

Okubara PA, Dickman MB, Blechl AE (2014) Molecular and genetic aspects of controlling the soilborne necrotrophic pathogens *Rhizoctonia* and *Pythium*. Plant Sci 228:61–70

Onate-Sanchez L, Singh KB (2002) Identification of *Arabidopsis* ethylene-responsive element binding factors with distinct induction kinetics after pathogen infection. Plant Physiol 128:1313–1322

Onate-Sanchez L, Anderson JP, Young J, Singh KB (2007) AtERF14, a member of the ERF family of transcription factors, plays a nonredundant role in plant defense. Plant Physiol 143:400–409

Pan XB, Chen ZX, Zhang Y, Zhu J, Ji XM (2001) Preliminary evaluation for breeding advancement of resistance to rice sheath blight. Chin J Rice Sci 15:218–220

Pannecoucque J, Hofte M (2009) Detection of rDNA ITS polymorphism in *Rhizoctonia solani* AG 2-1 isolates. Mycologia 101:26–33

Park D-S, Sayler RJ, Hong Y-G, Nam M-H, Yang Y (2008) A method for inoculation and evaluation of rice sheath blight disease. Plant Dis 92:25–29

Parmeter JRJ, Whitney HS (1970) Taxonomy and nomenclature of the imperfect state. In: Parmeter JR (ed) *Rhizoctonia solani*: biology and pathology. University of California Press, Berkeley, CA, pp 7–19

Parmeter JR, Sherwood RT, Platt WD (1969) Anastomosis grouping among isolates of *Thanatephorus cucumeris*. Phytopathology 59:1270–1278

Penmetsa RV, Uribe P, Anderson J, Lichtenzveig J, Gish JC, Nam YW, Engstrom E, Xu K, Sckisel G, Pereira M, Baek JM, Lopez-Meyer M, Long SR, Harrison MJ, Singh KB, Kiss GB, Cook DR (2008) The *Medicago truncatula* ortholog of *Arabidopsis EIN2*, *sickle*, is a negative regulator of symbiotic and pathogenic microbial associations. Plant J 55:580–595

Perl-Treves R, Foley RC, Chen W, Singh KB (2004) Early induction of the *Arabidopsis GSTF8* promoter by specific strains of the fungal pathogen *Rhizoctonia solani*. Mol Plant Microbe Interact 17:70–80

Piquerez SJ, Harvey SE, Beynon JL, Ntoukakis V (2014) Improving crop disease resistance: lessons from research on *Arabidopsis* and tomato. Front Plant Sci 5:671

Salazar O, Julian MC, Rubio V (2000) Primers based on specific rDNA-ITS sequences for PCR detection of *Rhizoctonia solani*, *R. solani* AG 2 subgroups and ecological types, and binucleate *Rhizoctonia*. Mycol Res 104:281–285

Schneebeli K, Mathesius U, Watt M (2015) *Brachypodium distachyon* is a pathosystem model for the study of the wheat disease *Rhizoctonia* root rot. Plant Pathol 64:91–100

Shah JM, Raghupathy V, Veluthambi K (2009) Enhanced sheath blight resistance in transgenic rice expressing an endochitinase gene from *Trichoderma virens*. Biotechnol Lett 31:239–244

Sharon M, Freeman S, Sneh B (2011) Assessment of resistance pathways induced in *Arabidopsis thaliana* by hypovirulent *Rhizoctonia* spp. isolates. Phytopathology 101:828–838

Sivalingam PN, Vishwakarma SN, Singh US (2006) Role of seed-borne inoculum of *Rhizoctonia solani* in sheath blight of rice. Indian Phytopathol 59:445–452

Smith JD, Kidwell KK, Evans MA, Cook RJ, Smiley RW (2003) Assessment of spring wheat genotypes for disease reaction to *Rhizoctonia solani* AG-8 in controlled environment and direct-seeded field evaluations. Crop Sci 43:694–700

Sneh B, Ichielevich-Auster M (1998) Induced resistance of cucumber seedlings caused by some non-pathogenic *Rhizoctonia* (np-R) isolates. Phytoparasitica 26:27–38

Sneh B, Burpee L, Ogoshi A (1991) Identification of *Rhizoctonia* species. American Phytopathological Society, St. Paul, MN

Solano R, Stepanova A, Chao Q, Ecker JR (1998) Nuclear events in ethylene signaling: a transcriptional cascade mediated by ETHYLENE-INSENSITIVE3 and ETHYLENE-RESPONSE-FACTOR1. Genes Dev 12:3703–3714

Song WY, Wang GL, Chen LL, Kim HS, Pi LY, Holsten T, Gardner J, Wang B, Zhai WX, Zhu LH, Fauquet C, Ronald P (1995) A receptor kinase-like protein encoded by the rice disease resistance gene, *Xa21*. Science 270:1804–1806

Sridevi G, Parameswari C, Sabapathi N, Raghupathy V, Veluthambi K (2008) Combined expression of chitinase and β-1,3-glucanase genes in indica rice (Oryza sativa L.) enhances resistance against *Rhizoctonia solani*. Plant Sci 175:283–290

Srinivasachary L, Willocquet L, Savary S (2011) Resistance to rice sheath blight (*Rhizoctonia solani* Kuhn) [Teleomorph: *Thanatephorus cucumeris* (A.B. Frank) Donk.] disease: current status and perspectives. Euphytica 178:1–22

Stretton HM, Baker KF, Flentje NT, McKenzie AR (1964) Formation of basidial stage of some isolates of *Rhizoctonia*. Phytopathology 54:1093–1095

Sweetingham MW, Cruickshank RH, Wong DH (1986) Pectic zymograms and taxonomy and pathogenicity of the Ceratobasidiaceae. Trans Br Mycol Soc 86:305–311

Thatcher LF, Anderson JP, Singh KB (2005) Plant defence responses: what have we learnt from *Arabidopsis*? Funct Plant Biol 32:1–19

Tisserat N, Fry J, Green D (1994) Managing *Rhizoctonia* large patch. Golf Course Manag 62: 58–61

Torres MA, Dangl JL, Jones JD (2002) *Arabidopsis* gp91phox homologues *AtrbohD* and *AtrbohF* are required for accumulation of reactive oxygen intermediates in the plant defense response. Proc Natl Acad Sci U S A 99:517–522

van Loon LC (1985) Pathogenesis-related proteins. Plant Mol Biol 4:111–116

Venu RC, Jia Y, Gowda M, Jia MH, Jantasuriyarat C, Stahlberg E, Li H, Rhineheart A, Boddhireddy P, Singh P, Rutger N, Kudrna D, Wing R, Nelson JC, Wang GL (2007) RL-SAGE and microarray analysis of the rice transcriptome after *Rhizoctonia solani* infection. Mol Genet Genomics 278:421–431

Wang L, Liu LM, Wang ZG, Huang SW (2013) Genetic structure and aggressiveness of *Rhizoctonia solani* AG1-IA, the cause of sheath blight of rice in Southern China. J Phytopathol 161: 753–762

Wang R, Lu L, Pan X, Hu Z, Ling F, Yan Y, Liu Y, Lin Y (2015) Functional analysis of OsPGIP1 in rice sheath blight resistance. Plant Mol Biol 87:181–191

Wen ZH, Zeng YX, Ji ZJ, Yang CD (2015) Mapping quantitative trait loci for sheath blight disease resistance in Yangdao 4 rice. Genet Mol Res 14:1636–1649

Wibberg D, Jelonek L, Rupp O, Hennig M, Eikmeyer F, Goesmann A, Hartmann A, Borriss R, Grosch R, Puhler A, Schluter A (2013) Establishment and interpretation of the genome sequence of the phytopathogenic fungus *Rhizoctonia solani* AG1-IB isolate 7/3/14. J Biotechnol 167:142–155

Wibberg D, Rupp O, Jelonek L, Krober M, Verwaaijen B, Blom J, Winkler A, Goesmann A, Grosch R, Puhler A, Schluter A (2015) Improved genome sequence of the phytopathogenic fungus *Rhizoctonia solani* AG1-IB 7/3/14 as established by deep mate-pair sequencing on the MiSeq (Illumina) system. J Biotechnol 203:19–21

Williams B, Kabbage M, Kim HJ, Britt R, Dickman MB (2011) Tipping the balance: *Sclerotinia sclerotiorum* secreted oxalic acid suppresses host defenses by manipulating the host redox environment. PLoS Pathog 7:e1002107

Xu SJ, Hong SJ, Choi W, Kim BS (2014) Antifungal activity of *Paenibacillus kribbensis* strain T-9 Isolated from soils against several plant pathogenic fungi. Plant Pathol J 30:102–108

Yin C, Hulbert SH, Schroeder KL, Mavrodi O, Mavrodi D, Dhingra A, Schillinger WF, Paulitz TC (2013) Role of bacterial communities in the natural suppression of *Rhizoctonia solani* bare patch disease of wheat (*Triticum aestivum* L.). Appl Environ Microbiol 79:7428–7438

Zheng A, Lin R, Zhang D, Qin P, Xu L, Ai P, Ding L, Wang Y, Chen Y, Liu Y, Sun Z, Feng H, Liang X, Fu R, Tang C, Li Q, Zhang J, Xie Z, Deng Q, Li S, Wang S, Zhu J, Wang L, Liu H, Li P (2013) The evolution and pathogenic mechanisms of the rice sheath blight pathogen. Nat Commun 4:1424

Belowground Defence Strategies Against *Verticillium* Pathogens

Eva Häffner and Elke Diederichsen

Abstract Plant pathogenic *Verticillium* species cause vascular infections in many dicot species and show a complex interaction with their hosts. The soil-borne fungi start infections on roots, traverse the root cortex to enter the xylem and spread systemically inside the vasculature. The disease symptoms include wilting, leaf necrosis, stem discoloration and/or premature senescence. Finally the host plant is systemically colonized, and resting structures are formed in the infected tissue. Control of this disease relies primarily on quantitative host resistance, and many studies have built a multifaceted picture of the many factors that are involved in defence on different levels. Once the first major barrier—the endodermis—has been overcome, defence reactions are primarily targeting the fungus in the vascular system and involve many components that have been described for pathogen-associated molecular pattern (PAMP)-triggered but also for effector-triggered immunity. Results from the recently described interaction between *Verticillium longisporum* and Brassicaceae hosts are reviewed more comprehensively, and own data on the gene expression pattern characterizing the defence response against systemic colonization in *Arabidopsis thaliana* are presented. Gene expression analysis in line with contrasting reactions revealed the absence of multiple defence gene induction in the susceptible line at the onset of systemic colonization. With respect to the available knowledge on *Verticillium* and its interactions, it should be possible to support the control of *Verticillium* by applying a plethora of science-based strategies that will more and more meet practical demands.

E. Häffner
Dahlem Centre of Plant Sciences, Institut für Biologie, Fachbereich Biologie, Chemie, Pharmazie, Freie Universität Berlin, Angewandte Genetik, Albrecht-Thaer-Weg 6, 14195 Berlin, Germany

Botanischer Garten und Botanisches Museum Berlin, Freie Universität Berlin, Königin-Luise-Str. 6-8, 14195 Berlin, Germany

E. Diederichsen (✉)
Dahlem Centre of Plant Sciences, Institut für Biologie, Fachbereich Biologie, Chemie, Pharmazie, Freie Universität Berlin, Angewandte Genetik, Albrecht-Thaer-Weg 6, 14195 Berlin, Germany
e-mail: elked@zedat.fu-berlin.de

C.M.F. Vos, K. Kazan (eds.), *Belowground Defence Strategies in Plants*, Signaling and Communication in Plants, DOI 10.1007/978-3-319-42319-7_6

1 Plant Pathogenic *Verticillium* Species and Their Impact on Different Crops

Verticillium pathogens are causing important vascular diseases that affect many plants in nearly all cropping areas in temperate and subtropical regions. They are members of the ascomycete genus *Verticillium*, which contains ten species that are regarded as plant pathogens. The taxonomy of these species has recently been revised based on DNA sequence information such as the ITS region (Inderbitzin and Subbarao 2014), and key facts characterizing the five major plant pathogenic *Verticillium* species are summarized in Table 1. *Verticillium* spp. have formerly been regarded as *fungi imperfecti*, and still no sexual stage has been observed in any of these species. Based on morphology and ITS sequences, two major lineages can be distinguished, the Flavexudans and the Flavnonexudans, a term that refers to the yellow colour of young, growing mycelium in vitro (Inderbitzin et al. 2011). All

Table 1 Major characteristics of the most relevant plant pathogenic *Verticillium* species (summarized after Inderbitzin and Subbarao 2014; Inderbitzin et al. 2011; Novakazi et al. 2015)

Species	Major hosts	Resting structures/ morphological features	Comment
V. albo-atrum	Potato (*Solanum tuberosum*), hop (*Humulus lupulus*)	Melanized resting mycelium and microsclerotia/growing mycelium white with yellow tinge	In literature, < 1990 *V. albo-atrum* and *V. dahliae* are not always correctly differentiated
V. tricorpus	Lettuce (*Lactuca sativa*)	Melanized mycelium, chlamydospores and microsclerotia/growing mycelium white with yellow tinge	Minor relevance as a pathogen, also described as endophyte
V. alfalfae	Alfalfa (*Medicago sativa*)	Melanized resting mycelium/growing mycelium white	
V. dahliae	Cotton (*Gossypium hirsutum*), olive (*Olea europaea*), tomato (*Lycopersicon esculentum*), potato (*Solanum tuberosum*), sunflower (*Helianthus annuus*); >200 hosts described	Microsclerotia/growing mycelium white Conidia are ca. 6 μm in length	Major relevance Defoliating and non-defoliating pathotypes, VCG groups
V. longisporum	Brassicaceae, such as oilseed rape (*Brassica napus*), cauliflower (*B. oleracea*), *A. thaliana*	Microsclerotia/growing mycelium white; long conidia (8.5 μm ± 2.5)	Diploid hybrid species with *V. dahliae* as one ancestor and two unknown ancestors, resp.; also pathogenic on non-Brassicaceae in pathogenicity tests

Verticillium pathogens are soil-borne; they survive for many years in soil as resting structures such as dark-resting mycelium, microsclerotia or, in some cases, chlamydospores. Host colonization is based on mycelium and conidiospores which are produced on phialides that arise from conidiophores in a whorl-like or *verticillate* structure or by budding from a hypha.

Verticillium infections cause a bunch of typical symptoms, such as stunted growth (Fig. 1a–c) and leaf chlorosis (Fig. 1a) that develops into necrosis and is often characterized by an asymmetric occurrence on individual leaves (Fig. 1a). Wilting of at least parts of the plant is a prominent symptom and often coincides with reduced growth and vascular discoloration. Not all *Verticillium* species induce identical symptoms; *V. longisporum* infections, for example, do not lead to wilting symptoms but are mainly characterized by premature seed ripening during the final growth stages of the host (Heale and Karapapa 1999; Fig. 1d, e). Certain highly aggressive strains of *V. dahliae* cause defoliation in cotton and, more recently, also in olive trees (Mercado-Blanco et al. 2002). Variation inside *V. dahliae* is mainly described by the vegetative compatibility grouping system (VCG, Joaquim and Rowe 1991), apart from race designations that can be made according to the pathogenicity on tomato or lettuce hosts (Klosterman et al. 2009). Chromosome variations due to rearrangements have been demonstrated by de Jonge et al. (2013) in *V. dahliae* and can be expected to contribute to variation in pathogenicity.

V. dahliae can be regarded as the most relevant pathogen in this genus due to its extremely broad host spectrum of more than 200 plant species, including major crops like cotton, tomatoes or potato (Pegg and Brady 2002). It also infects trees and causes significant damage in olives or maple trees (Goud et al. 2004; López-Escudero and Mercado-Blanco 2011). In cotton, yield losses of 0.5–3.5 % have been reported for the USA (Blasingame and Patel 2005); in Turkey, an average yield loss of 7 % in cotton has been found in a cultivar survey (Karademir et al. 2012). In potato crops, yield losses are commonly in the range of 10–15 % but may reach up to 50 %, whereas in lettuce, complete losses of a crop are regularly reported (Klosterman et al. 2009). A more recently described plant pathogenic *Verticillium* species is *V. longisporum,* which is the only species having a significant impact on Brassicaceae, such as oilseed rape (*Brassica napus*) or cauliflower (*B. oleracea*, Karapapa et al. 1997; Zeise and von Tiedemann 2002). Yield effects of *V. longisporum* in oilseed rape have been estimated in single-plant experiments to reach up to 80 % depending on disease severity (Dunker et al. 2008).

Epidemiology of *Verticillium* is characterized by the high persistence of the resting structures and its usually monocyclic nature. Spatial spread can occur by seed transmission, as it has been shown for *V. dahliae* on spinach (Spek 1973), cotton (Göre et al. 2011), lettuce (Vallad et al. 2005) and olives (Karajeh 2006). Only for *V. albo-atrum* wind dispersal of conidiospores has been described (Jiménez-Díaz and Millar 1988). Weeds can be assumed to play a significant role for the multiplication and rejuvenation of inoculum (Vallad et al. 2005). Severe disease symptoms and yield losses seem to depend on very high inoculum levels,

Fig. 1 *Verticillium longisporum*: symptoms and fungal structures. (**a–c**) Stunting caused by *V. longisporum* in greenhouse inoculation assays in *Brassica napus* (**a**) and *A. thaliana* (**b**, **c**). *Left side*: mock-inoculated controls, *right side*: *V. longisporum* inoculated plants. (**a**, **b**) Chlorosis of inoculated *B. napus* and *A. thaliana* plants (*right side* of each panel). The *arrow* (**a**) denotes asymmetric chlorosis in *B. napus*. *Bar panel* (**b**): 5 cm. (**d**) Premature senescence caused by *V. longisporum* in the susceptible oilseed rape cultivar 'Falcon'. (**e**) Protection from premature senescence in a breeding line carrying partial *V. longisporum* resistance. (**f**) *V. longisporum* hyphae in malt agar forming microsclerotia (*bar* = 1 mm). (**g**) *V. longisporum* microsclerotia on an oilseed rape stem. (**h**, **i**) *V. longisporum* outgrowth from apical stem segments of infected *A. thaliana* plated on malt agar medium. (**h**) Colonization-susceptible ecotype 'Landsberg *erecta*' (L*er*), (**i**) colonization-resistant ecotype 'Burren' (Bur)

which was indicated by the study of Dunker et al. (2008) for *V. longisporum* and described as a general attribute of *Verticillium* diseases by Schnathorst (1981).

Chemical control of *Verticillium* is not possible once the pathogen has established itself inside the host; other means of control such as reducing the amount of inoculum in the soil are still more of academic value and not yet established in cropping systems. Therefore, host resistance is a major control means, and resistant accessions have been described in many crops or related species; many of these are described by Pegg and Brady (2002). Different types of resistance have been reported, such as the race-specific and monogenic resistance conferred by the *Ve1* gene in tomato (Kawchuk et al. 2001) or on the other hand many reports on quantitative resistance/quantitative trait loci in several crops (Bolek et al. 2005; Jakse et al. 2013; Rygulla et al. 2008; Simko et al. 2004) and in *Arabidopsis thaliana* (Häffner et al. 2010, 2014; Veronese et al. 2003). The molecular basis of resistance to *Verticillium* has been studied intensively, and excellent reviews summarizing in particular the defence responses towards *V. dahliae* have been provided by Daayf (2015), Fradin and Thomma (2006) and Klosterman et al. (2009). During the last decade, a substantial knowledge increase has been generated on host reactions to control infections by *V. longisporum*; hence, this will be a focus of this review.

2 Life Cycle and Pathogenesis

The life cycle of *Verticillium* can be divided into a dormant phase, a parasitic phase and a saprophytic phase. The dormant phase is initiated by the formation of resting structures that are characterized by melanization and condensed accumulation of hyphal contents in either resting mycelium (*V. albo-atrum*) or microsclerotia (*V. dahliae* and *V. longisporum*, Fig. 1f, g), see Table 1. Unlike *V. dahliae*, *V. longisporum* produces also short melanized hyphae in between microsclerotia (Fig. 1f). Microsclerotia of *V. dahliae* (and most likely also those of *V. longisporum*) stay viable in soil for up to 15 years (Wilhelm 1955), while the resting mycelium of *V. albo-atrum* loses its germination capacity after 4 years (Fradin and Thomma 2006). Germination of fungal resting structures in the soil is inhibited until root exudates stimulate the germination of the melanized mycelium or the microsclerotia.

2.1 *Infection Process and Disease Progression*

Excess carbon and nitrogen amounts in root exudates seem to be the chemical stimulus that induces germination (Huisman 1982; Mol 1995; Olsson and Nordbring-Hertz 1985; Schreiber and Green 1963). Microsclerotia can germinate multiple times in a cell-by-cell manner to increase the number of successful infections. Hyphae that grow out of the resting structures can traverse only a limited distance (ca. 300 μm, Huisman 1982) to reach the host root.

Verticillium enters the parasitic stage by infecting either close to the root tip or at the sites of lateral root formation (Bishop and Cooper 1983). After establishing its mycelium on the rhizodermis, the fungus needs to grow through the cortex and the endodermis to reach its major niche, which is the xylem part of the vascular system. To get to the xylem, the hyphae can either grow inter- or intracellularly (or both) to trespass the cortical zone. The endodermis has been described as a physical barrier against infection in many interactions, and when comparing *V. dahliae* with *V. longisporum* infections on oilseed rape, this barrier seemed to explain non-host resistance towards *V. dahliae* (Eynck et al. 2007; Eynck et al. 2009). Often, crossing the endodermis may only be achieved when it is not yet fully developed or when it is damaged by wounding or nematode infection (Bowers et al. 1996; Eynck et al. 2007; Huisman 1982; Pegg 1974; Reusche et al. 2014; Schnathorst 1981). After crossing the endodermis, the fungus enters the xylem vessels. Usually, only a few vessels are initially affected, and horizontal spread into adjacent xylem vessels can start from here. Eynck et al. (2009) observed only a limited number of vessels colonized by *V. longisporum* in oilseed rape and concluded that this could explain the absence of wilting symptoms in this interaction. From the initial xylem colonization, disease progress is primarily based on acropetal spread inside the host. The fungus spreads inside the vasculature by hyphal growth (short distance) or by the formation of conidiospores. Conidia are carried with the sap stream and can be trapped in pit cavities or at vessel ends, where they can germinate and grow into adjacent vessel elements in order to continue colonization. Heinz et al. (1998) reported that the colonization during the systemic spread on *V. dahliae* in the vasculature appeared to occur in cycles of fungal spread and fungal elimination, which might reflect the struggle between defence responses and fungal attack. For *V. longisporum*, it has been demonstrated that the onset of systemic spread into the upper stem depended either on the onset of host flowering (Dunker et al. 2008; Häffner et al. 2010; Zhou et al. 2006) or on susceptibility—only very susceptible hosts seem to be colonized systemically at early stages (Dunker et al. 2008).

The final infection stage is characterized by the onset of host tissue necrosis and the saprophytic growth of *Verticillium* into the dead host tissue. The fungus grows into the adjacent necrotic parts of the host, proliferates extensively and finalizes its development at these sites by the formation of resting structures (Fig. 1g). This can be restricted to single leaves or happen on all infected plant tissues but is usually most profound on lower parts of the stem. For *V. longisporum*, this is the stage where the most typical symptom becomes apparent, the premature ripening. Infected plants undergo precocious senescence which is thought to affect the yield by shortening the seed-filling phase (Gladders 2009). The newly formed resting structures are released to the soil after decomposition of plant materials. In perennial hosts, the mycelium can also overwinter within the plant or in propagative organs such as tubers, bulbs or seeds, if the maternal part of the seed coat is infected.

2.2 Pathogenicity and Virulence Factors

Verticillium spp. employ a variety of pathogenicity and virulence factors such as enzymes, toxins and elicitors to successfully establish in the host and to manipulate host physiology to meet their own requirements. After the definition of Sacristán and García-Arenal (2008), pathogenicity refers to the capacity of a pathogen to cause disease, whereas virulence refers to the degree of damage caused in the host. A detailed description of *Verticillium* pathogenicity and virulence factors is given, for example, by Fradin and Thomma (2006) and by Luo et al. (2014). Here, a short summary is given to illustrate major pathogenicity and virulence mechanisms against which some of the defence responses described below are directed. Among the pathogenicity factors first detected to influence host colonization capacity of *Verticillium* are cell wall-degrading enzymes (CWDE). Most prominently, pectinases are produced by *Verticillium*. Their role seems plausible since *Verticillium* spp. have to penetrate cell walls to grow intracellularly in the root cortex and to overcome pit membranes between xylem vessels. Indeed, a CWDE secretion mutant strain caused less symptoms and had very low colonization capacity in tomato (Durrands and Cooper 1988). Among the pathogenicity factors are also all those enzymes that allow survival under the low-nutrient conditions of xylem sap. Examples are genes mediating cross-pathway control (CPC), a mechanism by which amino acid synthesis is activated if external supply is scarce. Impairment of CPC has been shown to reduce *V. longisporum* proliferation in the host (Timpner et al. 2013). A *V. dahliae* mutant deficient in thiamine synthesis is unable to cause disease in tomato (Hoppenau et al. 2014). Confirmed virulence factors that are also elicitors of defence responses known from *Verticillium* spp. include necrosis and ethylene-inducing proteins (NEP; Wang et al. 2004) and *Ave1*, a plant-type natriuretic peptide possibly interfering with host ion homeostasis (de Jonge et al. 2012). *V. dahliae* SPECIFIC SECRETED PROTEIN 1 (VdSSP1) increased virulence of *V. dahliae* in cotton and has a function in cell wall degradation (Liu et al. 2013). Isochorismate hydrolase is an enzyme putatively involved in host defence suppression. It is expressed in many pathogenic fungi and was characteristic for a highly aggressive *V. dahliae* isolate in a proteomic study (El-Bebany et al. 2010). Isochorismate is the immediate precursor of salicylic acid (SA), an important defence phytohormone.

3 Belowground Defence Mechanisms Against *Verticillium*

3.1 Tolerance and Resistance

There are two fundamental ways of hosts to defend themselves against an infection: resistance and tolerance. Here, the definition of Roy and Kirchner (2000) is used, defining ‘resistance’ as all host strategies limiting infection, while ‘tolerance’ does not limit infection itself but reduces its fitness consequences for the host. In pathogenesis caused by *Verticillium* spp., both strategies can be observed, often within the same host species. For olive, it was repeatedly reported that symptoms caused by *V. dahliae* were strongly correlated with the extent of systemic colonization (Markakis et al. 2010; Mercado-Blanco et al. 2003). However, Arias-Calderón et al. (2015) found no correlation between root or stem colonization and the disease intensity in the olive progenies tested. Some genotypes showing mild symptoms were strongly colonized with respect to intensity and extent, supporting the concept of tolerance for the olive—*V. dahliae* pathosystem. Reduced defence gene activation seems to be a common principle in tolerant genotypes (Robb et al. 2007; Tai et al. 2013). Defence responses can be inadequate and lead to susceptibility rather than resistance (Robb et al. 2012). Robb (2007) interpreted *Verticillium* tolerance as a step on the way to a mutualistic relationship. The molecular mechanisms underlying tolerance are less understood than active defence responses leading to the elimination of pathogens. Tai et al. (2013) found a pronounced up-regulation of chlorophyll biosynthesis genes in a tolerant potato clone as compared to a resistant clone. Resistance and tolerance against *Verticillium* are both quantitative traits that rely on a multitude of genes and mechanisms. In the following, examples for both types of defence will be given. While tolerance to *Verticillium* is reported for some species, most hosts depend on pathogen restriction or elimination to maintain plant health.

3.2 Vascular Defence in Root and Hypocotyl

Verticillium spp. invade roots of susceptible and resistant host genotypes equally (e.g. Eynck et al. 2009; Robb et al. 2007; Vallad and Subbarao 2008). Major differences in host resistance exist in the extent of systemic colonization and symptom development in various hosts such as olive (López-Escudero and Mercado-Blanco 2011; Mercado-Blanco et al. 2003), lettuce (Vallad and Subbarao 2008), cotton (Cui et al. 2000), oilseed rape (Eynck et al. 2009) and model plant *Arabidopsis thaliana* (Häffner et al. 2010; Johansson et al. 2006; Veronese et al. 2003; Fig. 1h, i). This leads to the conclusion that defence mechanisms against *Verticillium* are focused on the xylem of the root, hypocotyl and shoot of hosts. Studies comparing defence reactions in susceptible and resistant hosts emphasize the significance of induced defences that are activated more quickly and more strongly in resistant hosts. Over the last decades, induced vascular defences against *Verticillium* spp. have been studied in various host–pathogen interactions, and molecular components mediating pathogen perception, signal transduction and execution of defence have been elucidated. Preventing systemic spread of *Verticillium* spp. has been associated with vessel occlusions of various kinds: tyloses, which are invaginations of adjacent xylem parenchyma cells, have been shown to occur as a consequence of *Verticillium* infection in various hosts such as hop (Talboys 1958), tomato (Dixon and Pegg 1969), chrysanthemum (Robb et al. 1979) and olive (Báidez et al. 2007). Cells respond to *Verticillium* with a marked ultrastructural reorganization involving changes of the cytoskeleton and the vacuole (Wang et al. 2011; Yao et al. 2011; Yuan et al. 2006). Vessels can also be blocked by compounds secreted by neighbouring xylem parenchyma cells (Benhamou 1995; Eynck et al. 2009). The benefit of vessel occlusion consists in preventing the fungus from further spread, but if vessels are blocked in excess, wilting can occur as a consequence (Fradin and Thomma 2006; Talboys 1972). A mechanism of escaping the deleterious effects of vessel obstruction is de novo xylem formation. *V. longisporum* was shown to cause xylem hyperplasia in *A. thaliana* and *B. napus*, which hardly occurred after *V. dahliae* infection. Transdifferentiation of xylem parenchyma cells into functional xylem vessels and reactivation of secondary cambium to produce new xylem elements occurred under the control of the transcription factor VASCULAR-RELATED NAC DOMAIN 7 (VND7; Reusche et al. 2012). This adaptation not only prevented wilting but even rendered the host more tolerant to drought stress (Reusche et al. 2012). It is further hypothesized that various processes contribute to fungal elimination from the xylem in resistant hosts and that this elimination is overcome in susceptible hosts (Heinz et al. 1998). There is experimental evidence for a diverse set of antifungal enzymes and substances to be involved in vascular defence against *Verticillium*. The most important compounds discussed are phenolic compounds such as lignin and soluble phenylpropanoids, terpenoids, glucosinolates and camalexin. Proteins involved in *Verticillium* defence are, for example, enzymes that degrade fungal cell walls or proteins inhibiting fungal enzymes. The defence strategies and the signalling events leading to their activation are reviewed for the most important pathosystems in the following sections.

3.2.1 Antimicrobial Compounds

Among the low molecular weight antimicrobial compounds, phytoalexins and phytoanticipins are distinguished depending on whether they are induced upon infection or constitutively present in the plant (Van Etten et al. 1994). The distinction does not refer to particular classes of substances. Compounds of one and the same class could either act as phytoanticipins or phytoalexins.

Phenylpropanoids

Many studies emphasize the role of phenolic compounds such as lignin and soluble phenylpropanoids in the restriction of systemic spread of *Verticillium* in the host. The built-up of these compounds has been shown to be quicker and stronger in resistant compared to susceptible hosts (e. g. Smit and Dubery 1997; Xu et al. 2011). Phenylpropanoids are synthesized from phenylalanine. The initial step leading to cinnamic acid via deamination is catalysed by phenylalanine ammonia lyase (PAL). Simple, soluble phenylpropanoids include, e.g. sinapic acid and the lignin precursors coniferyl alcohol and sinapyl alcohol. Via polymerization or condensation, more complex compounds like lignin, tannins or flavonoids are formed (Dixon et al. 2002). The signalling molecule SA, also a phenolic compound, is related to phenylpropanoids because it shares the precursor chorismic acid with phenylalanine (Wildermuth et al. 2001). The involvement of lignin in response to *Verticillium* spp. has long been known. So-called lignitubers or papillae are known to form at sites of attempted hyphal penetration in root epidermis and cortical cells (Bishop and Cooper 1983; Griffiths 1971; Talboys 1958). Beckman (2000) attributed a role to phenolic storage cells in facilitating rapid lignification.

Xu et al. (2011) determined expression profiles of enzymes involved in phenylpropanoid biosynthesis in roots of resistant sea-island cotton (*Gossypium barbadense*) and susceptible upland cotton (*G. hirsutum*). *G. barbadense* showed a quicker and stronger induction of enzymes in the lignin biosynthesis pathway like PAL, cinnamate 4-hydroxylase, cinnamoyl-CoA reductase and cinnamyl alcohol dehydrogenase. The activities of PAL and peroxidase, the enzyme required for polymerization of lignin monomers, were much higher in roots of *G. barbadense* following infection compared to *G. hirsutum*. Intraspecies variability of resistance in *G. hirsutum* has also been attributed to differences in lignification of hypocotyl tissue (Smit and Dubery 1997). In olive, various flavonoids with antifungal activity like rutin, luteolin glucoside, oleuropein and tyrosol were detected in the vascular tissue of stems (Báidez et al. 2007). Phenolics also played a role in resistance of *B. napus* to *V. longisporum* (Eynck et al. 2009). While fungal entry into roots was similar for a resistant and a susceptible accession, systemic colonization of the shoot system was inhibited in the resistant genotype 'SEM 05-500256'. Microscopic analyses of the hypocotyl revealed a much higher extent of vessel occlusions as well as cell wall reinforcements with lignin and cell wall-bound phenolics as compared to the susceptible accession 'Falcon'. The resistant genotype also accumulated more soluble phenolics after infection compared to the susceptible genotype, and phenolic storage cells were more abundant (Eynck et al. 2009). Metabolomic analyses have shown that soluble phenylpropanoids accumulated in *A. thaliana* after *V. longisporum* challenge (König et al. 2014). Correspondingly, genes encoding enzymes in the phenylpropanoid pathway were induced upon infection, and the sinapate-deficient *fah1-2* mutant was more susceptible to the fungus. Moreover, the soluble phenylpropanoids sinapoyl glucose, coniferyl alcohol and coniferin inhibited fungal growth in vitro. Although metabolomic analyses have been performed in leaves, it is conceivable that phenylpropanoids also play a role in lower parts of *A. thaliana* since hypocotyls and petioles of infected plants exhibited stronger lignification of the xylem (König et al. 2014). Natural variation in genes controlling the phenylpropanoid pathway may well account for observed *V. longisporum* resistance QTL: the *vec3* QTL that controlled systemic colonization to *V. longisporum* in *A. thaliana* co-localized with the phenylpropanoid biosynthesis genes encoding cinnamyl alcohol dehydrogenase (*Cad5*, *Cad8*) and UDP-glycosyltransferase (*Ugt84a3*; Häffner et al. 2014). These genes were induced upon infection (König et al. 2014). In *B. napus, V. longisporum* resistance QTL were also found to co-localize with QTL for contents of some phenylpropanoids in the hypocotyls of the host plants (Obermeier et al. 2013).

Increased lignification in response to *Verticillium* is not restricted to roots and hypocotyls, but is also reported for stem tissue of various hosts (Báidez et al. 2007; Cui et al. 2000).

Terpenoids

Terpenoid phytoalexins synthesized via the isoprenoid pathway have been shown to be potent inhibitors of *V. dahliae* growth in *G. barbadense*. Four major hemigossypol derivatives in cotton stems killed conidia and mycelium in vitro, and one of them (desoxyhemigossypol) reached fungicidal concentrations in the cotton stele and had the required water solubility to act in xylem sap (Mace et al. 1985). But also in roots of *G. barbadense*, sesquiterpene aldehydes (e.g. hemigossypol) were correlated with resistance, as could be shown with plants that were silenced for (+)-δ-cadinene synthase (Gao et al. 2013), an important enzyme in the gossypol biosynthetic pathway (Chen et al. 1995). The same enzyme has been induced in roots of *G. barbadense* by *V. dahliae* infection (Wang et al. 2011). Terpenoids also seemed to play a role in root defence of a resistant *G. hirsutum* genotype compared to a susceptible genotype, as transcription of a respective biosynthesis gene was up-regulated after infection only in the resistant genotype (Zhang et al. 2012a).

Glucosinolates and Camalexin

Glucosinolates are a class of defensive compounds that are characteristic for Brassicaceae. Glucosinolates are amino acid derivatives and consist of glucose which is bound via a sulphur atom to a (Z)-*N*-hydroximinosulfate ester. A variable side chain renders considerable chemical diversity to this class of compounds (Halkier and Gershenzon 2006). The antibiotic effect is not exerted by glucosinolates themselves but by their degradation products: nitriles, epithionitriles and isothiocyanates that are produced after hydrolysis of glucosinolates by specific β-glucosidases (myrosinases). Most myrosinases are located in the vacuole, and glucosinolates are only cleaved upon tissue damage, for example, after insect herbivory. However, the atypical myrosinase PENETRATION 2 (PEN2) has been shown to cleave indole glucosinolates derived from tryptophan (Trp) also in living cells to produce potent antimicrobial glucosinolate degradation products (Bednarek et al. 2009). Glucosinolates are mostly regarded as phytoanticipins, but the pattern has been shown to change as a consequence of infection (Witzel et al. 2015). The idea that glucosinolates and their degradation products contribute to defence of crucifers against *Verticillium* has been investigated in recent studies. Iven et al. (2012) could show that genes involved in converting Trp to secondary metabolites like indole glucosinolates and camalexin were up-regulated in

A. thaliana roots after infection with *V. longisporum*. Likewise, transcription of the PEN2 homologue PEN2-LIKE 1 (PEL1) was increased. The authors found that a double mutant lacking the enzymes CYP79B2 and CYP79B3, which catalyse the bottleneck biosynthesis step from Trp to indole glucosinolates and camalexin, was more susceptible to *V. longisporum* and contained higher amounts of fungal biomass. However, deficiency in camalexin or indole glucosinolates alone did not significantly increase susceptibility. Witzel et al. (2013) investigated whether resistance against *V. longisporum* was correlated with glucosinolate profiles in different ecotypes of *A. thaliana*. They found a correlation between the presence of alkenyl glucosinolates in leaf extracts and fungal growth inhibition. A degradation product of 2-propenyl glucosinolate, 2-propenyl isothiocyanate, proved to be a potent inhibitor of *V. longisporum* growth in vitro. Further analyses (Witzel et al. 2015) showed that concentrations of glucosinolates and their breakdown products responded to *V. longisporum* infection in an organ- and genotype-specific manner. In the ecotype 'Burren' (Bur), which has been shown to be highly resistant against systemic colonization by *V. longisporum* (Häffner et al. 2010), glucosinolate contents in the roots increased after infection. In the colonization-susceptible genotype 'Landsberg *erecta*' (L*er*), this was not the case (Witzel et al. 2015). These findings suggest that glucosinolates at least contribute to attacking *V. longisporum* in the root, while other mechanisms are active as well.

3.2.2 Antifungal Proteins and Enzymes

Pathogenesis-related (PR) proteins are inducible proteins with antimicrobial activity that have been classified according to their structure and function (van Loon et al. 2006). Root transcriptomics and proteomics following *V. dahliae* infection have been most intensively studied in cotton. Up-regulated defence proteins in the cotton root include peroxidase (Dong and Cohen 2002; Hanson and Howell 2004; Zhang et al. 2012a), beta-glucanase (PR2; Zhang et al. 2013a), chitinase (Wang et al. 2011), Bet v1 protein (PR10), whose mode of action is not yet elucidated (Wang et al. 2011; Zhang et al. 2012b, 2013a), thaumatin-like protein (PR5; Zhang et al. 2013a) and polygalacturonase-inhibiting protein (PGIP), which can inactivate fungal cell wall-degrading enzymes (James and Dubery 2001). Furthermore, lectins have been shown to respond to infection in the cotton root (Wang et al. 2011). Root-specific lectins also played a role in hop (*Humulus lupulus*) resistance to *V. albo-atrum*. They were present in high concentrations in a resistant hop cultivar but absent from a susceptible cultivar (Mandelc et al. 2013). Interestingly, the susceptible cultivar showed a marked induction of PR proteins like chitinase, beta-glucanase and thaumatin-like proteins, which the resistant cultivar did not. This situation is reminiscent of tolerance, and indeed both genotypes were colonized to a comparable degree (Mandelc et al. 2013). In studies with biocontrol agents, the induction of PR proteins like PR1, a protein of yet unknown mode of action which is typically induced via the SA pathway, PR2 (beta-glucanase) and PR4 (chitinase) correlated well with increased resistance (Angelopoulou et al. 2014; Tjamos et al. 2005, Sect. 3.3.3). PR proteins also play a role in the *V. longisporum*–Brassicaceae pathosystem. While PR1 and PLANT DEFENSIN1.2 (PDF1.2), a peptide with antifungal activity responsive to jasmonic acid (JA) and ethylene, were not up-regulated in *A. thaliana* roots shortly after infection (Iven et al. 2012), both genes have been found to respond to *V. longisporum* infection locally in *Brassica* hypocotyls at defined infection stages (Kamble et al. 2013). Johansson et al. (2006) deduced PR2 induction in *A. thaliana* roots from promoter-GUS studies, and Iven et al. (2012) showed that chitinases, peroxidases, germin-like proteins and protease inhibitors were up-regulated in *A. thaliana* roots upon *V. longisporum* infection. PR5, a thaumatin-like protein presumably attacking cell membranes of pathogens, was up-regulated by *V. longisporum* in hypocotyls of *A. thaliana* (see Sect. 3.3.3).

3.3 Pathogen Perception and Defence Signalling

To mount an effective defence response against *Verticillium* involving the abovementioned and potentially further unknown mechanisms, pathogen recognition and subsequent defence signalling are indispensable. While the signalling events leading to immunity in leaves are well characterized, defence signalling in the roots or the hypocotyl is less investigated. However, several studies have addressed this topic recently (de Coninck et al. 2015; Millet et al. 2010; Yadeta and Thomma 2013).

3.3.1 Immune Receptors Mediating Defence Responses Against *Verticillium*

Receptor-mediated immunity has traditionally been divided into two fundamental processes: pathogen-associated-molecular-pattern (PAMP)-triggered immunity (PTI) that occurs upon perception of widespread molecular patterns of pathogens like flagellin or chitin by pattern recognition receptors (PRR) and effector-triggered immunity (ETI) acting specifically against certain pathogens by perceiving effectors, or their effects on hosts, through resistance genes (R-genes; Jones and Dangl 2006). Recently, this strict division has been challenged, since PAMPs and effectors and their specificity cannot always be clearly separated (Thomma et al. 2011). A good example for the sometimes unclear distinction between R-genes and PRR is the receptors involved in the interaction between *Verticillium* spp. and their hosts. Among the receptors induced by *Verticillium* are definitive PRR like chitin receptors (Sect. 3.3.3) but, for example, also the *Ve*-genes that recognize the effector *Ave1*, which is, however, surprisingly widespread among pathogenic basidiomycetes. Furthermore, the resistance conferred is quantitative and relatively weak compared to the effect of typical R-genes (de Jonge et al. 2012). The experimental evidence reviewed in the following suggests that other still uncharacterized receptors take part in the recognition of *Verticillium*.

Ve-Genes

Ve-genes have first been cloned from tomato (Kawchuk et al. 2001) and later been identified in other hosts as well (Chai et al. 2003; Fei et al. 2004; Vining and Davis 2009; Zhang et al. 2011). *Ve1* and its homologue *Ve2* from tomato have been characterized as receptor-like proteins with an N-terminal hydrophobic signal peptide, extracellular leucine-rich repeats (LRR) containing potential glycosylation sites, a membrane-associated domain, and an intracellular endocytosis signal (Kawchuk et al. 2001). Both genes mediated resistance against race 1 of *V. albo-atrum* in potato (Kawchuk et al. 2001). Fradin et al. (2009) found that only *Ve1*, but not *Ve2* mediated resistance against *V. dahliae* in tomato. *Ve1* has been shown to respond to the fungal effector *Ave1* that was most likely acquired by pathogens through horizontal gene transfer (de Jonge et al. 2012). *Ve*-genes were induced upon *V. dahliae* infection, while resistant accessions with active alleles responded more quickly. Downstream signalling involved ENHANCED DISEASE SUSCEPTIBILITY 1 (EDS1) and NON-RACE-SPECIFIC DISEASE RESISTANCE 1 (NDR1) as well as the NB-LRR protein NRC1, the F-box protein ACIF, the mitogen-activated protein kinase (MAPK) MEK2 and SOMATIC EMBRYOGENESIS RECEPTOR KINASE 3 (SERK3)/BRASSINOSTEROID-ASSOCIATED KINASE 1 (BAK1) as deduced from virus-induced gene silencing (Fradin et al. 2009). The resulting defence response has been shown to include induction of hydrogen peroxide, PAL and peroxidase in roots of resistant tomato cultivars. The concentration of selected metabolites from the phenylpropanoid pathway increased more quickly and more strongly in roots of a resistant tomato line compared to a susceptible line (Gayoso et al. 2010). *Ve* homologues from other hosts include *Stve* from *Solanum torvum* (Fei et al. 2004) and *SlVe* from *Solanum lycopersicoides* (Chai et al. 2003). Yet another *Ve1* ortholog has been found in *Nicotiana glutinosa* (Zhang et al. 2013b). *Ve* homologues have been identified outside the Solanaceae as well: for *mVe1* from *Mentha* spp. (Vining and Davis 2009) a resistance effect against *V. dahliae* is likely, while *Gbve1* from island cotton *G. barbadense* has been shown to mediate resistance against *V. dahliae* race 1 (Zhang et al. 2012c). The resistance effect exerted by *Gbve1* was expression-dependent, and promoter activity was shown to be highest in the vasculature of roots and stems (Zhang et al. 2012c). Although *Ve*-genes have not been identified in Brassicaceae, it has been shown that expression of *Ve1* and *Gbve1* in *A. thaliana* mediated resistance against race 1 of *V. dahliae* (Fradin et al. 2011; Zhang et al. 2011). This shows that the molecular machinery for *Ve*-mediated resistance is present in *A. thaliana*. An interesting difference in the immune response between different hosts consists in the occurrence of a hypersensitive response (HR). While HR occurred in tomato and *Nicotiana tabacum* plants where *Ave1* and *Ve1* were co-expressed (de Jonge et al. 2012; Zhang et al. 2013b), *Ve1*-mediated resistance in transgenic *N. benthamiana* and in *A. thaliana* was independent of an HR (Zhang et al. 2013b, c).

Other *Verticillium* Receptors

While *Ve*-genes are the most extensively characterized *Verticillium* immune receptor genes, other *Verticillium*-specific receptors have been found especially in cotton. The *Gbvdr5* gene codes for a membrane-localized receptor-like protein in *G. barbadense*. A putative loss-of-function mutation in *Gbvdr5* was found in all *Verticillium*-susceptible island cotton genotypes (Yang et al. 2014). *Gbvdr5* promoter activity was observed in all tissues in a reporter gene approach in *A. thaliana*, but expression was strongest in roots and shoot apices. *Gbvdr5* was induced by some *V. dahliae* isolates in *G. barbadense* but interestingly was unaffected or even suppressed in susceptible *G. hirsutum*. *Gbvdr5* was also induced by the stress phytohormones JA, abscisic acid (ABA) and ethylene. Silencing of *Gbvdr5* compromised resistance, and as shown for *Ve*-genes, resistance could be transferred to *A. thaliana* by expressing *Gbvdr5* in transgenic plants. *Gbvdr5*-mediated resistance was race-specific (Yang et al. 2014). In *G. raimondii*, another *V. dahliae*-resistant cotton species, resistance gene analogues were found to be arranged in clusters. Within these clusters, *V. dahliae* response loci were identified using RNA sequencing (RNA-seq) of root tissue. Some of these response loci were located in the vicinity of known *V. dahliae* resistance QTL (Chen et al. 2015). This leads to the conclusion that more as yet uncharacterized immune receptor genes mediating *V. dahliae* resistance are present in the cotton genome.

3.3.2 Root Defence Signalling in the Cotton–*V. dahliae* Pathosystem

Phytohormones play an important role in the defence response of plants to pathogens. Complex and highly cross-linked signalling cascades affecting many biological processes in the plant are triggered by relatively few phytohormones. The most important defence-related phytohormones are ethylene, jasmonic acid and salicylic acid. All three hormones participate in PTI. During more specific defence reactions, the SA-signalling pathway and the defence response triggered by JA and ethylene are mutually antagonistic: SA signalling mediates defence against biotrophic pathogens, while JA and ethylene together are required to fight necrotrophic pathogens (Glazebrook 2005). Signalling pathways involved in the response of the cotton root to *V. dahliae* have been identified in various cotton genotypes with different methods. The ethylene-signalling pathway has been found to respond in most studies, but its role is ambiguous: ethylene biosynthesis and response genes were induced in roots of resistant *G. barbadense* and susceptible *G. hirsutum* but with different patterns (Xu et al. 2011). A quick up-regulation of aminoacyl-cyclopropane oxidase (ACO), the enzyme catalysing the last step in ethylene biosynthesis, seems typical and important for resistance of *G. barbadense* (Wang et al. 2011). An interesting new mechanism involving an element of the ethylene-signalling cascade has recently been discovered by Yang et al. (2015a): cotton major latex protein 28 (GhMLP28) enhanced the transcription factor activity of ETHYLENE RESPONSE FACTOR 6 (ERF6) and led to enhanced transcription of some GCC-box genes that are responsive to ERFs. Cotton plants silenced for *Ghmlp28* showed increased susceptibility towards *V. dahliae*, and transgenic tobacco plants overexpressing *Ghmlp28* were more resistant. *Ghmlp28* had the highest expression levels in the root and was inducible by *V. dahliae*, ethylene, JA and SA (Yang et al. 2015a).

JA signalling contributes to early defence against *V. dahliae* in cotton roots (Gao et al. 2013; Zhang et al. 2013a). The expression of the key JA biosynthesis enzyme allene oxide synthase (AOS) was much higher in roots of a resistant *G. barbadense* genotype as compared to a susceptible *G. hirsutum* genotype (Zhang et al. 2013a). Although the gene expression study of Xu et al. (2014) was not specific for root tissue, the role of JA signalling in *V. dahliae* defence was confirmed. Li et al. (2014) discovered an interesting regulatory node influencing the defence–growth equilibrium while confirming the role of JA in *V. dahliae* defence of cotton. The transcription factor GbWRKY1 negatively regulated JA-mediated defences against *V. dahliae* in cotton roots. Interestingly, it is induced by *V. dahliae* and methyl jasmonate, possibly as an element of negative feedback control. In accordance with the antagonism between JA/ethylene- and SA-mediated defence responses, cotton plants over-accumulating SA and reactive oxygen species due to silencing of the *Gbssi2* gene were more susceptible to *V. dahliae* in a leaf-inoculation assay (Gao et al. 2013). However, these plants also accumulated the SA-induced PR proteins PR1, PR2 and PR5 that have been associated with increased resistance in roots (see Sect. 3.2.2). It may be concluded that these hormones act synergistically rather than antagonistically in early belowground defences against *V. dahliae*.

Experimental evidence exists that brassinosteroids contribute to cotton resistance against *V. dahliae*. Brassinosteroid-signalling components like the receptor BRASSINOSTEROID INSENSITIVE 1 (BRI1) and the response factor BRASSINAZOLE RESISTANT 1 (BZR1) were up-regulated upon infection in cotton roots, and exogenously applied brassinolide reduced *V. dahliae* symptoms and activated JA signalling (Gao et al. 2013). A resistance-promoting role of brassinosteroids has also been reported by Roos et al. (2014) in the *A. thaliana*–*V. longisporum* pathosystem (see Sect. 3.3.3).

3.3.3 Defence Signalling and Gene Expression in Cruciferous Hosts After *V. longisporum* Infection

V. longisporum is recognized by *A. thaliana* roots within less than an hour after spore germination and before hyphal penetration as evidenced by gene expression studies (Tischner et al. 2010). Ten minutes after the first contact with spores, the phosphorylation pattern of proteins changed not only in the root but also in the shoot. This suggested a highly mobile signal. A transient nitric oxide (NO) burst occurred 35 min after spore contact, which was possibly the initial signal for root-to-shoot communication. After 50 min, the expression pattern of 732 genes in the root and 474 genes in the shoot had changed. As expected, many genes related to signalling such as receptor-like kinases (RLKs), genes related to calcium signalling and transcription factors changed their expression pattern but also genes related to the cell wall, to proteolysis, to defence and to secondary metabolism. Whereas most of the differentially expressed genes in the root were transcription factors or associated with the cell wall or with proteolysis, the focus in the shoot was on regulation of genes related to defence, proteolysis and signalling (Tischner et al. 2010).

The transcriptional response of *A. thaliana* roots in the phase of *V. longisporum* spore germination (1 day post-infection, dpi) and of hyphal penetration into the root (3 dpi) was studied by Iven et al. (2012). 269 genes were differentially expressed at 1 dpi, 490 at 3 dpi with only minor overlap. Again, transcription factors, genes related to defence and stress response and genes encoding apoplastic proteins dominated. Over-representation of the gene ontology terms 'indole phytoalexin biosynthetic process', 'camalexin biosynthetic process', and 'tryptophan metabolic process' suggested that these metabolites may play a crucial role in early defences against *V. longisporum* as mentioned above. In the first 8 days after inoculation, phytohormone levels did not change significantly upon *V. longisporum* infection.

Mutant analysis revealed a role of additional signalling molecules in the *A. thaliana–V. longisporum* interaction. A prominent role of the F-box protein CORONATINE INSENSITIVE 1 (COI1) in *V. longisporum* pathogenesis was discovered by Ralhan et al. (2012). COI1 is a central component of the JA-signalling pathway. The authors showed that functional COI1 in *A. thaliana* roots is required for symptom development such as stunting and early senescence. This function of COI1, however, was shown to be JA-independent. It was concluded that *V. longisporum* exploits COI1 for the induction of early senescence, which allows the fungus to grow necrotrophically on senescent tissue. Consequently, wild-type *A. thaliana* showed much stronger *V. longisporum* colonization at the late stage of infection compared to *coi1* mutants. However, *coi1*-mediated resistance did not prevent host colonization. A similar disease-promoting function of COI1 was also detected in *Fusarium oxysporum* pathogenesis (Thatcher et al. 2009). Rab GTPase-ACTIVATING PROTEIN 22 (*RabGAP22*) is another signalling component that was found to promote *V. longisporum* resistance (Roos et al. 2014). It was expressed in root meristems, vascular tissue and stomata and showed increased expression after infection. The authors provide evidence that suggests a role of RabGAP22 in brassinosteroid signalling. Moreover, brassinosteroid treatment could reduce *V. longisporum* colonization of the host (Roos et al. 2014).

Transcriptional Response to *V. longisporum* in the Hypocotyl–Shoot Transition Zone of a Susceptible and a Resistant *A. thaliana* Line

In order to monitor molecular processes that underlie ecotype-specific resistance to systemic colonization, we performed a microarray analysis on tissue of the hypocotyl and the shoot basis. Two developmental stages were chosen for analysis: at the onset of flowering, systemic colonization has been shown to start (Häffner et al. 2010; Zhou et al. 2006), whereas at the onset of silique maturity, extensive fungal proliferation occurred (Häffner et al. 2010). The analysis was performed with two (Bur×L*er*) near-isogenic lines (NILs) differing for the major colonization resistance QTL *vec1* (Häffner et al. 2010, 2014). NIL9 contained only alleles of the colonization-susceptible parent L*er* in the variable region, whereas tmNIL130 contained a maximum 530 kb introgression of the resistant parent Bur (Fig. 2a, Häffner et al. 2015). Both NILs showed significantly different shoot colonization by *V. longisporum* at the onset of silique maturity (Fig. 2b). The transcriptional response of the more resistant tmNIL130 at the onset of flowering was comparable to the aforementioned studies: 295 genes were differentially expressed in infected plants compared to mock-inoculated plants (Fig. 2c). Among them, 117 genes were related to biotic stress based on their annotations. Many of them could be attributed to processes characteristic for innate immune response (Fig. 3): genes associated with pathogen recognition, such as chitin receptor genes, with calcium signalling, MAPK signalling or production of reactive oxygen species (ROS), were up-regulated. Specifically, transcripts of ENHANCED DISEASE SUSCEPTIBILITY 1 (EDS1) and PHYTOALEXIN-DEFICIENT 4 (PAD4) were up-regulated, which is typical for pathogen-associated molecular pattern (PAMP)-triggered immunity but also for effector-triggered immunity (ETI). While there was no evidence for phytohormone action at the early stages of infection (Tischner et al. 2010; Iven et al. 2012), there was a clear indication that ethylene and salicylic acid played a role in the systemic phase of the infection. Genes involved in ethylene biosynthesis were up-regulated, and SARD1, the main activator of isochorismate synthase, the key enzyme in SA biosynthesis, was up-regulated sixfold. A role of SA in *V. longisporum* defence was also observed by Ratzinger et al. (2009) who demonstrated that SA was present in the xylem sap of *B. napus* after infection and that disease symptoms were negatively correlated with the levels of SA and its glucoside in the shoot. WRKY transcription factors played a major role in *V. longisporum* defence signalling, as has also been demonstrated by Tischner et al. (2010) and Iven et al. (2012). In the present study, WRKY33, which is essential for an effective immune response against necrotrophic pathogens (Zheng et al. 2006), was up-regulated in the resistant NIL. Some of the induced genes were shown or hypothesized to play a role in glucosinolate metabolism: the transcription factor MYB51 was shown to control indole glucosinolate synthesis in roots and shoots of *A. thaliana* (Frerigmann and Gigolashvili 2014), and the cytochrome P450 protein CYP83B1 catalyses the formation of aromatic and indole glucosinolates (Bak et al. 2001). Interestingly, the jacalin-lectin domain containing protein JAL4 and the β-glucosidase BGLU11 were also up-regulated. Proteins of both families have been shown to be involved in glucosinolate degradation to produce antimicrobial compounds (Nagano et al. 2008). These findings support the idea that indole glucosinolates are involved in fighting *V. longisporum* in the vascular phase. The most striking finding, however, was the almost complete absence of a defence response in the susceptible NIL9 at the onset of flowering. Only 18 genes responded to *V. longisporum* infection in the hypocotyl and the shoot basis during this stage, and only three of them were defence-related (Häffner

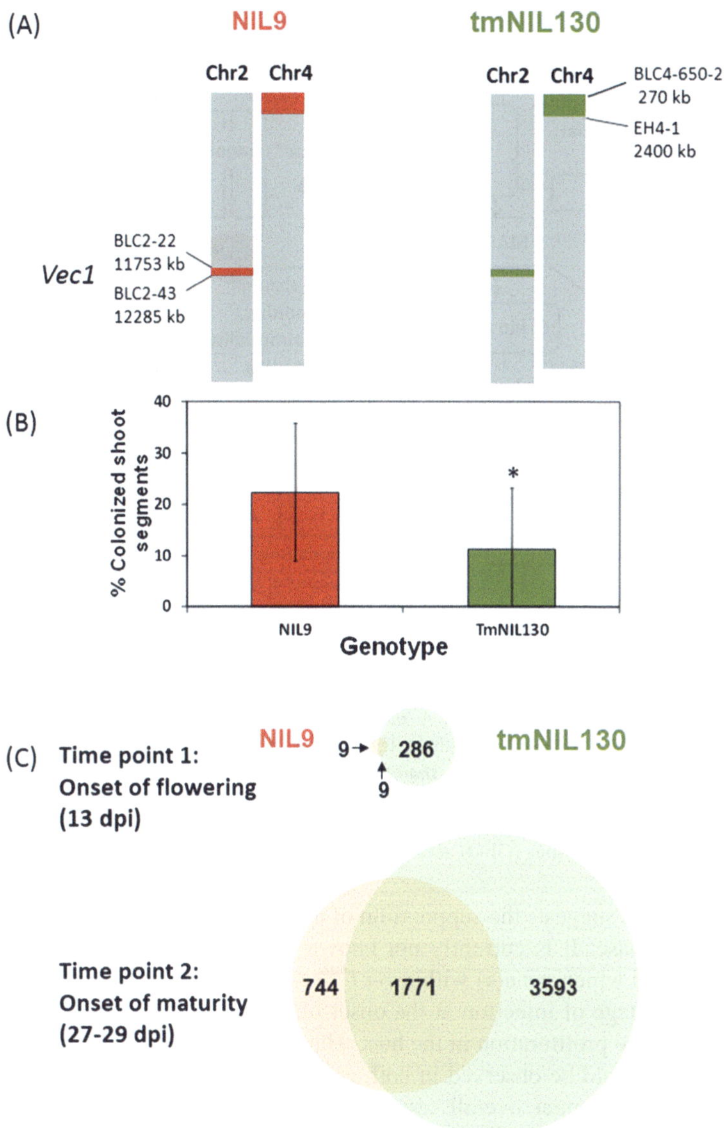

Fig. 2 Differential gene expression and systemic colonization after *V. longisporum* infection in two (Bur × L*er*) near-isogenic lines (NILs) differing in a region within the colonization resistance QTL *vec1*. (**a**) Genotype of NIL9 and tmNIL130. *Red parts* stand for L*er* alleles, *green parts* for Bur alleles in the variable regions. *Grey parts* are isogenic with respect to the tested marker loci. Names and physical positions in kilobases (kb) of markers delimiting variable regions are given next to the *bars* representing chromosomes. (**b**) Systemic colonization of NIL9 and tmNIL130 at the onset of silique maturity. $N = 12$, *t*-test. Samples were taken from 30 plants per replicate, among which were also the plants sampled for microarray analysis. (**c**) Genes differentially expressed by *V. longisporum* infection in the hypocotyl–shoot transition zone of NIL9 and tmNIL130 at two time points after infection. For growth, inoculation and RNA extraction protocol, see GEO accession GSE70021. Modified from Häffner et al. (2015)

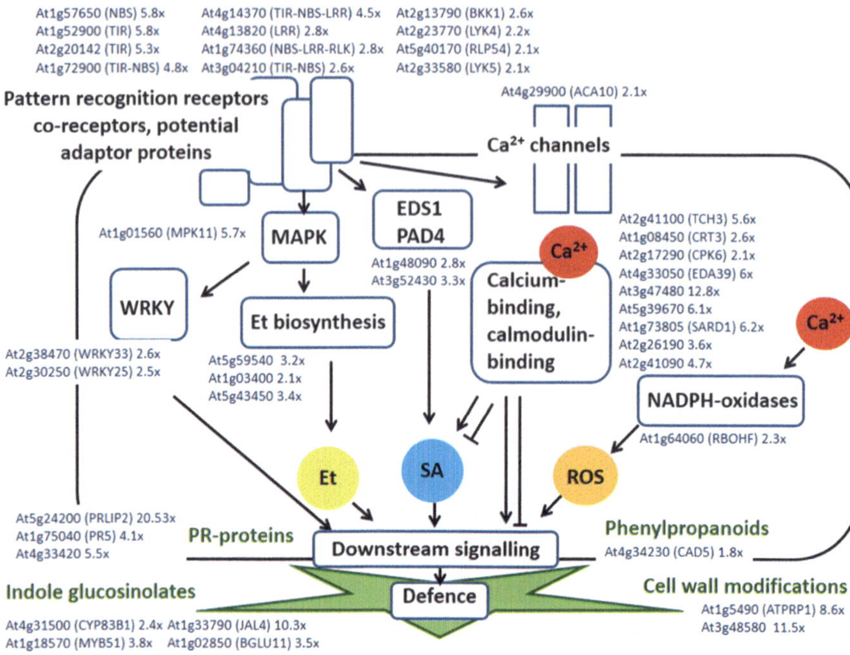

Fig. 3 A model for defence responses triggered by *V. longisporum* in the hypocotyl–shoot transition zone of colonization-resistant *A. thaliana* tmNIL130 at the onset of flowering. Genes and their up-regulation (fold change) upon *V. longisporum* infection are shown in *blue*. Gene assignment to biological roles in the *A. thaliana*–*V. longisporum* interaction is hypothetical and based on information from MAPMAN (Thimm et al. 2004) and The Arabidopsis Information Resource (TAIR). For a full record of differentially expressed genes and experimental procedures, see Gene Expression Omnibus (GEO) accession GSE70021

et al. 2015). This suggests the suppression of a defence reaction by the pathogen in the systemic phase. It is currently not known which fungal effector caused this suppression and which gene(s) within *vec1*^{Bur} could counteract it.

At the late stage of infection at the onset of silique maturity, when the fungus showed extensive proliferation in the host (Häffner et al. 2010), massive transcriptional changes could be observed in both genotypes. Still, the resistant tmNIL130 showed a much stronger overall response (Fig. 2c). Especially genes related to auxin metabolism, signalling and response and to the mitigation of oxidative stress responded much more strongly in the resistant NIL. This was interpreted as the capacity to exert a stricter control on damaging senescence-like processes that would benefit the pathogen and to keep up tissue viability and pathogen defence (Häffner et al. 2015).

3.3.4 RNA Silencing and Defence Signalling

Regulation of gene activity by small RNAs (sRNAs) is increasingly recognized as a mechanism that controls responses to pathogens (Voinnet 2008). sRNAs occur either as small interfering RNAs (siRNAs) or as microRNAs (miRNAs) mediating transcriptional or post-transcriptional gene silencing (TGS or PTGS; Pumplin and Voinnet 2013). Sequencing of sRNAs in cotton roots following *V. dahliae* infection showed that a resistant *G. barbadense* genotype had a different sRNA response pattern compared to a susceptible *G. hirsutum* genotype (Yin et al. 2012). PTGS has been shown to play an important role in defence against *V. dahliae* in *A. thaliana*, as mutants defective in PTGS were much more susceptible to the pathogen compared to wild type (Ellendorff et al. 2009). Interestingly, resistance to other necrotrophic and hemibiotrophic pathogens such as *Fusarium oxysporum* or *Alternaria brassicicola* was not compromised in the mutants. This suggests that a *Verticillium*-specific defence mechanism depends on PTGS (Ellendorff et al. 2009). Evidence exists that miRNAs are not only involved in defence but also in promoting the disease. For example, microRNA 482e (miR482e) of potato targets a CC–NBS–LRR resistance protein involved in mediating resistance to *V. dahliae*. Overexpression of miR482e greatly compromised resistance to *V. dahliae*. In wild-type plants, miR482e was downregulated in defence against *V. dahliae*, which led to the accumulation of the target resistance gene (Yang et al. 2015b). In other cases, disease-promoting microRNAs are manipulated by the pathogen to counteract host defence: *B. napus* miR168 has been shown to be strongly downregulated in *V. longisporum*-infected roots. This led to the induction of its target ARGONAUTE 1 (AGO1) which is required for *V. longisporum* development in the host (Shen et al. 2014). Presumably, AGO1 helps in suppressing host innate immunity by delivering sRNAs to targets with a role in pathogen defence. The necrotrophic pathogen *Botrytis cinerea* has even been shown to deliver such sRNAs into the host as pathogenicity factors (Weiberg et al. 2013).

3.4 Defence Strategies Based on Microbial Biocontrol Agents

Resistance to *Verticillium* can be greatly enhanced by beneficial microorganisms in the rhizosphere of the host. The meta-analysis of Bonanomi et al. (2010) confirmed that suppressiveness of soil amendments is most strongly correlated with the composition of the microbial community and especially with the presence of fluorescent pseudomonads and *Trichoderma* fungi. Biocontrol using selected microorganisms for host inoculation is therefore a promising approach to support plant health. In most cases where the biocontrol mechanism has been studied at the molecular level, the effect was rather due to the induction of host defences instead of a direct inhibitory effect on the pathogen. Diverse fungal and bacterial microorganisms have been shown to increase resistance of different hosts to *Verticillium*. The studies described in the following illustrate some facets of biocontrol agents (BCA) as an important belowground defence strategy.

An enormous diversity of potential biocontrol organisms against *V. dahliae* has been described for solanaceous hosts including the fungal root endophytes *Heteroconium chaetospira*, *Phialocephala fortinii* and species of *Penicillium*, *Fusarium* and *Trichoderma* (Narisawa et al. 2002). Colonization of eggplant and tomato roots with the arbuscular mycorrhizal fungus *Glomus mosseae* prevented fresh weight loss caused by *V. dahliae* from hosts (Karagiannidis et al. 2002). Pepper (*Capsicum annuum*) colonized by *Glomus deserticola* showed induction of acidic chitinases, superoxide dismutases and peroxidases and, after *V. dahliae* infection, also an increase of PAL and peroxidase activity in roots (Garmendia et al. 2006). Non-pathogenic *Fusarium oxysporum* 47 (Fo47) prevented fresh weight- and dry weight loss caused by *V. dahliae* from pepper plants (Veloso and Díaz 2012). This was associated with the increased induction of three defence genes (a PR1 protein, a sesquiterpene cyclase and a chitinase) in roots following *V. dahliae* infection compared to plants that were not colonized by Fo47 (Veloso and Díaz 2012). In potato, *Pseudomonas fluorescens* Biotype F isolate DF37 and *Bacillus pumilus* isolate M1 were successful in controlling *V. dahliae* wilt symptoms depending on the host genotype (Uppal et al. 2008). Colonization with these biocontrol agents was associated with accumulation of phenylpropanoids, especially the flavonol glycoside rutin. In eggplant, the biocontrol agents *Paenibacillus alvei* K165 and non-pathogenic *Fusarium oxysporum* F2, which reduced *V. dahliae* symptoms, induced PR1 and PR4 in the stems in a manner that depended on the rhizosphere size of the BCA population (Angelopoulou et al. 2014). Non-pathogenic *V. dahliae* Dvd E6 had a protective effect on tomato plants infected with pathogenic *V. dahliae* isolate VD1. When applied in advance of or together with VD1 infection, Dvd E6 almost completely excluded the pathogen from host roots. When applied after infection, both isolates competed at an equal basis (Shittu et al. 2009). Gene expression analysis suggests that Dvd E6 induced defence genes that were efficient in inhibiting VD1 colonization of the host.

A class of lipopeptides, iturins, shows high antifungal activity. A *Bacillus amyloliquefaciens* strain endophytic to cotton showed a high biocontrol efficacy against *V. dahliae* based on iturin production. Iturins not only had a direct toxic effect on *V. dahliae* but also induced PTI in cotton roots (Han et al. 2015). *Trichoderma viride*, another *Verticillium* BCA of cotton, led to increased terpenoid concentrations and peroxidase activity in seedling radicles (Hanson and Howell 2004).

In olive, the use of BCA is an important measure to control *V. dahliae*, complementing resistance breeding in an integrated approach that is needed to control its most important soil-borne pathogen (López-Escudero and Mercado-Blanco 2011). Aranda et al. (2011) have isolated rhizosphere microorganisms from wild olive and assessed the isolates for their biocontrol potential. About 14 % of the isolates had an antagonistic effect on *V. dahliae*. Typical compounds produced by the antagonists were indoleacetic acid (IAA) and siderophores, which are generally associated with growth promotion of the host and growth inhibition of pathogens, respectively (Arshad and Frankenberger 1993; Scher and Baker 1982). Furthermore, chitinolytic, lipolytic and proteolytic enzymes were produced that can potentially attack pathogenic fungi (Aranda et al. 2011). In an inoculation experiment with nursery material, root endophytic pseudomonads have proven to be effective BCA of *V. dahliae*. Growth promotion under *V. dahliae* challenge and symptom reduction was highest with the *Pseudomonas fluorescens* isolate PICF7. Pseudomonads even exerted an antagonistic effect on *V. dahliae* in vitro, which was, however, not correlated with the effect *in planta* (Mercado-Blanco et al. 2004). Microscopic studies with fluorescent *V. dahliae* and PICF7 showed that endophytic growth of PICF7 greatly inhibited root and xylem colonization by *V. dahliae*. Studying the underlying mechanisms in the model plant *A. thaliana* revealed that siderophore production was not required for the biocontrol effect. At least part of the effect was systemic, as root colonization by PICF7 also promoted resistance against *Botrytis cinerea* applied to leaves. This led to the conclusion that induced resistance contributes to biocontrol by PICF7.

Verticillium has been reported to be successfully controlled by BCA in Brassicaceae. Nejad and Johnson (2000) identified bacterial isolates that promoted growth and at the same time reduced symptoms from a Swedish *Verticillium* isolate from oilseed rape. *Paenibacillus alvei* K165, a plant growth promoting rhizobacterium, significantly reduced chlorosis caused by *V. longisporum* in *A. thaliana* (Tjamos et al. 2005). Since the BCA did not have a direct antagonistic effect on *V. longisporum*, induced resistance is the likely cause for the biocontrol effect. Molecular components which were necessary for induction of resistance were identified by mutant analysis and included SID1/EDS5, SID2/EDS16 and NPR1, which all act in the salicylic acid pathway. Consequently, the defence genes *Pr1*, *Pr2* and *Pr5* were most strongly activated in *V. longisporum* infected plants that were pretreated with K165 (Tjamos et al. 2005). Apart from bacteria, endophytic fungi also exerted a biocontrol effect against *V. longisporum*: the dark septate endophytic (DSE) fungi contain several potent BCAs. Two isolates of the DSE *Phialocephala fortinii* and a third unidentified DSE fungus reduced *V. longisporum* symptoms in Chinese cabbage up to 88 % (Narisawa et al. 2004). *Piriformospora indica*, which also belongs to the DSE fungi, is well known for its manifold beneficial effects on plant growth and health (Pham et al. 2008; Varma et al. 1999). *P. indica* protected *A. thaliana* from disease development through *V. dahliae*. Interestingly, *P. indica*-colonized plants that were infected with *V. dahliae* did not show the same degree of phytohormone accumulation and defence gene expression as infected plants without *P. indica* (Sun et al. 2014). This suggests that *P. indica* exerts its biocontrol effect via other mechanisms than induced resistance, possibly by a direct antagonistic effect. Indeed, *P. indica* inhibited growth of *V. dahliae* on agar plates (Sun et al. 2014).

4 Conclusions

The interaction of *Verticillium* spp. with their host plants is characterized by complexity in every respect. A great variety of symptoms is met by a diversity of defence mechanisms constituting quantitative resistance that relies on a complex genetic basis. Nevertheless, some common principles about belowground defences against *Verticillium* can be deduced from the research reviewed in this article:

(1) Hyphae of pathogenic *Verticillium* spp. are always capable of entering the host root cortex. Penetration resistance has not been observed so far. However, root colonization can be prevented or strongly reduced by beneficial endophytic or rhizosphere microorganisms. (2) The mature and intact endodermis is an impenetrable barrier for *Verticillium* spp., and infection of vascular tissue occurs via injuries or in young root tissues where the endodermis is not yet fully developed. (3) Induced defences are relying on a wide variety of signalling processes and lead to extensive proteomic and metabolomic changes that mostly take place in the xylem, ideally resulting in the elimination of the fungus from the xylem. Defence mechanisms are most strongly expressed in roots and hypocotyl, but are not restricted to these tissues. (4) In some cases, *Verticillium* is tolerated, and the host benefits from constrained or even suppressed defences.

Two main approaches in the control of *Verticillium* based on biological knowledge are resistance breeding and biocontrol. Many encouraging results have been obtained from experiments with biocontrol agents in various hosts. Researchers have started to study host prerequisites for biocontrol effects with experiments on defined mutants. Studying natural genetic variation of hosts with respect to their response towards biocontrol agents might lead to the identification of synergistic effects. By far, the most molecular knowledge about genetic resources of *Verticillium* resistance has been gained in cotton. There is a rich basis for combining different genes or QTL conferring *Verticillium* resistance in future breeding efforts. *Ve*-genes make an important contribution to quantitative resistance especially in solanaceous hosts and in cotton, but their effect needs to be complemented by other sources of resistance. In *Brassicaceae*, the host–pathogen interaction is well understood at the molecular level, mainly owing to numerous studies in the model plant *A. thaliana*. However, unlike in cotton, genetic variation leading to natural differences in *Verticillium* resistance is still poorly understood at the molecular level. *Ve*-like genes do not exist in crucifers, and only few QTL have been elucidated at the gene level. A more thorough understanding of how genetic variation leads to *Verticillium* resistance will greatly stimulate resistance breeding. Generally, translational approaches where homologues of known resistance genes from *A. thaliana* or cotton are studied in other crops should be extended. They may contribute to enhancing *Verticillium* resistance in crops where its genetic basis is still poorly understood.

Future research and applications can build upon a plethora of evidence from *Verticillium* research, which has received great impetus from molecular biology research within the last years. The high number of studies is more than justified to keep up with the complexity of the defence mechanisms. To achieve a high level of resistance, several defence mechanisms have to add up in each host. This illustrates the necessity of an integrated approach to achieve *Verticillium* control.

Acknowledgements We thank Sophia Harrand and Rebecca Werner for the excellent technical assistance. Funding from German Research Foundation (DFG, DI1502/3-1 grant) and the breeding company NPZ Norddeutsche Pflanzenzucht Hans-Georg Lembke KG is gratefully acknowledged.

References

Angelopoulou DJ, Naska EJ, Paplomatas EJ et al (2014) Biological control agents (BCAs) of Verticillium wilt: influence of application rates and delivery method on plant protection triggering of host defence mechanisms and rhizosphere populations of BCAs. Plant Pathol 63:1062–1069

Aranda S, Montes-Borrego M, Jiménez-Díaz RM et al (2011) Microbial communities associated with the root system of wild olives (*Olea europaea* L subsp *europaea* var *sylvestris*) are good reservoirs of bacteria with antagonistic potential against *Verticillium dahliae*. Plant Soil 343:329–345

Arias-Calderón R, León L, Bejarano-Alcázar J et al (2015) Resistance to Verticillium wilt in olive progenies from open-pollination. Sci Hortic 185:34–42

Arshad M, Frankenberger WT (1993) Microbial production of plant growth regulators. In: Blaine F, Metting JR (eds) Soil microbial ecology. Marcel and Dekker, New York, pp 307–347

Báidez AG, Gómez P, Del Río JA et al (2007) Dysfunctionality of the xylem in *Olea europaea* L plants associated with the infection process by *Verticillium dahliae* Kleb. Role of phenolic compounds in plant defense mechanism. J Agric Food Chem 55:3373–3377

Bak S, Tax FE, Feldmann KA et al (2001) CYP83B1 a cytochrome P450 at the metabolic branch point in auxin and indole glucosinolate biosynthesis in Arabidopsis. Plant Cell 13:101–111

Beckman CH (2000) Phenolic-storing cells: keys to programmed cell death and periderm formation in wilt disease resistance and in general defence responses in plants? Physiol Mol Plant Pathol 57:101–110

Bednarek P, Piślewska-Bednarek M, Svatoš A et al (2009) A glucosinolate metabolism pathway in living plant cells mediates broad-spectrum antifungal defence. Science 323:101–106

Benhamou N (1995) Ultrastructural and cytochemical aspects of the response of eggplant parenchyma cells in direct contact with *Verticillium*-infected xylem vessels. Physiol Mol Plant Pathol 46:321–338

Bishop CD, Cooper RM (1983) An ultrastructural study of root invasion in three vascular wilt diseases. Physiol Plant Pathol 22:15–27

Blasingame D, Patel MV (2005) Cotton disease loss estimate committee report. In: Proceedings of Beltwide cotton conference, pp 259–262

Bolek Y, El-Zik KM, Pepper AE et al (2005) Mapping of verticillium wilt resistance genes in cotton. Plant Sci 168:1581–1590

Bonanomi G, Antignani V, Capodilupo M et al (2010) Identifying the characteristics of organic soil amendments that suppress soilborne plant diseases. Soil Biol Biochem 42:136–144

Bowers JH, Nameth ST, Riedel RM et al (1996) Infection and colonization of potato roots by *Verticillium dahliae* as affected by *Pratylenchus penetrans* and *P. crenatus*. Phytopathology 86:614–621

Chai Y, Zhao L, Liao Z et al (2003) Molecular cloning of a potential *Verticillium dahliae* resistance gene SlVe 1 with multi-site polyadenylation from *Solanum licopersicoides*. DNA Seq 14:375–384

Chen XY, Chen Y, Heinstein P et al (1995) Cloning expression and characterization of (+)-δ-cadinene synthase: a catalyst for cotton phytoalexin biosynthesis. Arch Biochem Biophys 324:255–266

Chen JY, Huang JQ, Li NY et al (2015) Genome-wide analysis of the gene families of resistance gene analogues in cotton and their response to Verticillium wilt. BMC Plant Biol 15:148

Cui Y, Bell AA, Joost O et al (2000) Expression of potential defense response genes in cotton. Physiol Mol Plant Pathol 56:25–31

Daayf F (2015) Verticillium wilts in crop plants: pathogen invasion and host defence responses. Can J Plant Pathol 37:8–20

De Coninck B, Timmermans P, Vos C et al (2015) What lies beneath: belowground defence strategies in plants. Trends Plant Sci 20:91–101

de Jonge R, van Esse HP, Maruthachalam K et al (2012) Tomato immune receptor Ve1 recognizes effector of multiple fungal pathogens uncovered by genome and RNA sequencing. Proc Natl Acad Sci USA 109:5110–5115

de Jonge R, Bolton MD, Kombrink A et al (2013) Extensive chromosomal reshuffling drives evolution of virulence in an asexual pathogen. Genome Res 23:1271–1282

Dixon GR, Pegg GF (1969) Hyphal lysis and tylose formation in tomato cultivars infected by *Verticillium albo-atrum*. Trans Br Mycol Soc 53:109–118

Dixon RA, Achnine L, Kota P et al (2002) The phenylpropanoid pathway and plant defence—a genomics perspective. Mol Plant Pathol 3:371–390

Dong H, Cohen Y (2002) Dry mycelium of *Penicillium chrysogenum* induces resistance against Verticillium wilt and enhances growth of cotton plants. Phytoparasitica 30:147–157

Dunker S, Keunecke H, Steinbach P et al (2008) Impact of *Verticillium longisporum* on yield and morphology of winter oilseed rape (*Brassica napus*) in relation to systemic spread in the plant. J Phytopathol 156:698–707

Durrands PK, Cooper RM (1988) The role of pectinases in vascular wilt disease as determined by defined mutants of *Verticillium albo-atrum*. Physiol Mol Plant Pathol 32:363–371

El-Bebany AF, Rampitsch C, Daayf F (2010) Proteomic analysis of the phytopathogenic soilborne fungus *Verticillium dahliae* reveals differential protein expression in isolates that differ in aggressiveness. Proteomics 10:289–303

Ellendorff U, Fradin EF, de Jonge R et al (2009) RNA silencing is required for *Arabidopsis* defence against *Verticillium* wilt disease. J Exp Bot 60:591–602

Eynck C, Koopmann B, Grunewaldt-Stoecker G et al (2007) Differential interactions of *Verticillium longisporum* and *V. dahliae* with *Brassica napus* detected with molecular and histological techniques. Eur J Plant Pathol 118:259–274

Eynck C, Koopmann B, Karlovsky P et al (2009) Internal resistance in winter oilseed rape inhibits systemic spread of the vascular pathogen *Verticillium longisporum*. Phytopathology 99:802–811

Fei J, Chai Y, Wang J et al (2004) cDNA cloning and characterization of the *Ve* homologue gene *StVe* from *Solanum torvum* Swartz. DNA Seq 15:88–95

Fradin EF, Thomma BP (2006) Physiology and molecular aspects of *Verticillium* wilt diseases caused by *V. dahliae* and *V. albo-atrum*. Mol Plant Pathol 7:71–86

Fradin EF, Zhang Z, Ayala JCJ et al (2009) Genetic dissection of *Verticillium* wilt resistance mediated by tomato *Ve1*. Plant Physiol 150:320–332

Fradin EF, Abd-El-Haliem A, Masini L et al (2011) Interfamily transfer of tomato *Ve1* mediates *Verticillium* resistance in *Arabidopsis*. Plant Physiol 156:2255–2265

Frerigmann H, Gigolashvili T (2014) MYB34, MYB51 and MYB122 distinctly regulate indolic glucosinolate biosynthesis in *Arabidopsis thaliana*. Mol Plant 7:814–828

Gao W, Long L, Zhu LF et al (2013) Proteomic and virus-induced gene silencing (VIGS) analyses reveal that gossypol, brassinosteroids and jasmonic acid contribute to the resistance of cotton to *Verticillium dahliae*. Mol Cell Proteomics 12:3690–3703

Garmendia I, Aguirreolea J, Goicoechea N (2006) Defence-related enzymes in pepper roots during interactions with arbuscular mycorrhizal fungi and/or *Verticillium dahliae*. BioControl 51:293–310

Gayoso C, Pomar F, Novo-Uzal E et al (2010) The *Ve*-mediated resistance response of the tomato to *Verticillium dahliae* involves H_2O_2, peroxidase and lignins and drives PAL gene expression. BMC Plant Biol 10:232

Gladders P (2009) Relevance of verticillium wilt (*Verticillium longisporum*) in winter oilseed rape in the UK. HGCA Res Rev 72

Glazebrook J (2005) Contrasting mechanisms of defense against biotrophic and necrotrophic pathogens. Annu Rev Phytopathol 43:205–227

Göre ME, Erdoğan O, Altin N et al (2011) Seed transmission of Verticillium wilt of cotton. Phytoparasitica 39:285–292

Goud JKC, Termorshuizen AJ, Blok WJ et al (2004) Long-term effect of biological soil disinfestation on *Verticillium* wilt. Plant Dis 88:688–694

Griffiths D (1971) The development of lignitubers in roots after infection by *Verticillium dahliae* Kleb. Can J Microbiol 17:441–444

Häffner E, Karlovsky P, Diederichsen E (2010) Genetic and environmental control of the *Verticillium* syndrome in *Arabidopsis thaliana*. BMC Plant Biol 10:235

Häffner E, Karlovsky P, Splivallo R et al (2014) ERECTA salicylic acid abscisic acid and jasmonic acid modulate quantitative disease resistance of *Arabidopsis thaliana* to *Verticillium longisporum*. BMC Plant Biol 14:85

Häffner E, Konietzki S, Diederichsen E (2015) Keeping Control: the role of senescence and development in plant pathogenesis and defence. Plants 4:449–488

Halkier BA, Gershenzon J (2006) Biology and biochemistry of glucosinolates. Annu Rev Plant Biol 57:303–333

Han Q, Wu F, Wang X et al (2015) The bacterial lipopeptide iturins induce *Verticillium dahliae* cell death by affecting fungal signalling pathways and mediate plant defence responses involved in pathogen-associated molecular pattern-triggered immunity. Environ Microbiol 17:1166–1188

Hanson LE, Howell CR (2004) Elicitors of plant defense responses from biocontrol strains of *Trichoderma virens*. Phytopathology 94:171–176

Heale JB, Karapapa VK (1999) The *Verticillium* threat to Canada's major oilseed crop: canola. Can J Plant Pathol 21:1–7

Heinz R, Lee SW, Saparno A et al (1998) Cyclical systemic colonization in *Verticillium*-infected tomato. Physiol Mol Plant Pathol 52:385–396

Hoppenau CE, Tran VT, Kusch H et al (2014) *Verticillium dahliae* VdTHI4 involved in thiazole biosynthesis stress response and DNA repair functions is required for vascular disease induction in tomato. Environ Exp Bot 108:14–22

Huisman OC (1982) Interrelations of root growth dynamics to epidemiology of root-invading fungi. Annu Rev Phytopathol 20:303–327

Inderbitzin P, Subbarao KV (2014) *Verticillium* systematics and evolution: how confusion impedes *Verticillium* wilt management and how to resolve it. Phytopathology 104:564–574

Inderbitzin P, Bostock RM, Davis RM et al (2011) Phylogenetics and taxonomy of the fungal vascular wilt pathogen *Verticillium* with the descriptions of five new species. PLoS One 6: e28341

Iven T, König S, Singh S et al (2012) Transcriptional activation and production of tryptophan-derived secondary metabolites in *Arabidopsis* roots contributes to the defence against the fungal vascular pathogen *Verticillium longisporum*. Mol Plant 5:1389–1402

Jakse J, Cerenak A, Radišek S et al (2013) Identification of quantitative trait loci for resistance to *Verticillium* wilt and yield parameters in hop (*Humulus lupulus* L). Theor Appl Genet 126:1431–1443

James JT, Dubery IA (2001) Inhibition of polygalacturonase from *Verticillium dahliae* by a polygalacturonase inhibiting protein from cotton. Phytochemistry 57:149–156

Jiménez-Díaz RM, Millar RL (1988) Sporulation on infected tissues and presence of airborne Verticillium albo-atrum in alfalfa fields in New York. Plant Pathol 37:64–70
Joaquim TR, Rowe RC (1991) Vegetative compatibility and virulence of strains of *Verticillium dahliae* from soil and potato plants. Phytopathology 81:552–558
Johansson A, Staal J, Dixelius C (2006) Early responses in the *Arabidopsis-Verticillium longisporum* pathosystem are dependent on NDR1 JA-and ET-associated signals via cytosolic NPR1 and RFO1. Mol Plant Microbe Interact 19:958–969
Jones JD, Dangl JL (2006) The plant immune system. Nature 444:323–329
Kamble A, Koopmann B, von Tiedemann A (2013) Induced resistance to *Verticillium longisporum* in *Brassica napus* by β-aminobutyric acid. Plant Pathol 62:552–561
Karademir E, Karademir C, Ekinci R et al (2012) Effect of *Verticillium dahliae* Kleb on cotton yield and fiber technological properties. Int J Plant Prod 6:387–407
Karagiannidis N, Bletsos F, Stavropoulos N (2002) Effect of *Verticillium* wilt (*Verticillium dahliae* Kleb) and mycorrhiza (*Glomus mosseae*) on root colonization, growth and nutrient uptake in tomato and eggplant seedlings. Sci Hortic 94:145–156
Karajeh MR (2006) Seed transmission of *Verticillium dahliae* in olive as detected by a highly sensitive nested PCR-based assay. Phytopathol Mediterr 45:5–23
Karapapa VK, Bainbridge BW, Heale JB (1997) Morphological and molecular characterization of *Verticillium longisporum* comb. nov. pathogenic to oilseed rape. Mycol Res 101:1281–1294
Kawchuk LM, Hachey J, Lynch DR et al (2001) Tomato *Ve* disease resistance genes encode cell surface-like receptors. Proc Natl Acad Sci USA 98:6511–6515
Klosterman SJ, Atallah ZK, Vallad GE et al (2009) Diversity Pathogenicity and management of *Verticillium* species. Annu Rev Phytopathol 47:39–62
König S, Feussner K, Kaever A et al (2014) Soluble phenylpropanoids are involved in the defence response of *Arabidopsis* against *Verticillium longisporum*. New Phytol 202:823–837
Li C, He X, Luo X et al (2014) Cotton WRKY1 mediates the plant defense-to-development transition during infection of cotton by *Verticillium dahliae* by activating JASMONATE ZIM-DOMAIN1 expression. Plant Physiol 166:2179–2194
Liu SY, Chen JY, Wang JL et al (2013) Molecular characterization and functional analysis of a specific secreted protein from highly virulent defoliating *Verticillium dahliae*. Gene 529:307–316
López-Escudero FJ, Mercado-Blanco J (2011) *Verticillium* wilt of olive: a case study to implement an integrated strategy to control a soil-borne pathogen. Plant Soil 344:1–50
Luo X, Xie C, Dong J (2014) Interactions between *Verticillium dahliae* and its host: vegetative growth, pathogenicity, plant immunity. Appl Microbiol Biotechnol 98:6921–6932
Mace ME, Stipanovic RD, Bell AA (1985) Toxicity and role of terpenoid phytoalexins in *Verticillium* wilt resistance in cotton. Physiol Plant Pathol 26:209–218
Mandelc S, Timperman I, Radišek S et al (2013) Comparative proteomic profiling in compatible and incompatible interactions between hop roots and *Verticillium albo-atrum*. Plant Physiol Biochem 68:23–31
Markakis EA, Tjamos SE, Antoniou PP et al (2010) Phenolic responses of resistant and susceptible olive cultivars induced by defoliating and nondefoliating *Verticillium dahliae* pathotypes. Plant Dis 94:1156–1162
Mercado-Blanco J, Rodríguez-Jurado D, Pérez-Artés E et al (2002) Detection of the defoliating pathotype of *Verticillium dahliae* in infected olive plants by nested PCR. Eur J Plant Pathol 108:1–13
Mercado-Blanco J, Collado-Romero M, Parrilla-Araujo S et al (2003) Quantitative monitoring of colonization of olive genotypes by *Verticillium dahliae* pathotypes with real-time polymerase chain reaction. Physiol Mol Plant Pathol 63:91–105
Mercado-Blanco J, Rodrıguez-Jurado D, Hervás A et al (2004) Suppression of *Verticillium* wilt in olive planting stocks by root-associated fluorescent Pseudomonas spp. Biol Control 30:474–486

Millet YA, Danna CH, Clay NK et al (2010) Innate immune responses activated in *Arabidopsis* roots by microbe-associated molecular patterns. Plant Cell 22:973–990
Mol L (1995) Formation of microsclerotia of *Verticillium dahliae* on various crops. NJAS Wageningen J Life Sci 43:205–215
Nagano AJ, Fukao Y, Fujiwara M et al (2008) Antagonistic jacalin-related lectins regulate the size of ER body-type β-glucosidase complexes in *Arabidopsis thaliana*. Plant Cell Physiol 49:969–980
Narisawa K, Kawamata H, Currah RS et al (2002) Suppression of *Verticillium* wilt in eggplant by some fungal root endophytes. Eur J Plant Pathol 108:103–109
Narisawa K, Usuki F, Hashiba T (2004) Control of *Verticillium* yellows in Chinese cabbage by the dark septate endophytic fungus LtVB3. Phytopathology 94:412–418
Nejad P, Johnson PA (2000) Endophytic bacteria induce growth promotion and wilt disease suppression in oilseed rape and tomato. Biol Control 18:208–215
Novakazi F, Inderbitzin P, Sandoya G et al (2015) The three lineages of the diploid hybrid verticillium longisporum differ in virulence and pathogenicity. Phytopathology 105:662–673
Obermeier C, Hossain MA, Snowdon R et al (2013) Genetic analysis of phenylpropanoid metabolites associated with resistance against *Verticillium longisporum* in *Brassica napus*. Mol Breed 31:347–361
Olsson S, Nordbring-Hertz B (1985) Microsclerotial germination of *Verticillium dahliae* as affected by rape rhizosphere. FEMS Microbiol Ecol 31:293–299
Pegg GF (1974) *Verticillium* diseases. Rev Plant Pathol 53:82
Pegg GF, Brady BL (2002) *Verticillium* wilts. CABI, Wallingford
Pham GH, Singh A, Malla R et al (2008) Interaction of *Piriformospora indica* with diverse microorganisms and plants. In: Varma A, Abbott L, Werner D, Hampp R (eds) Plant surface microbiology. Springer, Heidelberg, pp 237–265
Pumplin N, Voinnet O (2013) RNA silencing suppression by plant pathogens: defence counter-defence and counter-counter-defence. Nat Rev Microbiol 11:745–760
Ralhan A, Schöttle S, Thurow C et al (2012) The vascular pathogen *Verticillium longisporum* requires a jasmonic acid-independent COI1 function in roots to elicit disease symptoms in *Arabidopsis* shoots. Plant Physiol 159:1192–1203
Ratzinger A, Riediger N, von Tiedemann A et al (2009) Salicylic acid and salicylic acid glucoside in xylem sap of *Brassica napus* infected with *Verticillium longisporum*. J Plant Res 122:571–579
Reusche M, Thole K, Janz D et al (2012) *Verticillium* infection triggers VASCULAR-RELATED NAC DOMAIN7-dependent de novo xylem formation and enhances drought tolerance in *Arabidopsis*. Plant Cell 24:3823–3837
Reusche M, Truskina J, Thole K et al (2014) Infections with the vascular pathogens *Verticillium longisporum* and *Verticillium dahliae* induce distinct disease symptoms and differentially affect drought stress tolerance of *Arabidopsis thaliana*. Environ Exp Bot 108:23–37
Robb J (2007) *Verticillium* tolerance: resistance susceptibility or mutualism? Botany 85:903–910
Robb J, Brisson JD, Busch L et al (1979) Ultrastructure of wilt syndrome caused by *Verticillium dahliae*. VII Correlated light and transmission electron microscope identification of vessel coatings and tyloses. Can J Bot 57:822–834
Robb J, Lee B, Nazar RN (2007) Gene suppression in a tolerant tomato-vascular pathogen interaction. Planta 226:299–309
Robb J, Shittu H, Soman KV et al (2012) Arsenal of elevated defense proteins fails to protect tomato against *Verticillium dahliae*. Planta 236:623–633
Roos J, Bejai S, Oide S et al (2014) RabGAP22 is required for defence to the vascular pathogen *Verticillium longisporum* and contributes to stomata immunity. PLoS One 9:e88187
Roy BA, Kirchner JW (2000) Evolutionary dynamics of pathogen resistance and tolerance. Evolution 54:51–63

Rygulla W, Snowdon RJ, Friedt W et al (2008) Identification of quantitative trait loci for resistance against *Verticillium longisporum* in oilseed rape (*Brassica napus*). Phytopathology 98:215–221

Sacristán S, García-Arenal F (2008) The evolution of virulence and pathogenicity in plant pathogen populations. Mol Plant Pathol 9:369–384

Scher FM, Baker R (1982) Effect of *Pseudomonas putida* and a synthetic iron chelator on induction of soil suppressiveness to *Fusarium* wilt pathogens. Phytopathology 72:1567–1573

Schnathorst WC (1981) Life-cycle and epidemiology of Verticillium. In: Mace ME, Bel AA, Beckman CH (eds) Fungal wilt diseases of plants. Academic, New York, pp 81–108

Schreiber LR, Green RJ (1963) Effect of root exudates on germination of conidia and microsclerotia of *Verticillium albo-atrum* inhibited by soil fungistatic principle. Phytopathology 53:260–264

Shen D, Suhrkamp I, Wang Y et al (2014) Identification and characterization of microRNAs in oilseed rape (*Brassica napus*) responsive to infection with the pathogenic fungus *Verticillium longisporum* using *Brassica* AA (*Brassica rapa*) and CC (*Brassica oleracea*) as reference genomes. New Phytol 204:577–594

Shittu HO, Castroverde DC, Nazar RN et al (2009) Plant-endophyte interplay protects tomato against a virulent *Verticillium*. Planta 229:415–426

Simko I, Costanzo S, Haynes KG et al (2004) Linkage disequilibrium mapping of a *Verticillium dahliae* resistance quantitative trait locus in tetraploid potato (*Solanum tuberosum*) through a candidate gene approach. Theor Appl Genet 108:217–224

Smit F, Dubery IA (1997) Cell wall reinforcement in cotton hypocotyls in response to a *Verticillium dahliae* elicitor. Phytochemistry 44:811–815

Spek J (1973) Seed transmission of *Verticillium dahliae*. Mededelingen van de Faculteit Landbouwwetenschappen Rijksuniversiteit Gent 38:1427–1434

Sun C, Shao Y, Vahabi K et al (2014) The beneficial fungus *Piriformospora indica* protects *Arabidopsis* from *Verticillium dahliae* infection by downregulation plant defense responses. BMC Plant Biol 14:268

Tai HH, Goyer C, De Koeyer D et al (2013) Decreased defense gene expression in tolerance versus resistance to *Verticillium dahliae* in potato. Funct Integr Genomics 13:367–378

Talboys PW (1958) Association of tylosis and hyperplasia of the xylem with vascular invasion of the hop by *Verticillium albo-atrum*. Trans Br Mycol Soc 41:249–260

Talboys PW (1972) Resistance to vascular wilt fungi. Proc R Soc Lond B Biol Sci 181:319–332

Thatcher LF, Manners JM, Kazan K (2009) *Fusarium oxysporum* hijacks COI1-mediated jasmonate signalling to promote disease development in Arabidopsis. Plant J 58:927–939

The Arabidopsis information resource (TAIR). http://www.arabidopsis.org. Accessed 21 Nov 2015

Thimm O, Bläsing O, Gibon Y et al (2004) MAPMAN: a user-driven tool to display genomics data sets onto diagrams of metabolic pathways and other biological processes. Plant J 37:914–939

Thomma BP, Nürnberger T, Joosten MH (2011) Of PAMPs and effectors: the blurred PTI-ETI dichotomy. Plant Cell 23:4–15

Timpner C, Braus-Stromeyer SA, Tran VT et al (2013) The Cpc1 regulator of the cross-pathway control of amino acid biosynthesis is required for pathogenicity of the vascular pathogen *Verticillium longisporum*. Mol Plant Microbe Interact 26:1312–1324

Tischner R, Koltermann M, Hesse H et al (2010) Early responses of *Arabidopsis thaliana* to infection by *Verticillium longisporum*. Physiol Mol Plant Pathol 74:419–427

Tjamos SE, Flemetakis E, Paplomatas EJ et al (2005) Induction of resistance to *Verticillium dahliae* in *Arabidopsis thaliana* by the biocontrol agent K-165 and pathogenesis-related proteins gene expression. Mol Plant Microbe Interact 18:555–561

Uppal AK, El Hadrami A, Adam LR et al (2008) Biological control of potato *Verticillium* wilt under controlled and field conditions using selected bacterial antagonists and plant extracts. Biol Control 44:90–100

Vallad GE, Subbarao KV (2008) Colonization of resistant and susceptible lettuce cultivars by a green fluorescent protein-tagged isolate of *Verticillium dahliae*. Phytopathology 98:871–885
Vallad GE, Bhat RG, Koike ST et al (2005) Weedborne reservoirs and seed transmission of *Verticillium dahliae* in lettuce. Plant Dis 89:317–324
Van Etten HD, Mansfield JW, Bailey JA et al (1994) Two classes of plant antibiotics: phytoalexins versus “phytoanticipins”. Plant Cell 6:1191
Van Loon LC, Rep M, Pieterse CMJ (2006) Significance of inducible defence-related proteins in infected plants. Annu Rev Phytopathol 44:135–162
Varma A, Verma S, Sahay N et al (1999) *Piriformospora indica*, a cultivable plant-growth-promoting root endophyte. Appl Environ Microbiol 65:2741–2744
Veloso J, Díaz J (2012) *Fusarium oxysporum* Fo47 confers protection to pepper plants against *Verticillium dahliae* and *Phytophthora capsici* and induces the expression of defence genes. Plant Pathol 61:281–288
Veronese P, Narasimhan ML, Stevenson RA et al (2003) Identification of a locus controlling *Verticillium* disease symptom response in *Arabidopsis thaliana*. Plant J 35:574–587
Vining K, Davis T (2009) Isolation of a *Ve* homolog *mVe1* and its relationship to *Verticillium* wilt resistance in *Mentha longifolia* (L) Huds. Mol Genet Genomics 282:173–184
Voinnet O (2008) Post-transcriptional RNA silencing in plant–microbe interactions: a touch of robustness and versatility. Curr Opin Plant Biol 11:464–470
Wang JY, Cai Y, Gou JY et al (2004) VdNEP an elicitor from *Verticillium dahliae* induces cotton plant wilting. Appl Environ Microbiol 70:4989–4995
Wang FX, Ma YP, Yang CL et al (2011) Proteomic analysis of the sea-island cotton roots infected by wilt pathogen *Verticillium dahliae*. Proteomics 11:4296–4309
Weiberg A, Wang M, Lin FM et al (2013) Fungal small RNAs suppress plant immunity by hijacking host RNA interference pathways. Science 342:118–123
Wildermuth MC, Dewdney J, Wu G et al (2001) Isochorismate synthase is required to synthesize salicylic acid for plant defence. Nature 414:562–565
Wilhelm S (1955) Longevity of the *Verticillium* wilt fungus in the laboratory and field. Phytopathology 45:180–181
Witzel K, Hanschen FS, Schreiner M et al (2013) *Verticillium* suppression is associated with the glucosinolate composition of *Arabidopsis thaliana* leaves. PLoS One 8:e71877
Witzel K, Hanschen FS, Klopsch R et al (2015) *Verticillium longisporum* infection induces organ-specific glucosinolate degradation in *Arabidopsis thaliana*. Front Plant Sci 6:508
Xu L, Zhu L, Tu L et al (2011) Lignin metabolism has a central role in the resistance of cotton to the wilt fungus *Verticillium dahliae* as revealed by RNA-seq-dependent transcriptional analysis and histochemistry. J Exp Bot 62:5607–5621
Xu L, Zhang W, He X et al (2014) Functional characterization of cotton genes responsive to *Verticillium dahliae* through bioinformatics and reverse genetics strategies. J Exp Bot 65:6679–6692
Yadeta KA, Thomma BP (2013) The xylem as battleground for plant hosts and vascular wilt pathogens. Front Plant Sci 4:97
Yang Y, Ling X, Chen T et al (2014) A Cotton *Gbvdr5* gene encoding a leucine-rich-repeat receptor-like protein confers resistance to *Verticillium dahliae* in transgenic *Arabidopsis* and Upland Cotton. Plant Mol Biol Rep. doi:10.1007/s11105-014-0810-5
Yang CL, Liang S, Wang HY et al (2015a) Cotton major latex protein 28 functions as a positive regulator of the ethylene responsive factor 6 in defense against *Verticillium dahliae*. Mol Plant 8:399–411
Yang L, Mu X, Liu C et al (2015b) Overexpression of potato miR482e enhanced plant sensitivity to *Verticillium dahliae* infection. J Integr Plant Biol 57(12):1078–1088. doi:10.1111/jipb.12348
Yao LL, Zhou Q, Pei BL et al (2011) Hydrogen peroxide modulates the dynamic microtubule cytoskeleton during the defence responses to *Verticillium dahliae* toxins in *Arabidopsis*. Plant Cell Environ 34:1586–1598

Yin Z, Li Y, Han X et al (2012) Genome-wide profiling of miRNAs and other small non-coding RNAs in the *Verticillium dahliae*-inoculated cotton roots. PLoS One 7:e35765

Yuan HY, Yao LL, Jia ZQ et al (2006) *Verticillium dahliae* toxin induced alterations of cytoskeletons and nucleoli in *Arabidopsis thaliana* suspension cells. Protoplasma 229:75–82

Zeise K, von Tiedemann A (2002) Host specialization among vegetative compatibility groups of *Verticillium dahliae* in relation to *Verticillium longisporum*. J Phytopathol 150:112–119

Zhang Y, Wang X, Yang S et al (2011) Cloning and characterization of a *Verticillium* wilt resistance gene from *Gossypium barbadense* and functional analysis in *Arabidopsis thaliana*. Plant Cell Rep 30:2085–2096

Zhang WW, Wang SZ, Liu K et al (2012a) Comparative expression analysis in susceptible and resistant *Gossypium hirsutum* responding to *Verticillium dahliae* infection by cDNA-AFLP. Physiol Mol Plant Pathol 80:50–57

Zhang WW, Jian GL, Jiang TF et al (2012b) Cotton gene expression profiles in resistant *Gossypium hirsutum* cv. Zhongzhimian KV1 responding to *Verticillium dahliae* strain V991 infection. Mol Biol Rep 39:9765–9774

Zhang B, Yang Y, Chen T et al (2012c) Island cotton *Gbve1* gene encoding a receptor-like protein confers resistance to both defoliating and non-defoliating isolates of *Verticillium dahliae*. PLoS One 7:e51091

Zhang Y, Wang XF, Ding ZG et al (2013a) Transcriptome profiling of *Gossypium barbadense* inoculated with *Verticillium dahliae* provides a resource for cotton improvement. BMC Genomics 14:637

Zhang Z, Fradin E, de Jonge R et al (2013b) Optimized agroinfiltration and virus-induced gene silencing to study *Ve1*-mediated *Verticillium* resistance in tobacco. Mol Plant Microbe Interact 26:182–190

Zhang Z, van Esse HP, Damme M et al (2013c) *Ve1*-mediated resistance against *Verticillium* does not involve a hypersensitive response in *Arabidopsis*. Mol Plant Pathol 14:719–727

Zheng Z, Qamar SA, Chen Z et al (2006) Arabidopsis WRKY33 transcription factor is required for resistance to necrotrophic fungal pathogens. Plant J 48:592–605

Zhou L, Hu Q, Johansson A et al (2006) *Verticillium longisporum* and *V dahliae*: infection and disease in *Brassica napus*. Plant Pathol 55:137–144

Belowground and Aboveground Strategies of Plant Resistance Against *Phytophthora* Species

Daigo Takemoto and Yuri Mizuno

Abstract The oomycete genus *Phytophthora* includes some of the most destructive plant pathogens in the world. Plant diseases caused by *Phytophthora* species have an extremely significant impact on a wide range of agriculturally important crops and plants in natural ecosystems such as trees and shrubs in forests. In this chapter, we will describe the infection processes and strategies of *Phytophthora* pathogens and the counter defence mechanisms of belowground and aboveground tissues of host plants.

1 Introduction

As the genus name implies, *Phytophthora* (phyto = plant and phthora = destroyer in Greek) species include a large number of the destructive plant pathogens. The most known pathogen in this genus is *P. infestans*, the potato late blight pathogen, causal agent of Irish potato famine in the 1840s (Fry 2008). Damage and associated control costs caused by potato late blight is estimated to be more than 1 billion € in Europe and $3 billion worldwide per year (Fry 2008; Haverkort et al. 2008). Root and stem rot of soybean caused by *P. sojae* is the most damaging and widespread disease of soybean, with an annual cost worldwide of $1–2 billion (Tyler 2007). Other *Phytophthora* species (e.g., *P. cactorum*, *P. cinnamomi*, *P. citrophthora*) cause root, crown, and collar rots on a wide range of fruit trees such as apples, citrus, cherries , peaches, pears, olives, and avocados (Erwin and Ribeiro 1996).

In addition to the impacts on agricultural production, many *Phytophthora* species are known as serious threats to trees and shrubs in natural ecosystems (Hansen et al. 2012). *P. cinnamomi* is an aggressive soilborne pathogen with an extremely wide host range, which includes over 3000 plant species (Hardham 2005). *P. cinnamomi* is the causal agent of ink disease in chestnuts, oak decline, little leaf disease in pines, dieback of eucalyptus, and many more. *P. ramorum* causes sudden oak death (or ramorum blight and dieback) in Tanoak (*Lithocarpus densiflorus*), Coast live oak (*Quercus agrifolia*), and other *Quercus* species

D. Takemoto (✉) • Y. Mizuno
Graduate School of Bioagricultural Sciences, Nagoya University, Nagoya, Japan
e-mail: dtakemo@agr.nagoya-u.ac.jp

C.M.F. Vos, K. Kazan (eds.), *Belowground Defence Strategies in Plants*, Signaling and Communication in Plants, DOI 10.1007/978-3-319-42319-7_7

(Grünwald et al. 2008). Destructive effects of *Phytophthora* species on natural trees affect other organisms in ecosystems such as animals and insects dependent on infected trees as foods and shelters.

Mechanisms of plant resistance against *Phytophthora* pathogens have been extensively investigated in pathosystems between Solanaceae plants and *P. infestans* and soybean and *P. sojae*. Recent advances on transcriptome, proteome, and metabolome analyses have opened up the opportunity for studies to understand the resistance mechanisms of trees to soilborne *Phytophthora* pathogens. In this chapter, we overview the infection processes and ingenious infection strategies employed by *Phytophthora* pathogens and the mechanisms of plant defence against infection by *Phytophthora* species.

2 Infection Process of *Phytophthora* Pathogens

Phytophthora species produce motile asexual spores, zoospores, which have two flagella to swim in flooded soil or on the wet surface of plant tissues (Fig. 1). Zoospores of *Phytophthora* species are attracted to amino acids (Deacon and Donaldson 1993); thus, the root exudates of any plant species can attract

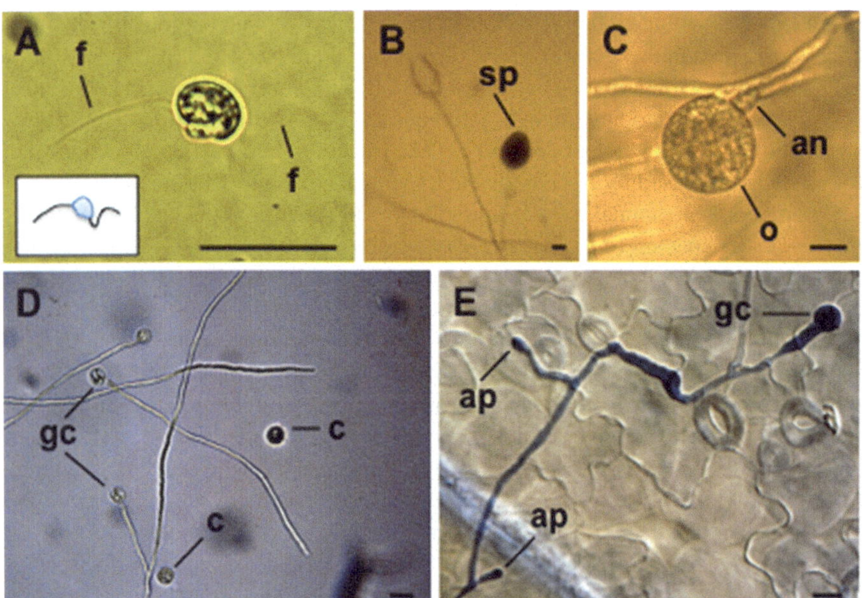

Fig. 1 (**a**) Zoospore of *Phytophthora sojae*. *f* flagellum. (**b**) Sporangia of *P. sojae*. *sp* sporangium. (**c**) Oospores production by *P. sojae*. *o* oogonium, *an* antheridium. (**d**) Encystment and germination of *P. sojae* cysts. *c* cyst, *gc* germinated cyst. (**e**) Penetration attempt by *P. sojae* on the leaf surface of *Arabidopsis thaliana*. *gc* germinated cyst, *ap* appressorium-like swelling. *Bars* = 10 μm

Phytophthora zoospores, though there are some reports showing that metabolites produced by particular plant species can attract specific *Phytophthora* species. For example, isoflavones in the root exudates of soybean, daidzein and genistein, specifically attract zoospores of *P. sojae* (Morris and Ward 1992; Tyler et al. 1996). Once zoospores reach the surface of plant roots, they rapidly produce cell walls, lose flagella, and become round shaped and adhesive (encystment, Fig. 1). On the surface of roots, attached cysts germinate and form appressorium-like swellings on the junction of the epidermal cells and penetrate mainly between the anticlinal walls of the root cells (Enkerli et al. 1997, Fig. 2). Recently, the *RAM2* gene of barrel medic (*Medicago truncatula*) was identified as an essential gene for the appressoria-mediated root infection of *P. palmivora* as well as root colonization by mycorrhizal fungi (Wang et al. 2012). *RAM2* encodes a glycerol-3-phosphate acyl transferase (GPAT) involved in the production of cutin monomers. Expression of potato *RAM2* was enhanced upon infection with *P. infestans*, suggesting that the cutin monomer acts as a plant signal that promotes the invasion of *Phytophthora* species into plant tissues (Kaschani et al. 2010; Wang et al. 2012).

In susceptible plants, intracellular hyphae produce a large number of haustoria in root cortical cells and the penetrating hypha further invade into the vascular tissues (Enkerli et al. 1997, Fig. 2). In resistant plants, thickening of the cortical cell walls, wall appositions, collapse of cortical cells, and accumulation of osmophilic granules are observed around penetrating hyphae (Oh and Hansen 2007).

Though the majority of *Phytophthora* species are soilborne pathogens, there are some airborne *Phytophthora* species (e.g., *P. infestans*) that utilize sporangia as the primary source of propagation. Sporangia are produced on the top of aerial hyphae of *P. infestans* and transferred via wind or insects. Sporangia can germinate directly (Fig. 1) or release zoospores on the wet surface of a host leaf, and germinated

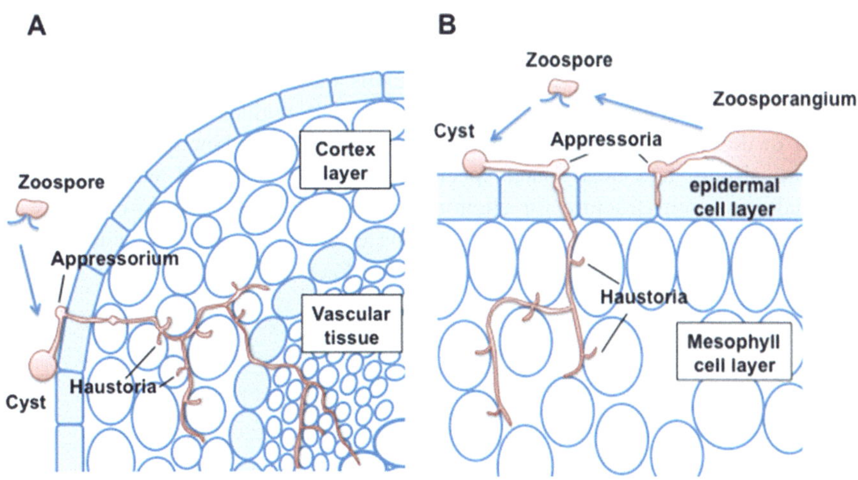

Fig. 2 Representative infection processes of *Phytophthora* pathogens to root (**a**) and leaf (**b**) tissues of host plant

sporangia or cysts form appressorium-like swelling on the surface of leaf epidermal cells (Figs. 1 and 2). Hyphae reaching to mesophyll cell layer form haustoria in contacting cells.

Most of the *Phytophthora* species are hemibiotrophic pathogens, which form haustoria to uptake nutrients from living plant cells in the early stages of infection (biotrophic phase) and become necrotrophic in the later stages of plant colonization. Haustoria also act as the center of production of virulence factors (effectors), which can suppress the defence mechanisms of host plants (see Sect. 5).

3 *Phytophthora*-Derived Molecules Recognized by Plant

One of critical process for plants to induce effective defence responses against pathogens is the recognition of molecules derived from microorganisms. Conserved molecules of potential pathogens, called pathogen-associated molecular patterns (PAMPs), are recognized by plant cells to induce the innate immunity of host plants (Jones and Takemoto 2004). Several molecules derived from mycelial walls or secretory proteins of *Phytophthora* and related species could act as PAMPs of oomycete pathogens.

Twenty-carbon poly-unsaturated fatty acids derived from the cell wall of *P. infestans*, eicosapentaenoic acid and arachidonic acid, elicit production of phytoalexins in potato (Bostock et al. 1981). Eicosapentaenoic acid and arachidonic acid are generally not found in plant tissues but are abundant in *Phytophthora* species. In the early stages of infection into a host plant, these fatty acids are released from spores of *Phytophthora* (Ricker and Bostock 1994). Glucans derived from cell walls of *Phytophthora* species have the activity to elicit or enhance defence responses of the host plant. Arachidonic acid alone can induce active defence reactions, but glucans from *P. infestans*, inactive as elicitors, enhanced accumulation of the sesquiterpenoid phytoalexins and defence response induced by arachidonic acid (Preisig and Kuć 1985). Cell wall β-glucan of *P. sojae* is the elicitor of defence responses in a wide range of Fabaceae plant species including soybean, alfalfa, and other plant species such as tobacco and sunflower. The essential minimum structure for elicitor activity of *P. sojae* glucan elicitor was determined as β-1,6-1,3 heptaglucan (Cheong et al. 1991). Such principal molecules in cell wall of *Phytophthora* species act as PAMPs for the induction of plant defence.

4 Apoplastic Elicitor Proteins Produced by *Phytophthora* Species

A large number of secretory proteins are produced by *Phytophthora* species in the apoplast of a host plant during the infection process, and some of them act as elicitor molecules for defence induction in host plants. Elicitins are the sterol-binding proteins secreted by *Phytophthora* and *Pythium* species. Studies determining the three-dimensional solution structure of elicitins from *P. cryptogea* (cryptogein) and *P. cinnamomi* (β-cinnamomin) revealed that the hydrophobic core of elicitins would have the capacity to capture sterols derived from the plasma membrane of host plant (Boissy et al. 1996, 1999). Given that *Phytophthora* species cannot produce sterols, elicitins are probably essential factors for their growth in host plants as the scavenger of phytosterols. *Phytophthora* species have multiple genes for elicitins and elicitin-like proteins (Tyler 2002). Class I elicitins (e.g., INF1 for *P. infestans* and cryptogein for *P. cryptogea*) are generally secreted most abundantly in culture and have robust elicitor activity for a limited range of plant species including most of the *Nicotiana* species, some cultivars of *Brassica*, *Raphanus* species, and a few *Solanum* species (Kamoun et al. 1993; Takemoto et al. 2005; Vleeshouwers et al. 2006). Usually, responsive plants can recognize elicitins from different *Phytophthora* species; thus, elicitins have a conserved molecular pattern of *Phytophthora* and *Pythium* species. Gene silencing of the elicitin *inf1*, which enhanced the virulence of *P. infestans* on the non-host *Nicotiana benthamiana*, indicated that elicitins are avirulence factors for responsive plant species (Kamoun et al. 1998). Recently, a gene for the receptor of elicitins, elicitin response (*ELR*), was identified from elicitin-responsive genotype of *Solanum microdontum* (Du et al. 2015). ELP is a receptor-like protein, structurally similar to the tomato R proteins Cf9 and Cf2 for resistance to *Cladosporium fulvum* and Ve1 for *Verticillium* resistance (Jones et al. 1994; Dixon et al. 1996; Kawchuk et al. 2001). Introduction of ELR to the highly susceptible potato cultivar Désirée enhanced resistance to *P. infestans*, indicating that ELR is an extracellular pattern recognition receptor for *Phytophthora* elicitins (Du et al. 2015).

NPP1 of *P. parasitica*, PsojNIP of *P. sojae*, and NPP1.1 of *P. infestans* are members of the Nep1-like proteins (NLPs), which induce cell death in dicotyledonous, but not in monocotyledonous plants (Fellbrich et al. 2002; Qutob et al. 2002). Induction of cell death by NLPs facilitates the virulence of some pathogens, including *P. parasitica* and *Pythium aphanidermatum*. As elicitors of plant defence, it is expected that the cell death-inducing activity of NLPs may induced the release of immunogenic damage-associated molecular patterns (DAMPs) from plant cells (Ottmann et al. 2009). Homologue of NLPs can be found in some species of bacteria, fungi, and oomycete, including plant symbiotic fungi, insect pathogens, and animal-related fungi (Oome and Van den Ackerveken 2014). A large number of *NLP* genes can be identified from the genome sequences of *Phytophthora* species. In the genome of *P. sojae*, 33 *NLP*-like genes were predicted, and the expression of 20 genes was detected. However, only 8 out of

19 *P. sojae* NLP have cell death induction activity (Dong et al. 2012). Expression of many nontoxic NLPs were induced during biotrophic stage, whereas genes for cell death-inducing NLPs were expressed during the necrotrophic phase, probably indicating the functional diversification of NLPs (Judelson et al. 2008; Dong et al. 2012).

A 42-kDa cell wall glycoprotein, GP42, was isolated from the cell wall of *P. sojae* as an elicitor protein of parsley suspension cells (Parker et al. 1991). GP42 is a calcium-dependent transglutaminase conserved among *Phytophthora* species. A sequence of C-terminal 13 amino acids, Pep-13, is highly conserved among *Phytophthora* species and is essential and sufficient for the elicitor activity. This indicates that this peptide is a conserved molecular pattern in *Phytophthora* species (Nürnberger et al. 1994).

Cellulose-binding elicitor lectin, CBEL, is another cell wall glycoprotein with elicitor activity isolated from root rot pathogen of tobacco *P. parasitica* (Séjalon-Delmas et al. 1997). CBEL has elicitor activity in a variety of plant species including tobacco (Solanaceae), *Arabidopsis* (Brassicaceae), French bean (Fabaceae), and Zinnia (Asteraceae) (Khatib et al. 2004). CBEL of *P. parasitica* has been shown to be required for the organization of the hyphal cell walls (Gaulin et al. 2002). Homologues of CBEL were identified from various *Phytophthora* species, and the highly conserved cellulose-binding domain (CBD) of CBEL is sufficient for the induction of defence responses (Gaulin et al. 2006). Therefore, CBD is considered as a PAMP in *Phytophthora* species.

P. parasitica OPEL was recently identified as an elicitor in *Nicotiana* species (Chang et al. 2015). OPEL can induce a series of defence responses such as HR-like cell death, callose deposition, and ROS production. Application of OPEL can induce the resistance of tobacco to a wide range of pathogens, including virus, bacteria, and oomycete. OPEL has a signature motif in active site of laminarinases, ExDxxE, which is probably essential for the enzymatic activity of OPEL. This conserved motif is also required for the elicitor activity (Chang et al. 2015). As OPEL is an oomycete-specific secretory protein, the laminarinases domain of OPEL is another conserved molecular pattern of oomycete, but elicitor activity of OPEL homologues from other oomycete species have not been tested.

5 RXLR Effectors of *Phytophthora* Species

Phytophthora species produce a large number of secretary proteins with a conserved RXLR-dEER motif, called RXLR effectors (Bozkurt et al. 2012). Approximately, 560, 400, and 350 genes for potential RXLR effectors are identified from the genome of *P. infestans*, *P. sojae*, and *P. ramorum*, respectively (Haas et al. 2009). RXLR effectors secreted from haustoria of *Phytophthora* are translocated from the extrahaustorial matrix into the cytoplasm of host cells and targeted to the site of their actin in plant cells. RXLR effectors suppress a wide range of plant mechanisms for disease resistance. *P. infestans* Avr3a stabilizes and modifies the activity of an E3 ligase of the host plant, while CMPG1 is required for the induction of cell death by the plant. Disease symptoms caused by *P. infestans* were significantly reduced by the suppression of *Avr3a* (Bos et al. 2010), indicating the crucial role of Avr3a in the pathogenicity of *P. infestans*. AVRblb1 and AVRblb2 are RXLR effectors highly conserved among strains of *P. infestans* (Vleeshouwers et al. 2008; Oh et al. 2009). The host target of AVRblb1 is the lectin receptor kinase LecRK-I.9, a putative mediator of cell wall–plasma membrane adhesions. The expected function of AVRblb1 as a virulence factor is the destabilization of the interaction between the host cell wall and plasma membrane continuum (Bouwmeester et al. 2011). AVRblb2 suppress the secretion of a host immune cysteine protease C14 at the haustorial interface to promote infection (Bozkurt et al. 2011). *P. infestans* PexRD2 is an interactor of MAPKKKε of Solanaceae plants, a positive regulator of cell death for plant immunity. Expression of *PexRD2* or gene silencing of *MAPKKKε* in *N. benthamiana* enhanced disease symptoms caused by *P. infestans* (King et al. 2014). Another RxLR effector of *P. infestans*, Pi03192, directly interacts with the host's NAC transcriptional factors NTP1 and NTP2 and inhibits their translocation from the ER membrane to the nucleus, which is required for disease resistance (McLellan et al. 2013).

P. sojae Avr3b is an ADP-ribose/NADH pyrophosphorylase, which suppresses the resistance reaction of *N. benthamiana*. Silencing of *Avr3b* compromised the virulence of *P. sojae* on susceptible soybean cultivar, suggesting that Avr3b is an essential virulence factor for *P. sojae* (Dong et al. 2011). *P. sojae* PSR1 and PSR2 (*Phytophthora* suppressors of RNA silencing) are inhibitors of the biogenesis of small RNAs (Qiao et al. 2013). PSR1 can bind to a host nuclear protein PINP1, which contains a RNA helicase domain. The localization of the dicer protein complex in the nucleus is impaired in *PSR1*-expressing or *PINP1*-silenced cells, indicating that PSR1 targets PINP1 to disturb the assembly of dicing complexes (Qiao et al. 2015).

Isolated avirulence proteins of *P. infestans* (e.g., Avr1, Avr2, Avr3a, Avr4, Avrblb1, and Avrblb2) and *P. sojae* (e.g., Avr1a, Avr1b, Avr3a/5, Avr3c, and Avr4/6) so far have been identified as RXLR effectors (Birch et al. 2009). Despite the diversity of functions of RXLR effectors as virulence factors, plant resistance (R) proteins for effector-induced defence are generally coiled coil domain nucleotide-binding site-leucine-rich repeat (CC-NBS-LRR) or Toll/interleukin-1 receptor domain (TIR)-NBS-LRR type proteins. Generally, *Phytophthora* resistances of potato and soybean determined by R proteins are effective in both aboveground and belowground tissues (Fig. 3).

Fig. 3 Resistance of potato leaf and tuber determined by *R* gene. (*Top*) Leaves of potato cv. Rishiri (*R1*) are inoculated with *Phytophthora infestans* isolate PI0-1 (race 0, *left panel*) or isolate PI1234-1 (race 1.2.3.4, *right panel*). (*Bottom*) Tubers of potato cvs. Rishiri (*R1*, Resistant) and Irish cobbler (*r*, Susceptible) are inoculated with *Phytophthora infestans* isolate PI0-1 (race 0)

6 Resistance Mechanisms of Potato Against *Phytophthora infestans*

Interactions between potato tubers and *P. infestans* have been used as a model system to investigate the plant defence responses against *Phytophthora* species. The susceptibility and resistance of potato tubers against *P. infestans* are principally determined by the combination of R proteins of potato and avirulence factors (effectors) of the pathogen (Fig. 3). There are several *R* genes, encoding NBS-LRR type resistance proteins, cloned from *Solanum* species, including *R1*, *R2*, *R3a*, and *R3b* from *S. demissum* for race-specific resistance and *RB*/*Rpi-blb1* and *Rpi-blb2* of *S. bulbocastanum* for broad-spectrum resistance to *P. infestans* (Ballvora et al. 2002; Song et al. 2003; van der Vossen et al. 2003, 2005; Huang et al. 2005; Lokossou et al. 2009; Li et al. 2011). Where a potato cultivar shows resistance to *P. infestans*, rapid responses of potato tuber cells are induced almost immediately after the invasion of the pathogen. One of rapid responses is the production of reactive oxygen species (ROS) (Doke 1983). Change of plant cytoplasmic streaming is also promptly induced at the invasion sites of *P. infestans*, resulting in the translocation of cellular components to the site of pathogen attack (Tomiyama 1956). Such quick defence responses are followed by induction of programmed cell death (hypersensitive cell death) and production of sesquiterpenoid phytoalexin, risitin (Kitazawa and Tomiyama 1969; Ishizaka et al. 1969). The death of infected cells and accumulation of phytoalexins together restrict the further growth of an invading pathogen.

Salicylic acid (SA) is recognized as an important signaling factor for the induction of plant disease resistance in a wide range of plant species (Vlot et al. 2009). In *Arabidopsis*, mutations of SID2/ICS1, which encodes an enzyme for SA production, or heterologous expression of *NahG* (gene for bacterial salicylate hydroxylase), reduces resistance against bacterial and oomycete pathogens (Delaney et al. 1994; Wildermuth et al. 2001). In contrast, potato plants expressing *NahG* didn't show any significant effect on the development of disease symptoms caused by *P. infestans*, although the expression of *NahG* increased the biomass of *P. infestans* in potato (Yu et al. 1997; Halim et al. 2007). Pep-13-induced resistance reactions such as hypersensitive cell death and ROS production are impaired in potato expressing *NahG*, indicating that SA is a key regulator for the induction of potato resistance to a PAMP of *Phytophthora*. Silencing of genes for jasmonic acid (JA) production, such as allene oxide cyclase and 12-oxophytodienoic acid reductase, compromised Pep-13-induced accumulation of ROS and hypersensitive cell death. Therefore, both SA and JA signaling are involved in PAMP responses and basal defence of potato against *P. infestans* (Halim et al. 2009).

Studies employing gene silencing or overexpression of target genes identified several potato genes involved in defence against *P. infestans*. Du et al. (2013) performed virus-induced gene silencing (VIGS) of candidate potato genes highly expressed during the infection of *P. infestans*. Several genes including a lipoxygenase and a suberization-associated anionic peroxidase were identified as genes involved in the resistance of potato against *P. infestans*. (Du et al. 2013). Transient expression of *StPRp27*, encoding a secreted protein, in potato as well as in *Nicotiana benthamiana* enhanced resistance to *P. infestans* indicating its potential contribution to disease resistance. However, gene silencing of *PRp27* homologues in *N. benthamiana* showed no effects on the resistance conferred by R proteins, suggesting that StPRp27 contributes to race-nonspecific resistance against *P. infestans* (Shi et al. 2012). Xyloglucan-specific endoglucanase inhibitors (XEIP) located in the extracellular regions of the plant are often embedded in the cell wall. Silencing of *XEIP* resulted in a significant increase in lesion size and water-soaked disease symptoms caused by *P. infestans* (Jones et al. 2006).

7 Resistance of *Nicotiana benthamiana* Against *Phytophthora infestans*

N. benthamiana is commonly used as a model Solanaceae host plant for *P. infestans*. Infection attempts by encysted zoospores or sporangia of *P. infestans* are generally stopped on the surface of mature *N. benthamiana* plants before penetration, through the induction of a few HR-like cell death events in the epidermal cells (Shibata et al. 2010). This is in contrast to the frequent penetration and induction of HR cell death on potato leaves inoculated with a zoospore suspension of *P. infestans* (Kitazawa and Tomiyama 1969, Fig. 3), implying that preinvasion resistance plays a key role in the resistance of *N. benthamiana* against *P. infestans*. *Nicotiana* species produce the sesquiterpenoid phytoalexins such as capsidiol (Bailey et al. 1975). Silencing of *NbEAS* (5-epi-aristrochen synthase) and *NbEAH* (5-epi-aristrochen dihydroxylase) genes of specialized enzymes for capsidiol production significantly compromises the resistance of *N. benthamiana* to *P. infestans* (Shibata et al. 2010). Silencing of *NbEIN2* (ethylene insensitive 2), a gene required for ethylene signaling, resulted in the suppression of *NbEAS* and *NbEAH* expression, and subsequent capsidiol production, indicating that the production of this phytoalexin is regulated by ethylene in *N. benthamiana* (Shibata et al. 2010; Ohtsu et al. 2014). A gene for plant-specific calreticulin *NbCRT3* was isolated as a required gene for resistance to *P. infestans*. *NbCRT3* encodes an endoplasmic reticulum (ER) quality control chaperone for the maturation of secreted glycoproteins. Several recent reports indicated that plant CRT3 is required for the maturation and stable accumulation of cell surface receptors; thus, it was expected that extracellular LRR receptor(s) are involved in the recognition of elicitors derived from *P. infestans* in *N. benthamiana* (Matsukawa et al. 2013). Consistently, the receptor-like kinase SERK3/BAK1 is also isolated as an essential factor for the resistance of *N. benthamiana* to *P. infestans* (Chaparro-Garcia et al. 2011). Functional analyses of *P. infestans* RXLR effectors identified several factors of *N. benthamiana* involved in disease resistance (as the virulence targets of effectors), including the E3 ligase CMPG1, the lectin receptor kinase LecRK-I.9, the cysteine protease C14, MAPKKKε, the NAC transcription factors NTP1 and NTP2, and machineries for biogenesis of small RNAs (see Sect. 5.5).

8 Resistance Mechanisms of Soybean Against Root Rot Pathogen *P. sojae*

Resistance of soybean to the root rot pathogen *P. sojae* is generally determined by R proteins encoded by *Rps* genes that provide effective resistance against *P. sojae* races with corresponding *Avr* genes. Fourteen *Rps* genes have been identified (Grau et al. 2004). Cloned *Rps* genes so far encode NBS-LRR type disease resistance proteins. Two functional Rps1k (Rps1k-1 and Rps1k-2) were identified from the

Rps1k locus, which encode CC-NBS-LRR class R proteins. (Bhattacharyya et al. 2005). Of these, *Rps2* encodes a TIR-NBS-LRR class R protein (Graham et al. 2002), whereas Rps4 (identified form the *Rps1* locus) is a CC-NBS-LRR resistance protein (Sandhu et al. 2004). The *Rps* gene-based resistance in soybean is usually effective for roots as well as aboveground tissues, but Rps2 confers incomplete resistance only in roots (Kilen et al. 1974). Some soybean cultivars have partial resistance determined by dominant *R* genes. Partial resistance is effective against all races of *P. sojae* (Dorrance et al. 2003).

Soybean roots and aboveground tissues produce isoflavonoid phytoalexin, glyceollin, after inoculation with *P. sojae*, or treatment with the β-glucan elicitor (Ayers et al. 1976; Ebel and Grisebach 1988). Production of glyceollin is positively and negatively regulated by ethylene and abscisic acid (ABA), respectively (Yoshikawa et al. 1990; Mohr and Cahill 2001). Application of norflurazon, an ABA biosynthesis inhibitor, to susceptible soybean enhanced the production of glyceollin and reduced the disease symptom caused by *P. sojae*, whereas ABA treatment to the resistant soybean cultivar reduced glyceollin accumulation and resistance to *P. sojae*. Given ABA treatment did not change the induction of HR in resistant soybean inoculated with *P. sojae*, glyceollin plays the most important role in the resistance of soybean against *P. sojae* (Mohr and Cahill 2001). Some fungal pathogens of soybean, such as *Colletotrichum truncatum* and *Rhizoctonia solani*, can detoxify glyceollin, but *P. sojae* cannot metabolize this phytoalexin. Consistently, the growth of *P. sojae* is significantly inhibited by glyceollin (Lygin et al. 2010). Silencing of genes for isoflavone synthase (*IFS*) or chalcone reductase (*CHR*), encoding enzymes for isoflavonoids production, compromised *Rps*-mediated resistance of soybean, further supporting the importance of isoflavonoids in the resistance of soybean to *P. sojae* (Graham et al. 2007).

Recently, the roles of small RNAs in soybean resistance against *P. sojae* were reported. MicroRNAs (miRNAs) miR393 and miR166 are induced by heat-treated *P. sojae* hyphae in soybean roots. Silencing of miR393 causes reduction of genes for glyceollin biosynthesis and enhances susceptibility of soybean roots to *P. sojae*. These data suggest that miR393 promotes soybean defence against *P. sojae*. Infection of *P. sojae* also increases the accumulation of phased siRNAs generated from genes encoding NB-LRR proteins and genes for pentatricopeptide repeat-containing proteins. Thus, specific miRNAs and phasiRNAs are involved in the regulation of defence genes in soybean during attack by *P. sojae* (Wong et al. 2014). Interestingly, RXLR effectors of *P. sojae* (PSR1 and PSR2) prevent the biogenesis of small RNAs (Qiao et al. 2013, see above). Given that homologous effectors of PSR2 can be identified from various *Phytophthora* species, regulation of defence genes by small RNAs is probably a common key event for the induction of plant resistance.

9 Resistance Mechanisms of Trees Against *Phytophthora* Species

In contrast to potato–*P. infestans* and soybean–*P. sojae* interactions, little is known about the resistance mechanisms of trees against *Phytophthora* species, but there are several reports indicating the importance of antimicrobial compounds produced by host plants. *P. citrophthora* is the causal agent of citrus collar and root rot. Citrus species resistant to *P. citrophthora* (e.g., macrophylla, trifoliate orange) produced much higher amount of scoparone, a phenylpropanoid phytoalexin, than susceptible species (e.g., rough lemon, shamouti). Mycelial growth of *P. citrophthora* was inhibited by scoparone. Treatment of resistant citrus with aminooxyacetic acid (AOA), an inhibitor of phenylpropanoid production, reduced the resistance to *P. citrophthora*, indicating that scoparone plays a crucial role for the resistance of citrus to *P. citrophthora* (Afek and Sztejnberg 1988). In the interaction between coast live oak and *P. ramorum*, productivity of ellagic acid, a phenolic compound, was associated with the resistance of oak to *P. ramorum* (Nagle et al. 2011). Ellagic acid has also been shown to inhibit the growth of *P. ramorum* (McPherson et al. 2014).

Recent advances made in omics-based approaches also have revealed new insights into the mechanisms of tree root resistance to *Phytophthora* species. For example, transcriptome analysis was performed for European chestnut, *Castanea sativa*, inoculated with the ink disease pathogen, *P. cinnamomi*. Gene ontology annotation and differential gene expression analysis for the root transcriptome of the susceptible *C. sativa* and the resistant *C. crenata* after inoculation with *P. cinnamomi* enabled the selection of candidate genes for ink disease resistance in *Castanea* species (Serrazina et al. 2015). Similar transcriptome analyses were performed for avocado–*P. cinnamomi* (Reeksting, et al. 2014), citrus–*P. parasitica* (Rosa et al. 2007), and tanoak–*P. ramorum* interactions (Hayden et al. 2014).

10 *Arabidopsis–P. parasitica* Pathosystem for Dissecting the Resistance of Plant Roots to *Phytophthora* Species

Arabidopsis thaliana is the most commonly used model plant to investigate all kinds of plant activities including plant–microbe interactions. *Arabidopsis* as model host has great advantages, because of available genetic and genomic resources, established research techniques, and a large collection of mutants. Recently, a model *Arabidopsis–P. parasitica* pathosystem has been established (Attard et al. 2010). In compatible interactions, *P. parasitica* forms appressoria on the surface of *Arabidopsis* roots to penetrate into the cortex layer of the root. *P. parasitica* produces a lot of haustoria during biotrophic phase of infection but becomes necrotrophic in later stage of the infection. *Arabidopsis* mutants with impaired salicylic acid (SA), jasmonic acid (JA), or ethylene (ET) signaling pathways are more susceptible than the wild type, indicating that SA, JA, and ET are all involved in the basal resistance of *Arabidopsis* to *P. parasitica*. Importantly, the interactions between *Arabidopsis* ecotypes and *P. parasitica* isolates have been tested, and there are natural variations in susceptibility and resistance between *Arabidopsis* ecotypes and *P. parasitica* isolates (Wang et al. 2011). Larroque et al. (2013) reported that the *Arabidopsis* mutant *bak1-4* (encode receptor coupled protein kinase) and *rbohD/F* (NADPH oxidases for ROS production) are significantly more susceptible to *P. parasitica* than the wild type, indicating that BAK1 and RBOH are required for the basal resistance of *Arabidopsis* against *P. parasitica*. Evangelisti et al. (2013) reported the function of an RXLR effector of *P. parasitica*, penetration-specific effector 1 (PSE1). Expression of *PSE1* in *Arabidopsis* altered the distribution of auxin efflux carriers and suppressed the induction of elicitor-induced cell death. *PSE1* expression in *Arabidopsis* also increases susceptibility to *P. parasitica*, and auxin treatment suppressed the disease symptom of *PSE1*-expressing *Arabidopsis*, indicating that PSE1 is an effector that modulates the local auxin content for the root infection of *P. parasitica*. Draft genome sequencing for *P. parasitica* was recently completed, and transcriptome analysis for the *Arabidopsis–P. parasitica* interaction was reported (*Phytophthora parasitica* assembly dev initiative, Broad Institute, Attard et al. 2014). Such new resources will further reveal the belowground mechanisms involved in plant defence against *Phytophthora* species.

Acknowledgments We thank Dr. Nicola Skoulding for English editing of the manuscript. This work was supported by a Grant-in-Aid for Young Scientists (B) (23780041), by a Grant-in-Aid for Scientific Research (B) (26292024), and by a Grant for Basic Science Research Projects from the Sumitomo Foundation.

References

Afek U, Sztejnberg A (1988) Accumulation of scoparone, a phytoalexin associated with resistance of citrus to *Phytophthora citrophthora*. Phytopathology 78:1678–1682

Attard A, Gourgues M, Callemeyn-Torre N, Keller H (2010) The immediate activation of defense responses in *Arabidopsis* roots is not sufficient to prevent *Phytophthora parasitica* infection. New Phytol 187:449–460

Attard A, Evangelisti E, Kebdani-Minet N, Panabières F, Deleury E, Maggio C, Ponchet M, Gourgues M (2014) Transcriptome dynamics of *Arabidopsis thaliana* root penetration by the oomycete pathogen *Phytophthora parasitica*. BMC Genomics 15:538

Ayers AR, Ebel J, Finelli F, Berger N, Albersheim P (1976) Host-pathogen interactions: IX. Quantitative assays of elicitor activity and characterization of the elicitor present in the extracellular medium of cultures of *Phytophthora megasperma* var. *sojae*. Plant Physiol 57:751–759

Bailey JA, Burden RS, Vincent GG (1975) Capsidiol: an antifungal compound produced in *Nicotiana tabacum* and *Nicotiana clevelandii* following infection with tobacco necrosis virus. Phytochemistry 14:597

Ballvora A, Ercolano MR, Weiß J, Meksem K, Bormann CA, Oberhagemann P, Salamini F, Gebhardt C (2002) The *R1* gene for potato resistance to late blight (*Phytophthora infestans*) belongs to the leucine zipper/NBS/LRR class of plant resistance genes. Plant J 30:361–371

Bhattacharyya MK, Narayanan NN, Gao H, Santra DK, Salimath SS, Kasuga T, Liu Y, Espinosa B, Ellison L, Marek L, Shoemaker R, Gijzen M, Buzzell RI (2005) Identification of a large cluster of coiled coil-nucleotide binding site-leucine rich repeat-type genes from the *Rps1* region containing Phytophthora resistance genes in soybean. Theor Appl Genet 111:75–86

Birch PR, Armstrong M, Bos J, Boevink P, Gilroy EM, Taylor RM, Wawra S, Pritchard L, Conti L, Ewan R, Whisson SC, van West P, Sadanandom A, Kamoun S (2009) Towards understanding the virulence functions of RXLR effectors of the oomycete plant pathogen *Phytophthora infestans*. J Exp Bot 60:1133–1140

Boissy G, de La Fortelle E, Kahn R, Huet JC, Bricogne G, Pernollet JC, Brunie S (1996) Crystal structure of a fungal elicitor secreted by *Phytophthora cryptogea*, a member of a novel class of plant necrotic proteins. Structure 4:1429–1439

Boissy G, O'Donohue M, Gaudemer O, Perez V, Pernollet J-C, Brunie S (1999) The 2.1 Å structure of an elicitin-ergosterol complex: a recent addition to the sterol carrier protein family. Protein Sci 8:1191–1199

Bos JI, Armstrong MR, Gilroy EM, Boevink PC, Hein I, Taylor RM, Zhendong T, Engelhardt S, Vetukuri RR, Harrower B, Dixelius C, Bryan G, Sadanandom A, Whisson SC, Kamoun S, Birch PR (2010) *Phytophthora infestans* effector AVR3a is essential for virulence and manipulates plant immunity by stabilizing host E3 ligase CMPG1. Proc Natl Acad Sci U S A 107:9909–9914

Bostock RM, Kuc JA, Laine RA (1981) Eicosapentaenoic and arachidonic acids from *Phytophthora infestans* elicit fungitoxic sesquiterpenes in the potato. Science 212:67–69

Bouwmeester K, de Sain M, Weide R, Gouget A, Klamer S, Canut H, Govers F (2011) The lectin receptor kinase LecRK-I.9 is a novel Phytophthora resistance component and a potential host target for a RXLR effector. PLoS Pathog 7:e1001327

Bozkurt TO, Schornack S, Win J, Shindo T, Ilyas M, Oliva R, Cano LM, Jones AM, Huitema E, van der Hoorn RA, Kamoun S (2011) *Phytophthora infestans* effector AVRblb2 prevents secretion of a plant immune protease at the haustorial interface. Proc Natl Acad Sci U S A 108:20832–20837

Bozkurt TO, Schornack S, Banfield MJ, Kamoun S (2012) Oomycetes, effectors, and all that jazz. Curr Opin Plant Biol 15:483–492

Chang YH, Yan HZ, Liou RF (2015) A novel elicitor protein from *Phytophthora parasitica* induces plant basal immunity and systemic acquired resistance. Mol Plant Pathol 16:123–136

Chaparro-Garcia A, Wilkinson RC, Gimenez-Ibanez S, Findlay K, Coffey MD, Zipfel C, Rathjen JP, Kamoun S, Schornack S (2011) The receptor-like kinase SERK3/BAK1 is required for basal resistance against the late blight pathogen *Phytophthora infestans* in *Nicotiana benthamiana*. PLoS One 6:e16608

Cheong JJ, Birberg W, Fügedi P, Pilotti A, Garegg PJ, Hong N, Ogawa T, Hahn MG (1991) Structure-activity relationships of oligo-β-glucoside elicitors of phytoalexin accumulation in soybean. Plant Cell 3:127–136

Deacon JW, Donaldson SP (1993) Molecular recognition in the homing responses of zoosporic fungi, with special reference to *Pythium* and *Phytophthora*. Mycol Res 97:1153–1171

Delaney TP, Uknes S, Vernooij B, Friedrich L, Weymann K, Negrotto D, Gaffney T, Gut-Rella M, Kessmann H, Ward E, Ryals J (1994) A central role of salicylic acid in plant disease resistance. Science 266:1247–1250

Dixon MS, Jones DA, Keddie JS, Thomas CM, Harrison K, Jones JDG (1996) The tomato *Cf-2* disease resistance locus comprises two functional genes encoding leucine-rich repeat proteins. Cell 84:451–459

Doke N (1983) Generation of superoxide anion by potato tuber protoplasts upon the hypersensitive response to hyphal wall components of Phytophthora infestans and specific inhibition of the reaction by suppressor of hypersensitivity. Physiol Plant Pathol 23:359–367

Dong S, Yin W, Kong G, Yang X, Qutob D, Chen Q, Kale SD, Sui Y, Zhang Z, Dou D, Zheng X, Gijzen M, Tyler BM, Wang Y (2011) *Phytophthora sojae* avirulence effector Avr3b is a secreted NADH and ADP-ribose pyrophosphorylase that modulates plant immunity. PLoS Pathog 7:e1002353

Dong S, Kong G, Qutob D, Yu X, Tang J, Kang J, Dai T, Wang H, Gijzen M, Wang Y (2012) The NLP toxin family in *Phytophthora sojae* includes rapidly evolving groups that lack necrosis-inducing activity. Mol Plant Microbe Interact 25:896–909

Dorrance AE, McClure SA, St. Martin SK (2003) Effect of partial resistance on *Phytophthora* stem rot incidence and yield of soybean in Ohio. Plant Dis 87:308–312

Du J, Tian Z, Liu J, Vleeshouwers VG, Shi X, Xie C (2013) Functional analysis of potato genes involved in quantitative resistance to *Phytophthora infestans*. Mol Biol Rep 40:957–967

Du J, Verzaux E, Chaparro-Garcia A, Bijsterbosch G, Keizer LCP, Zhou J, Liebrand TWH, Xie C, Govers F, Robatzek S, van der Vossen EAG, Jacobsen E, Visser RGF, Kamoun S, Vleeshouwers VGAA (2015) Elicitin recognition confers enhanced resistance to *Phytophthora infestans* in potato. Nat Plants 1:15034

Ebel J, Grisebach H (1988) Defense strategies of soybean against the fungus *Phytophthora megasperma* f.sp. *glycinea*: a molecular analysis. Trends Biochem Sci 13:23–27

Enkerli K, Hahn MG, Mims CW (1997) Ultrastructure of compatible and incompatible interactions of soybean roots infected with the plant pathogenic oomycete *Phytophthora sojae*. Can J Bot 75:1494–1508

Erwin DC, Ribeiro OK (1996) *Phytophthora* diseases worldwide. APS, St Paul, MN

Evangelisti E, Govetto B, Minet-Kebdani N, Kuhn ML, Attard A, Ponchet M, Panabières F, Gourgues M (2013) The *Phytophthora parasitica* RXLR effector penetration-specific effector 1 favours *Arabidopsis thaliana* infection by interfering with auxin physiology. New Phytol 199:476–489

Fellbrich G, Romanski A, Varet A, Blume B, Brunner F, Engelhardt S, Felix G, Kemmerling B, Krzymowska M, Nürnberger T (2002) NPP1, a *Phytophthora*-associated trigger of plant defense in parsley and *Arabidopsis*. Plant J 32:375–390

Fry W (2008) *Phytophthora infestans*: the plant (and *R* gene) destroyer. Mol Plant Pathol 9:385–402

Gaulin E, Jauneau A, Villalba F, Rickauer M, Esquerré-Tugayé MT, Bottin A (2002) The CBEL glycoprotein of *Phytophthora parasitica* var. *nicotianae* is involved in cell wall deposition and adhesion to cellulosic substrates. J Cell Sci 115:4565–4575

Gaulin E, Dramé N, Lafitte C, Torto-Alalibo T, Martinez Y, Ameline-Torregrosa C, Khatib M, Mazarguil H, Villalba-Mateos F, Kamoun S, Mazars C, Dumas B, Bottin A, Esquerré-Tugayé MT, Rickauer M (2006) Cellulose binding domains of a *Phytophthora* cell wall protein are novel pathogen-associated molecular patterns. Plant Cell 18:1766–1777

Graham MA, Marek LF, Shoemaker RC (2002) Organization, expression and evolution of a disease resistance gene cluster in soybean. Genetics 162:1961–1977

Graham TL, Graham MY, Subramanian S, Yu O (2007) RNAi silencing of genes for elicitation or biosynthesis of 5-deoxyisoflavonoids suppresses race-specific resistance and hypersensitive cell death in *Phytophthora sojae* infected tissues. Plant Physiol 144:728–740

Grau CR, Dorrance AE, Bond J, Russin J (2004) Fungal diseases. In: Boerma HR, Specht JE (eds) Soybeans: improvement, production and uses, 3rd edn, Agronomy monograph. American Society of Agronomy, Madison, WI, pp 679–763

Grünwald NJ, Goss EM, Press CM (2008) *Phytophthora ramorum*: a pathogen with a remarkably wide host range causing sudden oak death on oaks and ramorum blight on woody ornamentals. Mol Plant Pathol 9:729–740

Haas BJ, Kamoun S, Zody MC, Jiang RH, Handsaker RE, Cano LM, Grabherr M, Kodira CD, Raffaele S, Torto-Alalibo T, Bozkurt TO, Ah-Fong AM, Alvarado L, Anderson VL, Armstrong

MR, Avrova A, Baxter L, Beynon J, Boevink PC, Bollmann SR, Bos JI, Bulone V, Cai G, Cakir C, Carrington JC, Chawner M, Conti L, Costanzo S, Ewan R, Fahlgren N, Fischbach MA, Fugelstad J, Gilroy EM, Gnerre S, Green PJ, Grenville-Briggs LJ, Griffith J, Grünwald NJ, Horn K, Horner NR, Hu CH, Huitema E, Jeong DH, Jones AM, Jones JD, Jones RW, Karlsson EK, Kunjeti SG, Lamour K, Liu Z, Ma L, Maclean D, Chibucos MC, McDonald H, McWalters J, Meijer HJ, Morgan W, Morris PF, Munro CA, O'Neill K, Ospina-Giraldo M, Pinzón A, Pritchard L, Ramsahoye B, Ren Q, Restrepo S, Roy S, Sadanandom A, Savidor A, Schornack S, Schwartz DC, Schumann UD, Schwessinger B, Seyer L, Sharpe T, Silvar C, Song J, Studholme DJ, Sykes S, Thines M, van de Vondervoort PJ, Phuntumart V, Wawra S, Weide R, Win J, Young C, Zhou S, Fry W, Meyers BC, van West P, Ristaino J, Govers F, Birch PR, Whisson SC, Judelson HS, Nusbaum C (2009) Genome sequence and analysis of the Irish potato famine pathogen *Phytophthora infestans*. Nature 461:393–398

Halim VA, Eschen-Lippold L, Altmann S, Birschwilks M, Scheel D, Rosahl S (2007) Salicylic acid is important for basal defense of *Solanum tuberosum* against *Phytophthora infestans*. Mol Plant Microbe Interact 20:1346–1352

Halim VA, Altmann S, Ellinger D, Eschen-Lippold L, Miersch O, Scheel D, Rosahl S (2009) PAMP-induced defense responses in potato require both salicylic acid and jasmonic acid. Plant J 57:230–242

Hansen EM, Reeser PW, Sutton W (2012) *Phytophthora* beyond agriculture. Annu Rev Phytopathol 50:359–378

Hardham AR (2005) *Phytophthora cinnamomi*. Mol Plant Pathol 6:589–604

Haverkort AJ, Boonekamp PM, Hutten R, Jacobsen E, Lotz LAP, Kessel GJT, Visser RGF, Vossen EAG (2008) Societal costs of late blight in potato and prospects of durable resistance through cisgenic modification. Potato Res 51:47–57

Hayden K, Garbelotto M, Knaus B, Cronn R, Rai H, Wright J (2014) Dual RNA-seq of the plant pathogen *Phytophthora ramorum* and its tanoak host. Tree Genet Genomes 10:489–502

Huang S, van der Vossen EA, Kuang H, Vleeshouwers VG, Zhang N, Borm TJ, vanEck HJ, Baker B, Jacobsen E, Visser RG (2005) Comparative genomics enabled the isolation of the *R3a* late blight resistance gene in potato. Plant J 42:251–261

Ishizaka N, Tomiyama K, Katsui N, Murai A, Masamune T (1969) Biological activities of rishitin. An antifungal compound isolated from diseased potato tubers, and its derivatives. Plant Cell Physiol 10:183–192

Jones DA, Takemoto D (2004) Plant innate immunity—direct and indirect recognition of general and specific pathogen-associated molecules. Curr Opin Immunol 16:48–62

Jones DA, Thomas CM, Hammond-Kosack KE, Balint-Kurti PJ, Jones JDG (1994) Isolation of the tomato *Cf-9* gene for resistance to *Cladosporium fulvum* by transposon tagging. Science 266:789–793

Jones RW, Ospina-Giraldo M, Deahl K (2006) Gene silencing indicates a role for potato endoglucanase inhibitor protein in germplasm resistance to late blight. Am J Potato Res 83:41–46

Judelson HS, Ah-Fong AM, Aux G, Avrova AO, Bruce C, Cakir C, da Cunha L, Grenville-Briggs L, Latijnhouwers M, Ligterink W, Meijer HJ, Roberts S, Thurber CS, Whisson SC, Birch PR, Govers F, Kamoun S, van West P, Windass J (2008) Gene expression profiling during asexual development of the late blight pathogen *Phytophthora infestans* reveals a highly dynamic transcriptome. Mol Plant Microbe Interact 21:433–447

Kamoun S, Klucher KM, Coffey MD, Tyler BM (1993) A gene encoding a host-specific elicitor protein of *Phytophthora parasitica*. Mol Plant Microbe Interact 6:573–581

Kamoun S, van West P, Vleeshouwers VGAA, de Groot KE, Govers F (1998) Resistance of *Nicotiana benthamiana* to *Phytophthora infestans* is mediated by the recognition of the elicitor protein INF1. Plant Cell 10:1413–1425

Kaschani F, Shabab M, Bozkurt T, Shindo T, Schornack S, Gu C, Ilyas M, Win J, Kamoun S, van der Hoorn RA (2010) An effector-targeted protease contributes to defense against

Phytophthora infestans and is under diversifying selection in natural hosts. Plant Physiol 154:1794–1804

Kawchuk LM, Hachey J, Lynch DR, Kulcsar F, van Rooijen G, Waterer DR, Robertson A, Kokko E, Byers R, Howard RJ, Fischer R, Prufer D (2001) Tomato *Ve* disease resistance genes encode cell surface-like receptors. Proc Natl Acad Sci U S A 98:6511–6515

Khatib M, Lafitte C, Esquerré-Tugayé MT, Bottin A, Rickauer M (2004) The CBEL elicitor of *Phytophthora parasitica* var. *nicotianae* activates defence in *Arabidopsis thaliana* via three different signalling pathways. New Phytol 162:501–510

Kilen TC, Hartwig EE, Keeling BL (1974) Inheritance of a second major gene for resistance to *Phytophthora* rot in soybeans. Crop Sci 14:260–262

King SR, McLellan H, Boevink PC, Armstrong MR, Bukharova T, Sukarta O, Win J, Kamoun S, Birch PR, Banfield MJ (2014) *Phytophthora infestans* RXLR effector PexRD2 interacts with host MAPKKKε to suppress plant immune signaling. Plant Cell 26:1345–1359

Kitazawa K, Tomiyama K (1969) Microscopic observations of infection of potato cells by compatible and incompatible races of *Phytophthora infestans*. J Phytopathol 66:317–324

Larroque M, Belmas E, Martinez T, Vergnes S, Ladouce N, Lafitte C, Gaulin E, Dumas B (2013) Pathogen-associated molecular pattern-triggered immunity and resistance to the root pathogen *Phytophthora parasitica* in *Arabidopsis*. J Exp Bot 64:3615–3625

Li G, Huang S, Guo X, Li Y, Yang Y, Guo Z, Kuang H, Rietman H, Bergervoet M, Vleeshouwers VG, van der Vossen EA, Qu D, Visser RG, Jacobsen E, Vossen JH (2011) Cloning and characterization of *R3b*; members of the *R3* superfamily of late blight resistance genes show sequence and functional divergence. Mol Plant Microbe Interact 24:1132–1142

Lokossou AA, Park TH, van Arkel G, Arens M, Ruyter-Spira C, Morales J, Whisson SC, Birch PR, Visser RG, Jacobsen E, van der Vossen EA (2009) Exploiting knowledge of *R/Avr* genes to rapidly clone a new LZ-NBS-LRR family of late blight resistance genes from potato linkage group IV. Mol Plant Microbe Interact 22:630–641

Lygin AV, Hill CB, Zernova OV, Crull L, Widholm JM, Hartman GL, Lozovaya VV (2010) Response of soybean pathogens to glyceollin. Phytopathology 100:897–903

Matsukawa M, Shibata Y, Ohtsu M, Mizutani A, Mori H, Wang P, Ojika M, Kawakita K, Takemoto D (2013) *Nicotiana benthamiana* calreticulin 3a is required for the ethylene-mediated production of phytoalexins and disease resistance against oomycete pathogen *Phytophthora infestans*. Mol Plant Microbe Interact 26:880–892

McLellan H, Boevink PC, Armstrong MR, Pritchard L, Gomez S, Morales J, Whisson SC, Beynon JL, Birch PRJ (2013) An RxLR effector from *Phytophthora infestans* prevents re-localisation of two plant NAC transcription factors from the endoplasmic reticulum to the nucleus. PLoS Pathog 9:e1003670

McPherson BA, Mori SR, Opiyo SO, Conrad AO, Wood DL, Bonello P (2014) Association between resistance to an introduced invasive pathogen and phenolic compounds that may serve as biomarkers in native oaks. For Ecol Manage 312:154–160

Mohr P, Cahill D (2001) Relative roles of glyceollin, lignin and the hypersensitive response and the influence of ABA in compatible and incompatible interactions of soybeans with *Phytophthora sojae*. Physiol Mol Plant Pathol 58:31–41

Morris PF, Ward EWB (1992) Chemoattraction of zoospores of the soybean pathogen, *Phytophthora sojae*, by isoflavones. Physiol Mol Plant Pathol 40:17–22

Nagle AM, McPherson BA, Wood DL, Garbelotto M, Bonello P (2011) Relationship between field resistance to Phytophthora ramorum and constitutive phenolic chemistry of coast live oak. For Pathol 41:464–469

Nürnberger T, Nennstiel D, Jabs T, Sacks WR, Hahlbrock K, Scheel D (1994) High affinity binding of a fungal oligopeptide elicitor to parsley plasma membranes triggers multiple defense responses. Cell 78:449–460

Oh E, Hansen EA (2007) Histopathology of infection and colonization of susceptible and resistant port-orford-cedar by *Phytophthora lateralis*. Phytopathology 97:684–693

Oh S-K, Young C, Lee M, Oliva R, Bozkurt TO, Cano LM, Win J, Bos JIB, Liu H-Y, van Damme M, Morgan W, Choi D, Van der Vossen EAG, Vleeshouwers VGAA, Kamoun S (2009) *In Planta* expression screens of *Phytophthora infestans* RXLR effectors reveal diverse phenotypes, including activation of the *Solanum bulbocastanum* disease resistance protein Rpi-blb2. Plant Cell 21:2928–2947

Ohtsu M, Shibata Y, Ojika M, Tamura K, Hara-Nishimura I, Mori H, Kawakita K, Takemoto D (2014) Nucleoporin 75 is involved in the ethylene-mediated production of phytoalexin for the resistance of *Nicotiana benthamiana* to *Phytophthora infestans*. Mol Plant Microbe Interact 27:1318–1330

Oome S, Van den Ackerveken G (2014) Comparative and functional analysis of the widely occurring family of Nep1-like proteins. Mol Plant Microbe Interact 27:1081–1094

Ottmann C, Luberacki B, Küfner I, Koch W, Brunner F, Weyand M, Mattinen L, Pirhonen M, Anderluh G, Seitz HU, Nürnberger T, Oecking C (2009) A common toxin fold mediates microbial attack and plant defense. Proc Natl Acad Sci U S A 106:10359–10364

Parker JE, Schulte W, Hahlbrock K, Scheel D (1991) An extracellular glycoprotein from *Phytophthora megasperma* f. sp. *glycinea* elicits phytoalexin synthesis in cultured parsley cells and protoplasts. Mol Plant Microbe Interact 4:19–27

Preisig CL, Kuć JA (1985) Arachidonic acid-related elicitors of the hypersensitive response in potato and enhancement of their activities by glucans from *Phytophthora infestans*. Arch Biochem Biophys 236:379–389

Qiao Y, Liu L, Xiong Q, Flores C, Wong J, Shi J, Wang X, Liu X, Xiang Q, Jiang S, Zhang F, Wang Y, Judelson HS, Chen X, Ma W (2013) Oomycete pathogens encode RNA silencing suppressors. Nat Genet 45:330–333

Qiao Y, Shi J, Zhai Y, Hou Y, Ma W (2015) *Phytophthora* effector targets a novel component of small RNA pathway in plants to promote infection. Proc Natl Acad Sci U S A 112:5850–5855

Qutob D, Kamoun S, Gijzen M (2002) Expression of a *Phytophthora sojae* necrosis-inducing protein occurs during transition from biotrophy to necrotrophy. Plant J 32:361–373

Reeksting BJ, Coetzer N, Mahomed W, Engelbrecht J, van den Berg N (2014) *De Novo* sequencing, assembly, and analysis of the root transcriptome of *Persea americana* (Mill.) in response to *Phytophthora cinnamomi* and flooding. PLoS One 9:e86399

Ricker KE, Bostock RM (1994) Eicosanoids in the *Phytophthora infestans*-potato interaction: lipoxygenase metabolism of arachidonic acid and biological activities of selected lipoxygenase products. Physiol Mol Plant Pathol 44:65–80

Rosa DD, Campos MA, Targon MLP, Souza AA (2007) *Phytophthora parasitica* transcriptome, a new concept in the understanding of the citrus gummosis. Genet Mol Biol 30:997–1008

Sandhu D, Gao H, Cianzio S, Bhattacharyya MK (2004) Deletion of a disease resistance nucleotide-binding-site leucine-rich repeat-like sequence is associated with the loss of the *Phytophthora* resistance gene *Rps4* in soybean. Genetics 168:2157–2167

Séjalon-Delmas N, Mateos FV, Bottin A, Rickauer M, Dargent R, Esquerré-Tugayé MT (1997) Purification, elicitor activity, and cell wall localization of a glycoprotein from *Phytophthora parasitica* var. *nicotianae*, a fungal pathogen of tobacco. Phytopathology 87:899–909

Serrazina S, Santos C, Machado H, Pesquita C, Vicentini R, Pais MS, Sebastiana M, Costa R (2015) Castanea root transcriptome in response to *Phytophthora cinnamomi* challenge. Tree Genet Genomes 11:1–19

Shi X, Tian Z, Liu J, van der Vossen EA, Xie C (2012) A potato pathogenesis-related protein gene, StPRp27, contributes to race-nonspecific resistance against *Phytophthora infestans*. Mol Biol Rep 39:1909–1916

Shibata Y, Kawakita K, Takemoto D (2010) Age-related resistance of *Nicotiana benthamiana* against hemibiotrophic pathogen *Phytophthora infestans* requires both ethylene- and salicylic acid-mediated signaling pathways. Mol Plant Microbe Interact 23:1130–1142

Song J, Bradeen JM, Naess SK, Raasch JA, Wielgus SM, Haberlach GT, Liu J, Kuang H, Austin-Phillips S, Buell CR, Helgeson JP, Jiang J (2003) Gene *RB* cloned from *Solanum bulbocastanum* confers broad spectrum resistance to potato late blight. Proc Natl Acad Sci U S A 100:9128–9133

Takemoto D, Hardham AR, Jones DA (2005) Differences in cell death induction by Phytophthora elicitins are determined by signal components downstream of MAP kinase kinase in different species of *Nicotiana* and cultivars of *Brassica rapa* and *Raphanus sativus*. Plant Physiol 138:1491–1504

Tomiyama K (1956) Cell physiological studies on the resistance of potato plant to *Phytophthora infestans*. Ann Phytopathol Soc Jpn 21:54–62

Tyler BM (2002) Molecular basis of recognition between *Phytophthora* pathogens and their hosts. Annu Rev Phytopathol 40:137–167

Tyler BM (2007) *Phytophthora sojae*: root rot pathogen of soybean and model oomycete. Mol Plant Pathol 8:1–8

Tyler BM, Wu M, Wang J, Cheung W, Morris PF (1996) Chemotactic preferences and strain variation in the response of *Phytophthora sojae* zoospores to host isoflavones. Appl Environ Microbiol 62:2811–2817

van der Vossen E, Sikkema A, Hekkert B, Gros J, Stevens P, Muskens M, Wouters D, Pereira A, Stiekema W, Allefs S (2003) An ancient *R* gene from the wild potato species *Solanum bulbocastanum* confers broad-spectrum resistance to *Phytophthora infestans* in cultivated potato and tomato. Plant J 36:867–882

van der Vossen EAG, Gros J, Sikkema A, Muskens M, Wouters D, Wolters P, Pereira A, Allefs S (2005) The *Rpi-blb2* gene from *Solanum bulbocastanum* is an *Mi-1* gene homolog conferring broad-spectrum late blight resistance in potato. Plant J 44:208–222

Vleeshouwers VG, Driesprong JD, Kamphuis LG, Torto-Alalibo T, Van't Slot KA, Govers F, Visser RG, Jacobsen E, Kamoun S (2006) Agroinfection-based high-throughput screening reveals specific recognition of INF elicitins in *Solanum*. Mol Plant Pathol 7:499–510

Vleeshouwers VG, Rietman H, Krenek P, Champouret N, Young C, Oh SK, Wang M, Bouwmeester K, Vosman B, Visser RG, Jacobsen E, Govers F, Kamoun S, van der Vossen EA (2008) Effector genomics accelerates discovery and functional profiling of potato disease resistance and *Phytophthora infestans* avirulence genes. PLoS One 3:e2875

Vlot AC, Dempsey DA, Klessig DF (2009) Salicylic Acid, a multifaceted hormone to combat disease. Annu Rev Phytopathol 47:177–206

Wang Y, Meng YL, Zhang M, Tong XM, Wang QH, Sun YY, Quan JL, Govers F, Shan WX (2011) Infection of *Arabidopsis thaliana* by *Phytophthora parasitica* and identification of variation in host specificity. Mol Plant Pathol 12:187–201

Wang E, Schornack S, Marsh JF, Gobbato E, Schwessinger B, Eastmond P, Schultze M, Kamoun S, Oldroyd GE (2012) A common signaling process that promotes mycorrhizal and oomycete colonization of plants. Curr Biol 22:2242–2246

Wildermuth MC, Dewdney J, Wu G, Ausubel FM (2001) Isochorismate synthase is required to synthesize salicylic acid for plant defence. Nature 414:562–565

Wong J, Gao L, Yang Y, Zhai J, Arikit S, Yu Y, Duan S, Chan V, Xiong Q, Yan J, Li S, Liu R, Wang Y, Tang G, Meyers BC, Chen X, Ma W (2014) Roles of small RNAs in soybean defense against *Phytophthora sojae* infection. Plant J 79:928–940

Yoshikawa M, Takeuchi Y, Horino O (1990) A mechanism for ethylene- induced disease resistance in soybean: enhanced synthesis of an elicitor-releasing factor, β-1,3-endoglucanase. Physiol Mol Plant Pathol 37:367–376

Yu D, Liu Y, Fan B, Klessig DF, Chen Z (1997) Is the high basal level of salicylic acid important for disease resistance in potato? Plant Physiol 115:343–349

Belowground Signaling and Defence in Host–*Pythium* Interactions

Patricia A. Okubara, Jin-Ho Kang, and Gregg A. Howe

Abstract Members of the genus *Pythium* interact with plants and microbial members of the rhizosphere using a variety of signaling mechanisms. Biochemical signaling has a role in pathogen–host specificity, host defence response induction, and antagonism between *Pythium* and biocontrol microorganisms. *Pythium irregulare*, *P. aphanidermatum*, and *P. arrhenomanes* are among the plant-pathogenic species that share a common mode of infection but vary in host range and virulence, possibly due to differences in nutrient acquisition and sensitivity to host and biocontrol interactions. Host innate immunity to *Pythium* is conferred by the jasmonic acid (JA) and ethylene (E) signal pathways in roots; triggers of these pathways include pathogen cell surface components, and metabolite and protein effectors. Roots also can mount chemical (metabolite-based) defences against specific *Pythium* spp., and, reciprocally, *Pythium* can degrade defence metabolites. In contrast, *P. oligandrum* is a mycoparasite of other *Pythium* species and also sends signals that trigger defence responses in plants. Interactions between plant-pathogenic *Pythium* and biocontrol bacteria have revealed additional complexities of belowground signaling. In this chapter, we summarize current knowledge about rhizosphere signaling between *Pythium* spp., other microbial community members, and plant roots in agricultural production venues, with emphasis on molecular mechanisms. We also report new findings for the role of JA-mediated defence in protection of tomato from *P. aphanidermatum*.

P.A. Okubara (✉)
United States Department of Agriculture-Agricultural Research Service, Root Disease Unit and Biological Control Research Unit, P.O. Box 6430, Pullman, WA 99164-6430, USA
e-mail: patricia.okubara@ars.usda.gov

J.-H. Kang
MSU-DOE Plant Research Laboratory, Department of Biochemistry and Molecular Biology, Michigan State University, East Lansing, MI 48824, USA

Graduate School of International Agricultural Technology, Seoul National University, Seoul, Republic of Korea

G.A. Howe
MSU-DOE Plant Research Laboratory, Department of Biochemistry and Molecular Biology, Michigan State University, East Lansing, MI 48824, USA

C.M.F. Vos, K. Kazan (eds.), *Belowground Defence Strategies in Plants*, Signaling and Communication in Plants, DOI 10.1007/978-3-319-42319-7_8

About 130–150 species of the genus *Pythium* have been characterized on the bases of spore morphology and ribosomal DNA intergenic transcribed spacer sequence (Benhamou et al. 2012; Schroeder et al. 2013). These species colonize plants, algae, fish, and mammals in soil, freshwater, and aboveground niches (Davis et al. 2006; Schroeder et al. 2013); generally are rapid growers; and are necrotrophic, acquiring nutrients from dying or dead cells. *Pythium* are oomycetes, related to diatoms and brown algae, and hence harbor cellulose and β-glucans in their cell walls, in contrast to the true fungi having chitin-containing cell walls. Many plant-pathogenic species are generalists that attack a wide range of crops and persist in wet, clay soils (Schroeder et al. 2006); others are more specialized in host and environmental niches. In a survey of *Pythium* in 80 cereal production sites of Washington, USA, 12 major species, including *P. irregulare* Buisman, *P. intermedium* de Bary, *P. abappressorium* Paulitz and M. Mazzola, and *P. ultimum* Trow, were grouped into six communities based on prevalence (Paulitz and Adams 2003). The most abundant and widespread species was the moderately pathogenic *P. abappressorium*, whereas the highly pathogenic, broad host range species *P. ultimum* was a minority in all but one of the six communities. Occurrence of certain *Pythium* species in a range of soil types, meteorological zones, and hosts indicate that species survival depends on a complex set of factors.

Soilborne *Pythium* causes *Pythium* root rot and damping-off and seed (embryo) and crown rot in agricultural soils throughout the world (Paulitz and Adams 2003; Schroeder et al. 2013). The pathogen attacks the seminal, crown, and lateral roots of young seedlings and interferes with root hair development (Schroeder and Paulitz 2006; Van Buyten and Höfte 2013). Hyphae of germinating oospores or zoospores penetrate the epidermis of host roots, likely due to production of cell wall degrading enzymes (see Schroeder et al. 2013). Extent of infection is an indicator of virulence. For example, the virulent rice pathogen *P. arrhenomanes* Drechsler invades the root inner cortex and stele, causing extensive cellular breakdown and disruption of the vascular system (Van Buyten and Höfte 2013). Infection of rice and tomato roots by less virulent isolates, such as *Pythium* group F, is slower and less invasive, such that the host can mount cell wall fortifications and other defences (Rey et al. 1998; Van Buyten and Höfte 2013). In contrast, the nonpathogen *Pythium uncinulatum* Plaäts-Nit. & I. Blok colonizes the outer cell layers of the tomato root and does not evoke a defence response (Rey et al. 1998). A comprehensive review of the *Pythium* life cycle, host recognition and infection, and disease management is available (Martin and Loper 1999).

1 Biochemical Aspects of Pathogenicity

The *Pythium*–host interaction is first evident when *Pythium* zoospores perceive and swim toward a prospective host root. Chemotaxis of zoospores to specific hosts appears to be governed by root exudate composition and zoospore perception in a manner that reflects the coevolution of the association. Zoospores of the cereal pathogens *P. arrhenomanes* and *P. graminicola* Subraman preferentially accumulated and encysted on the roots of field-grown grasses and cultivated cereals relative to dicot weeds, whereas those of the generalists *P. ultimum* and *P. aphanidermatum* (Edson) Fitzp. accumulated on both monocot and dicot hosts (Mitchell and Deacon 1985). In the *Pythium*–cucumber system, zoospore attraction was correlated with pathogenicity. Zoospores of pathogenic *P. aphanidermatum* and *Pythium* group F accumulated behind the root tips of cucumber at higher densities compared to those from nonpathogenic *P. oligandrum* Drechsler (Wulff et al. 1998). Electrical fields generated by ion pumps at the root tip and zone of elongation of wheat, ryegrass, and cress roots attracted *P. aphanidermatum* zoospores in an exudate-independent manner (van West et al. 2002). While the charge differential and direction of the electrical field was predicted to vary with host species, physiology, and environment, this mechanism could account for the nonspecific migration of zoospores from multiple *Pythium* spp. to the roots of the same host.

During the encystment stage of infection, motile zoospores of *P. aphanidermatum* become embedded in host-derived surface glycoproteins and cell wall polysaccharides of cress roots (Estrada-Garcia et al. 1990). A fraction of host mucilage containing 5 % fucose and low uronic acid triggered the encystment in vitro. Encystment also was observed using the lectin concanavalin A and a monoclonal antibody, PA1, that binds to the zoospore surface and flagella (Estrada-Garcia et al. 1990), suggesting active roles of both zoospore and host surface components. Encystment was not necessarily correlated to pathogenicity, however. When zoospores of *P. aphanidermatum*, a generalist on dicots, were exposed to roots of tomato, alfalfa, or sugar beet (*Beta vulgaris* L.), the density of encysted zoospores at the zone of elongation was correlated to extent of root pruning only on alfalfa (Raftoyannis and Dick 2006). Varietal and isolate differences in encystment also were observed. *P. dissimile* Vaartaja, moderately pathogenic to wheat and oat, encysted less densely on roots of those hosts relative to *P. aphanidermatum* on alfalfa roots, but similar extents of root pruning were seen on all hosts. The data indicated that *Pythium* pathogenicity is mediated by factors or processes downstream of encystment.

Root exudates provide carbon and nitrogen that attract and support rhizosphere microbial members, so it is not surprising that certain hosts preferentially interact with *Pythium*. A stable isotope ($^{13}CO_2$) pulse labeling approach was used to trace exudates from switchgrass roots into bacteria and fungi (Mao et al. 2014). On the assumption that microbes in strong association with host roots were more enriched for ^{13}C than transiently or distantly associated microbes, the authors concluded that *Pythium* was a major genus on switchgrass roots. Further characterization will be needed to determine the species of *Pythium*. Extensive utilization of exudates can account for more aggressive and persistent associations. The virulence of rice pathogens was attributed to their ability to metabolize a wide range of amino acids, including the host defence compounds L-threonine and hydroxyl-L-proline (Van Buyten and Höfte 2013).

Root exudates also can include secondary metabolites that have antimicrobial activities; several examples of anti-*Pythium* metabolites have been reported

(Fig. 1). Roots of oat produce families of glycosylated triterpenes and steroid aglycone derivatives, called saponins, having antifungal activity against a number of soilborne pathogens, including *Pythium* spp. (Deacon and Mitchell 1985). The activity of the saponin avenacin A-1 (Fig. 1a) is attributed to its ability to permeabilize fungal and oomycete membranes (see Osbourn et al. 2011). As preformed or constitutively produced root exudates, the saponins represent one of the first lines of defence against soilborne pathogens, but direct interaction with soil biota predisposes these compounds to biodegradation, especially by microbes that are susceptible to their action (Bouarab et al. 2002). Susceptibility or resistance to avenacin A-1 was found to be host-dependent in three oat–wheat rotation regimens (Carter et al. 1999). Of 47 morphologically distinct fungi that were isolated in the continuous oat regimen, 44 (94 %) were resistant. However, if wheat, which does not produce avenacin A-1, was planted after two seasons of oat, only 6 of 14 (43 %) associated fungi were resistant. No resistance was observed in 18 isolates collected from wheat that followed an oat–wheat rotation. Resistance was correlated with the removal (detoxification) of glucose residues of avenacin and to disruption of host defence signaling by the products of detoxification (Bouarab et al. 2002). While none of the isolates were *Pythium*, these findings demonstrate the importance of signaling and the transient and host-dependent composition of rhizosphere inhabitants.

Roots of American ginseng (*Panax quinquefolius* L.) produce ginsenosides (Fig. 1b), a class of antifungal triterpenoid saponins having anti-inflammatory and other health benefits (Leung and Wong 2010). Ginsenosides are synthesized from mevalonate via the dammarenediol synthase pathway (Oh et al. 2014). These secondary metabolites reduced the growth of nonpathogenic fungi and a foliar pathogen about 20-fold, compared to a 3–8-fold reduction of soilborne root and crown pathogens (Nicol et al. 2002). Degradation or enzymatic detoxification of ginsenosides by the latter group was implicated but not quantified. In a separate set of studies, isolates of *P. irregulare* were found to secrete glycosidases and other enzymes, collectively called ginsenosidases, that partially or completely detoxified ginsenosides (Ivanov and Bernards 2012). The enzymes were induced in *P. irregulare* by the substrates, and detoxification was correlated with greater disease severity (decreased root vigor and increased chlorophyll fluorescence, a stress response) in ginseng seedlings. The findings indicated that specific isolates of *P. irregulare* can overcome biochemical defences mounted by its host, with consequences to pathogenicity. In the case of the *P. irregulare*–ginseng interaction, additional growth benefits of ginsenosides for the pathogen were observed in vitro (Nicol et al. 2003).

Collagen served as a substrate for *Pythium* in culture, indicating an alternative nutrient source to cell wall polysaccharides. This prompted a study of protease secretion by the cereal pathogen *P. graminicola*, the algae pathogen *P. grandisporangium* Fell and Master, and the mammalian pathogen *P. insidiosum* De Cock, L. Mend., A. A. Padhye, Ajello, and Kaufman (Davis et al. 2006). Despite their distinctive ecological niches, all three species secreted serine proteases in vitro. However, proteases could act on wall-associated proteins

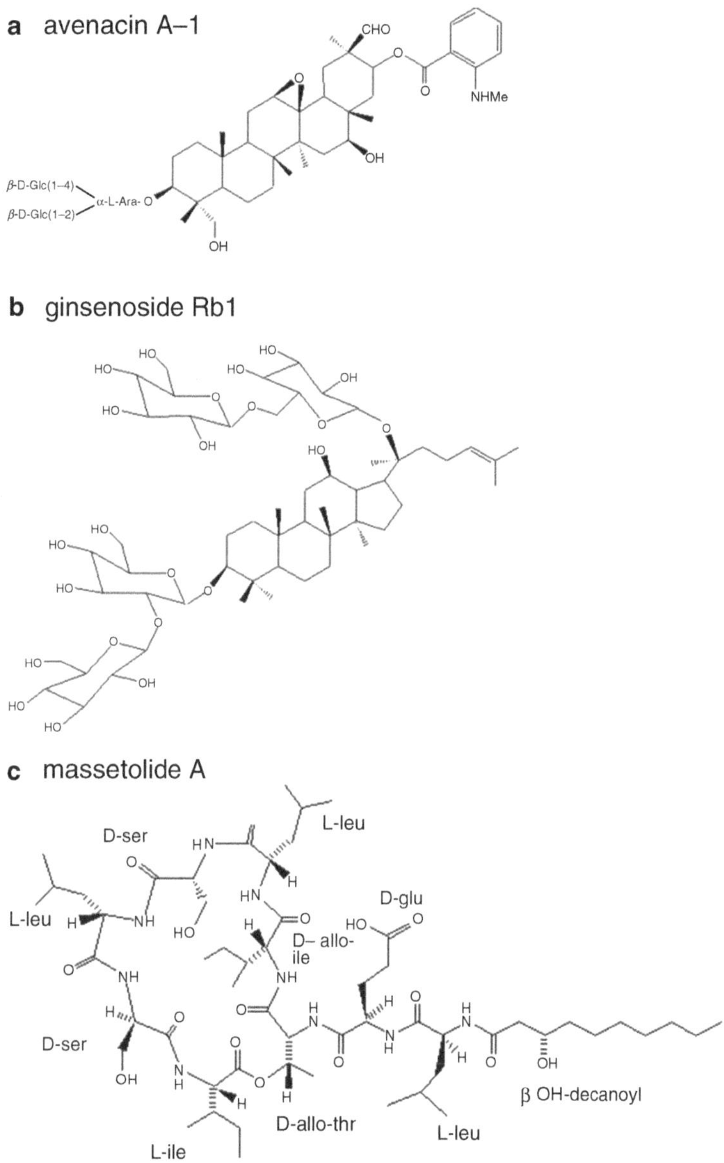

Fig. 1 Examples of metabolites having activity against *Pythium* spp. (**a**) The saponin avenacin A-1 produced by oat roots (Osbourn et al. 2011); (**b**) a JA-responsive (PPD type) saponin, ginsenoside Rb1, from American ginseng roots (Ivanov and Bernards 2012; Oh et al. 2014); (**c**) the cyclic lipopeptide surfactant massetolide A produced by the biocontrol strain *Pseudomonas fluorescens* SS101 (de Souza et al. 2003b; de Bruijn et al. 2008). Structures were generated using ChemDraw Pro vers. 4.0.1 (Cambridge Soft Corp., Cambridge, Massachusetts, USA)

to weaken the integrity of plant and fungal cell walls. The role of proteases in *Pythium* pathogenicity and rhizosphere persistence requires further testing.

2 Role of the Jasmonic Acid Pathway in Host Defences Against *Pythium*

The jasmonic acid (JA) signal pathway confers broad spectrum, innate immunity against insect damage, wounding, and abiotic stress; mediates induced systemic resistance elicited by biocontrol bacteria (Pieterse et al. 2002; van Loon 2007); and is required for pollen development in *Arabidopsis* and seed maturation in tomato (reviewed in Campos et al. 2014). Molecular components of the JA pathway have been well-characterized, and inducers of the pathway, such as cell surface components and microbial effectors, have been identified over the past 20 years. The JA pathway is stimulated by JA, methyl JA (MeJA), and an isoleucine conjugate of JA (JA-Ile), the bioactive and mobile form of JA in plant (Staswick and Tiryaki 2004). Applied MeJA induced the production of ginsenosides (Fig. 1b) in the stele of American ginger (Oh et al. 2014), indicating ginsenoside synthesis was regulated by the JA pathway. Ginsenoside accumulation was the result of reduced flux through cycloartenol synthase branch of the mevalonate pathway leading to sterol production and increased flux through dammarenediol synthase, the first committed step in ginsenoside production.

In belowground interactions, roots of nearly all host species are susceptible to pathogenic *Pythium*, but the JA pathway provides a degree of protection. *Arabidopsis* mutants deficient in JA biosynthesis and in CORONATINE INSENSITIVE1 (COI1), the key component of JA perception and signal transduction, displayed more chlorosis and foliar wilting than wild-type plants in the presence of a soilborne pathogen later identified as *P. mastophorum* (Vijayan et al. 1998). The wilting phenotype and low expression of the JA-regulated defensin gene *PDF1.2* in the mutants were rescued by MeJA. In addition, *Arabidopsis jar1* mutants deficient in the accumulation of JA-Ile (Staswick and Tiryaki 2004; Thines et al. 2007; Sheard et al. 2010), due the absence of a functional Ile-conjugating enzyme, displayed more severe disease symptoms after challenge with *P. irregulare* (Staswick et al. 1998, 2002). Since the *jar1* mutants are insensitive to JA, native roots did not display the typical growth inhibition observed in wild type after JA treatment. The role of the JA defence pathway in tomato roots is presented in the following section.

Resulting root-localized defence responses include accumulation of defence metabolites and proteins (van Loon 2007; Campos et al. 2014). The inducible nature of these responses has implications for host vigor, fitness of specific community members, and composition of the rhizosphere microbial community. However, as part of the stress response, this pathway also might promote cell disruption and death favored by *Pythium* and other necrotrophic pathogens, thereby offsetting the benefits of partial protection. For instance, the rapid colonization of rice root tissues by the virulent species *P. arrhenomanes* is accompanied by production of reactive oxygen species and necrosis-associated induction of *JA-Myb*, a stress response transcription factor gene (Van Buyten and Höfte 2013).

3 JA-Mediated Protection Against *Pythium aphanidermatum* in Tomato Roots

3.1 Rationale and Hypothesis

Leaves of tomato plants harboring mutations in *COI1*, the co-receptor for JA perception and signaling, were more susceptible to insect feeding (Li et al. 2004). These mutants, called *jasmonic acid insensitive1* (*jai1*), also displayed substantial wilting, chlorosis, roots stunting, and mortality compared in wild-type plants when grown in a field at Michigan State University, East Lansing, Michigan. The symptoms were typical of *Pythium* root rot, and the soil was diagnosed for *P. ultimum* (Campos et al. 2014). The JA defence pathway conferred partial protection to *Pythium* spp. in *Arabidopsis* and maize (Staswick et al. 1998; Vijayan et al. 1998; Yan et al. 2012). Here, we hypothesized that the *jai1* mutants also would be more susceptible to the tomato pathogen *P. aphanidermatum* and to other soilborne necrotrophic pathogens of tomato, such as *Rhizoctonia solani* AG-8 and *R. solani* AG-2-1 (data not shown).

3.2 Materials and Methods

3.2.1 Tomato Plants, Field Experiments, and Sample Collection

Tomato cultivar Castlemart has a functional JA signal pathway and served as a wild-type control for responses to *Pythium*. The *jai1-1* mutant of tomato, which was isolated in the cv. Micro-Tom genetic background, harbors two copies of the null allele (*jai1-1*) of the tomato *COI1* gene and is deficient in JA signaling (Li et al. 2004). *jai1-1* homozygous plants display a number of developmental phenotypes, including reduced fruit weight, decreased pollen fertility, and defective seed maturation (Li et al. 2004); hence, the *jai1-1* mutation was maintained in the heterozygous state. Our experiments were done with a (BC_2F_5) line in which the *jai1-1* mutant (cv. Micro-Tom) was backcrossed twice to cv. Castlemart followed by self-pollination. Homozygous *jai1* individuals were distinguished from *Jai1* homozygotes and from *Jai1*/*jai1-1* heterozygotes on the basis of PCR product size (Li et al. 2004) using genomic DNA from a single cotyledon (Lin et al. 2001).

Twenty-eight-day-old Castlemart and *jai1-1* homozygotes were planted in alternating rows, two rows per genotype and 15 plants per row (Fig. 2a) in a field plot

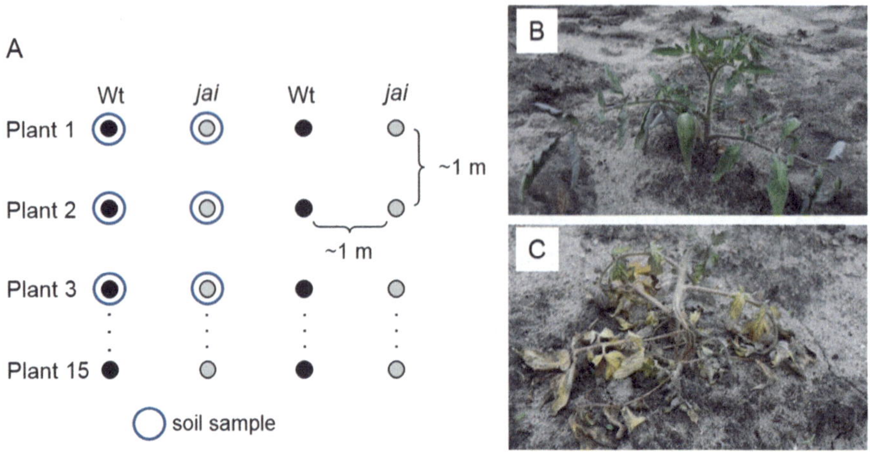

Fig. 2 Planting and sampling scheme used for field experiments (**a**). *Blue circles* indicate the location of soils collected for greenhouse experiments done at Michigan State University. Foliar symptoms of wild-type Castlemart (**b**) and homozygotes of *jai1-1* (**c**) 9 days after planting in a *Pythium*-infested field in 2010

located at Michigan State University, East Lansing, MI. The field was known to cause early foliar necrosis and root stunting, symptoms typical of *Pythium* root rot, and to harbor *P. ultimum* (Campos et al. 2014). Plants were spaced about 1 m apart at the time of planting. In 2009, the planting date was June 8, and plants were harvested on June 16 and June 23, 8, and 15 days post inoculation (dpi). The field experiment was repeated in 2010 using similar planting and harvest dates. Root samples were dried in a laminar flow hood for 2–3 h prior to DNA extraction. Rhizosphere soil also was collected by shaking the roots over clean bench paper. At the time of planting in 2009, field soil was collected at sites of planting (Fig. 2a). The soil was used to grow 28-day-old Castlemart and *jai1-1* plants in the greenhouse at Michigan State University. Greenhouse-grown plants were harvested after 7 days.

3.2.2 Extraction of Total DNA from Soil and Root Samples

Soil extracts were obtained from triplicate 0.8-g dried samples, and each dried root mass was extracted in one to four batches of 100–400 mg per batch. To improve DNA extraction efficiency, both soil and root samples were subjected to 15 cycles of ambient pressure for 10 s alternated with 35,000 psi (235 MPa) for 20 s using the Barocycler™ NEP 3229 (Pressure BioSciences, Inc., Bridgewater, Massachusetts, USA) as described in Okubara et al. (2007). Pressure cycling was performed in FT500-ND PULSE Tubes™ (Pressure BioSciences, Inc.) containing premeasured lysis solution, 120 μL of S1 (sodium dodecyl sulfate solution), 400 μL of inhibitor removal solution, and 600 μL of guanidine thiocyanate bead solution (UltraClean

Soil DNA Kit, MO BIO Laboratories, Solana Beach, California, USA). Clarified supernatants were incubated with 400 μL S2 acetate solution and 1.8 mL S3 guanidine HCl/isopropanol solution, passed through spin filter columns, and washed with 300 μL S4 ethanol solution as recommended by the manufacturer. Total soil or root DNA was eluted in 60 μL S5 TRIS buffer solution into a clean Eppendorf tube containing 5 mg of insoluble polyvinylpyrrolidone (PVP; Sigma Chemical Co., St. Louis, Missouri, USA) to remove residual low molecular weight fluorescent compounds. The PVP was dispensed as 50 μL aliquots of a 10 % (w/v) aqueous suspension. Excess water was removed from the PVP by centrifugation prior to adding column-eluted DNA (Okubara et al. 2007). DNA extracts containing PVP were clarified by centrifugation immediately before real-time PCR.

3.2.3 Real-Time PCR Quantification of *Pythium ultimum* and *P. aphanidermatum*

Real-time PCR primers were designed to amplify the intergenic transcribed spacer (ITS) regions of the nuclear ribosomal DNA of *Pythium aphanidermatum* or *P. ultimum*. PCR primers for *P. ultimum* were ULT1F (5′) GACACTGGAACGGGAGTCAGC (3′) and ULT4R (5′) AAAGGACTCGACAGATTCTCGATC (3′) (Schroeder et al. 2006); primers for *P. aphanidermatum* were PaphF2 (5′) GGGCTGCTTAATTGTAGTCTGCC (3′) and PaphR2 (5′) CTAACCGAAGTCGCCCAAATG (3′) (P. Okubara, this study). Each PCR reaction consisted of 5.8 μL nanopure water, 1 μL FastStart DNA Master SYBR Green I reagent (Roche Applied Science, Indianapolis, Indiana, USA), 1.2 μL 25 mM $MgCl_2$, 5 pmol of each primer, and 1 μL of DNA extract in a total volume of 10 μL. Samples were amplified in duplicate using the Roche Light Cycler (Roche Applied Science) and the following amplification protocol: 95 °C for 10 min; 95 °C for 10 s/70 °C at 5 s/72 °C at 10 s for 50 cycles; and 40 °C for 30 s. Amplicon melting and fluorescence data were transformed as described earlier (Okubara et al. 2008). *Pythium* DNA (pg) was calculated from average Ct values (y) using the equation $y = -3.734 \log(x) + 24.741$ (Schroeder et al. 2006). Pathogen DNA in each soil sample was the average of three extracts per sample normalized to a gram of soil (pg g^{-1}). Pathogen DNA in each root was the sum of all extracts from a single root (pg $root^{-1}$).

3.2.4 *Pythium* Isolates, Inocula, and Greenhouse Pathogenicity Assays

Pythium ultimum isolate 0900119 and *P. irregulare* group I isolate 0900101 were obtained from no-till plots in Garfield, Washington (Schroeder et al. 2006), and maintained on potato dextrose agar. An isolate of *P. aphanidermatum* was isolated from pepper in Florida (Chellemi et al. 2000). *Pythium* inocula consisted of colonized oat particles inoculated with cubes of fungi from agar cultures. Greenhouse pathogenicity assays were performed essentially as described in Okubara and

Jones (2011). *Pythium* on oats was enumerated and used to infest soil at rates of 0, 100, 250, and 500 propagules g^{-1} soil (ppg). Seven-day-old tomato seedlings were transferred to 10-cm^2 plastic pots containing infested soil and grown for 14 days at 15 ± 1 °C with 12-h daily supplemental lighting (66–90 $\mu mol\ m^{-2}\ s^{-1}$). Six to eight plants of each genotype were used per treatment. Disease severity was assessed on the basis of root fresh weight and total root length. The latter was quantified using digital scans of roots and WinRHIZO 5.0 (Regents Instruments, Inc., Quebec, Canada). To normalize for endogenous differences in root mass among the Castlemart and *jai1-1* genotypes, the root variables were expressed as ratios of the means of inoculated to non-inoculated plants. Experiments with *P. aphanidermatum* and *P. ultimum* were done twice.

3.2.5 Statistical Analyses

Mean pathogen DNA values were calculated from three independent soil or root samples; mean root fresh weight and total root length were the averages of six to eight plants per treatment. Fisher's protected least significant difference (LSD) test at $P < 0.05$ was used to compare mean values from Castlemart and *jai1-1* plants in all field and greenhouse experiments (Statistix 8.1, Analytical Software, Tallahassee, Florida, USA). Significant differences among the means were indicated by different letters.

3.3 *Results and Discussion*

3.3.1 *jai1-1* Homozygotes Showed Enhanced Susceptibility to *Pythium ultimum* in the Field and Greenhouse

Roots and rhizosphere soils of homozygous *jai1-1* plants harbored substantially more *P. ultimum* DNA than those of Castlemart after growth in naturally infested soil (Table 1). Differentials of about 300- and 60-fold were observed in *jai1-1* roots in 2009 and 2010, respectively, and about 5–120-fold in *jai1-1* rhizosphere soils in 2009 and 2010. Our findings indicated that a deficiency in JA signaling enhanced the susceptibility of tomato to the pathogen.

A single amplicon was obtained in PCR assays, indicating that *P. ultimum* was the sole or predominant species in field-grown roots and rhizosphere soils. The differential in *P. ultimum* DNA was observed in roots of the two genotypes when they were grown in the greenhouse using soil taken from the 2009 field plot. The roots of *jai1-1* plants harbored an average of 321 pg DNA $root^{-1}$ compared to those of Castlemart, at 2.4 pg DNA $root^{-1}$ (data not shown).

The roots of these *jai1-1* plants displayed an additional PCR product, likely a second *Pythium* species. To test the hypothesis that the second species was the common tomato pathogen *P. aphanidermatum*, primers were designed for the ITS

Table 1 Real-time PCR quantification of *Pythium ultimum* DNA (pg)[a] in roots and rhizosphere soils of wild-type Castlemart and homozygous *jai1-1* tomato plants in 2009 and 2010 field plots

	Harvest point			
	2009[b]		2010[b]	
Genotype	pg root^{-1}	pg g^{-1} soil	pg root^{-1}	pg g^{-1} soil
8 dpi in field				
Castlemart	12.7 b	0.5 b	53.3 b	29.7 b
jai1-1	4226 a	57.9 a	3195 a	1572 a
15 dpi in field				
Castlemart	5.8 b	2.9 b	29.3 b	129 b
jai1-1	1730 a	19.4 a	1854 a	470 a

[a]Letters indicate significant ($P < 0.05$) differences between means of three independent root or soil samples from wild-type and *jai1* plants at each harvest point
[b]Twenty-eight-day-old plants were transferred to field plots and harvested 8 and 15 days after planting (dpi)

region of this pathogen and found to amplify total DNA samples from the roots. The *P. aphanidermatum* primers did not amplify DNA from *P. ultimum* or nine other *Pythium* species and detected *P. irregulare* group IV DNA with 10^4–10^5 less sensitivity than *P. aphanidermatum* DNA (data not shown).

3.3.2 *jai1-1* Was Susceptible to *P. aphanidermatum* in Greenhouse Pathogenicity Assays

The BC_2F_5 population in the Castlemart genetic background segregated for the *jai1-1* mutation, The BC_2F_5 seedlings resembled the Castlemart parental line. Nevertheless, we compared the root variables of *Pythium*-infected seedlings relative to noninfected seedlings for each genotype, to normalize for subtle inherent differences in root development.

In greenhouse assays, root dry weights of Castlemart were reduced about 40 % after 14 days of growth in 500 ppg of *P. aphanidermatum*, whereas root weights of *jai1-1* homozygotes dropped about 85 % (Table 2), supporting the observation that loss of the JA signal pathway resulted in enhanced susceptibility.

Roots of Castlemart and homozygous *jai1-1* generally were indistinguishable in the absence of the pathogen, but roots of the latter were more severely stunted with 100–500 ppg of *P. aphanidermatum* (Fig. 3). As expected, *Jai1* homozygotes and wild-type Castlemart showed similar reductions in root dry weight and total root length, and *Jai1*/*jai1-1* heterozygotes were somewhat less sensitive to the pathogen than the *jai1-1* homozygotes (Table 3). The latter genotype also showed enhanced susceptibility to *P. irregulare* group I (data not shown). Our data demonstrates that the absence of a functional JA signal pathway in tomato results in enhanced root susceptibility to *Pythium* species and supports observations reported in other plant species.

Table 2 Root dry weights (mg)[a] of Castlemart and homozygous *jai1-1* tomato plants after 14 days of growth in non-infested soil or in soil infested with *Pythium aphanidermatum* in the greenhouse

	Expt 1		Expt 2	
Inoculum	Castlemart	*jai1-1*	Castlemart	*jai1-1*
0 ppg	57.0 ± 9.3	59.0 ± 7.1	65.8 ± 4.9	69.0 ± 5.0
500 ppg	33.9 ± 1.9	8.0 ± 1.3*	39.5 ± 2.3	9.7 ± 0.8*
Ratio[b]	0.59	0.13	0.60	0.14

[a]Means and standard errors of 6–8 control (0 ppg) and *Pythium*-treated roots (500 ppg). *Asterisks* indicate significant ($P < 0.05$) differences between means of the genotypes at a given inoculum density

[b]Ratio of average weights of pathogen infected and noninfected roots for each genotype

Fig. 3 Roots of wild-type Castlemart (**a**) and *jai1-1* homozygotes (**b**) after 14 days of growth in soil infested with 0, 100, 250, and 500 propagules g^{-1} soil (ppg) of *Pythium aphanidermatum*

Table 3 Mean root weight (mg)[a] and total root length (cm)[a] of wild-type Castlemart and *Jai1* tomato genotypes after 14 days of growth in soil infested with *Pythium aphanidermatum*

Genotype	Inoculum (ppg)	Root dry wt (mg)	Weight ratio[b]	Root length (cm)	Length ratio[b]
Castlemart	0	207 bcd	0.78	127 ab	0.60
	500	161 cd		77 bc	
Jai1/*Jai1*	0	230 abc	0.79	157 a	0.50
	500	181 cd		78 bc	
Jai1/*jai1*	0	254 ab	0.66	157 a	0.54
	500	168 cd		85 bc	
jai1/*jai1*	0	301 a	0.38	165 a	0.33
	500	115 d		55 c	

[a]Letters indicate mean significance classes determined using Fischer's protected LSD test ($P < 0.05$) for all values within the column
[b]Ratios of values at 500 ppg relative 0 ppg for each genotype

4 Tritrophic Signaling in *Pythium oligandrum* Biocontrol Interactions

Among the 130 recognized species of *Pythium*, several are distinctive for their disease-suppressive properties. The best characterized is *Pythium oligandrum* (*PO*), a ubiquitous mycoparasite of *Phytophthora*, *Trichoderma*, and other *Pythium* spp. The mycoparasite uses a number of signaling strategies to interact with target fungi in the soil and to induce defence responses in tomato, wheat, sugar beet, and other plants (reviewed in Benhamou et al. 2012; Gerbore et al. 2014). Mycoparasitism is manifest as a coiling of the hyphae of *PO* around that of its target, followed by proliferation of *PO* hyphae and cytoplasmic disorganization and loss within target cells (Benhamou et al. 2001, 2012). This species is not an endophyte, as the hyphae decline after an initial rapid colonization of host roots (Picard et al. 2000; Takenaka et al. 2008), possibly due to the inability of *PO* to tolerate host defences or of the host to support fungal replication. *PO* responds to an uncharacterized chitin complex from the cell wall of the target *Fusarium oxysporum* f. sp. *radicis-lycopersici* and undergoes adhesion to the target surface. Induction of cellulases and proteases in *PO* by the *Fusarium* is proposed to be involved in adhesion (Horner et al. 2012).

Certain pathogen-suppressive *PO* appears to modulate rhizosphere phytohormone levels, leading to plant growth promotion. In one case, *PO* produced tryptamine and low levels of an auxin-like metabolite in vitro if the precursor tryptophan was added at specific concentrations to the culture media. Furthermore, tomato roots exposed to the culture media appeared to take up the auxin-like metabolite and accumulated more biomass than roots grown in medium without tryptophan (reviewed in Benhamou et al. 2012). In contrast, an auxin-producing isolate of moderately pathogenic *Pythium* group F, which caused yield reduction without visible symptoms, produced abnormal root morphology and browning lesions (Le Floch et al. 2003), indicating that the activity of plant growth promoting factor is threshold-sensitive and likely conditions host defence reactions. *PO* colonization was not accompanied by the hypersensitive response that is common in pathogenic host interactions (Picard et al. 2000).

PO reduced populations of *P. dissotocum* Drechsler in a hydroponic tomato growing system (Vallance et al. 2009), indicating that its activity is based in part on external signals. Isolates of *PO* produce small secreted protein and peptide signal molecules that trigger systemic resistance and reduce disease symptoms caused by a range of foliar and soilborne pathogens (Picard et al. 2000; Hase et al. 2006; Takenaka et al. 2003, 2006, 2008). Oligandrin, a 10 kDa secreted peptide found in the supernatant fraction of *PO* cultures, was translocated from the site of application at the petiole or excised leaf to intact leaves, indicating its potential for inducing systemic resistance. In an interesting variation, oligandrin applied to tomato stems without leaves was able to elicit root defences against the soilborne pathogen *Fusarium oxysporum* f. sp. *radicis-lycopersici* (Benhamou et al. 2001).

A second group of proteinaceous signal molecules, called POD, are present in cell wall protein fractions of *PO* and have been shown to induce defence responses in host roots. POD proteins harbor elicitin domains initially found as conserved motifs in *Phytophthora* effectors, but the POD form a phylogenetic cluster distinct from the elicitins ELI and ELL of *Phytophthora* (Takenaka et al. 2006; Masunaka et al. 2010). The effects of POD on defence and protection vary with both host and pathogen species (Benhamou et al. 2012). Unlike the elicitins, oligandrin and POD do not trigger a hypersensitive response in host plants or in *Nicotiana benthamiana* leaf assays (Picard et al. 2000; Takenaka et al. 2006; Masunaka et al. 2010).

PO genotypes produced different structural isoforms of POD, including POD-1 and POD-2, which varied in ability to induce defence proteins. In sugar beet roots, POD-1 and POD-2 differentially regulated phenylalanine ammonia lyase (PAL), chitinase and cell wall-associated ferulic acid, and defence genes encoding oxalate oxidase and glutathione S-transferase (Takenaka et al. 2003, 2006). When roots of tomato seedlings were treated with hexameric (bioactive) forms of POD-1, mRNAs encoding PR-6, proteinase inhibitor II, PR-2b, a basic glucanase, and LeCAS, an enzyme in the hydrogen cyanide detoxification pathway, were induced in the roots (Takenaka et al. 2011). The induction of *PR-6* and *LeCAS* implicated the involvement of the JA and ethylene (ET) defence pathways, respectively. Mycelial homogenates and cell wall proteins extracts of *PO* induced the accumulation of ET in tomato roots (Hase et al. 2006; Takenaka et al. 2011) and induced *ETR4* (E receptor), *ERF2* (ET-responsive transcription factor), and three pathogenesis-related mRNAs known to regulated by the ET pathway (Hase et al. 2006). Using a similar system, Hase et al. (2008) demonstrated that the JA-responsive *PR-6* (Kunitz trypsin inhibitor) gene was induced by *PO* extracts in wild-type but not in *jai1-1* tomato plants. In *Arabidopsis* mutants *coi1*, *jar1*, *ein2*, and *etr1* that were deficient in JA or E signaling, cell wall proteins fractions of *PO* failed to induce defence gene expression (Kawamura et al. 2009). *PO*-mediated defence was also systemic, as indicated by defence gene induction in leaves and suppression of foliar pathogens (Hase et al. 2006, 2008; Kawamura et al. 2009). The induction of ET by secreted peptides of *PO* distinguishes this defence signaling pathway from that of induced systemic resistance following *Pseudomonas* root colonization (Sect. 5), in which JA and ET levels remain constant (Pieterse et al. 2000). Host receptors of the *PO* peptides remain unidentified.

Despite having activity against soilborne plant pathogens, *PO* appears to have minimal impact on rhizosphere microbial populations per se. For instance, growth of the pathogen *Rhizoctonia solani* Kühn was not substantially suppressed in vitro or in the rhizosphere, and niche competition was not indicated by the transient nature of *PO* populations (Takenaka et al. 2003, 2008). Production of diffusible and stable antimicrobial compounds by *PO* has not been documented, with the exception of possible volatiles (Gerbore et al. 2014). The diversity of rhizoplane bacterial communities from hydroponically grown tomato roots shifted over an 8-month sampling period, but the changes were not consistently associated with the presence or absence of *PO* (Vallance et al. 2012). The effectiveness of *PO* as a biocontrol organism might lie with its nonspecific but localized and transient activities.

5 Signaling Between Pathogenic *Pythium*, Plants, and Biocontrol Bacteria

Plant pathogenic species of *Pythium* are subject to suppression by biocontrol bacteria, such as *Pseudomonas* and *Bacillus* spp. Suppression results from the action of antifungal metabolites, nutrient competition, iron chelation, phytohormone (growth hormone) production, and induced systemic resistance in the host (reviewed in Martin and Loper 1999; Lugtenberg and Kamilova 2009; Mavrodi et al. 2006). The antifungal metabolites 2,4-diacetylphloroglucinol (DAPG), phenazine-1-carboxylic acid and derivatives, cyclic and straight-chain lipopeptide surfactants, hydrogen cyanide (HCN), and other volatile compounds disrupt hyphal integrity, cytoplasmic organization, or cellular functions, or interfere with the life cycle of *Pythium* spp. For instance, *Bacillus cereus* UW85 and the germ tube elongation inhibitor zwittermicin A reduced movement of *P. torulosum* zoospores around roots of tomato, although a higher degree of disease suppression was conferred by intact *B. cereus* cells (Shang et al. 1999). Applied DAPG was particularly effective against zoospores of the sugar beet pathogen *P. ultimum* var. *sporangiiferum* Drechsler, causing rapid zoospore immobility and disintegration. Hyphae of the pathogen displayed abnormal plasma membrane morphology, cytoplasmic vesiculation, and disorganization of cellular contents (de Souza et al. 2003a). Siderophores pyoluteorin, pyoverdin, and pyochelin sequester iron to the detriment of *Pythium* (Buysens et al. 1996). Inhibition of hyphal growth by bacterium-derived antifungal metabolites often has been observed on Petri plates, in which the bacterium is grown adjacent to the target pathogen. However, metabolite activity in the rhizoplane or field depends on biotic and abiotic factors that favor niche establishment of the bacterium and production, dispersal, and stability of the metabolite in order to attain bioactive thresholds. Furthermore, *Pythium* spp.

and isolates exhibit differential sensitivity to antifungal metabolites (e.g., Nielsen et al. 2002; de Souza et al. 2003a; Mazzola et al. 2007).

Quantity and quality of native root exudates differ among hosts of the same pathogen and can be modulated by biocontrol metabolites (Martin and Loper 1999; Phillips et al. 2004), so it is not surprising that host genotype is a major driver of rhizosphere microbial activity. Two sugar beet cultivars harboring different rhizosphere *Pseudomonas* differentially regulated *P. aeruginosa* transcripts in a cultivar-specific manner (Mark et al. 2005). Many transcripts were identified as having metabolic functions, allowing the bacterium to adapt to quality and quantity of host exudates. *Pseudomonas fluorescens* isolates from the sugar beet rhizosphere controlled *P. ultimum* on barley and sugar beet roots in vitro and in planta (Nielsen et al. 1998; Jousset et al. 2011). In the sugar beet interaction, disease suppression was correlated to growth in high glucose medium and attributed to DAPG (Nielsen et al. 1998). In the barley interaction, the DAPG biosynthetic locus was upregulated in the bacterium (Jousset et al. 2011).

Systemic induction of defence by *P. fluorescens* strain CHA0 was demonstrated in barley using a split root system, in which roots were physically separated into two portions; the proximal half was treated with the bacterium and the distal half was treated with the pathogen. *Pythium* infection was associated with increased root exudation of the secondary metabolites vanillic, fumaric, and *p*-coumaric acids in the distal portion. Application of these compounds to roots induced *PhlA* expression (Jousset et al. 2011). In this case, DAPG might be the systemic signal, as has been observed in induced systemic resistance in *Arabidopsis* (Weller et al. 2012), but the signal pathway remains unknown. However, these findings demonstrate both indirect and direct effects of strain CHA0 on *Pythium* disease suppression.

Cyclic lipopeptide surfactants (CLP) represent a diverse structural class of anti-*Pythium* metabolites that also are involved in motility and biofilm formation. These compounds are synthesized by non-ribosomal peptide synthesis and polyketide synthesis loci in bacteria and vary in the numbers and types of amino acids in the peptide ring backbone and in the composition of the fatty acid side chains (Raaijmakers et al. 2006). The amphipathic nature of the peptide ring and lipid side chain renders the CLP somewhat soluble, with potential for membrane and cell wall disruption (Schneider et al. 2014). A novel CLP, named viscosinamide, produced by *Pseudomonas fluorescens* strain DR54A, suppressed damage by the sugar beet pathogen *P. ultimum* (Nielsen et al. 1998, 1999). Viscosinamide caused abnormal hyphal morphology in and encystment of zoospores. A more extensive survey of pseudomonads from sugar beet revealed additional *Pseudomonas* spp. that controlled *P. ultimum* (Nielsen et al. 2002). The isolates grouped into two biovars based on CLP production and carbon utilization profiles; isolates active against the pathogen all produced a common CLP having an 11-amino acid peptide ring and a 3-hydroxydecanoyl side chain. In this collection, HCN did not appear to be the primary active metabolite in *Pythium* suppression. A CLP produced by *P. fluorescens* strain SS101 caused rapid lysis of zoospores of *P. ultimum* var. *sporangiiferum* and *P. intermedium*, causal agents of *Pythium* root rot of hyacinth (de Souza et al. 2003b), and was identified as massetolide A (de Bruijn et al. 2008)

(Fig. 1c). However, a transposon mutation in the massetolide biosynthetic locus of strain SS101 did not compromise suppression of *P. irregulare*, *P. sylvaticum*, and *P. ultimum* var. *ultimum* nor systemic resistance in wheat or apple seedlings (Mazzola et al. 2007). Since these *Pythium* spp. are not prolific zoospore producers, factors other than CLP might be involved in suppression. The data suggest that a biocontrol strain that produces multiple factors having different mechanisms of disease suppression and utilizes non-overlapping biosynthetic pathways for production of the factors is the most competitive (Xu et al. 2011). Synergy between phenazines and rhamnolipid biosurfactants was observed against *P. splendens* Hans Braun of bean and *P. myriotylum* Drechsler of cocoyam (*Xanthosoma sagittifolium* L. Schott) (Perneel et al. 2008).

Biocontrol bacteria harbor type III secretion systems (TTSS), as do symbiotic rhizobia and plant-pathogenic bacteria. In the latter, the TTSS plays a role in the delivery of virulence proteins to host cells, leading to disease. If the host has adapted to recognize the virulence protein and protein recognition has been linked to a defence pathway, then the outcome can be disease resistance. The role of the TTSS in biocontrol interactions generally is understudied, and it is not clear whether it conditions interactions with the host, or with the target pathogen, or both. In the case of *P. fluorescens* strain KD, which protects cucumber seedlings against *P. ultimum*, several lines of evidence indicate that the TTSS is involved in pathogen rather than host interactions (Rezzonico et al. 2005). The expression of the TTSS locus, monitored using the *hrpJ′:inaZ* reporter construct, was induced in vitro by *P. ultimum* but not by autoclaved cucumber seedlings. Expression was also induced in the rhizosphere if the pathogen was present. An insertional mutation in the TTSS gene *hrcV* of strain KD did not affect cucumber seedling growth and vigor, or bacterial rhizosphere populations in absence of *P. ultimum*. However, the mutant was reduced in suppressiveness when the pathogen was present, and activity of the pathogenicity factor pectinase polygalacturonase in *Pythium* was reduced more in wild type compared to the mutant. The findings provide a framework for future signaling studies between pathogen, biocontrol bacteria, and plants.

6 Concluding Remarks

The JA and ET pathways have been recruited for defence signaling in roots during interactions with other types of microbes, including rhizobia and *Trichoderma*, and it is natural to ask whether pathway components can be modulated for defence against *Pythium*. One unique JA-dependent signaling of innate immunity to *Pythium* involves endogenous host peptides (Huffaker et al. 2006; Huffaker and Ryan 2007). The propeptides are induced by JA, the ET mimic ethephon, and wounding and also are auto-induced. Overexpression of the propeptides results in increased expression of JA-responsive defence genes and root biomass in the presence of *P. irregulare* in *Arabidopsis* (Huffaker and Ryan 2007). This intriguing signal pathway has yet to be explored in roots of *Pythium*-susceptible crops.

Few clues regarding *Pythium* defence signaling in plants have been obtained from disease-resistant genotypes of small grain cereals (Okubara and Jones 2011). Resistance or tolerance is considered to be multigenic in most cases, as is the case for *P. ultimum* resistance in bean, attributed to seed coat color, seedling emergence, and vigor (Campa et al. 2010). One exception is the *CzR1* locus for resistance to *P. aphanidermatum* in wild turmeric (*Curcuma zedoaria* Loeb.) which encodes a protein structurally similar to the barley powdery mildew resistance proteins Mla1 and MLO (coiled-coil nucleotide binding site leucine-rich repeat domain protein, or CC-NBS-LRR) and other proteins conferring race-specific resistance to biotrophic pathogens (Joshi et al. 2013; Kar et al. 2013). Structural modeling indicated that six amino acid residues in the folded protein potentially can form hydrogen bonds with a β-1,3-D-glucan ligand from the cell wall of *P. aphanidermatum* (causal agent of rhizome rot of ginger), possibly leading to enzymatic cleavage of wall polymers (Joshi et al. 2013). Recognition of structural components of microbes by plants also is a feature of pathogen-triggered immunity. Genomic approaches are being used to identify *Pythium* genes involved in pathogenicity (Horner et al. 2012; Lévesque et al. 2010) and might provide leads to host-induced gene silencing (HIGS) for control of specific *Pythium* spp.

Finally, *Pythium* in native and agroecosystems is one genus in a complex and dynamic community of organisms interconnected by different signals. Metagenomic profiling is beginning to shed light on community composition, but time and effort is required to understand the biological function of genes and of rhizosphere community members relative to *Pythium* disease and management. Expanded knowledge about signals used by other mycoparasitic and nonpathogenic *Pythium* spp. will expand our understanding of the perception and responses of host plants and target pathogens, and, possibly, the evolution of mycoparasitic *Pythium* interactions.

Acknowledgments The authors thank Nathalie Walter and Joe Hulbert for expert technical assistance, and Erin Rosskopf for the gift of *P. aphanidermatum*. This work was supported by grants 3019-3019-4564 and 3019-3564 from the Washington Wheat Commission and by USDA ARS Project Number 2090 22000 016 00D (P.O.) and the Chemical Sciences, Geosciences, and Biosciences Division, Office of Basic Energy Sciences, Office of Science, US Department of Energy (grant no. DE-FG02-91ER20021 (G.A.H.). G.A.H. also acknowledges support from the Michigan AgBioResearch Project MICL02278. References to a company and/or product by the USDA are only for the purposes of information and do not imply approval or recommendation of the product to the exclusion of others that may also be suitable. USDA is an equal opportunity employer.

References

Benhamou N, Bélanger RR, Rey P, Tirilly Y (2001) Oligandrin, the elicitin-like protein produced by the mycoparasite *Pythium oligandrum*, induces systemic resistance to *Fusarium* crown and root rot in tomato plants. Plant Physiol Biochem 39:681–698

Benhamou N, le Floch G, Vallance J, Gerbore J, Grizard D, Rey P (2012) *Pythium oligandrum*: an example of opportunistic success. Microbiology 158:2679–2694

Bouarab K, Melton R, Pearl J, Baulcombe D, Osbourn A (2002) A saponin-detoxifying enzyme mediates suppression of plant defences. Nature 418:889–892

Buysens S, Heungens K, Poppe J, Höfte M (1996) Involvement of pyochelin and pyoverdin in suppression of *Pythium*-induced damping-off of tomato by *Pseudomonas aeruginosa* 7NSK2. Appl Environ Microbiol 62:865–871

Campa A, Pérez-Vega E, Pascual A, Ferreira JJ (2010) Genetic analysis and molecular mapping of quantitative trait loci in common bean against *Pythium ultimum*. Phytopathology 100:1315–1320

Campos ML, Kang J-H, Howe GA (2014) Jasmonate-triggered plant immunity. J Chem Ecol 40:657–675

Carter JP, Spink J, Cannon PF, Daniels MJ, Osbourn AE (1999) Isolation, characterization, and avenacin sensitivity of a diverse collection of cereal-root-colonizing fungi. Appl Environ Microbiol 65:3364–3372

Chellemi DO, Mitchell DJ, Kannwischer-Mitchell ME, Rayside PA, Rosskopf EN (2000) *Pythium* associated with bell pepper production in Florida. Plant Dis 84:1271–1274

Davis DJ, Lanter K, Makselan S, Bonati C, Asbrock P, Ravishankar JP, Money NP (2006) Relationship between temperature optima and secreted protease activities of three *Pythium* species and pathogenicity toward plant and animal hosts. Mycol Res 110:96–103

Deacon JW, Mitchell RT (1985) Toxicity of oat roots, oat root extracts, and saponins to zoospores of *Pythium* spp. and other fungi. Trans Br Mycol Soc 84:479–487

de Bruijn I, de Kock MJD, de Waard P, van Beek TA, Raaijmakers JM (2008) Massetolide A biosynthesis in *Pseudomonas fluorescens*. J Bacteriol 190:2777–2789

de Souza JT, Arnould C, Deulvot C, Lemanceau P, Gianinazzi-Pearson V, Raaijmakers JM (2003a) Effect of 2,4-diacetylphloroglucinol on *Pythium*: cellular responses and variation in sensitivity among propagules and species. Phytopathology 93:966–975

de Souza JT, de Boers M, de Waard P, van Beek TA, Raajimakers JM (2003b) Biochemical, genetic, and zoosporicidal properties of cyclic lipopeptide surfactants produced by *Pseudomonas fluorescens*. Appl Environ Microbiol 69:7161–7172

Estrada-Garcia T, Ray TC, Green JR, Callow JA, Kennedy JF (1990) Encystment of *Pythium aphanidermatum* zoospores is induced by root mucilage polysaccharides, pectin and a monoclonal antibody to a surface antigen. J Exp Bot 41:693–699

Gerbore J, Benhamou N, Vallance J, Le Floch G, Grizard D, Regnault-Roger C, Rey P (2014) Biological control of plant pathogens: advantages and limitations seen through the case study of *Pythium oligandrum*. Environ Sci Pollut Res 21:4847–4860

Hase S, Shimizu K, Nakaho K, Takenaka S, Takahashi H (2006) Induction of transient ethylene and reduction in severity of tomato bacterial wilt by *Pythium oligandrum*. Plant Pathol 55:537–543

Hase S, Takahashi S, Takenaka S, Nakaho K, Arie T, Seo S, Ohashi Y, Takahasi H (2008) Involvement of jasmonic acid signalling in bacterial wilt disease resistance induced by biocontrol agent *Pythium oligandrum* in tomato. Plant Pathol 57:870–876

Horner NR, Grenville-Briggs LJ, van West P (2012) The oomycete *Pythium oligandrum* expresses putative effectors during mycoparasitism of *Phytophthora infestans* and is amenable to transformation. Fungal Biol 116:24–41

Huffaker A, Ryan CA (2007) Endogenous peptide defense signals in Arabidopsis differentially amplify signaling for the innate immune response. Proc Nat Acad Sci USA 104:10732–10736

Huffaker A, Pearce G, Ryan CA (2006) An endogenous peptide signal in Arabidopsis activates components of the innate immune response. Proc Natl Acad Sci U S A 103:10098–10103

Ivanov DA, Bernards MA (2012) Ginsenosidases and the pathogenicity of *Pythium irregulare*. Phytochemistry 78:44–53

Joshi RK, Nanda S, Rout E, Kar B, Naik PK, Nayak S (2013) Molecular modeling and docking characterization of *CzR1*, a CC-NBS-LRR *R*-gene from *Curcuma zedoaria* Loeb. that confers resistance to *Pythium aphanidermatum*. Bioinformation 9:560–564

Jousset A, Rochat L, Lanoue A, Bonkowski M, Keel C, Scheu S (2011) Plants respond to pathogen infection by enhancing the antifungal gene expression of root-associated bacteria. Mol Plant Microbe Interact 24:352–358

Kar B, Nanda S, Nayak PK, Nayak S, Joshi RJ (2013) Molecular characterization and functional analysis of CzR1, a coiled-coil-nucleotide-binding-site-leucine-rich repeat R-gene from *Curcuma zedoaria* Loeb. that confers resistance to *Pythium aphanidermatum*. Physiol Mol Plant Pathol 83:59–68

Kawamura Y, Takenaka S, Hase S, Kubota M, Ichinose Y, Kanayama Y, Nakaho K, Klessig D, Takahashi H (2009) Enhanced defense responses in Arabidopsis induced by the cell wall protein fractions from Pythium oligandrum require SGT1, RAR1, NPR1 and JAR1. Plant Cell Physiol 50:924–934

Le Floch G, Rey P, Benizri E, Benhamou N, Tirilly Y (2003) Impact of auxin-compounds produced by the antagonistic fungus *Pythium oligandrum* or the minor pathogen *Pythium* group F on plant growth. Plant Soil 257:459–470

Leung KW, Wong AS-T (2010) Pharmacology of ginsenosides: a literature review. Chin Med 5:20

Lévesque CA, Brouwer H, Cano L et al (2010) Genome sequence of the necrotrophic plant pathogen *Pythium ultimum* reveals original pathogenicity mechanisms and effector repertoire. Genome Biol 11:R73

Li L, Zhao Y, McCaig BC, Wingerd BA, Wang J, Whalon ME, Pichersky E, Howe GA (2004) The tomato homolog of CORONATINE-INSENSITIVE1 is required for the maternal control of seed maturation, jasmonate-signaled defense responses, and glandular trichome development. Plant Cell 16:126–143

Lin R-C, Ding Z-S, Li L-B, Kuang T-Y (2001) A rapid and efficient DNA minipreparation suitable for screening transgenic plants. Plant Mol Biol Rep 19:379a–379e

Lugtenberg B, Kamilova F (2009) Plant-growth-promoting rhizobacteria. Annu Rev Microbiol 63:541–556

Mao L, Li X, Smyth EM, Yannarell AC, Mackie RI (2014) Enrichment of specific bacterial and eukaryotic microbes in the rhizosphere of switchgrass (*Panicum virgatum* L.) through root exudates. Environ Microbiol Rep 6:293–306

Mark GL, Dow JM, Kiely PD, Higgins H, Haynes J, Baysse C, Abbas A, Foley T, Franks A, Morrissey J, O'Gara F (2005) Transcriptome profiling of bacterial responses to root exudates identifies genes involved in microbe–plant interactions. Proc Natl Acad Sci U S A 102:17454–17459

Martin FN, Loper JE (1999) Soilborne plant diseases caused by *Pythium* spp.: ecology, epidemiology, and prospects for biological control. Crit Rev Plant Sci 18:111–181

Masunaka A, Sekiguchi H, Takahashi H, Takenaka S (2010) Distribution and expression of elicitin-like protein genes of the biocontrol agent *Pythium oligandrum*. J Phytopathol 158:417–426

Mavrodi DV, Blankenfeldt W, Thomashow LS (2006) Phenazine compounds in fluorescent *Pseudomonas* spp. biosynthesis and regulation. Annu Rev Phytopathol 44:417–445

Mazzola M, Zhao X, Cohen MF, Raaijmakers JM (2007) Cyclic lipopeptide surfactant production by *Pseudomonas fluorescens* SS101 is not required for suppression of complex *Pythium* spp. populations. Phytopathology 97:1348–1355

Mitchell RT, Deacon JW (1985) Differential (host-specific) accumulation of zoospores of *Pythium* on roots of graminaceous and non-graminaceous plants. New Phytol 102:113–122

Nicol RW, Traquair JA, Bernards MA (2002) Ginsenosides as host resistance factors in American ginseng (*Panax quinquefolius*). Can J Bot 80:557–562

Nicol RW, Yousef L, Traquair JA, Bernards MA (2003) Ginsenosides stimulate the growth of soilborne pathogens of American ginseng. Phytochemistry 64:257–264

Nielsen MN, Sørensen J, Fels J, Pedersen HC (1998) Secondary metabolite- and endochitinase-dependent antagonism toward plant-pathogenic microfungi of *Pseudomonas fluorescens* isolates from sugar beet rhizosphere. Appl Environ Microbiol 64:3563–3569

Nielsen TH, Christophersen C, Anthoni U, Sørensen J (1999) Viscosinamide, a new cyclic depsipeptide with surfactant and antifungal properties produced by *Pseudomonas fluorescens* DR54. J Appl Microbiol 86:80–90

Nielsen TH, Sørensen D, Tobiasen C, Andersen JB, Christophersen C, Givskov M, Sørensen J (2002) Antibiotic and biosurfactant properties of cyclic lipopeptides produced by fluorescent *Pseudomonas* spp. from the sugar beet rhizosphere. Appl Environ Microbiol 68:3416–4323

Oh JY, Kim Y-J, Jang M-G, Joo SC, Kwon W-S, Kim S-Y, Jung S-K, Yang D-C (2014) Investigation of ginsenosides in different tissues after elicitor treatment in *Panax ginseng*. J Ginseng Res 38:270–277

Okubara PA, Jones SS (2011) Seedling resistance to *Rhizoctonia* and *Pythium* in wheat chromosome group 4 addition lines from *Thinopyrum* spp. Can J Plant Pathol 33:415–422

Okubara PA, Li C, Schroeder KL, Schumacher RT, Lawrence NP (2007) Improved extraction of *Rhizoctonia* and *Pythium* DNA from wheat roots and soil samples using pressure cycling technology. Can J Plant Pathol 29:304–310

Okubara PA, Schroeder KL, Paulitz TC (2008) Identification and quantification of *Rhizoctonia solani* and *R. oryzae* using real-time polymerase chain reaction. Phytopathology 98:837–847

Osbourn A, Goss RJM, Field RA (2011) The saponins—polar isoprenoids with important and diverse biological activities. Nat Prod Rep 28:1261–1268

Paulitz TC, Adams K (2003) Composition and distribution of *Pythium* communities in wheat fields in eastern Washington state. Phytopathology 93:867–873

Perneel M, D'hondt L, De Maeyer K, Adiobo A, Rabaey K, Höfte M (2008) Phenazines and biosurfactants interact in the biological control of soil-borne diseases caused by *Pythium* spp. Environ Microbiol 10:778–788

Phillips DA, Fox TC, King MD, Bhuvaneswari TV, Teuber LR (2004) Microbial products trigger amino acid exudation from plant roots. Plant Physiol 136:2887–2894

Picard K, Ponchet M, Blein J-P, Rey P, Tirilly Y, Benhamou N (2000) Oligandrin. A proteinaceous molecule produced by the mycoparasite *Pythium oligandrum* induces resistance to *Phytophthora parasitica* infection in tomato plants. Plant Physiol 124:379–395

Pieterse CMJ, Van Pelt JA, Ton J, Parchmann S, Mueller MJ, Buchala AJ, Métraux J-P, Van Loon LC (2000) Rhizobacteria-mediated induced systemic resistance (ISR) in *Arabidopsis* requires sensitivity to jasmonate and ethylene but is not accompanied by an increase in their production. Physiol Mol Plant Pathol 57:123–134

Pieterse CMJ, Van Wees SCM, Ton J, Van Pelt JA, Van Loon LC (2002) Signalling in rhizobacteria-induced systemic resistance in *Arabidopsis thaliana*. Plant Biol 4:535–544

Raaijmakers JM, de Bruijn I, de Kock MJD (2006) Cyclic lipopeptide production by plant-associated *Pseudomonas* spp.: diversity, activity, biosynthesis, and regulation. Mol Plant Microbe Interact 7:699–710

Raftoyannis Y, Dick MW (2006) Effect of oomycete and plant variation on zoospore cover and disease severity. J Plant Pathol 88:95–101

Rey P, Benhamou N, Tirilly Y (1998) Ultrastructural and cytochemical investigation of asymptomatic infection by *Pythium* spp. Phytopathology 88:234–244

Rezzonico F, Binder C, Défago G, Möenne-Loccoz Y (2005) The type III secretion system of biocontrol *Pseudomonas fluorescens* KD targets the phytopathogenic Chromista *Pythium ultimum* and promotes cucumber protection. Mol Plant Microbe Interact 18:991–1001

Schneider T, Müller A, Miess H, Gross H (2014) Cyclic lipopeptides as antibacterial agents—potent antibiotic activity mediated by intriguing mode of actions. Int J Med Microbiol 304:37–43

Schroeder KL, Paulitz TC (2006) Root diseases of wheat and barley during the transition from conventional tillage to direct seeding. Plant Dis 90:1247–1253

Schroeder KL, Okubara PA, Tambong JT, Lévesque CA, Paulitz TC (2006) Identification and quantification of pathogenic *Pythium* spp. from soils in eastern Washington using real-time PCR. Phytopathology 96:637–647

Schroeder KL, Martin FN, de Cock AWAM, Lévesque CA, Spies CFJ, Okubara PA, Paulitz TC (2013) Molecular detection and quantification of *Pythium* species: evolving taxonomy, new tools, and challenges. Plant Dis 97:4–20

Shang H, Chen J, Handelsman J, Goodman RM (1999) Behavior of *Pythium torulosum* zoospores during their interaction with tobacco roots and *Bacillus cereus*. Curr Microbiol 38:199–204

Sheard LB, Tan X, Mao H, Withers J, Ben-Nissan G, Hinds TR, Kobayashi Y, Hsu F-F, Sharon M, Browse J, He SY, Rizo J, Howe GA, Zheng N (2010) Jasmonate perception by inositol-phosphate-potentiated COI1-JAZ co-receptor. Nature 468:400–407

Staswick PE, Tiryaki I (2004) The oxylipin signal jasmonic acid is activated by an enzyme that conjugates it to isoleucine in Arabidopsis. Plant Cell 16:2117–2127

Staswick PE, Yuen GY, Lehman CC (1998) Jasmonate signaling mutants of *Arabidopsis* are susceptible to the soil fungus *Pythium irregulare*. Plant J 15:747–754

Staswick PE, Tiryaki I, Rowe ML (2002) jasmonate response locus *JAR1* and several related Arabidopsis genes encode enzymes of the firefly luciferase superfamily that show activity on jasmonic, salicylic, and indole-3-acetic acids in an assay for adenylation. Plant Cell 14:1405–1415

Takenaka S, Nishio Z, Nakamura Y (2003) Induction of defense reactions in sugar beet and wheat by treatment with cell wall protein fractions from the mycoparasite *Pythium oligandrum*. Phytopathology 93:1228–1232

Takenaka S, Nakamura Y, Kono T, Sekiguchi H, Masunaka A, Takahashi H (2006) Novel elicitin-like proteins isolated from the cell wall of the biocontrol agent *Pythium oligandrum* induce defence related genes in sugar beet. Mol Plant Pathol 7:325–339

Takenaka S, Sekiguchi H, Nakaho K, Tojo M, Masunaka A, Takahashi H (2008) Colonization of *Pythium oligandrum* in the tomato rhizosphere for biological control of bacterial wilt disease analyzed by real-time pcr and confocal laser-scanning microscopy. Phytopathology 98:187–195

Takenaka S, Yamaguchi K, Masunaka A, Hase S, Inoue T, Takahashi H (2011) Implications of oligomeric forms of POD-1 and POD-2 proteins isolated from cell walls of the biocontrol agent *Pythium oligandrum* in relation to their ability to induce defense reactions in tomato. J Plant Physiol 168:1972–1979

Thines B, Katsir L, Melotto M, Niu Y, Mandaokar A, Liu G, Nomura K, He SY, Howe GA, Browse J (2007) JAZ repressor proteins are targets of the SCFCOI1 complex during jasmonate signaling. Nature 448:661–666

Vallance J, Le Floch G, Déniel F, Barbier G, Lévesque CA, Rey P (2009) Influence of *Pythium oligandrum* biocontrol on fungal and oomycete population dynamics in the rhizosphere. Appl Environ Microbiol 75:4790–4800

Vallance J, Déniel F, Barbier G, Guerin-Dubrana L, Benhamou N, Rey P (2012) Influence of *Pythium oligandrum* on the bacterial communities that colonize the nutrient solutions and the rhizosphere of tomato plants. Can J Microbiol 58:1124–1134

Van Buyten E, Höfte M (2013) Pythium species from rice roots differ in virulence, host colonization and nutritional profile. BMC Plant Biol 13:203

Van Loon LC (2007) Plant responses to plant growth-promoting rhizobacteria. Eur J Plant Pathol 119:243–254

van West P, Morris BM, Reid B, Appiah AA, Osborne MC, Campbell TA, Shepherd TA, Gow NAR (2002) Oomycete plant pathogens use electric fields to target roots. Mol Plant Microbe Interact 15:790–798

Vijayan P, Shockey J, Lévesque C, Cook RJ, Browse J (1998) A role for jasmonate in pathogen defense of *Arabidopsis*. Proc Natl Acad Sci USA 95:7209–7214

Weller DM, Mavrodi DV, van Pelt JA, Pieterse CJM, van Loon LC, Bakker PAHM (2012) Induced systemic resistance in *Arabidopsis thaliana* against *Pseudomonas syringae* pv. *tomato* by 2,4-diacetylphloroglucinol-producing *Pseudomonas fluorescens*. Phytopathology 102:403–412

Wulff EG, Pham ATH, Chérif M, Rey P, Tirilly Y, Hockenhill J (1998) Inoculation of cucumber roots with zoospores of mycoparasitic and plant pathogenic *Pythium* species: differential zoospore accumulation, colonization ability and plant growth response. Eur J Plant Pathol 104:69–76

Xu X-M, Jefferies P, Pautasso M, Jeger MJ (2011) Combined use of biocontrol agents to manage plant diseases in theory and practice. Phytopathology 101:1024–1031

Yan Y, Christensen S, Isakeit T, Engelberth J, Meeley R, Hayward A, Emery RJN, Kolomiets MV (2012) Disruption of OPR7 and OPR8 reveals the versatile function of jasmonic acid in maize development and defense. Plant Cell 24:1420–1436

Belowground Defence Strategies Against Clubroot (*Plasmodiophora brassicae*)

Jutta Ludwig-Müller

Abstract The clubroot disease is one of the most devastating root-borne diseases of brassica crops. While breeding of resistant cultivars is still a method of choice, the control of clubroot by either biocontrol agents or even plant strengtheners could be improved. More environmentally friendly alternatives or additional means to make the resistance response of crop plants more durable are needed. Chemical control of clubroot is in many cases not successful; only liming has been used traditionally with good success. In some cases, the model plant *Arabidopsis thaliana* has been used; a plethora of work however has been done on oilseed rape/canola (in this chapter, the common name for *Brassica napus* will be chosen according to the name in the respective publications, mainly canola in Canada and oilseed rape in Europe) (*Brassica napus*). The clubroot pathogen is called *Plasmodiophora brassicae* and constitutes an obligate biotrophic protist that lives in close relationship with its host cell. The roots of the host plants are colonized, and the plant growth is altered upon infection. While shoots can be stunted and show wilt symptoms after longer infection periods, the root system is converted to a tumorous root tissue, called "clubroot" by alterations of plant hormones and metabolic pathways essential for pathogen nutrition. In this chapter, the major focus will, however, be on biocontrol of clubroot by either endophytic organisms or by plant strengtheners or plant growth regulators; and some mechanisms behind it, independent of which host plant was employed, will be discussed.

1 Introduction

The clubroot disease is caused by the obligate biotrophic protist *Plasmodiophora brassicae* on roots of the host plants mainly from the Brassicaceae. This disease affects economically important crops and can be considered a worldwide threat to brassica crop farming (Dixon 2009, 2014). Many recent review articles have dealt

J. Ludwig-Müller (✉)
Institut für Botanik, Technische Universität Dresden, 01062 Dresden, Germany
e-mail: Jutta.Ludwig-Mueller@tu-dresden.de

C.M.F. Vos, K. Kazan (eds.), *Belowground Defence Strategies in Plants*, Signaling and Communication in Plants, DOI 10.1007/978-3-319-42319-7_9

with the economic problem of this worldwide disease by publishing the proceedings of the various recent clubroot meetings and workshops. For instance, one series of review articles has been published in 2014 in the Canadian Journal of Plant Pathology and another one several years earlier (2009) in the Journal of Plant Growth Regulation. The roots of clubroot work go back to the discovery of its causal agent *P. brassicae* by the Russian scientist Woronin (Woronin 1878). Since then, the main structures of *P. brassicae* as well as the major parts of the complex intracellular life cycle have been elucidated (Kageyama and Asano 2009). Nevertheless, there are still many open questions concerning specific stages of development and colonization. The major problem when dealing with this root pathogenic protist is its obligate biotrophic lifestyle. Despite many efforts, there has been no progress in cultivating the pathogen outside of its host until now. For instance, Arnold et al. (1996) reported the cultivation of *P. brassicae* in *Escherichia coli*, but the resulting amoeba failed to infect host plants. As noted by Dixon (2014), the protist "*P. brassicae* exists in a highly protected environment for the majority of its life cycle. Here, *P. brassicae* has immediate access to all the nutrition that is required for growth and reproduction."

1.1 Disease Cycle

The disease starts by infection of host root hairs (Kageyama and Asano 2009). The resting spores germinate in the vicinity of host roots and produce biflagellate zoospores, which then move through the capillary water of the soil and penetrate a root hair mechanically (Aist and Williams 1971). The root hair elongates, and eventually, the plasmodia produce zoospores again which are either released into the soil or enter the cortex by yet unknown mechanisms (Kageyama and Asano 2009). Donald and Porter (2004) observed what they called "secondary zoospores drifting within root hairs," indicating that the movement could occur directly from the root hair to the cortex. In the cortex, the first structure visible within the host cell is a binucleate myxamoeba (Kobelt 2000), which develops into a so-called secondary multinucleate plasmodium (Mithen and Magrath 1992). This secondary plasmodium reorganizes host metabolism and ultimately host tissues (Ludwig-Müller et al. 2009). Cell division of plasmodia might occur concomitantly with the respective host cell (Kageyama and Asano 2009), leading to cell clusters that all contain large secondary plasmodia. These infected cells are then induced to undergo hypertrophic growth (Fig. 1). Once the disease symptoms are fully developed, the vasculature is partially destroyed, and therefore, the upper plant parts suffer from drought stress symptoms (Ludwig-Müller 2009). Using the plant hormonal network, the plasmodia induce cell divisions and cell elongation in their host which is dependent on auxin, cytokinin, and brassinosteroids (Siemens et al. 2006; Ludwig-Müller et al. 2009; Jahn et al. 2013; Schuller et al. 2014). These events ultimately lead to the development of visible clubroot symptoms on a

Fig. 1 (**a**) An example of *Plasmodiophora brassicae*-infected roots of *Brassica rapa* 6 weeks after inoculation with resting spores of the pathogen. (**b**) Thin section stained with methylene blue/ Azure II/basic fuchsine (Buczacki and Moxham 1979) through an *Arabidopsis thaliana* root 4 weeks after inoculation with *P. brassicae*. *RS* resting spores, *PL* secondary plasmodia. The *bar* represents 50 µm. Microscopic picture was taken by Claudia Seidel, Technische Universität Dresden, Germany

cellular and organ level (Fig. 1). Finally, the mature plasmodia develop into millions of resting spores which are released into the soil.

Many mechanisms have been described on how the clubroot pathogen can benefit from the changes in host metabolism and also host hormone homeostasis. However, using the findings of such approaches to control clubroot is difficult due to the effect of these approaches on the overall growth and development of the host plant, in particular, if the hormone homeostasis or essential metabolites are altered (Siemens et al. 2006, 2011; Schuller et al. 2014). Dwarfed plants have been described for cytokinin- and brassinosteroid-deficient mutants, although they showed a resistance phenotype against clubroot (Siemens et al. 2002, 2006; Schuller et al. 2014). Thus, clubroot is mainly controlled by using resistant cultivars. The challenge is to find environmentally friendly means to control clubroot. While in most cases the administered treatment is equally effective for different hosts, in other cases it was shown that treatments were effective for one, but not for the other organism, which will be explained in more detail below. Also, in many cases, successful treatments under different environmental conditions have been reported, but the question remains whether the treatment would also be effective in the field.

Belowground control methods against clubroot that will only be briefly described include pH and liming, fungicides, and biofumigation, because excellent reviews exist to which references will be made. Biocontrol agents (i.e., bacteria and fungi), as well as plant-strengthening formulations and growth regulators, will be covered in more detail. It will also be tried to give the most likely point(s) in the life cycle of the pathogen, where the method might be more effective and where possible mechanistic insights will be presented. In the end, some remarks on integrated control of clubroot will be made.

2 Liming, pH, and Ca^{2+}

The clubroot pathogenesis is successful in the field at lower pH values (Einhorn and Bochow 1990). Therefore, liming is a good method to increase the pH of the soil, but this is not the only effect of this treatment on disease development since calcium ions could also directly affect *P. brassicae* growth. One of the most effective products against clubroot is calcium cyanamide (Donald et al. 2004), which also acts as fertilizer. Treatments with cyanamide can lead to reduced resting spore germination (Fig. 2) and diminished infections (Naiki and Dixon 1987; Niwa et al. 2008). However, the release of the active compound depends on soil type and moisture as well as pH, and nitrogen input also needs to be considered (Diederichsen et al. 2014). Moreover, the form of lime, its particle size, mixing with the soil, and finally the time point of application are other important factors (Donald and Porter 2009). However, alkaline pH does not result in the reduction of clubroot if other conditions are still conducive for infection (Gossen et al. 2014).

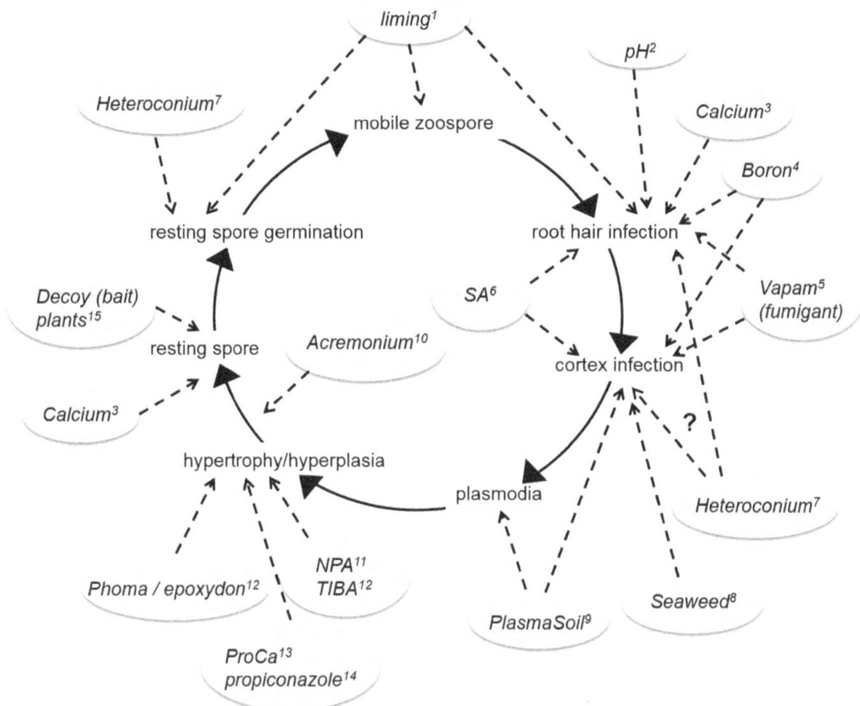

Fig. 2 Brief life cycle of *Plasmodiophora brassicae* with indications where the different control methods might be most effective. While all control mechanisms may ultimately reduce clubroot symptoms, the reduction of the hypertrophied tissue is given as a possible point where a control agent might interfere with. Not all control mechanisms described in the text have been included. The numbers refer to the selected references given here in the legend. [1]Donald et al. (2004), Naiki and Dixon (1987); [2]Einhorn and Bochow (1990); [3]Webster and Dixon (1991a), Murakami et al. (2001); [4]Deora et al. (2011); [5]Hwang et al. (2014a); [6]Agarwal et al. (2011), Lovelock et al. (2013); [7]Narisawa et al. (2005), Lahlali et al. (2014); [8]Wite et al. (2015); [9]Kammerich et al. (2014); [10]Jäschke et al. (2010); [11]Devos and Prinsen (2006); [12]Arie et al. (1999); [13]Päsold and Ludwig-Müller (2013); [14]Schuller et al. (2014); [15]Murakami et al. (2002), Ahmed et al. (2011)

Calcium ions have been considered to act against *P. brassicae* together with high pH (Webster and Dixon 1991a). Ca^{2+} alone inhibited either sporangial dehiscence at higher inoculum pressures or development of sporangia (here this will be called sporulating plasmodia) at low inoculum pressures (Fig. 2). However, it was noted that application of salts that simultaneously increase Ca^{2+} and pH had a stronger effect on reducing clubroot symptoms than those raising only the pH (Webster and Dixon 1991a). Other studies have suggested that calcium and magnesium ions in lime have additional effects on disease control that are independent of pH (Murakami et al. 2002). However, this interrelation between minerals and soil is complex (Myers and Campbell 1985; Donald and Porter 2009; Dixon 2014). In addition, fertilizer treatments can potentially alter soil microbes (Dixon 2014), so that there might be an indirect beneficial effect of this treatment on biocontrol agents.

To elucidate at which point in the life cycle of the pathogen the respective treatment has the major influence may require the establishment of specific in vitro growth systems that allow the direct observation of growth stages of *P. brassicae*. Since the pathogen is an obligate biotroph, it is difficult to observe its growth stages in the soil. To overcome this problem, Donald and Porter (2004) designed a sand-solution cultivation technique that enabled them to observe the effect of Ca^{2+} and pH on root hair and cortical infection. The system was later adapted for other experimental approaches, i.e., transcriptome analyses of early stages in the life cycle using *Arabidopsis thaliana* (Agarwal et al. 2011).

When the effect of Ca^{2+} on clubroot is discussed, this is mainly attributed to the factors mentioned above. However, Ca^{2+} is also a signal in various pathways that regulate biotic stress responses (Lecourieux et al. 2006). Backing this thought up is a publication by Takahashi et al. (2002) showing that endogenous Ca^{2+} is required for transient induction of phenylalanine ammonia lyase after *P. brassicae* infection in resistant turnip cells. Whether the exogenous Ca^{2+} from soil might also have this effect is not clear, but it is an alternative to think about the role of calcium as a signal for defence induction rather than an inhibitor of spore germination or other direct effects on *P. brassicae*.

In addition to Ca^{2+}, boron was able to reduce clubroot on Chinese cabbage (*Brassica rapa* ssp. *pekinensis*) (Webster and Dixon 1991b), where boron had a better effect at higher pH values compared to lower pH. A similar effect was observed on canola (*Brassica napus*) (Deora et al. 2011). Through boron application, development of the infection within root hair and cortex was reduced (Fig. 2) as well as the incidence and severity of the disease (Deora et al. 2011). Therefore, the authors concluded that boron can be used as a component of an integrated management program (see Sect. 7).

3 Chemical Control

As noted by Donald and Porter (2009), the use of the term "fungicide" in clubroot control is misleading since *P. brassicae* is a protist and not a true fungus. Nevertheless, several compounds were reported to be applied against clubroot, i.e., cyazofamid (Zhou et al. 2014) or pentachloronitrobenzene, the latter is stable and persists in the soil for a long time (Arie et al. 1999). Some of the chemicals effective on clubroot are not allowed by regulatory authorities due to their potentially undesirable effects. Consistent control of clubroot in the field was reported only for a few fungicides (Donald and Porter 2009). One example that should be mentioned is mercurous chloride (Calomel™), but its high toxicity and persistence in the environment has led to the withdrawal of the compound from the market. However, up to now no other comparable chemical effective against clubroot has been reported (Donald and Porter 2009). A summary of fungicides that have been

Table 1 Compilation of fungicides used against clubroot mainly based on the review by Donald and Porter (2009)

Compound	(Trade) name	Clubroot control	Problem/remark
Cyazofamid	Ranman™	Yes	Specificity against oomycetes
Pentachloronitrobenzene		Yes	Persistence in environment Hexachlorobenzene as impurity
Mercurous chloride	Calomel™	Highly efficient	Toxic to mammals Persistence in environment
Dithiocarbamate Sodium *N*-methyldithiocarbamate[a]	Vapam	Efficient	Fumigant
Benzimidazoles (benomyl derivatives)	Methyl benzimidazol-2-ylcarbamate	Efficient only when incorporated into the soil	Precursor used which is converted to active compound in soil Some compounds toxic to plant
Alkylene bisdithiocarbamates	Maneb, mancozeb, zinep	Yes	
N-(1-alkoxy-2,2,2-trichloroethyl)-2-hydroxybenzamides	Trichlamide	Yes	High concentrations needed for efficient control
4-chloro-*N*-(2-chloro-4-nitrophenyl)-*a*,*a*,*a*-trifluoro-*m*-toluene sulfonamide	Flusulfamide	Inhibits spore germination	
3-chloro-*N*-(3-chloro-5-trifluoro-methyl-2-pyridyl)-*a*,*a*,*a*-trifluoro-2,6-dinitro-*p* toluidine	Fluazinam (Shirlan™ or Omega™)	Yes	Interrupts the production of energy in fungal pathogens by an uncoupling effect on oxidative phosphorylation
AG3 phosphonate		Yes	

[a]From Hwang et al. (2014a). For more information, see Sect. 3

used against clubroot is given in Table 1, which is mainly based on the extensive review published by Donald and Porter (2009). Alternatively, or in addition to these fungicides, surfactants have been used in clubroot control (Hildebrand and McRae 1998; Donald and Porter 2009).

In Canada, the fumigant Vapam (dithiocarbamate; sodium *N*-methyldithiocarbamate) was used to investigate its effect on *P. brassicae* primary and secondary infection, clubroot severity, and growth of canola under greenhouse and field conditions (Hwang et al. 2014a). The effect of Vapam has been mainly attributed to its conversion to methyl isothiocyanate, a volatile compound that diffuses as a gaseous form through the soil after application (Smelt and Leistra 1974). Both primary and secondary infection could be reduced as well as the overall clubroot symptom severity through the use of Vapam (Fig. 2). Concomitantly, the seed yield was increased. The authors suggest its use in brassica vegetable production, for example, in transplant propagation beds, as well as for controlling clubroot in small patches of clubroot incidents (Hwang et al. 2014a).

4 Biocontrol Agents

Biocontrol aims to use natural enemies to reduce the population size of a plant's pest to a level where the host is not or less strongly affected by the pest. Two main mechanisms are proposed so far: antimicrobial compounds and/or induction of plant defence mechanisms. Some biocontrol agents mentioned here are summarized in Table 2. The pests could be insects, nematodes, fungi, or bacteria. In a molecular

Table 2 Summary of biocontrol organisms presented here together with their possible point in the life cycle where they exert their function (see also Fig. 2) and if known possible mechanism of clubroot control

Organism	Name	Product	Point and mechanism of clubroot control
Actinomycetes	*Microbispora rosea* ssp. *rosea*		n.d.
	Streptomyces olivochromogenes		n.d.
	Streptomyces griseoviridis		n.d
	Streptomyces lydicus		n.d.
Other bacteria	*Bacillus subtilis* QST713	Serenade®	Suppression of root hair and cortical infection; induction of defence
	Bacillus subtilis XF-1		Chitosanase production
	Lysobacter antibioticus		Release of antimicrobial compound?
	Bacillus megaterium		n.d.
	Clostridium tyrobutyricum		n.d
Fungi	*Acremonium alternatum*		Reduction of resting spore production; induction of defence
	Clonostachys rosea f. *catenulate*	Prestop®	Suppression of root hair and cortical infection; antibiosis; induction of defence
	Gliocladium catenulatum		n.d.
	Heteroconium chaetospira		Resting spore germination, root hair and cortex infection; induction of defence
	Phoma glomerata		Synthesis of epoxydon
	Trichoderma harzianum		Slight control

n.d., mechanism not yet determined. For more information, see Sect. 4

sense, biocontrol agents could also induce the defence response of a plant by triggering systemic acquired resistance (SAR) or induced systemic resistance (ISR). The latter mechanism is also called priming (Conrath et al. 2001) and can not only be induced by live organisms but also by elicitor molecules.

Work on biocontrol of clubroot has possibly started with the observation that some soils were suppressive toward the pathogen (reviewed in Dixon 2014). In some cases, the suppressiveness was retained after autoclaving the soil, while in other cases, it was not. In light of the knowledge gained nowadays on the mechanisms induced by biocontrol agents, this observation could mean that either heat-stable antimicrobial compounds were still present in the soil or that even autoclaved spores could induce plant defence as elicitors (see Sect. 4.2). Chitosan is also an elicitor, and it was shown that the compound was able to reduce clubroot symptoms (Wang et al. 2012). However, the authors only showed a direct effect of chitosan on resting spores (i.e., spore germination was inhibited). Therefore, it is unknown whether defence pathways in the plant were also induced.

At which stage should an effective biocontrol agent (BCA) against clubroot be affecting the pathogen? At best, the biofungicide treatment should target the release of the zoospores, which can occur shortly after sowing into infested and moist soils (Peng et al. 2014). In addition, there should be the possibility to control a later step if the zoospores were too numerous to be completely controlled. However, it is not trivial to find novel biocontrol organisms. While it was reported recently from China that novel organisms were found (Zhou et al. 2014), the screen of more than 5000 soil microbial isolates from the Canadian prairies showed no promising candidate for clubroot control (Peng et al. 2014).

The first trials with potential BCA certainly need to be carried out under controlled environmental conditions, i.e., in a temperature-controlled chamber or greenhouse, where not only the environment but also the inoculum density can be controlled. However, often the efficacy of the biofungicides varies among trials when moved to the field conditions. This is true especially across crops and test sites as well as application methods (Peng et al. 2014). Therefore, finding a biocontrol agent that is reducing clubroot in the greenhouse is only the beginning in finding a cure in the field. Commercial BCAs include Serenade® (*Bacillus subtilis*), Prestop® (*Clonostachys rosea* f. *catenulate*), Mycostop® (*Streptomyces griseoviridis*), and RootShield® (*Trichoderma harzianum* Rifai) (Peng et al. 2014), but not all have been tested against clubroot. Furthermore, formulations need to be developed that can be used easily in fields. This has been done as granular and seed treatment formulations for canola (Peng et al. 2014).

4.1 Bacteria

Antagonistic bacteria used to control clubroot include *Bacillus subtilis* (Lahlali et al. 2011; Guo et al. 2013), *Lysobacter antibioticus* (Zhou et al. 2014),

Streptomyces sp. (Cheah et al. 2001; Joo et al. 2004), or various actinomycetes including *Microbispora rosea* ssp. *rosea* and *Streptomyces* species such as *S. olivochromogenes* (Lee et al. 2008), *S. griseoviridis*, and *S. lydicus* (Peng et al. 2011).

Results from Zhou et al. (2014) indicated that 6 out of 14 bacterial strains that were isolated from the soil around the roots of vegetables reduced disease severity of Chinese cabbage by more than 50 % under greenhouse conditions, but no mechanism was elucidated even though the authors speculated antibiotic factors might be responsible. Also field trials were performed and resulted in similar data concerning the disease reduction. The authors compared the efficacy of BCAs, e.g., *L. antibioticus*, to a fungicide (cyazofamid) and found comparable results. Interestingly, the treatment of seeds with the biocontrol strain also reduced clubroot severity later in the greenhouse, albeit to a lesser extent than the soil drench method (Zhou et al. 2014).

One of the biocontrol agents already in the market is *B. subtilis*, and respective products have already been tested successfully against clubroot on canola in Canada, albeit so far only in the greenhouse (Lahlali et al. 2011). The effect of the commercial biocontrol agent Serenade® (*B. subtilis* QST713) on reducing clubroot incidents had been attributed to suppressing root hair and cortical infection by *P. brassicae* (Fig. 2), because resting spore germination was only marginally affected (Lahlali et al. 2011). In addition, Serenade® and another biofungicide Prestop® suppressed the disease on canola via antibiosis and induced host resistance under controlled-environment conditions (see also Sect. 4.2). Lahlali et al. (2013) showed the induction of a set of defence genes involved in phenylpropanoid, jasmonic acid (JA), and ethylene (ET) pathways upon treatment with the BCA. They also positively correlated the amount of *P. brassicae* DNA with the reduction of clubroot symptoms (Lahlali et al. 2013). Granular and seed treatment formulations were developed to facilitate the delivery of biofungicides in field trials (Peng et al. 2014). Other bacteria tested were *S. griseoviridis* and *S. lydicus* which also showed some control potential against clubroot (Peng et al. 2011).

Another *B. subtilis* strain, XF-1, which showed high potential to suppress *P. brassicae*, was sequenced, and it was shown that a gene cluster involved in the synthesis of chitosanase is related to the suppression of clubroot (Guo et al. 2013). This might indicate that the chitin in the resting spores could be a target. Gao and Xu (2014) used a cocktail of three different biocontrol organisms (*Bacillus megaterium*, *Clostridium tyrobutyricum*, and *Saccharomyces cerevisiae*) to analyze their potential to control clubroot. It was shown that their mixture of organisms could diminish clubroot symptoms.

4.2 Fungal Endophytes

The initial reports on the possible use of fungal endophytes came from reports published in the 1990s, where endophytic fungi such as *Heteroconium chaetospira* were isolated from the rhizosphere (Narisawa et al. 1998, 2000). In these initial experiments, no mechanism was postulated, even though reduction of clubroot incidence was shown. Some data pointed to the germination of resting spores as a target site (Fig. 2). Later, Lahlali et al. (2014) showed that the clubroot resistance induced by *Heteroconium chaetospira* can be related to the induction of plant defence pathways via jasmonic acid (JA), ethylene (ET), and auxin (indole-3-acetic acid; IAA) in canola, but not via salicylic acid (SA).

Using the fungus *Phoma glomerata*, Arie et al. (1999) showed reduction of clubroot on various brassica crops. They were able to attribute the effect to a compound, epoxydon, that was isolated as active principle from fungal cultures. They found that the compound could neither exert antifungal activity against a variety of plant pathogenic fungi in vitro nor induce acquired resistance (Arie et al. 1999). However, the compound was reported to display antiauxin activity (Sakai et al. 1970), and it was shown that another antiauxin (2,3,5-triiodobenzoic acid; TIBA) had similar effects on clubroot control (Arie et al. 1999). The authors concluded that the control of clubroot was most likely conferred via an alteration of auxin levels or distribution, since TIBA is an auxin transport inhibitor (see Sect. 6).

Other endophytes were tested as BCA against clubroot. Doan et al. (2010), Jäschke et al. (2010), and Auer and Ludwig-Müller (2014) evaluated the fungus *Acremonium alternatum* for its potential to control clubroots of Chinese cabbage, oilseed rape, and *Arabidopsis*. While for Chinese cabbage (Doan et al. 2010) and *Arabidopsis* (Jäschke et al. 2010) a good biocontrol effect was observed, the effects on oilseed rape were not very strong (Auer and Ludwig-Müller 2014), but maybe the conditions that are needed to exert the full biocontrol potential have yet to be identified for the latter species.

In *Arabidopsis*, the endophyte *A. alternatum* slowed down the development of *P. brassicae* (Fig. 2), because the major form found in infected roots were secondary plasmodia (Jäschke et al. 2010). This was confirmed by the observation that genes of *P. brassicae* expressed at different time points during the disease cycle were upregulated at later time points under the influence of the endophyte. The resting spore germination, however, was not inhibited. Since autoclaved spores of *A. alternatum* were also able to induce the tolerance against clubroot, it was speculated that the defence mechanism of the plant was induced. This assumption was confirmed by microarray analyses which showed that several defence genes were upregulated more in the co-inoculation with *A. alternatum* and *P. brassicae* than in the inoculation with only one of the two organisms (S. Auer and J. Ludwig-Müller, unpublished results). Contrary to *H. chaetospira* (Lahlali et al. 2014), the endophyte *A. alternatum* seems to induce SA-dependent defence pathways, because some typical pathogen-associated molecular pattern genes were upregulated (S. Auer and J. Ludwig-Müller, unpublished results).

Formulations containing the biocontrol agent *Gliocladium catenulatum* reduced clubroot severity as well (Peng et al. 2011). *G. catenulatum* reduced clubroot severity by more than 80 % relative to controls only inoculated with *P. brassicae* on a highly susceptible canola cultivar. This efficacy was comparable to that of the fungicides fluazinam and cyazofamid (Peng et al. 2011). In this study, *Trichoderma harzianum* was also tested, which was somewhat less efficient in clubroot control compared to *G. catenulatum*. Based on experiments with cell-free filtrates which also suppressed clubroot, it was concluded that there might be also antimicrobial compounds present (Peng et al. 2010).

A biocontrol agent available as commercial product, *Clonostachys rosea* (Prestop®), reduces clubroot symptoms via induced host resistance (Lahlali and Peng 2014). Pathways probably involved, as identified by gene expression analyses, included the phenylpropanoid pathway and JA and ET signaling, but not SA (Lahlali and Peng 2014). Since the authors found that this biofungicide did not reduce the germination or viability of *P. brassicae* resting spores, they concluded that the suppression of clubroot disease probably results from the reduction of root hair and/or cortical infection (Lahlali and Peng 2014). To elucidate the functional principle, they partitioned the key product components and found that the whole product gave the most efficient clubroot control compared to *C. rosea* spore suspension or product filtrate. They also observed that high treatment doses were necessary for full efficacy, which might be a problem for the application in the field (Lahlali and Peng 2014).

4.3 Other Organisms

Studies on the interaction of the clubroot pathogen and earthworms were carried out since it was reasoned that earthworms can alter soil properties by changing minerals and/or microbial communities and thereby may also change the outcome of specific diseases (Winding et al. 1997; Clapperton et al. 2001). While the effect might be more indirect via (biocontrol) microbes, a treatment with several earthworm species on clubroot incidence was carried out. It was considered that since the galls disintegrate at the end of the disease cycle and the spores are liberated into the soil, these might be consumed by soil grazers feeding on microbes (Friberg et al. 2008). The fate of plant pathogen propagules during the passage through the gut of earthworms can vary from complete survival to complete digestion (Moody et al. 1996). Therefore, it is not predictable what would happen to the very resistant resting spores of *P. brassicae*. In their experiments, Nakamura et al. (1995) found that the presence of the earthworm *Pheretima hilgendorfi* reduced clubroot disease severity in experimental pots, but not the number of resting spores. It was therefore suggested that the effect was based on a chemical inactivation of the resting spores, resulting in reduced ability of the pathogen to infect the host plants. Contrary to these promising results, the presence of the earthworm *Aporrectodea caliginosa* did not change clubroot disease severity in *Brassica rapa* var. *pekinensis* in various treatment combinations (Friberg et al. 2008).

4.4 *Biofumigation*

As for chemical fumigants (see Sect. 3), toxic compounds leaching out from plant material can be used to control diseases in the soil. Alternatively, the plant materials containing these compounds are composted in the soil, thereby it is assumed that the volatile toxic compounds diffuse through the soil (Gimsing and Kirkegaard 2009). This process is called biofumigation. Treatments with high glucosinolate-containing plants, for example, *B. napus* and *B. rapa* cultivars, were shown to reduce soil inoculum of *P. brassicae* (Cheah et al. 2001, 2006). It is important to note that for clubroot, the second technique using composted plant material is more promising, because it avoids host plants for *P. brassicae* in the field (Donald and Porter 2009).

5 Plant Growth Stimulants

The application of plant growth stimulants in the field or greenhouse should increase plant performance in general and often under abiotic stress conditions (Metting et al. 1990; Mancuso et al. 2006). They can be grouped according to their ingredients such as inorganic mixtures (e.g., sodium or potassium hydrogen carbonate), organic mixtures (e.g., algal extracts, humic acids, plant extracts, animal products), and microbial extracts or components (Kühne et al. 2006). Recently, members of such compounds have also been noticed to increase plant resistance against pathogens (Kofoet and Fischer 2007). Thus, they may also be considered as biocontrol agents. While the composition of fungicides is better documented, the specific components within strengthening formulations are sometimes not completely freely available (Kammerich et al. 2014).

Despite these possible drawbacks, the use of such strengtheners for clubroot control has been tested. Kammerich et al. (2014) tested the liquid strengthener formulation Frutogard® that consists essentially of algal extract, amino acids, and phosphonate and a similar product on the basis of a granulate formulation, PlasmaSoil®, on possible clubroot control. They showed that both mixtures reduced clubroot symptoms on Chinese cabbage and oilseed rape, but the granulate formulation was more effective. In addition, light microscopy has indicated reduction of pathogen structures, especially plasmodia, in treated root sections as well as several anatomical changes compared to untreated controls and infected roots (Kammerich et al. 2014). These anatomical changes induced by PlasmaSoil® were summarized as follows: "(i) strengthening of the vascular cylinder to prevent *P. brassicae* from entering the vasculature; (ii) larger cortex cells, which could absorb and transport more nutrients; and (iii) a suberin layer, which is only one cell layer in clubroot infected roots, but is at least two cell layers thick in controls and PlasmaSoil®-treated roots" (Kammerich et al. 2014).

Seaweed extracts are another prominent class of plant strengtheners (Metting et al. 1990). Wite et al. (2015) used a commercial seaweed extract (Seasol Commercial®) containing two different brown algal species, *Durvillaea potatorum* and *Ascophyllum nodosum*, to control clubroot in broccoli (*Brassica oleracea* var. *italica*). The seaweed extract had a better effect on the suppression of the secondary infection phase than on the reduction of root hair colonization (Wite et al. 2015). The authors speculated that the seaweed extract might induce the plant's defence mechanisms possibly due to their laminarin content or growth regulators present. These results are not in agreement with the observation made by Kammerich et al. (2014) that one single component of the plant strengthener used in their study, the seaweed constituent consisting of *Ascophyllum nodosum* and *Laminaria* species, could not reduce clubroot symptoms alone. Furthermore, it was not possible to reduce clubroot symptoms of Chinese cabbage in the greenhouse using a commercial seaweed extract (Afrikelp® LG-1) containing the giant brown seaweed *Ecklonia maxima* (J. Ludwig-Müller, unpublished results). Clearly, different host plants, cultivation conditions, and algal species could be the reason for this discrepancy, and this needs more research in the future.

6 Plant Growth Regulators

Plant growth regulators can be used to regulate the performance of a plant. Often they directly target the biosynthesis, perception, or transport of plant hormones. Many of them inhibit gibberellin (GA) biosynthesis (Rademacher 2000) and thus act as antagonists of the plant's growth response. Such compounds might therefore be successfully employed against the clubroot pathogen, because the plant hormonal system is dramatically altered in these infected roots (Ludwig-Müller et al. 2009; Diederichsen et al. 2014). While the mutation of a specific pathway most likely results in unwanted growth changes, treatments with inhibitors could circumvent this problem by applying them only when needed. Thus, the unwanted effects on plant growth and development might be reduced.

Since it was shown that flavonoids accumulated in clubroots (Päsold et al. 2010), it was tested whether an inhibitor for enzymes belonging to the class of oxoglutaric acid-dependent dioxygenases, prohexadione-calcium (ProCa), would have an influence on the development of the clubroot symptoms (Päsold and Ludwig-Müller 2013). The compound does not only inhibit an enzyme from the flavonoid biosynthetic pathway but also enzymes occurring in GA synthesis (Rademacher 2000). To investigate the specificity of the results, another growth regulator chlorcholinechloride (CCC) that targets specifically the GA biosynthetic pathway was also employed (Rademacher 2000). Evaluation of clubroot symptoms showed that the effect was surprisingly specific for ProCa, but not for CCC, since a reduction of *Arabidopsis* root symptoms was observed only with the former compound (Päsold and Ludwig-Müller 2013). This also demonstrates that GAs are not involved in the hypertrophy symptoms after *P. brassicae* infection. However, whether the observed accumulation of the flavanone naringenin is responsible for the suppression of clubroot symptoms could not be determined. So it cannot be ruled out that the inhibition of the flavonoid pathway results in other defects in the plant.

Auxin homeostasis plays a role for club development (e.g., Jahn et al. 2013). Auxin transport inhibitors such as TIBA and naphthylphthalamic acid (NPA) (reviewed in Muday and Murphy 2002) seem to suppress clubroot symptoms. Arie et al. (1999) called TIBA an antiauxin and showed its suppressive effect on clubs of various brassicas (see Sect. 4.2). Later, it was shown that application of the polar auxin transport inhibitor NPA reduced root galls (Devos and Prinsen 2006). However, it seems important during which period of infection the inhibitor was applied. Application during later time points, when the disease was already established in the roots, did not result in the reduction of clubroot symptoms even though the treated plants showed a dwarfed phenotype (Päsold et al. 2010). Treatment with the auxin influx inhibitor 1-naphthoxyacetic (NOA) acid resulted in somewhat reduced disease symptoms (Päsold et al. 2010). The reduction of auxin by means of auxin transport inhibition could directly result in reduced gall size, because it has been assumed that the increase in auxin is one prerequisite for hypertrophied cells (Ludwig-Müller et al. 2009).

An effect of the potassium channel blocker tetraethylammonium (TEA) was reported on clubroot disease symptoms of *Arabidopsis* (Jahn et al. 2013). While the overall phenotype of the treated plants was surprisingly normal, the reduction of clubroot incidence of treated roots compared to untreated ones was reduced by about 50 %. Also the green plant parts were as healthy as uninoculated plants (Jahn et al. 2013). This effect was attributed to the inhibition of the K^+-mediated cell elongation process in which auxin also is involved. To be more specific, K^+ channels are needed for the auxin-mediated cell elongation response (Christian et al. 2006). If the cells in infected roots can no longer perform the cell elongation via increase of turgor pressure, galls will remain small, and *P. brassicae* cannot develop into large sporulating plasmodia which would then result in the reduction of resting spore numbers as well.

Recently, a role for brassinosteroids (BR) for the development of clubroots, in addition to auxin and cytokinin, was reported (Schuller et al. 2014). Propiconazole, an inhibitor targeting the BR biosynthetic pathway (Hartwig et al. 2012), reduced clubroot symptoms of *Arabidopsis* substantially. While the growth of the treated plants was reduced, this phenotype was not as dramatic as found for the dwarfed biosynthesis mutants in the BR pathway (reviewed in, e.g., Choe 2004). Brassinosteroids are also involved in cell elongation, and therefore, their reduction has a direct effect, like auxin, on gall size.

In a sense, SA is also a plant (growth) regulator. It is definitely considered a regulator of the induction of plant resistance against pathogens and also involved in systemic acquired resistance (SAR). Direct treatment with SA during infection stages where the pathogen was already established in the plant could not be used to reduce clubroot symptoms of Chinese cabbage (Ludwig-Müller et al. 1995). Based on microarray data for early (root hair) infection, Agarwal et al. (2011)

identified a downregulation of the SA pathway in infected roots. They pretreated *Arabidopsis* plants with a SA solution and found a significant reduction of clubroot symptoms. This is in contrast to the observations by Ludwig-Müller et al. (1995) where the infected plants were treated with SA when infection by *P. brassicae* had already occurred. In line with the results of Agarwal et al. (2011), Lovelock et al. (2013) have shown that early treatment of broccoli roots with SA could reduce clubroot symptoms significantly. Concomitantly, the gene expression for two PR genes was upregulated already 24 h after SA treatment (Lovelock et al. 2013). A possible explanation for these different results concerning SA comes from recent work where it was shown that *P. brassicae* possesses a methyltransferase that can methylate SA (Fig. 3) (Ludwig-Müller et al. 2015). It was also shown that the respective gene was expressed as early as day 4 after inoculation. Thus, treatment with SA at a time point where *P. brassicae* is already established in the plant could lead to methylation of SA in infected roots, and the methyl ester of SA is better transported than SA in *Arabidopsis* plants from the roots to the leaves (Ludwig-Müller et al. 2015). It was concluded that methylation of SA by *P. brassicae* is one possibility to suppress the plant's defence response (Fig. 3a). If SA is administered at an early time point, as in the work of Agarwal et al. (2011) and Lovelock et al. (2013), then the *P. brassicae* methyltransferase would not yet be active to reduce SA concentrations, so that exogenous SA can induce resistance (Fig. 3b).

7 Integrated Clubroot Control

Main factors considered in integrated clubroot control management include a combination of soil treatment with fertilizers and lime, resistant cultivars, and hygiene measures in field plots and greenhouses (summarized in great detail by Donald and Porter 2009). Biocontrol agents, in conjunction with soil factors, are also being considered (Narisawa et al. 2005; Peng et al. 2011), whereas plant growth regulators and plant strengtheners are not included into thoughts about integrated control as yet. While most of the experiments with BCAs have been performed in controlled environmental conditions, their field performance is yet to be tested. In Australia, integrated clubroot control was shown to work effectively (Donald and Porter 2009, 2014), and also in Canada, integrated clubroot management was investigated for canola (Strelkov et al. 2011; Hwang et al. 2014b; Peng et al. 2014). In China, mainly resistant cultivars and BCAs are being investigated against clubroot (Chai et al. 2014).

For the integrated approach, also environmental factors have to be taken into account (Dixon 2014). Besides the pH value of the soil (see Sect. 2), temperature, rain, wind, etc. can play important roles in the outcome of the disease symptoms (e.g., Einhorn and Bochow 1990; Dixon 2009; Gossen et al. 2014; Hwang et al. 2014b). Temperature, in contrast to soil-related factors, is a factor often neglected because it cannot be controlled in the field. In a screen of *Arabidopsis*

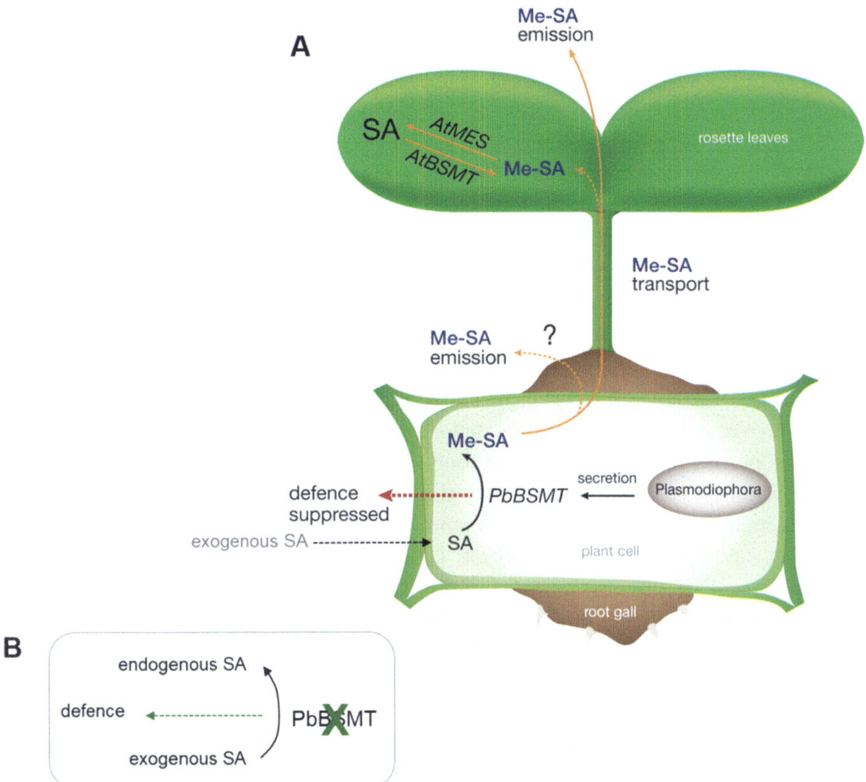

Fig. 3 Model on the role of SA and a methyltransferase from *Plasmodiophora brassicae* that can methylate SA (PbBSMT). The model is based on data from Ludwig-Müller et al. (2015). *P. brassicae* could secrete PbBSMT into the host cell, then the enzyme would methylate the defence signal SA. Since MeSA is not activating plant defence responses, the upregulation of the respective reaction in the host root would be suppressed. Also, MeSA is a better transport substance in clubroot-infected *Arabidopsis thaliana* plants than SA and is ultimately emitted from the leaves (or possibly converted back to SA). Whether MeSA can also be emitted from the root has not yet been determined. Ultimately, the SA levels in roots can be at least partially downregulated by this strategy of the protist. This model would also explain why addition of SA at a time point where *P. brassicae* is already established in the host root does not lead to the induction of defence responses. However, if SA is administered early enough at a time point where PbBSMT is not yet made, then the SA-dependent defence pathways in the plant can be induced as shown by Agarwal et al. (2011) and Lovelock et al. (2013). Both datasets together point to a role of PbBSMT as a possible important pathogenicity factor. In the *gray box*, the situation for exogenous SA without PbBSMT is shown. The cartoon in (**a**) was created by Donna Gibson, Institute for Plant & Food Research, Christchurch, New Zealand

mutants, Siemens et al. (2002) showed a reduction in colonization at lower (18 °C) compared higher temperature (24 °C) under controlled conditions. In crop plants, for example, canola, temperatures below 17 °C also reduced the development of *P. brassicae* at all life stages (Gossen et al. 2014), and growth at 10 °C completely suppressed clubroot symptoms (Sharma et al. 2011). Based on laboratory experiments and literature reviews, Dixon (2014) concluded that temperatures required for symptom development and expression are lower than those needed for movement and penetration of zoospores.

A range of alternative management strategies have been evaluated for their usefulness in clubroot suppression, including manipulation of sowing time (Gossen et al. 2012; Hwang et al. 2012) and the use of bait crops (Kroll et al. 1983; Ikegami 1985; Murakami et al. 2001; Ahmed et al. 2011). Moreover, the distribution of clubroot from infested field plots is also a problem that needs to be considered (Gossen et al. 2014). In this context, controlling farm and nursery hygiene is very important (Donald and Porter 2009, 2014), since *P. brassicae* spores can be easily spread, for example, by wind (Rennie et al. 2015) or through irrigation water (Gossen et al. 2014). It was shown that these resting spores remain viable in water for over 30 months, and repeated irrigation with water containing as few as 10 spores per ml resulted in clubbed roots (Donald 2005).

Crop rotation can also help to keep the disease manageable (Robak 1994), but the growers need to follow the recommended schemes, meaning at least only once a brassica crop within a 4-year rotation (Diederichsen et al. 2014). In Canada, a 2-year interval of nonhosts was recommended, but only when resistant canola cultivars are employed, to reduce *P. brassicae* resting spore load (Peng et al. 2014). Wallenhammar already determined in 1996 a half-life for *P. brassicae* resting spores of 3.6 years which indicates the need for even longer periods between brassica crops. Dixon (2014) calculated that it would take 18 years, in the absence of a suitable host, for a field population of *P. brassicae* to decrease to less than 10 % of the original spore population. The avoidance of host plants is, however, difficult to achieve when considering the presence of volunteers of oilseed rape or cruciferous weeds (Diederichsen et al. 2014). Weed control might therefore be another—indirect—factor that could lead to successful control of clubroot in an integrated approach. Many weeds are hosts for *P. brassicae* and need to be spotted in the given environment (Howard et al. 2010).

In general, these management approaches hold some potential, but they are still not cost-effective in many areas where clubroot is a problem on crops. For example, the most cost-effective method to control clubroot on canola is by using resistant cultivars (Strelkov et al. 2011). A major problem however here is that the number of clubroot resistance genes available for breeding is low (Hirai 2006) and that single-gene-dependent resistance can be broken down rather quickly by the development of more virulent pathotypes of *P. brassicae* (Kuginuki et al. 1999). Therefore, it is recommended to complement the cultivation of resistant cultivars by at least one or two other methods to reduce clubroot in the field (Fig. 4). On the other hand, it was reported that when the soil resting spore load was too high, neither biocontrol nor chemical control agents could be effective in reducing disease development. Therefore, resistant cultivars and crop rotation need to be employed in conjunction with other measures (Peng et al. 2014). In Brazil, the clubroot control of cauliflower (*Brassica oleracea* var. *botrytis*) and Chinese cabbage was possible on highly infested fields using a combination of liming, fungicide (flusulfamide), and

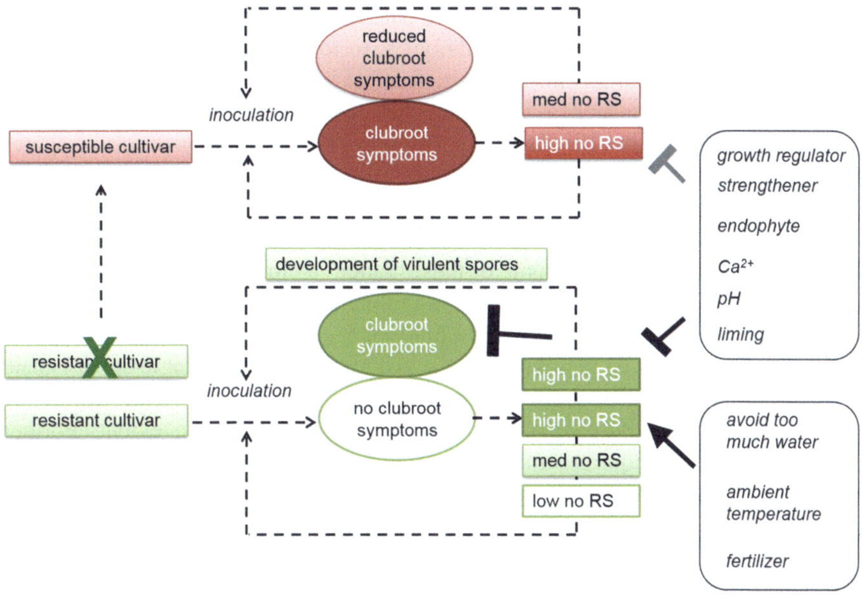

Fig. 4 Scheme for integrated clubroot control comparing the effect of a susceptible and resistant cultivar on resting spore (RS) numbers (no). In the case of a susceptible cultivar (in *red*), the inoculation leads in a linear chain of events to the development of the (severe) clubroot symptoms. The club produces a high number of resting spores which are liberated into the soil for another infection cycle. The spore load will stay high, unless measures (displayed in *box*) for reducing infection will take place. These ultimately reduce the spore numbers to medium (med) which in turn leads to reduced clubroot symptoms, eventually the spore load will gradually get lower. In the case of a resistant cultivar (in *green*), the club development is blocked either at the root hair or cortex infection, so that the clubroot symptoms are very small or nonexisting. That will reduce the spore load via medium to low numbers. However, eventually more virulent spores can develop which can now infect the resistant cultivar. After some time, the resistant cultivar turns into a susceptible one, producing high spore loads of the more virulent form. If at any stage before this happens the spore load can be reduced by methods to control clubroot (*boxes*), then the time frame that a resistant cultivar can retain the resistance mechanism is quite high

solarization treatment (hydrothermal treatment employing solar radiation to heat the soil under a transparent plastic film), showing that high temperatures can be employed against *P. brassicae* (Kowata-Dresch and May-De Mio 2012).

In this context, a good prediction of clubroot formation can be important as well. A good knowledge of the site, the severity of disease in the most recent brassica crop, the rotational history, soil properties, and treatments applied in previous crops would be very helpful (Donald and Porter 2009). However, for the evaluation of success during various treatments, the determination of the clubroot pathogen in the soil is necessary. Many (q)PCR-based methods for detecting spores of *P. brassicae* in water and soils have been developed over the years (Faggian et al. 1999; Faggian and Strelkov 2009) as sensitive as detecting 1000 spores per gram soil (Wallenhammar et al. 2012). These can be used to determine not only actual soil spore load but also detect heavily infested patches within a field. So far, it has not been easy to determine individual pathotypes, to follow up on new more virulent *P. brassicae* strains, but this may become feasible within the next few years. In the last century, methods such as PCR on clubroot spores were unthinkable, and researchers were trying to develop methods for improved detection of resting spores in soil (Takahashi and Yamaguchi 1987) followed by methods to distinguish between viable and dead spores and antibody-linked assays (Wakeham and White 1996). In fact, one disadvantage of PCR is that it cannot give any information on the viability of resting spores, so some of these "older" methods are still important for some applications. Nevertheless, given the tremendous advances in the molecular methods, detection of pathotype and viability of spores seems to be just around the corner or may already be facilitated by the genome draft of the single-spore isolate e3 (Schwelm et al. 2015). In conclusion, clubroot, a disease with worldwide importance to agriculture, can be controlled at least to some extent. Further progress in this area requires strong collaboration among agronomists, plant pathologists, breeders, farmers, and molecular biologists.

Acknowledgments Work in the author's laboratory was funded over the years by the State of Saxony (SMWK and SAB), the Deutsche Forschungsgemeinschaft (DFG), Bundesministerium für Bildung und Forschung (BMBF), Bundesministerium für Landwirtschaft und Ernährung (BLME), and the Deutsche Akademische Austauschdienst (DAAD). Apologies to all colleagues whose work could not be included even though they contributed significantly to the clubroot world.

References

Agarwal A, Kaul V, Faggian R, Rookes JE, Ludwig-Müller J, Cahill DM (2011) Analysis of global host gene expression during the primary phase of the *Arabidopsis thaliana–Plasmodiophora brassicae* interaction. Funct Plant Biol 38:462–478

Ahmed HU, Hwang SF, Strelkov SE, Gossen BD, Peng G, Howard RJ, Turnbull GD (2011) Assessment of bait crops to reduce inoculum of clubroot (*Plasmodiophora brassicae*) of canola. Can J Plant Sci 91:545–551

Aist JR, Williams PH (1971) The cytology and kinetics of cabbage root hair penetration by *Plasmodiophora brassicae* Wor. Can J Bot 49:2023–2034

Arie T, Kobayashi Y, Kono Y, Gen O, Yamaguchi I (1999) Control of clubroot of crucifers by *Phoma glomerata* and its product epoxydon. Pestic Sci 55:602–604

Arnold DL, Blakesley D, Clarkson JM (1996) Evidence for the growth of *Plasmodiophora brassicae* in vitro. Mycol Res 100:535–540

Auer S, Ludwig-Müller J (2014) Effects of the endophyte *Acremonium alternatum* on oilseed rape (*Brassica napus*) development and clubroot progression. Albanian J Agric Sci 13:15–20

Buczacki ST, Moxham SE (1979) A triple stain for differentiating resin-embedded sections of *Plasmodiophora brassicae* in host tissues under light microscope. Trans Br Mycol Soc 72:311–347

Chai AL, Xie XW, Shi YX, Li BJ (2014) Research status of clubroot (*Plasmodiophora brassicae*) on cruciferous crops in China. Can J Plant Pathol 36(S1):142–153

Cheah L-H, Kent G, Gowers S (2001) Brassica crops and a *Streptomyces* sp. as potential biocontrol for clubroot of brassicas. N Z Plant Prot 54:80–83

Cheah L-H, Gowers S, Marsh AT (2006) Clubroot control using Brassica break crops. Acta Hortic 706:329–332
Choe S (2004) Brassinosteroid biosynthesis and metabolism. In: Davies PJ (ed) Plant hormones: biosynthesis, signal transduction, action! Kluwer Academic, Dordrecht, pp 156–178
Christian M, Steffens B, Schenck D, Burmester S, Böttger M, Lüthen H (2006) How does auxin enhance cell elongation? Roles of auxin-binding proteins and potassium channels in growth control. Plant Biol 8:346–352
Clapperton MJ, Lee NO, Binet F, Conner RL (2001) Earthworms indirectly reduce the effects of take-all (*Gaeumannomyces graminis* var. *tritici*) on soft white spring wheat (*Triticum aestivum* cv. Fielder). Soil Biol Biochem 33:1531–1538
Conrath U, Thulke O, Katz V, Schwindling S, Kohler A (2001) Priming as a mechanism in induced systemic resistance of plants. Eur J Plant Pathol 107:113–119
Deora A, Gossen BD, Walley F, McDonald MR (2011) Boron reduces development of clubroot in canola. Can J Plant Pathol 33:475–484
Devos S, Prinsen E (2006) Plant hormones: a key in clubroot development. Commun Agric Appl Biol Sci 71(3 Pt B):869–872
Diederichsen E, Frauen M, Ludwig-Müller J (2014) Clubroot disease management challenges from a German perspective. Can J Plant Pathol 36(S1):85–98
Dixon GR (2009) The occurrence and economic impact of *Plasmodiophora brassicae* and clubroot disease. J Plant Growth Regul 28:194–202
Dixon GR (2014) Clubroot (*Plasmodiophora brassicae* Woronin)—an agricultural and biological challenge worldwide. Can J Plant Pathol 36(S1):5–18
Doan TT, Jäschke D, Ludwig-Müller J (2010) An endophytic fungus induces tolerance against the clubroot pathogen *Plasmodiophora brassicae* in *Arabidopsis thaliana* and *Brassica rapa* roots. Acta Hortic 867:173–180
Donald EC, Porter IJ (2004) A sand–solution culture technique used to observe the effect of calcium and pH on root hair and cortical stages of infection by *Plasmodiophora brassicae*. Australas Plant Pathol 33:585–589
Donald EC (2005) The influence of abiotic factors and host plant physiology on the survival and pathology of Plasmodiophora brassicae of vegetable brassicas. PhD thesis, University of Melbourne
Donald C, Porter I (2009) Integrated control of clubroot. J Plant Growth Regul 28:289–303
Donald EC, Porter IJ (2014) Clubroot in Australia: the history and impact of *Plasmodiophora brassicae* in Brassica crops and research efforts directed towards its control. Can J Plant Pathol 36(S1):66–84
Donald EC, Lawrence JM, Porter IJ (2004) Influence of particle size and application method on the efficacy of calcium cyanamide for control of clubroot of vegetable brassicas. Crop Prot 23:297–303
Einhorn G, Bochow H (1990) Influence of temperature and soil pH value on the pathogenesis of clubroot *Plasmodiophora brassicae* Wor. Arch Phytopathol Pflanzen 26:131–138
Faggian R, Strelkov SE (2009) Detection and Measurement of *Plasmodiophora brassicae*. J Plant Growth Regul 28:282–288
Faggian R, Bulman SR, Lawrie AC, Porter IJ (1999) Specific polymerase chain reaction primers for the detection of *Plasmodiophora brassicae* in soil and water. Phytopathology 89:392–397
Friberg H, Lagerlöf J, Hedlund K, Rämert B (2008) Effect of earthworms and incorporation of grass on *Plasmodiophora brassicae*. Pedobiologia 52:29–39
Gao Y, Xu G (2014) Development of an effective nonchemical method against *Plasmodiophora brassicae* on Chinese cabbage. Int J Agron 2014:Article ID 307367, http://dx.doi.org/10.1155/2014/307367
Gimsing AL, Kirkegaard JA (2009) Glucosinolates and biofumigation: fate of glucosinolates and their hydrolysis products in soil. Phytochem Rev 8:299–310
Gossen BD, Adhikari KKC, McDonald MR (2012) Effect of seeding date on development of clubroot in short-season brassica crops. Can J Plant Pathol 34:516–523

Gossen BD, Deora A, Peng G, Hwang S-F, McDonald MR (2014) Effect of environmental parameters on clubroot development and the risk of pathogen spread. Can J Plant Pathol 36 (S1):37–48

Guo S, Mao Z, Wu Y, Hao K, He P, He Y (2013) Genome sequencing of *Bacillus subtilis* strain XF-1 with high efficiency in the suppression of *Plasmodiophora brassicae*. Genome Announc 1(2):e00066-13. doi:10.1128/genomeA.00066-13

Hartwig T, Corvalan C, Best NB, Budka JS, Zhu J-Y, Choe S, Schulz B (2012) Propiconazole is a specific and accessible brassinosteroid (BR) biosynthesis inhibitor for Arabidopsis and maize. PLoS One 7:e36625

Hildebrand PD, McRae KB (1998) Control of clubroot caused by *Plasmodiophora brassicae* with nonionic surfactants. Can J Plant Pathol 20:1–11

Hirai M (2006) Genetic analysis of clubroot resistance in Brassica crops. Breeding Sci 56:223–229

Howard RJ, Strelkov SE, Harding MW (2010) Clubroot of cruciferous crops—new perspectives on an old disease. Can J Plant Pathol 32:43–57

Hwang SF, Cao T, Xiao Q, Ahmed HU, Manolii VP, Turnbull GD, Gossen BD, Peng G, Strelkov SE (2012) Effects of fungicide, seeding date and seedling age on clubroot severity, seedling emergence and yield of canola. Can J Plant Sci 92:1175–1186

Hwang SF, Ahmed HU, Zhou Q, Strelkov SE, Gossen BD, Peng G, Turnbull GD (2014a) Efficacy of Vapam fumigant against clubroot (*Plasmodiophora brassicae*) of canola. Plant Pathol 63:1374–1383

Hwang S-F, Howard RJ, Strelkov SE, Gossen BD, Peng G (2014b) Management of clubroot (*Plasmodiophora brassicae*) on canola (*Brassica napus*) in western Canada. Can J Plant Pathol 36(S1):49–65

Ikegami H (1985) Decrease of clubroot fungus by cultivation of different crops in heavily infested soil. Res Bull Fac Agric Gifu Univ 50:19–32

Jahn L, Mucha S, Bergmann S, Horn C, Siemens J, Staswick P, Steffens B, Ludwig-Müller J (2013) The clubroot pathogen (*Plasmodiophora brassicae*) influences auxin signaling to regulate auxin homeostasis. Plants 2:726–749

Jäschke D, Dugassa-Gobena D, Karlovsky P, Vidal S, Ludwig-Müller J (2010) Suppression of clubroot (*Plasmodiophora brassicae*) development in *Arabidopsis thaliana* by the endophytic fungus *Acremonium alternatum*. Plant Pathol 59:100–111

Joo GJ, Kim YM, Kim JW, Kim WC, Rhee IK, Choi YH, Kim JH (2004) Biocontrol of cabbage clubroot by the organic fertilizer using *Streptomyces* sp. AC-3. Kor J Microbiol Biotechnol 32:173–178

Kageyama K, Asano T (2009) Life cycle of *Plasmodiophora brassicae*. J Plant Growth Regul 28:203–211

Kammerich J, Beckmann S, Scharafat I, Ludwig-Müller J (2014) Suppression of the clubroot pathogen *Plasmodiophora brassicae* by plant growth promoting formulations in roots of two *Brassica* species. Plant Pathol 63:846–857

Kobelt P (2000) Die Verbreitung von sekundären Plasmodien von *Plasmodiophora brassicae* (Wor.) im Wurzelgewebe von *Arabidopsis thaliana* nach immunhistologischer Markierung des plasmodialen Zytoskeletts. Dissertation, Institut für Angewandte Genetik, Freie Universität Berlin, Germany

Kofoet A, Fischer K (2007) Evaluation of plant resistance improvers to control *Peronospora* destructor, *P. parasitica*, *Bremia lactucae* and *Pseudoperonospora cubensis*. J Plant Dis Protect 114:54–61

Kowata-Dresch LS, May-De Mio LL (2012) Clubroot management of highly infested soils. Crop Prot 35:47–52

Kroll TK, Lacy GH, Moore LD (1983) A quantitative description of the colonization of susceptible and resistant radish plants by *Plasmodiophora brassicae*. J Phytopathol 108:97–105

Kuginuki Y, Hiroaki Y, Hirai M (1999) Variation in virulence of *Plasmodiophora brassicae* in Japan tested with clubroot-resistant cultivars of Chinese cabbage (*Brassica rapa* L. ssp. *pekinensis*). Eur J Plant Pathol 105:327–332

Kühne S, Burth U, Marx P (2006) Biologischer Pflanzenschutz im Freiland. Eugen Ulmer, Stuttgart

Lahlali R, Peng G (2014) Suppression of clubroot by *Clonostachys rosea* via antibiosis and induced host resistance. Plant Pathol 63:447–455

Lahlali R, Peng G, McGregor L, Gossen BD, Hwang SF, McDonald MR (2011) Mechanisms of the biofungicide Serenade (*Bacillus subtilis* GST713) in suppressing clubroot. Biocontrol Sci Technol 21:1351–1362

Lahlali R, Peng G, Gossen BD, McGregor L, Yu FQ, Hynes RK, Hwang S-F, McDonald MR, Boyetchko SM (2013) Evidence that the biofungicide Serenade (*Bacillus subtilis*) suppresses clubroot on canola via antibiosis and induced host resistance. Phytopathology 103:245–254

Lahlali R, McGregor L, Song T, Gossen BD, Narisawa K, Peng G (2014) *Heteroconium chaetospira* induces resistance to clubroot via upregulation of host genes involved in jasmonic acid, ethylene, and auxin biosynthesis. PLoS One 9(4):e94144

Lecourieux D, Ranjeva R, Pugin A (2006) Calcium in plant defence-signalling pathways. New Phytol 171:249–269

Lee SO, Choi GJ, Choi YH et al (2008) Isolation and characterization of endophytic actinomycetes from Chinese cabbage roots as antagonists to *Plasmodiophora brassicae*. J Microbiol Biotechnol 18:1741–1746

Lovelock DA, Donald CE, Conlan XA, Cahill DM (2013) Salicylic acid suppression of clubroot in broccoli (*Brassicae oleracea* var. *italica*) caused by the obligate biotroph *Plasmodiophora brassicae*. Australas Plant Pathol 42:141–153

Ludwig-Müller J (2009) Plant defence—what can we learn from clubroots? Australas Plant Pathol 38:318–324

Ludwig-Müller J, Kasperczyk N, Schubert B, Hilgenberg W (1995) Identification of salicylic acid in Chinese cabbage and its possible role during root infection with *Plasmodiophora brassicae*. Curr Top Phytochem 14:39–45

Ludwig-Müller J, Prinsen E, Rolfe S, Scholes J (2009) Metabolism and plant hormone action during the clubroot disease. J Plant Growth Regul 28:229–244

Ludwig-Müller J, Jülke S, Geiß K, Richter F, Sola I, Rusak G, Mithöfer A, Keenan S, Bulman S (2015) A novel methyltransferase from the intracellular pathogen *Plasmodiophora brassicae* methylates salicylic acid. Mol Plant Pathol 16:349–364

Mancuso S, Azzarello E, Mugnai S, Briand X (2006) Marine bioactive substances (IPA extract) improve foliar ion uptake and water tolerance in potted *Vitis vinifera* plants. Adv Hortic Sci 20:156–161

Metting W, Zimmerman J, Crouch IJ, Van Staden J (1990) Agronomic uses of seaweed and microalgae. In: Akatsuka I (ed) Introduction to applied phycology. SPB, The Hague, pp 269–307

Mithen R, Magrath R (1992) A contribution to the life history of *Plasmodiophora brassicae*: secondary plasmodia development in root galls of *Arabidopsis thaliana*. Mycol Res 96:877–885

Moody SA, Piearce TG, Dighton J (1996) Fate of some fungal spores associated with wheat straw decomposition on passage through the guts of *Lumbricus terrestris* and *Aporrectodea longa*. Soil Biol Biochem 28:533–537

Muday GK, Murphy AS (2002) An emerging model of auxin transport regulation. Plant Cell 14:293–299

Murakami H, Tsushima S, Akimoto T, Shishido Y (2001) Reduction of spore density of *Plasmodiophora brassicae* in soil by decoy plants. J Gen Plant Pathol 67:85–88

Murakami H, Tsushima S, Kuroyanagi Y, Shishido Y (2002) Reduction in resting spore density of *Plasmodiophora brassicae* and clubroot severity by liming. Soil Sci Plant Nutr 48:685–691

Myers DF, Campbell RN (1985) Lime and the control of clubroot of crucifers: effects of pH, calcium, magnesium and their interactions. Phytopathology 75:670–673

Naiki T, Dixon GR (1987) The effects of chemicals on developmental stages of *Plasmodiophora brassicae* clubroot. Plant Pathol 36:316–327

Nakamura Y, Itakura J, Matsuzaki I (1995) Influence of the earthworm *Pheretima hilgendorfi* (Megascolecidae) on *Plasmodiophora brassicae* clubroot galls of cabbage seedlings in pot. Edaphologia 54:39–41

Narisawa K, Tokumasu S, Hashiba T (1998) Suppression of clubroot formation in Chinese cabbage by the root endophytic fungus, *Heteroconium chaetospira*. Plant Pathol 47:206–210

Narisawa K, Ohiki KT, Hashiba T (2000) Suppression of clubroot and Verticillium yellows in Chinese cabbage in the field by the root endophytic fungus, *Heteroconium chaetospira*. Plant Pathol 49:141–146

Narisawa K, Shimura M, Usuki F, Fukuhara S, Hashiba T (2005) Effects of pathogen density, soil moisture, and soil pH on biological control of clubroot in Chinese cabbage by *Heteroconium chaetospira*. Plant Dis 89:285–290

Niwa R, Nomura Y, Osaki M, Ezawa T (2008) Suppression of clubroot disease under neutral pH caused by inhibition of spore germination of *Plasmodiophora brassicae* in the rhizosphere. Plant Pathol 57:445–452

Päsold S, Ludwig-Müller J (2013) Clubroot (*Plasmodiophora brassicae*) formation in Arabidopsis is reduced after treatment with prohexadione-calcium, an inhibitor for oxoglutaric acid dependent dioxygenases. Plant Pathol 62:1357–1365

Päsold S, Siegel I, Seidel C, Ludwig-Müller J (2010) Flavonoid accumulation in *Arabidopsis thaliana* root galls caused by the obligate biotrophic pathogen *Plasmodiophora brassicae*. Mol Plant Pathol 11:545–562

Peng G, Hwang SF, McDonald MR, Lahlali R, Gossen BD, Hynes RH, Boyetchko SM (2010) Management of clubroot (*Plasmodiophora brassicae*) with microbial and synthetic fungicides. Phytopathology 100:S99

Peng G, McGregor L, Lahlali R, Gossen BD, Hwang SF, Adhikari KK, Strelkov SE, McDonald MR (2011) Potential biological control of clubroot on canola and crucifer vegetable crops. Plant Pathol 60:566–574

Peng G, Lahlali R, Hwang S-F, Pageau D, Hynes RK, McDonald MR, Gossen BD, Strelkov SE (2014) Crop rotation, cultivar resistance, and fungicides/biofungicides for managing clubroot (*Plasmodiophora brassicae*) on canola. Can J Plant Pathol 36(S1):99–112

Rademacher W (2000) Growth retardants: effects on gibberellin biosynthesis and other metabolic pathways. Annu Rev Plant Physiol Plant Biol 51:501–531

Rennie DC, Holtz MD, Turkington TK, Leboldus JM, Hwang S-F, Howard RJ, Strelkov SE (2015) Movement of *Plasmodiophora brassicae* resting spores in windblown dust. Can J Plant Pathol. doi:10.1080/07060661.2015.1036362

Robak J (1994) Crop rotation effect on clubroot disease decrease. Acta Hortic 371:223–226

Sakai R, Sato R, Niki H, Sakamura S (1970) Biological activity of phyllosinol, a phytotoxic compound isolated from a culture filtrate of *Phyllosticta* sp. Plant Cell Physiol 11:907–920

Schuller A, Kehr J, Ludwig-Müller J (2014) Laser microdissection coupled to transcriptional profiling of Arabidopsis roots inoculated by *Plasmodiophora brassicae* indicates a role for brassinosteroids in clubroot formation. Plant Cell Physiol 55:392–411

Schwelm A, Fogelqvist J, Knaust A, Jülke S, Lilja T, Bonilla-Rosso G, Karlsson M, Shevchenko A, Choi SR, Dhandapani V, Kim HG, Park JY, Lim YP, Ludwig-Müller J, Dixelius C (2015) The genome sequence of the plant pathogenic Rhizarian *Plasmodiophora brassicae* reveals insights in its biotrophic life cycle and ancestry of chitin synthases. Sci Rep. doi:10.1038/srep11153

Sharma K, Gossen BD, McDonald MR (2011) Effect of temperature on primary infection by *Plasmodiophora brassicae* and initiation of clubroot symptoms. Plant Pathol 60:830–838

Siemens J, Nagel M, Ludwig-Müller J, Sacristán MD (2002) The interaction of *Plasmodiophora brassicae* and *Arabidopsis thaliana*: parameters for disease quantification and screening of mutant lines. J Phytopathol 150:592–605

Siemens J, Keller I, Sarx J, Kunz S, Schuller A, Nagel W, Schmülling T, Parniske M, Ludwig-Müller J (2006) Transcriptome analysis of *Arabidopsis* clubroots indicate a key role for cytokinins in disease development. Mol Plant Microbe Interact 19:480–494

Siemens J, Gonzales M-C, Wolf S, Hofmann C, Greiner S, Du Y, Rausch T, Roitsch T, Ludwig-Müller J (2011) Extracellular invertase is involved in the regulation of the clubroot disease in *Arabidopsis thaliana*. Mol Plant Pathol 12:247–262

Smelt JH, Leistra M (1974) Conversion of metham-sodium to methyl isothiocyanate and basic data on the behaviour of methyl isothiocyanate in soil. Pestic Sci 5:401–407

Strelkov SE, Hwang SF, Howard RJ, Hartman M, Turkington TK (2011) Progress towards the sustainable management of clubroot (*Plasmodiophora brassicae*) of canola on the Canadian Prairies. Prairie Soils Crops J 4:114–121

Takahashi K, Yamaguchi T (1987) An improved method for estimating the number of resting spores of *Plasmodiophora brassicae* in soil. Ann Phytopathol Soc Jpn 53:507–515

Takahashi H, Takita K, Kishimoto T, Mitsui T, Hori H (2002) Ca^{2+} is required by clubroot resistant turnip cells for transient increases in PAL activity that follow inoculation with *Plasmodiophora brassicae*. J Phytopathol 150:529–535

Wakeham AJ, White JG (1996) Serological detection in soil of *Plasmodiophora brassicae* resting spores. Physiol Mol Plant Pathol 48:289–303

Wallenhammar AC (1996) Prevalence of *Plasmodiophora brassicae* in a spring oilseed rape growing area in central Sweden and factors influencing soil infestation levels. Plant Pathol 45:710–719

Wallenhammar A-C, Almquist C, Söderström M, Jonsson A (2012) In-field distribution of *Plasmodiophora brassicae* measured using quantitative real-time PCR. Plant Pathol 61:16–28

Wang L, Li B, Chen X, Shi Y, Zhou Q, Qiu H, Ibrahim M, Xie GL, Sun GC (2012) Effect of chitosan on seed germination, seedling growth and the clubroot control in Chinese cabbage. J Food Agric Environ 10:673–675

Webster MA, Dixon GR (1991a) Calcium, pH and inoculum concentration influencing colonization by *Plasmodiophora brassicae*. Mycol Res 95:65–73

Webster MA, Dixon GR (1991b) Boron, pH and inoculum concentration influencing colonization by *Plasmodiophora brassicae*. Mycol Res 95:74–79

Winding A, Rønn R, Hendriksen NB (1997) Bacteria and protozoa in soil microhabitats as affected by earthworms. Biol Fertil Soils 24:133–140

Wite D, Mattner SW, Porter IJ, Arioli T (2015) The suppressive effect of a commercial extract from *Durvillaea potatorum* and *Ascophyllum nodosum* on infection of broccoli by *Plasmodiophora brassicae*. J Appl Phycol. doi:10.1007/s10811-015-0564-y

Woronin M (1878) *Plasmodiophora brassicae* Urheber der Kohlpflanzen—Hernie. Jahrb Wiss Bot 11:548–574 [translated by Chupp C (1934) Phytopathological classics no. 4. American Phytopathological Society, St. Paul, MN]

Zhou L, Zhang L, He Y, Liu F, Li M, Wang Z, Ji G (2014) Isolation and characterization of bacterial isolates for biological control of clubroot on Chinese cabbage. Eur J Plant Pathol 140:159–168

Belowground Defence Strategies Against Sedentary Nematodes

Marta Barcala, Javier Cabrera, Carmen Fenoll, and Carolina Escobar

Abstract Plant parasitic nematodes (PPN) represent a major threat to agriculture as they produce high economic losses. Among them, the sedentary endoparasites (root-knot nematodes, RKNs, and cyst nematodes) complete their life cycle inside the host roots where they induce a special feeding site for nutrient uptake, namely, giant cells for RKNs and syncytia for cyst nematodes. The root system represents the first physical barrier for nematode penetration. Cell wall hardening strategies used against many pathogens are not very effective against them, as they use a robust stylet during penetration or migration to apply mechanical force and/or to secrete a mixture of cell wall degrading enzymes from the subventral esophageal glands. Plant defences against endoparasitic nematodes include mechanisms as pattern-triggered immunity (PTI) and effector-triggered immunity (ETI), the last one leading to the hypersensitive response. The development of sensitive "omics" techniques, sometimes combined with feeding cell isolation, allowed global analysis of gene expression during this interaction. Hence, transcriptional changes associated to compatible and incompatible interactions of different plant species such as *Arabidopsis*, soybean, tomato, *Medicago*, etc. with different species of either cyst or RKN nematodes brought up a vast amount of genes induced or repressed during both interactions. Some of them will be useful for future applications on nematode control, as functional studies indicated their role in nematode resistance. Information on the molecular effectors used by nematodes during the cross talk with susceptible or resistant plants leading to plant defence responses is continuously increasing. Furthermore, in the recent years, some effectors that suppress plant defences were described, increasing the complexity of this particular plant–pathogen interaction.

M. Barcala • J. Cabrera • C. Fenoll • C. Escobar (✉)
Facultad de Ciencias Ambientales y Bioquímica, Universidad de Castilla-La Mancha, Av. Carlos III s/n, 45071 Toledo, Spain
e-mail: carolina.escobar@uclm.es

C.M.F. Vos, K. Kazan (eds.), *Belowground Defence Strategies in Plants*, Signaling and Communication in Plants, DOI 10.1007/978-3-319-42319-7_10

1 Plant Parasitic Nematodes: Introduction to Life Style

PPNs are roundworms within the Nematoda phylum. They are obligate parasites with a simple body structure and can be isolated from almost every vascular plant (crops, ornamental plants, and trees). They represent a major threat to agriculture, as yearly economic losses due to crop infestation by PPN have been estimated in more than $100 billion (Bird et al. 2009).

According to their lifestyle, PPN are classified into sedentary or migratory (either ectoparasites or endoparasites). Sedentary endoparasitic PPN represent one of the most important groups in terms of agricultural damage, economical losses, and cost of pest eradication. So far, more than 2500 species have been described (Zhang 2013), and they affect relevant staple crops such as potato, rice, corn, or wheat.

Sedentary endoparasitic nematodes complete their life cycle inside the host roots where they trigger the formation of a special feeding site for nutrient uptake (Fig. 1A). They deprive plant–host from food resources and water what in turn results in stunted and dwarf plants, lowering yields. The most representative members of this group are the root-knot nematodes (RKN, *Meloidogyne* spp.) and the cyst nematodes (represented mainly by *Heterodera* spp. and *Globodera* spp.), which are named based on the typical structures that can be observed in the host root system after nematode infection: the gall and the cyst, respectively (Fig. 1B, D). Both RKN and cyst nematodes display morphological similarities, as unicellular esophageal glands, sexual dimorphism, and the presence of a stylet. The esophageal glands are essential for parasitism, during penetration and migration stages and during establishment of the feeding site (nematode feeding site, NFS) (reviewed in Mitchum et al. 2013). During the first stages of parasitism, two subventral esophageal glands produce secretory granules containing a cocktail of cell wall (CW)-modifying enzymes that will help the nematode to penetrate and move through the host root. Among the CW-degrading enzymes, several endoglucanases and pectate lyases have been described to be actively secreted during invasion and/or migration (Davis et al. 2011), facilitating the parasitism. Similar CW-modifying enzymes have been reported from bacteria, suggesting a putative acquisition via gene horizontal transfer (Davis et al. 2011). At later stages, the dorsal esophageal gland enlarges and becomes more relevant in detriment of the subventral glands, producing nematode effectors involved in NFS formation and maintenance. The composition of the gland secretions varies among nematode species and throughout their developmental stages (reviewed in Hussey and Davis 2004; Rosso and Grenier 2011; Gardner et al. 2015; Truong et al. 2015).

The stylet is a distinctive PPN morphological feature, although not exclusive of them. It is a structure in the anterior part of the body in the shape of a needle. This can be protruded outside in repetitive thrusts providing mechanical force that, together with the enzymatic activity of secreted cellulases and endoglucanases, facilitates penetration and migration in the host root. The stylet is also used for

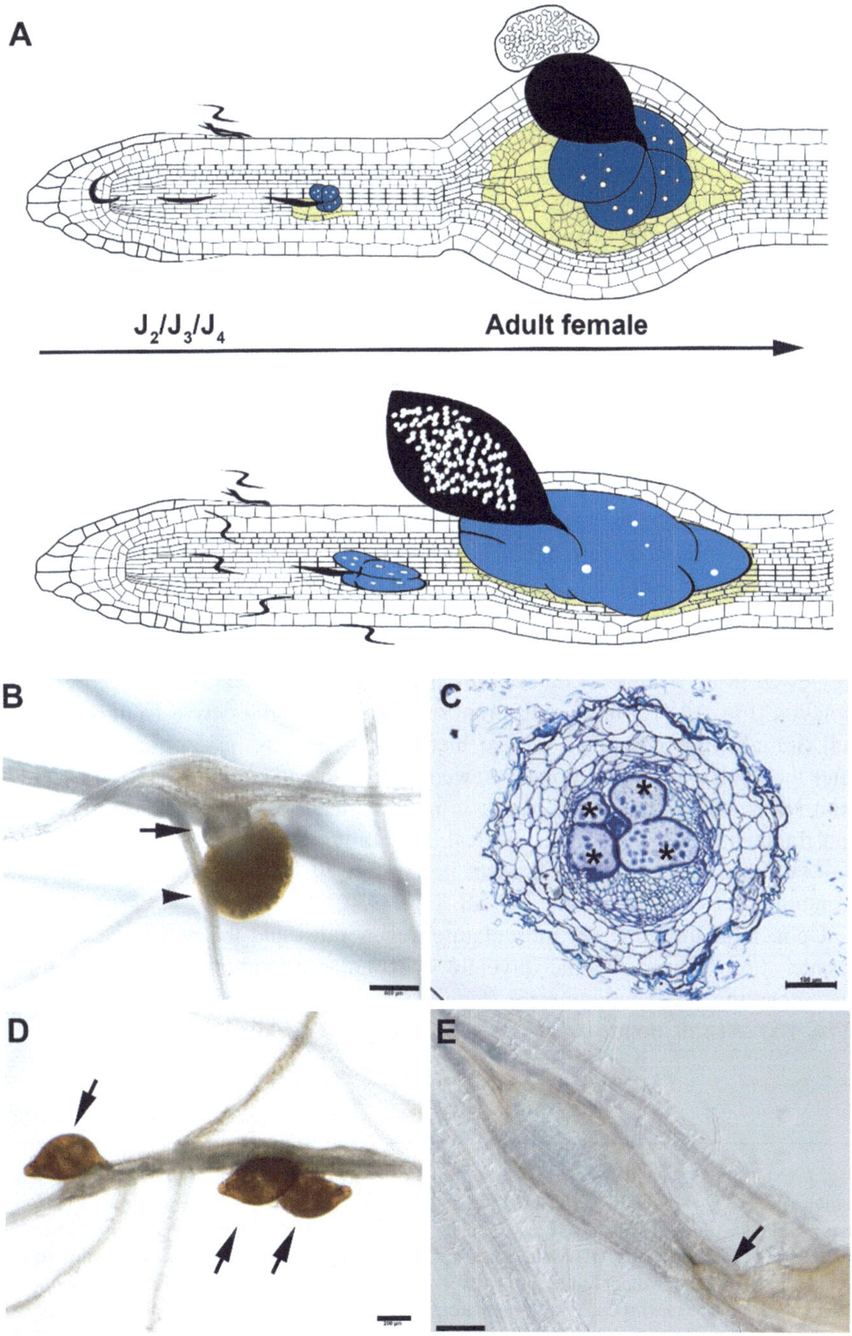

Fig. 1 Overview of life cycles of plant sedentary endoparasitic nematodes inside roots. (**A**) Schematic representation of the course of the interaction during invasion, migration, and feeding site (NFS) formation with a root-knot nematode (*upper panel*) and a cyst nematode (*lower panel*). From *left* to *right*, see preparasitic J2 larvae (*black*) invading the roots, initial stages of GC, and

piercing host cells, injecting secretions and withdrawing nutrients (Perry and Moens 2011).

Typically, the life cycle of both types of nematodes begins once infective juveniles (J2) hatch from eggs (reviewed in Escobar et al. 2015). Freshly hatched J2s are attracted to young roots, which penetrate using different strategies. Cyst nematodes possess a robust stylet that allow them to mechanically break cell walls and migrate directly from the point of entrance to the vascular cylinder trespassing the Casparian strip (Wyss and Zunke 1986). RKNs show a more elaborate behavior and they enter mainly in the elongation zone very close to the root meristem. Then, they first move intracellularly toward the root tip where they turn 360° to enter the vascular cylinder where they will establish and initiate feeding site formation (Wyss et al. 1992). NFS development is a crucial step; so far, the J2s have survived using their own lipid reservoir, but unless they succeed on feeding site development, they will starve. During evolution, RKNs and cyst nematodes have developed different strategies for inducing feeding sites, in both cases involving selection of root cells to reprogram their gene expression (through injected nematode secretions) which eventually lead to morphological and metabolic changes. A RKN punctures several cells (around 5–8) in the vascular tissues and injects its secretions. The most distinctive morphological change in these cells is their enlargement; thus, they are known as giant cells (GC, Fig. 1A, C). GC formation involves several rounds of mitosis with partial cytokinesis and endoreduplication events (reviewed in de Almeida Engler et al. 2015). Cyst nematode feeding cells, called syncytia (Fig. 1A, E), have a different ontogeny. Syncytia derive from one single cell (initial syncytial cell, ISC) that increases its size by fusion of adjacent cells after their cell wall dissolution (reviewed in Sobczak and Golinowski 2011). Thus, both NFSs, GC and syncytia, result in multinucleate cells, with dense cytoplasm and fragmented vacuoles. Root cortical cells around NFSs hypertrophy, forming a root swelling that is more prominent in the case of RKN, originating the typical structure after RKN infection, the gall (Fig. 1A–C).

Upon NFS initiation, J2 musculature degenerates and the nematode becomes sessile. From then on, the juvenile will enlarge and suffer several molts encompassing the developmental stages J3 and J4, until reaching the adult stage (Fig. 1A). At this point, the differences in the life cycle between both nematode

---

Fig. 1 (continued) syncytia development within the vascular cylinder and mature NFS with adult females. The eggs are deposited inside the females in the cyst nematodes (*lower panel*), and egg masses filled with eggs from RKN are deposited outside the female body (*upper panel*). See the main text for further details. (**B**) Mature gall in *Cucumis sativus* after *M. javanica* infection. Adult female body protrudes outside the gall (*black arrow*) with the gelatinous matrix containing the eggs (*black arrowhead*). (**C**) Semithin section of a gall 7 days postinfection from *A. thaliana* stained with toluidine. GCs are indicated with *asterisks*. (**D**) *Sinapis alba* root infected with *H. schachtii*. (**E**) *Arabidopsis thaliana* syncytia close-up micrograph. Adult females containing egg masses inside its body (cysts) are indicated by *black arrows*. *Scale bars, (C, E) 100μm; (B) 500μm; (D) 200μm*

groups are remarkable. Contrary to cyst nematodes that reproduce sexually, RKN present apomixis and reproductive individuals are all female, albeit a few adult males that appear under stress conditions (i.e., low nutrient availability). RKN adult females show a pear-like shape (Fig. 1B), and their head remains attached to the feeding cells while the posterior part of the body is exposed to the outside. These females deposit parthenogenetic eggs embedded in a gelatinous matrix (Fig. 1B), where the larvae will develop into J1 and J2 that will hatch to complete a new infective cycle. In cyst nematodes, the adult female is fertilized by a free-moving male, while its head stays inside the host root and the majority of its body protrudes outside the root. The fertilized female shows a lemonlike shape and deposits the eggs inside its body. Once the female dies, its body hardens to provide extra protection for eggs and serve as a resistance form called the cyst (Fig. 1D). Inside the eggs, the nematode will pass through the developmental stages J1 and J2, and it will remains inside the egg until favorable conditions for hatching. Nematodes remain viable inside the cyst for long periods, which makes their eradication from fields very difficult . The narrow host range that show the cyst nematodes, restricted mainly to Solanaceae (as for *Globodera* spp.) or Poaceae (for *Heterodera* spp.), could be a consequence of having this resistance from what allows them to remain quiescent for long periods and eliminates the urgent necessity to find a suitable host upon hatching as for RKN (Lambert and Bekal 2002).

2 Belowground Defences Against Endoparasitic Nematodes

Resistance to nematode infection is usually referred to as the result of a low capacity of the nematode to reproduce, so that no substantial increase in final nematode population can be observed. In most resistant plants, nematodes are able to penetrate and migrate inside the root, and even to initiate a NFS, although its development is usually blocked by plant defence responses after a while. Thus, the root system represents the first barrier for nematode penetration, and by hardening the CW, e.g., increasing the lignin content, plants try to protect themselves from nematode attack through preinfective mechanical defences. Chemical preinfective defences have been described for some plants whose exudates contain compounds with repellent or nematicidal activity (Tomczak et al. 2009). However, this hardening strategy that can be reliable for pathogens such as fungi or bacteria is not very effective against endoparasitic nematodes, as they enter very young roots using a stylet and/or a mixture of CW-degrading enzymes to loosen cell walls, thus facilitating penetration. Among CW-degrading enzymes, several endoglucanases or pectate lyases are actively secreted during cyst nematode invasion and/or migration (reviewed by Bohlmann and Sobczak 2014). Polygalacturonases have also been found in RKN secretions (Jaubert et al. 2002).

Plants have a battery of strategies to protect themselves against pathogen and to achieve plant immunity. These general responses are usually sufficient to avoid a broad range of pathogen attacks and are comprised in what has been called nonhost resistance. Basically, after any attempt of invasion, plants can recognize either conserved pathogen-derived molecules (i.e., flagellin, flg22, from bacteria), known as pathogen-associated molecular patterns (PAMPs), or molecules derived from the infection process itself. This process is usually executed by transmembrane pattern recognition receptors (PRRs) that activate pattern-triggered immunity or innate immunity (PTI; Macho and Zipfel 2014). Degradation products from CW-degrading enzymes (i.e., cellulases and pectate lyases (Lozano and Smant 2011)) or peptides derived from cleaved and degraded proteins that are produced as a result of the activities of the invading pathogen could also act as PAMPs (Tomczak et al. 2009; Albert 2013; Malinovsky et al. 2014). So far, no nematode PAMP that could induce plant defence responses has been reported (Mantelin et al. 2015). After PAMP recognition, plants initiate a response that involves transcriptional activation of defence genes encoding oxidases, peroxidases, or MAPKs and genes involved in the synthesis of defence compounds such as phytoalexin, flavonoids, or reactive oxygen species (ROS), etc. However, since nematodes and other pathogens deliver effectors that suppress PTI, some plants have a second layer of immune receptors encoded by resistance (R) genes. These proteins recognize effectors leading to effector-triggered immunity (ETI; Jones and Dangl 2006). After ETI signaling, plants initiate a cascade of events (i.e., synthesis of pathogenesis-related proteins (PR) or defence compounds such as phytoalexins) leading to the hypersensitive response (HR) (Smant and Jones 2011; Mantelin et al. 2015). Then, a cross talk among attack–defence strategies is initiated by pathogen and plant, and depending on the ability of the pathogen to evade or overcome these plant responses, it results in a compatible interaction, where the pathogen is able to feed and/or reproduce, or an incompatible interaction where plant is resistant.

The gene-for-gene resistance is so far the most successful plant tool to counteract PTI evasion and to fight nematode infection. This type of resistance is elicited by an effector from the pathogen, known as avirulence (Avr) factor, recognized by a plant plasma membrane or cytosolic receptor, the R protein, in a gene- and allele-specific fashion. For this reason, gene-for-gene resistance is highly specific for nematode pathotypes or races and specific plant species and cultivars, which represents a disadvantage for breeding. Identification of Avr or putative Avr genes from PPN is still scarce (e.g., *Mj-Cg-1*, *Gr-VAP1*, or *Gp-RBP1*; reviewed in Kaloshian et al. 2011; Rosso and Grenier 2011; Lozano-Torres et al. 2012). In the cases studied, the rapid resistance response (termed hypersensitive response, HR) involves the production of ROS and changes in the phosphorylation state of proteins and Ca^{2+} uptake (Jones and Dangl 2006; Lozano and Smant 2011), and it resembles an exacerbated and rapid PTI. This HR has been described in tomato plants carrying the *Hero A* gene after infection with the potato cyst nematode *G. rostochiensis* (Sobczak et al. 2005) or in tomato plants carrying the *Mi-1.2* gene after infection with *Meloidogyne incognita* (reviewed in Williamson and Roberts 2009). *R*-gene defence in most cases cannot prevent nematode infection/penetration, and it drives nematode population decrease by interfering with NFS development or promoting male formation, so that only in few cases some females complete their life cycle.

Besides these examples of direct interaction between the R protein and the Avr effector, there are evidences suggesting also indirect interactions (reviewed in Bogdanove 2002). For instance, the *Cf-2* resistance gene from *Solanum pimpinellifolium* guards an apoplastic papain-like cysteine protease (Rcr3) that recognizes the GrVAP1 effector from *G. rostochiensis* (Lozano-Torres et al. 2012; Mitchum et al. 2013). In this respect, the fungal effector Avr2 was first reported to activate defence signaling by perturbing Rcr3, sensed by Cf-2 that eventually triggers cell death (Krüger et al. 2002). Similarly, the *G. pallida Gp-RBP-1* gene encodes a secreted protein which induces effector-triggered immunity (ETI) mediated by the *Solanum tuberosum* disease resistance gene *Gpa2*, which encodes a nuclotide binding-leucin rich (NB-LRR) protein, but requires the Ran GTPase-activating protein 2 (RanGAP2), a protein known to interact with the Gpa2 N terminus (Sacco et al. 2009). Similarly, the potato (*Solanum tuberosum*) disease resistance protein Rx, which mediates resistance to the potato virus X (PVX), also interacts with the cofactor RanGAP2 for effective immune signaling (Tameling et al. 2010). Hence, it has been suggested that pathogens may target a limited number of host proteins that act as essential regulators of the plant defence responses (Mantelin et al. 2015). This agrees with the fact that several *R* genes provide protection against diverse pathogens, such as *Mi-1.2* that protects against several *Meloidogyne* species, aphids and whiteflies (Nombela et al. 2003) and *Cf-2* conferring resistance to *G. rostochiensis* and to the fungus *Cladosporium fulvum* (Lozano-Torres et al. 2012).

2.1 Defences Against Cyst Nematodes

The first *R* gene against cyst nematodes that was cloned is the sugar beet $Hs1^{pro\text{-}1}$ (Cai et al. 1997), which conferred resistance against *Heterodera schachtii*. Since then, other sources of resistance have been described, but few have been cloned. In plants harboring *R* genes such as $Hs^{pro\text{-}1}$ in sugar beet, *Hero A* in tomato or *Gpa2* in potato, the defence response is delayed and syncytia are initiated. Necrotic zones around and/or in syncytia, when observed, are not evident until a further developmental stage when syncytia degenerate. Sometimes nematode development is allowed but mainly males are produced. Syncytia collapse takes place at different times depending on different R genes (reviewed in Sobczak and Golinowski 2011). In barley carrying the *Rha2* gene (cv. Chebec), syncytia degenerate later than in cv. Galleon carrying the *Rha4* gene, suggesting that different defence cascades are elicited depending on *R* gene upon *Heterodera avenae* infection (Aditya et al. 2015).

Most reported *R* genes encode receptors of the nucleotide-binding–leucine-rich repeat (NB-LRR) family. This large family has been described in defence responses of other plant–pathogen interactions such as viruses, bacteria, fungi, or oomycetes (Grant et al. 2006; Kamoun 2006; Hogenhout et al. 2009; Stergiopoulos and de Wit

2009; Kaloshian et al. 2011). Activation of ETI by NB-LRR type of R proteins initiates defence cascades through MAP kinases what will probably activate in turn the jasmonic acid/salicylic (JA/SA) defence signaling pathway (Hammond-Kosack and Parker 2003). *R* genes such as *Gro1.4* from potato against *G. rostochiensis* are toll/interleukin 1 receptor (TIR) NB-LRR, whereas *Hero A* from tomato and *Gpa2* from potato against *G. rostochiensis* and *G. pallida* are members of the coiled-coil-NB-LRR (CC-NB-LRR), respectively (Caromel and Gebhardt 2011). All of them are located in the cytoplasm, contrary to another structural type of NB-LRR proteins that carry an extracellular LRR domain such as those encoded by $Hs1^{pro\text{-}1}$ from sugar beet or the *Cf* from tomato against *H. schachtii* and *G. rostochiensis*, respectively (reviewed in Fosu-Nyarko and Jones 2015).

Genome-wide mapping revealed that monogenic *R* genes are usually located in clusters of homologous genes, called "hot spots." *Hero A* is located within a region of 14 homologous genes (Ernst et al. 2002), *Gro1* is clustered with another 13 genes (Paal et al. 2004) and *Gpa2* locates in a small cluster (van der Vossen et al. 2000). Unfortunately, only for the *Gpa2* cluster, other resistance genes have been functionally identified, as the *Rx1* gene which confers resistance to PVX (van der Voort et al. 1997).

Although these *R*-encoded plant receptors are probed to be sufficient to confer complete resistance, in many cases, plants show intermediate resistance phenotypes that have been ascribed to polygenic resistance loci instead of single dominant *R* genes. In fact, quantitative trait loci (QTLs) are responsible for most of the resistances. Most studied QTL in soybean resistance against *Heterodera glycines* is the *Rhg* locus. Initially, three recessive loci were described (*Rhg1-Rhg3*; Caldwell et al. 1960), but *Rhg1* on chromosome 18 had the greatest impact on soybean cyst nematode (SCN) resistance, providing resistance against a broad range of SCN in the soybean cv. PI88788 (Melito et al. 2010). However, it is the combination of a resistance allele of *Rhg1*, with the corresponding dominant action *Rhg4* allele, which provides full resistance against some *Heterodera* races in resistant cv. Peking and Forrest (Melito et al. 2010). Mapping of this QTL located *Rhg1* in a 67 kb segment containing also 11 predicted genes. Among them, there is a gene encoding an LRR-receptor kinase (LRR-RK) (Glyma18g02680, GmRLK18-1); unfortunately, functional assays have revealed that this LRR-RK is not the major source of resistance for this locus (Melito et al. 2010). On the other hand, in vitro binding assays indicate the capacity of the GmRLK18-1 LRR domain to bind both nematode and plant signal peptides from the CLE protein family with high affinity (Afzal et al. 2013), in addition to other major effectors/molecules such as cyclophilins and methionine synthase. In a fine mapping of the *Rhg1* locus, a narrower region of 31 kb containing noncanonical *R* genes was described (Cook et al. 2012). Among them is the previously reported *α-SNAP* (Glyma18g02590), whose contribution to the resistance phenotype has been demonstrated by Matsye et al. (2012) and that is likely involved in vesicle trafficking and may influence exocytosis of products that alter feeding site development or nematode physiology. Another of these genes (Glyma18g02580) encodes a predicted amino acid transporter from the tryptophan/tyrosine permease family, which may affect auxin levels or distribution, yet another one (Glyma18g02610) encodes the WI12 protein that may participate in the production of compounds toxic to the nematodes. These three genes contribute to the resistance against SCN described for the *Rhg1-b* allele in the SCN-resistant soybean line PI 88788. Interestingly, higher resistance levels correlate with higher number of copies for this cluster. Fiber-FISH assays revealed a single cluster copy in the susceptible cultivar Williams82, whereas resistant cultivars Peking and Fayette had 10 and 4 copies, respectively (Cook et al. 2012).

The *Rhg4* QTL was mapped and cloned in soybean cv. Forrest by Liu et al. (2012). Within the cloned genes, some unrelated to the typical LRR R proteins were identified: one encoding a serine hydroxyl methyl transferase (*SHMT*) and the other a subtilisin-like protease (*SUB1*). Functional assays indicated that *SHMT* is responsible for the *Rhg4* resistance although its molecular mechanism of action is unknown. SHMT is involved in folate one-carbon metabolism, so its role during SCN resistance could be related to folate deficiency, by triggering HR and subsequent cell death or leading to nurturing deficiencies that eventually starve the SCN (Liu et al. 2012).

Global transcriptomic analysis of plants infected by cyst nematodes representing compatible and incompatible interactions revealed differential expression patterns of defence-related genes (Table 1). Many of those were involved in the basal defence, usually common to both interaction types, but a group of genes was more specific to plant resistance, mostly analyzed in crops of agronomical interest as soybean and tomato (Khan et al. 2004; Klink et al. 2005, 2007a, b, 2010a; Alkharouf et al. 2006; Ithal et al. 2007a; Ithal et al. 2007b; Puthoff et al. 2007; Uehara et al. 2010). Some studies were centered in soybean transcriptomes from different cultivars, one susceptible and the other resistant to a particular nematode race (Klink et al. 2011; Matsye et al. 2011; Mazarei et al. 2011; Wan et al. 2015). Alternatively, the same cultivar was tested against virulent and avirulent nematode races (Klink et al. 2007a, b, 2009a, b, 2010b; Hosseini and Matthews 2014). In some cases, near-isogenic lines (NILs; Klink et al. 2010a; Kandoth et al. 2011) were compared, and differences in responses to a certain pathogen can be presumably attributed to a narrow defined region of the genome. Similar results during incompatible interactions were observed throughout different host–cyst nematode interactions, showing upregulation of general plant disease and defence genes. Kandoth et al. (2011) used laser microdissection coupled with comparative microarray profiling of syncytia isolated from soybean-resistant and susceptible NILs differing at the locus *Rhg1* after infection with *H. glycines* type 0. They reported that the resistant NIL overexpressed genes encoding proteins related to plant defence or oxidative stress, like a homolog to the BCL-2-ASSOCIATED ATHANOGENE 6 (AtBAG6; its overexpression causes cell death in *Arabidopsis* and yeast), heat-shock proteins and factors, PRs, WRKY family transcription factors, proteins associated with HR, apoptotic cell death and the SA-mediated resistance pathway, or members of the canonical resistance family of CC-NB-LRR proteins. Interestingly, 23 NBS-LRR resistance genes and one LRR-receptor-like kinase from the biotic stress category were constitutively expressed in resistant lines, suggesting a

Table 1 Transcriptomic assays involving incompatible interactions of cyst nematodes with different plant species

Reference	Technique	Type of interaction	Nematode species	Plant species	Biological material
Puthoff et al. (2003)	Microarray	Compatible/ incompatible	*H. schachtii* *H. glycines*	*A. thaliana*	Roots 3 days postinoculation (dpi)
Klink et al. (2007a)	Microarray	Compatible/ incompatible	*H. glycines NL1-RHg* *H. glycines TN8*	*G. max cv. Peking (PI 548402)*	Roots 3 and 8 dpi
Klink et al. (2007b)	LCM and microarray	Compatible/ incompatible	*H. glycines NL1-RHg* *H. glycines TN8*	*G. max cv. Peking (PI 548402)*	LCM syncytia 3 and 8 dpi
Klink et al. (2009a)	Microarray	Compatible/ incompatible	*H. glycines NL1-RHg* *H. glycines TN8*	*G. max cv. Peking (PI 548402)*	Roots 12 h postinoculation (hpi), 3 or 8 dpi
Klink et al. (2009b)	LCM and microarray	Compatible/ incompatible	*H. glycines NL1-RHg* *H. glycines TN8*	*G. max cv. Peking (PI 548402)*	LCM syncytia 3, 6, and 9 dpi
Klink et al. (2010a)	LCM and microarray	Incompatible	*H. glycines NL1-RHg (HG type 7)*	*G. max (PI 88788)*	LCM syncytia 3, 6, and 9 dpi
Klink et al. (2010b)	LCM and microarray	Compatible/ incompatible	*H. glycines NL1-RHg (HG type 7)* *H. glycines TN8 (HG type 1.3.6.7)*	*G. max cv. Peking (PI 548402)*	LCM syncytia 3 dpi and 8 dpi
Klink et al. (2011)	LCM and microarray	Incompatible	*H. glycines NL1-RHg (HG type 7)*	*G. max cv. Peking (PI 548402)* *G. max (PI 88788)*	LCM syncytia 3, 6, and 9 dpi
Kandoth et al. (2011)	LCM and microarray	Compatible/ incompatible	*H. glycines* PA3 (HG type 0) *H. glycines* TN19 (HG type 1–7)	*G. max* cv. Williams 82 *G. max* cv. (PI 437654)	LCM syncytia 5 and 8 dpi
Matsye et al. (2011)	Microarray	Compatible/ incompatible	*H. glycines NL1-RHg (HG type 7)* *H. glycines TN8*	*G. max* cv. Peking (PI548402) *G. max* (PI 88788)	Roots 3, 6, and 9 dpi

(continued)

Table 1 (continued)

Reference	Technique	Type of interaction	Nematode species	Plant species	Biological material
			(HG type 1.3.6.7)		
Mazarei et al. (2011)	Microarray	Compatible/ incompatible	*H. glycines (race 2)*	*G. max TN02-275* *G. max TN02-226*	Roots 3, 6, and 9 dpi
Hosseini and Matthews (2014)	mRNA sequencing	Compatible/ incompatible	*H. glycines TN8 (race 14)* *H. glycines NL1-RHg (race 3)*	*G. max cv. Peking (PI 548402)*	Roots 6 and 8 dpi
Wan et al. (2015)	Microarray	Compatible/ incompatible	*Heterodera glycines Ichinohe*	*G. max cv. Magellan* *G. max (PI 437654)* *G. max (PI 567516C)*	Roots 3 and 8 dpi

possible role in regulating resistance to multiple SCN races in these soybean lines (Wan et al. 2015). In this respect, the *KR3* gene (encoding a TIR-NBS-LRR protein) was exclusively induced in a soybean-resistant cultivar at 3 and 6 dpi (Mazarei et al. 2011). Unique differentially expressed genes induced in resistant cultivars upon *H. glycines* infection were also those coding PR10, peroxidases, or cytochrome p450 proteins, as well as the β-1,4-glucanases and defence proteins belonging to the thioredoxin reductase family (RnDR and PDI, respectively; Hosseini and Matthews 2014; Klink et al. 2010a). Thioredoxin could act mediating SA defence through interaction with nonexpressor of PR genes 1 (NPR1). Moreover, transcriptomic studies pointed out the importance of the pathway of phenylpropanoids in the defence response to nematodes, not only because of their known role in lignin biosynthesis but because it is a branch for the synthesis of SA and flavonols. Thus, Hosseini and Matthews (2014) reported the induction of transcripts of key genes of phenylpropanoid and flavonoids pathway, such as chalcone synthase (*ChS*), chalcone isomerase (*ChI*), or chalcone reductase (*ChR*), and also two lipoxygenases (*A-8 LOX* and *LOX2*) in a resistant line upon *H. glycines*. Similar induction patterns of key genes from these former pathways were described in soybean-resistant cultivar transcriptomes after SCN infection (Klink et al. 2010a, b; Mazarei et al. 2011). Surprisingly, Matthews et al. (2013) observed no alteration of the reproductive index, female index (FI), as compared to the control when enzymes of this route were overexpressed, except for a moderate decrease in the case of *ChS* and *C4H* (cinnamate-4-hydroxylase).

Other genes related to defence responses as those encoding lipoxygenases (*LOX*s) involved in the biosynthesis of oxylipins and JA were induced in multiple SCN-resistant *Glycine max* genotypes (Ithal et al. 2007b; Klink et al. 2007a; Klink and Matthews 2009; Hosseini and Matthews 2014) and also in NFS and surrounding cells in resistant pea plants after infection with *H. goettingiana* (Veronico et al. 2006). According to this, overexpression of the lipoxygenase Glyma08g14550.1 in soybean significantly reduced FI to 42 % after SCN infection (Matthews et al. 2013). Other genes with a functional implication in nematode resistance and relevant for plant defence responses/syncytia formation are a cell wall modifier (Gm endo-b-1,4-glucanase), the ascorbate peroxidase 2, a lipoxygenase or the momilactone-A synthase implicated in the phytoalexin synthesis. Those were identified among 100 differentially expressed genes in a soybean-resistant cultivar as compared to those susceptible and tested for FI (Matthews et al. 2013). Peroxidases (PRXs) have a relevant role in defence response, as they are responsible for ROS production (Sharma et al. 2012). PRXs were induced after *Heterodera* spp. infection in resistant plants as compared to susceptible ones (Alkharouf and Matthews 2004; Simonetti et al. 2010; wheat and soybean respectively), contributing to the increase of ROS. ROS are not only toxic compounds but also trigger activation of MAPKs in downstream defence signaling cascades to elicit an HR. However, different types of these PRX, such as an ascorbate peroxidase 2 and a cationic peroxidase, are not acting in the same pathway for nematode resistance, as when overexpressed in soybean, the former reduced up to 70 % the FI, whereas the latter substantially increased FI (160 % to the control) (Matthews et al. 2014). In conclusion, a general trend in transcriptomes of resistance cultivars after cyst nematode infection, as compared to the compatible interaction, is the induction of defence-related genes, such as those encoding peroxidases, thioredoxins, and genes related to secondary metabolism (i.e., phenylpropanoids pathway) or lipid metabolism (i.e., phytoalexins or JA synthesis pathway).

The transcription factor category is usually numerous among the global transcriptomic analysis and are crucial regulators of entire transduction cascades. This category was also overrepresented in soybean-resistant lines such as those from the *WRKY* (e.g., *ZAP1*), *AP2*, or *MYB* families as compared to susceptible lines (Hosseini and Matthews 2014; Wan et al. 2015). Regarding cyst infection, WRKYs are either repressors of basal defence or positive regulators of ETI in compatible interactions (Eulgem and Somssich 2007).

2.1.1 Plant Susceptibility Factors with Functions in Resistance to Cyst Nematodes

Transcriptomic analysis of compatible interactions has contributed to the increase in knowledge about the battle between nematodes and their host during pathogenesis. The identification of plant genes required for a successful infection either induced or repressed (plant susceptibility factors) provides novel sources of resistance, as their loss of function or overexpression may effectively compromise nematode progression. Experimental approaches to identify susceptibility factors include the use of root samples enriched in syncytia, as well as techniques such as microaspiration or laser capture microdissection (LCM) to increase the enrichment in syncytia content and thus the sensitivity of the transcriptomic analysis. A general downregulation of defence genes was reported on microaspirated syncytia in *Arabidopsis* at 5 dpi (Szakasits et al. 2009), e.g., genes coding peroxidases or the *RAP2.6* ethylene response transcription factor. Overexpression of the *RAPD 2.6* gene in *Arabidopsis* plants leads to enhanced resistance, probably by increasing callose deposition, albeit in loss-of-function *rapd2.6* mutant lines, susceptibility to *H. schachtii* was not altered (Ali et al. 2013). Interestingly, other transcriptomic analysis performed with either whole root or isolated syncytia in soybean or *Arabidopsis* indicate that, similar to the incompatible interaction, genes related to defence and to secondary metabolism, in particular, the phenylpropanoids pathways, are induced (Alkharouf et al. 2006; Ithal et al. 2007b). This last route is involved in the biosynthesis of cell wall components such as lignin, and it is the pathway used for biosynthesis of phytoalexin and other defence compounds. The upregulation of these defence-related genes may also be a collateral response induced by the wounding damage performed during nematode migration rather than an expression of susceptibility (Escobar et al. 2011).

In *Arabidopsis*, the biosynthesis of the phytoalexin camalexin is initiated after activation of a MAPK signaling cascade, and it is dependent of WRKY33. This transcription factor is one of the most downregulated genes in the microaspirated syncytia transcriptome (Szakasits et al. 2009). Infection studies of a mutant from an interactor of WRKY33 (PHYTOALEXIN-DEFICIENT3; *pad3*) indicated a higher susceptibility to cyst nematode infection, with larger syncytia and larger nematodes as compared to the controls (Ali et al. 2014). Overexpression of the gene *AtPAD4*, corresponding to other *pad* mutant (Glazebrook et al. 1997) also showed an impact in *H. glycines* resistance in soybean roots (Youssef et al. 2013). Other WRKYs involved in biotic stress responses (WRKY6) or acting as negative regulators of basal resistance to *Pseudomonas syringae* (WRKY11 and WRKY17) were also downregulated in syncytia (Szakasits et al. 2009). WRKY6 overexpression decreased the number of females, whereas mutant lines for either WRKY 11 or 17 were more susceptible, displaying a higher number of females (Ali et al. 2014) suggesting that their loss of function or overexpression influence nematode resistance.

Contrary to Szakasits et al. (2009), other transcriptomic analysis at the initial stages of parasitism revealed induction of plant defences (Puthoff et al. 2003; Alkharouf et al. 2006; Ithal et al. 2007b; Mazarei et al. 2011). Accordingly, an increase in the promoter activity of two lipoxygenases (LOX) *LOX3* and *4* after cyst nematode infection in *Arabidopsis* was reported (Ozalvo et al. 2014). However, opposite phenotypes were observed in functional assays for both proteins. Whereas lack of the LOX4 activity increased plant susceptibility, the knockdown of LOX3 decreased plant susceptibility, suggesting that both elicit different defence pathways (Ozalvo et al. 2014). Other assays supporting the role of LOX in plant defence (Matthews et al. 2013) indicate their participation in signaling during basal defence.

Lipoxygenases are also linked to the JA signaling pathway (Kammerhofer et al. 2015). In this respect, hormone levels from infected roots were compared to uninfected roots at initial stages of *H. schachtii* parasitism in *A. thaliana* (Kammerhofer et al. 2015). An increase in JA and in the ethylene precursor 1-aminocyclopropane-1-carboxylic acid (ACC) were observed that was corroborated by qPCR of hormone marker genes such as some LOXs (*LOX3, LOX4, LOX6*) in accordance to Ozalvo et al. (2014). In the tomato compatible interaction, *G. rostochiensis* triggered the suppression of SA-mediated defences, as no increase in *PR1 (P4)* marker was observed upon nematode infection. On the contrary, *PR-6*, a JA response gene, showed higher levels of expression in the compatible interaction as compared to the resistant one, as did other JA-associated genes (Uehara et al. 2007, 2010). In this respect, when the genes *AtNPR1*, *AtGA2*, and *AtPR-5*, encoding specific components involved in SA regulation, synthesis, and signaling, were overexpressed in soybean roots, resistance to SCN was highly enhanced (decreasing infection by 60 %; Matthews et al. 2014). However, overexpression of other JA-related genes such as *AtAOS*, *AtAOC*, and *AtJAR1* did not influence nematode reproduction (Matthews et al. 2014). In contrast, Kammerhofer et al. (2015) did not find significant transcriptional changes in genes related to SA signaling.

2.2 Defences Against Root-Knot Nematodes

Several resistance genes or loci have been described in different plant species that confer resistance against species of RKNs (reviewed in Williamson and Roberts 2009). The physiological mechanisms involved in the incompatible interactions with RKNs have been at least partially characterized for a few plant species like tomato, pepper, or *Prunus* spp.

The resistance response in tomato, consisting in an early HR in the area of infection, is the best characterized (Milligan et al. 1998; Williamson et al. 1994; Martinez De Ilarduya et al. 2001; Ammiraju et al. 2003; Martinez de Ilarduya et al. 2004; Bhattarai et al. 2007; Schaff et al. 2007; Atamian et al. 2012; Mantelin et al. 2013; Iberkleid et al. 2014; Molinari et al. 2014; Zhou et al. 2015). Several accessions of *Solanum peruvianum* (formerly *Lycopersicon peruvianum*) possessed natural resistance to RKN what was used by genetic crosses to obtain resistant lines of the common tomato *S. lycopersicum* (Smith 1944; Watts 1947). However, the molecular mechanisms underlying this resistance have not been yet fully understood, although during the last years, molecular biology techniques helped to describe some of the transduction cascades involved in this complex process. In tomato, two main resistance genes were cloned, named *Mi-1.2* and, *Mi-9*, both encoding proteins belonging to the largest class of R proteins with central NB and C-terminal LRR domains. Only *Mi-1.2* confers resistance against the three main species of RKNs (*M. javanica, M. incognita, M. arenaria*) but also against potato aphids (Rossi et al. 1998) and the potato whitefly

(Nombela et al. 2003); however, it does not confer resistance against other species of RKNs, such as *M. hapla* (Brown et al. 1997) or *M. enterolobii* (Kiewnick et al. 2009). Additionally, it is inactive at soil temperatures above 28 °C (Dropkin 1969). Interestingly, the only putative Avr gene described for *Mi-1* is *Cg-1* that is present in an avirulent *M. javanica* strain but not in the virulent strain. It encodes a small nematode protein (the longest open reading frame of 32 amino acids) that is required for *Mi-1* resistance (Gleason et al. 2008). *Mi-9*-mediated resistance in *Solanum arcanum* L. is functional at soil temperatures as high as 32 °C conferring resistance only to *M. incognita* and *M. javanica* (Veremis et al. 1999; Ammiraju et al. 2003; Jablonska et al. 2007). Although most of the seven remaining *Mi-1.2* paralogs (Seah et al. 2004) are transcribed to detectable levels, their ability to confer resistance remains unknown. The *Rme1* locus is required for the early steps of the resistance response mediated by *Mi-1.2* (Martinez De Ilarduya et al. 2001, 2004), as the *rme1* mutant was compromised in resistance to *M. javanica* and to the potato aphid. *Rme1* acts early in the *Mi-1.2* pathway, either at the same step as the *Mi-1* protein product or upstream of *Mi-1* (Martinez de Ilarduya et al. 2004). However, the precise function and molecular nature of *RMe1* in the incompatible interaction needs to be further analyzed. Another protein involved in the signal transduction pathway mediated by *Mi-1.2* gene is HSP90, a chaperone capable to form hetero-multimeric recognition complexes together with R proteins, which remain in an inactive but signaling-competent state (Kaloshian et al. 2011). Silencing of *HSP90-1* by virus-induced gene silencing (VIGS) in *Nicotiana benthamiana* demonstrated the role of this gene in *Mi-1.2*-mediated resistance as it was attenuated (Bhattarai et al. 2007). Transcriptomic approaches have been used as well to decipher the differences between the resistant (*Mi-1*+) and susceptible (*Mi-1*–) tomato plants (Schaff et al. 2007). In the absence of nematodes, they showed only one differentially expressed gene corresponding to a glycosyltransferase (Table 2; Schaff et al. 2007), and, strikingly, silencing this gene restored the susceptibility to nematode infection of the resistant line (Schaff et al. 2007). Three key transcription factors mediating the signaling cascade for *Mi-1*-mediated resistance have been described so far, *SlWRKY70*, *SlWRKY72a*, and *SlWRKY72b*, whose silencing in resistant plants restored also the infection by *M. incognita* (Bhattarai et al. 2010; Atamian et al. 2012). The comparison between the transcriptomes of resistant and susceptible lines indicated that the JA signaling pathway had a role in basal defence but not in *Mi-1*-mediated resistance to RKNs, while low SA levels might be sufficient because *Mi-1*-1 resistance to RKN was not compromised in *Mi-1 NahG* tomato lines that fail to accumulate SA (Table 2; Bhattarai et al. 2008). The participation of ethylene in the *Mi-1*-mediated resistance to RKN seems also minor, as impairing ethylene biosynthesis or perception using VIGS, the ethylene-insensitive mutant *Never ripe*, or 1-methylcyclopropene treatment, did not attenuate the resistance (Mantelin et al. 2013).

Contrasting to the restricted resistance conferred by *Mi* genes, in *Prunus* spp. the *Ma* gene identified in *P. cerasifera* confers resistance to over 30 RKN species and isolates (Esmenjaud et al. 1994, 1996; Lecouls et al. 1997). *Ma* has been cloned

Table 2 Transcriptomic assays involving incompatible interactions of root-knot nematodes with different plant species

Reference	Technique	Type of interaction	Nematode species	Plant species	Biological material
Potenza et al. (1996)	Bidimensional electrophoresis	Compatible/ incompatible	*M. incognita*	*M. sativa*	3 days postinoculation roots
Callahan et al. (1997)	Bidimensional electrophoresis	Compatible/ incompatible	*M. incognita*	*G. hirsutum*	Infected roots at 8 days after inoculation
Lambert et al. (1999)	cDNA Library	Incompatible	*M. javanica*	*S. lycopersicum*	12 h infected root tips
Zhang et al. (2002)	cDNA Library	Incompatible	*M. incognita*	*G. hirsutum*	10 days after infection galls
Schaff et al. (2007)	Microarray	Compatible/ incompatible	*M. incognita*/*M. hapla*	*S. lycopersicum*	Root segments containing galls at 12, 36, and 72 h after inoculation and galls at 4 weeks after inoculation
Bhattarai et al. (2008)	Microarray	Compatible/ incompatible	*M. incognita*/*M. javanica*	*S. lycopersicum*	24 h after inoculation, 1 cm of the infected root tips
Das et al. (2010)	Microarray	Compatible/ incompatible	*M. incognita*	*V. unguiculata*	Nematode-infected root tissue at 3 and 9 days postinoculation
Franco et al. (2010)	Bidimensional electrophoresis	Incompatible	*M. paranaensis*/ *M. incognita*	*G. hirsutum and C. canephora*	Roots were collected at 6 and 10 days after inoculation
Tirumalaraju et al. (2011)	cDNA Library	Compatible/ incompatible	M. arenaria	*A. hypogaea*	Infected roots at 12, 24, 48, and 72h postinoculation
de Sa et al. (2012)	cDNA Library	Incompatible	*M. javanica*	*G. max*	Infected roots at 6, 12, 24, 48, 96, 144, and 192 h postinoculation
Bagnaresi et al. (2013)	Microarray	Compatible/ incompatible	*M. incognita*	*S. torvum*/*S. melongena*	Infected roots at 14 days after inoculation
Beneventi et al. (2013)	mRNA sequencing	Incompatible	*M. javanica*	*G. max*	Root sections at 0, 6, 12 h, 1, 2, 4, 6, and 8 days postinoculation
Postnikova et al. (2015)	mRNA sequencing	Compatible/ incompatible	*M. incognita*	*M. sativa*	Infected roots at 10 days after inoculation
Villeth et al. (2015)	Bidimensional electrophoresis	Incompatible	*M. incognita*	*V. unguiculata*	Infected roots at 3, 6, and 9 days after inoculation

as the gene *TNL1* (a TIR-NBS-LRR gene), and it conferred the same complete-spectrum and high-level resistance using its genomic sequence and native promoter region in *Agrobacterium rhizogenes*-transformed hairy roots and composite plants (Claverie et al. 2004, 2011).

In pepper, the resistance to RKN is driven by several heat-stable resistance genes (named *Me* genes; Djian-Caporalino et al. 1999, 2001) and the N-gene (Hare 1957), which cluster together in the P9 chromosome (Djian-Caporalino et al. 2007). Other important crop species present natural resistance against RKN infection, although the molecular knowledge of these sources of resistance is still scarce. However, the "omics" techniques like proteomic and transcriptomic have greatly contributed to the study of host defence responses related to the plant resistance (Table 2). The study of the transcriptome of a resistant cotton line uncovered a 14 kDa protein (Callahan et al. 1997), later identified in a cDNA library from 10 dpi *M. incognita* galls that was induced in a nematode-resistant line, *Meloidogyne*-Induced Cotton3 (*MIC-3* ; Zhang et al. 2002). *MIC-3* overexpression in a susceptible line reduced egg production by 60–75 % (Callahan et al. 1997, 2004; Wubben et al. 2015). In resistant cotton cultivars, there was a negative correlation between the presence of MIC-3 in the roots and the developing of nematode-induced galls (Callahan et al. 2004). Since then, 15 *MIC*-like cDNAs have been identified in cotton roots, showing their maximum induction before the appearance of visible signs of resistance (Wubben et al. 2008). Therefore, the *MIC* gene family has been proposed as part of a root-specific defence response mechanism in cotton (Wubben et al. 2008). The proteomic profiling of cotton and coffee resistant lines were further studied, showing differential expression of proteins related to disease resistance like a chitinase, a pathogenesis-related protein and a quinone reductase 2 (Franco et al. 2010).

The comparison of cDNA libraries from infected roots at 12, 24, 48, and 72 h post-inoculation with *M. arenaria* of the resistant peanut cultivar NemaTAM and a susceptible one showed expression of a higher number of stress-related genes in NemaTAM, including specific transcripts as those encoding PR proteins, patatin-like proteins and universal stress proteins (USP; Tirumalaraju et al. 2011).

In cowpea, the *Rk* locus drives the resistance against RKNs in a different way of the early response mediated by *Mi-1* in tomato, as it confers a later response without early hypersensitive signs but blocking the nematode reproduction. When genes expressed in *RK* plants were compared to those from a susceptible nearly isogenic line, the typical defence response was partially suppressed in resistant roots, even at 9 days postinoculation, allowing development of juvenile nematode stages. Differences in ROS concentrations, induction of toxins, and other defence-related genes seem to play a role in this unique resistance mechanism (Das et al. 2010). Other aproaches based on proteomics on cowpea resistant lines revealed 13 unique proteins including some related to oxidative stress (e.g., a multicatalytic endopeptidase complex (proteasome), hydroxyacid oxidase, gamma-type carbonic anhydrase family protein, ferredoxin-NADP reductase isozyme 2, and glutathione S-transferase), those proteins increased or were unique to the highly resistant CE 31 and could be involved in nematode defence and resistance mechanisms (Villeth et al. 2015). A complementary proteomics study comparing different resistant lines described the induction of enzymatic activities such as superoxide dismutase, chitinase, b-1,3-glucanase, peroxidase, and cysteine proteinase inhibitor in the highly resistant line as compared to the rest. This suggests that these activities may contribute to the high resistance of this cowpea cultivar against infection and colonization by *M. incognita* (Oliveira et al. 2012).

The resistance of alfalfa cv. Moapa 69 to *M. incognita* does not rely on apoptotic cell death, but may occur due to the inability of the RKN to enter the developing vascular cylinder of the root, as J2s remain at the root apex as early as 48–72 h after inoculation (Potenza et al. 1996). Massive sequencing of the transcriptomes of *M. sativa* susceptible and resistant lines (cv. Lahontan and cv. Moapa 69, respectively) at 10 days postinoculation with *M. incognita* showed the contribution of a high number of unique R genes in both interactions and identified nearly a thousand of differentially expressed genes that are presumably involved in basal defence responses (cv. Lahontan) and in resistance pathways (cv. Moapa). Interestingly, a number of transcripts potentially associated with resistance to nematodes, in particular two R genes (Medtr3g056585, an LRR and NB-ARC domain disease resistance protein and Medtr0277s0020.3, a disease resistance protein of TIR-NBS-LRR class) are upregulated during infection in cv. Moapa and repressed in cv. Lahontan, supporting typical gene-for-gene interaction (Postnikova et al. 2015). The data also suggest that the R genes could have a dual role as part of a general defence in the susceptible lines and as part of the resistance reaction in the incompatible lines (Postnikova et al. 2015).

In soybean, PR genes such as *PR-1*, *PR-2*, *PR-5*, or *PR-14* were upregulated in infected roots of a resistant line, being *PR-14* exclusive of the incompatible interaction (Beneventi et al. 2013). From this transcriptomic study, a complex model was proposed integrating putative crosstalk mechanisms between plant hormones, mainly gibberellins and auxins, which can be crucial to modulate the levels of ROS in the resistance reaction to nematode invasion (Beneventi et al. 2013). Furthermore, the ectopic expression of the *Arabidopsis* gene *PAD4* that encodes a lipase-like protein that plays a role in SA signaling and is required for the expression of multiple defence responses such as camalexin biosynthesis, resulted in a 77 % decrease in gall number (Youssef et al. 2013). Additionally, the analysis of soybean infected with *M. incognita* revealed six responsive genes encoding heat-shock proteins from the *HSP20* family (*GmHSP20* genes) by qRT-PCR. Some of them were downregulated in a susceptible line but upregulated in the resistant genotype (Lopes-Caitar et al. 2013).

2.2.1 Plant Susceptibility Factors with Functions in Resistance to Root-Knot Nematodes

As discussed for cyst nematodes, knowledge of factors involved in the susceptibility to RKN has provided important insights in the mechanisms of infection as well as molecular candidates for developing resistance. A good example is the analysis of the transcriptomes of early developing isolated GCs from *Arabidopsis* and tomato (Barcala et al. 2010; Portillo et al. 2013). In both transcriptomes, a massive downregulation of genes was observed at early infection stages, particularly in GCs. Those genes repressed in GCs were robustly conserved between *Arabidopsis* and tomato (Portillo et al. 2013). Many of these genes were related to stress, particularly to secondary metabolism as the phenylpropanoid pathway that was significantly overrepresented among the repressed genes in both tomato and *Arabidopsis* GCs. Among them are genes involved in lignin biosynthesis, such as those coding a group of peroxidases, together with genes from a biotic stress subcategory encoding protease inhibitors (Portillo et al. 2013). In this respect, infection tests with a tomato line overexpressing the TPX-1 peroxidase, highly repressed in *Arabidopsis* and tomato GCs, showed a 35 % reduction in the number of galls formed. In contrast, a remarkable induction of *TPX-1* (above ninefold) was observed in a resistant cultivar carrying the *Mi-1* gene, *S. lycopersicum* cv. Motelle (Mi-1/Mi-1) as compared to the susceptible near-isogenic line *S. lycopersicum* cv. Moneymaker (Portillo et al. 2013) both infected with *M. javanica*. Similarly, downregulation of secondary metabolism and defence-related genes as compared to the neighboring cells was also observed in *Medicago* spp. GCs. Therefore, defence and secondary metabolism (as the phenylpropanoids biosynthetic pathway) repression seems to be a hallmark of the GC transcriptome (Barcala et al. 2010; Damiani et al. 2012; Ji et al. 2013; Portillo et al. 2013).

Lipoxygenases are also crucial enzymes for the biosynthesis of oxylipins, which have an important function in the plant defence response against wounding and pathogen attack. Interestingly, significant roles during RKN interaction has been reported for some of the gene members encoding LOXs. Maize *lox3-4* mutants displayed increased attractiveness to RKN and an increased number of juveniles and eggs (Gao et al. 2008), and in *Arabidopsis*, *lox4* mutants were more susceptible to RKNs than control plants, but *lox3* mutants showed less susceptibility (Ozalvo et al. 2014). Additionally, the expression of six PAL genes related to the phenylpropanoids pathway, in three maize genotypes that were good, moderate, and poor hosts for *M. incognita* showed that *ZmPAL4* was most strongly expressed in the most resistant maize line (Starr et al. 2014), suggesting a role for this pathway in the defence against the RKNs.

The ability of nematodes to suppress local defence pathways (mainly ethylene and SA-related pathways) in feeding sites during the compatible interaction has been further confirmed in monocots as rice. Similar to what was described in *Arabidopsis* (Barcala et al. 2010), genes involved in the phenylpropanoid pathway, responsible for the biosynthesis of different metabolites, such as lignin precursors, flavonoids, and hydroxycinnamic acid esters and salicylic acid, most of them involved in plant defences, were strongly suppressed in 3 dai galls in rice (Kyndt et al. 2012). The downregulation of thionins, peptides involved in plant defence, in galls induced by *M. graminicola* at early and medium stages of development was shown to have a functional role during infection as overproducing OsTHI7 decreased susceptibility to *M. graminicola* infection (Ji et al. 2015). The proteins encoded by these genes are promising targets for developing crop varieties that are better adapted to biotic and abiotic stresses.

3 An Overview of Nematode Effectors Involved in Suppression of Plant Defences

In this section, we will only center in the most relevant effectors with described functions in suppression of plant defences. However, it is important to mention that there are many effectors with different roles in the plant–nematode interaction, for a complete review (Gardner et al. 2015; Truong et al. 2015; Mantelin et al. 2015).

The PTI and the ETI mechanisms protect the plant against nematode attack. Nematodes try to suppress these two immunity responses with molecules coating their surface or through the secretion of effector molecules. In the first case, it has been shown that *G. rostochiensis* presents a peroxiredoxin in the surface that could act as an inhibitor of the plant oxidative burst response (Robertson et al. 2000; Robertson et al. 1999). Similarly, glutathione peroxidases and superoxide dismutase (SOD) have been identified in *G. rostochiensis* (Robertson et al. 1999; Jones et al. 2004). *G. pallida* and *M. javanica* coat their surfaces with the fatty-acid- and retinol-binding proteins Gp-FAR-1 and Mj-FAR-1, respectively, that bind precursors of plant defence compounds and JA-related defensive molecules (Prior et al. 2001; Iberkleid et al. 2013). The silencing of Mj-FAR-1 in tomato hairy roots expressing a complementary double-stranded RNA rendered a decrease in the number of infections, while plants overexpressing this protein were most susceptible to nematode attack (Iberkleid et al. 2013).

Similar results to those described for Mj-FAR-1 were obtained when silencing by RNA interference or overexpressing a calreticulin (CRT) secreted by *M. incognita* (Jaubert et al. 2005; Dubreuil et al. 2009; Jaouannet et al. 2012). CRT are calcium-binding proteins highly conserved in plants and animals. Mi-CRT overexpression in *A. thaliana* suppressed the induction of defence marker genes, as well as callose deposition after treatment with the pathogen-associated molecular pattern elf18 (Dubreuil et al. 2009). GpRBP-1 from *G. pallida*, an effector of the SPRYSEC family, triggers Gpa2-mediated cell death in *N. benthamiana* (Sacco et al. 2009; see Sect. 2 on this chapter); however, SPRYSEC-19 enables the supression of programmed cell death and disease resistance mediated by several CC-NB-LRR proteins in plants (Postma et al. 2012). R gene-mediated cell death was also suppressed by the effector GrUBCEP12 from *G. rostochiensis*. Transgenic potato lines expressing GrUBCEP12 showed increased susceptibility to *G. rostochiensis*, and its suppression by RNAi led to a decrease in the infection. The gene *GrUBCEP12* encodes two functional units separated once secreted into the cells, one acting to suppress plant immunity (GrCEP12) and the other potentially affecting the host 26S proteasome, to promote feeding cell formation (Chronis et al. 2013). It is also known that the overexpression of the effector Hs10A06 from *H. schachtii* increases the number of infections in *Arabidopsis*. The model proposed involved 10A06 through the interaction with a spermidine synthase (SPDS2), thereby increasing spermidine content and consequently polyamine oxidase activity that will stimulate the induction of the cellular antioxidant machinery in syncytia.

Hs10A06 seems also to interfere with SA defence signaling (Hewezi et al. 2010). Similarly, the expression of Hg30C02 in *Arabidopsis* increased plant susceptibility, probably interfering with the function of a β-1,3-endoglucanase. The 30C02 protein also interacted with a β-1,3-endoglucanase in both yeast and plant cells, possibly interfering with its activity during pathogenesis (Hamamouch et al. 2012).

The annexin-like effector 4F01 (Gao et al. 2008) from *Heterodera* spp. interacted in a yeast two-hybrid screening with an oxidoreductase (Patel et al. 2010) previously described in the response to *Hyaloperonospora parasitica* (Van Damme et al. 2008). Constitutive expression of Hs4F01 in *Arabidopsis* increased susceptibility to *H. schachtii* infection. Experiments with *Arabidopsis* plants expressing double-stranded RNA complementary to *Hs4F01* resulted in a decrease in the number of infections and in the transcript levels in the nematode (Patel et al. 2010).

Both CNs and RKNs secrete effectors homologous to plant chorismate mutases (Bekal et al. 2003; Jones et al. 2003; Huang et al. 2005; Vanholme et al. 2009) which could prevent the SA-mediated host defence by competing for the chorismate (Doyle and Lambert 2003).More recently, it has been demonstrated that RKN and CN produce ascarosides, a group of conserved nematode pheromones (Manosalva et al. 2015). Pretreatment of *Arabidopsis* roots with 10 nM ascr#18, which is found in the exo-metabolome of *M. incognita* and *H. schachtii*, significantly reduced infection by these nematodes and induced the expression of the defence-related genes *PHI1*, *FRK1*, and *WRKY53* (Manosalva et al. 2015). The activation of PTI components such as mitogen-activated protein kinases, as well as SA- and JA-mediated defence signaling pathways by this ascaroside suggests that plants recognized this pheromone as a feature of these nematodes (Manosalva et al. 2015). Other effectors as Avr genes were already mentioned along the chapter.

By all means, the signaling molecules from the nematode that participate in the complex cross talk with the plant during the interaction provide a wide open and promising field. Although several plant genes conferring various levels of natural resistance to different endoparasitic nematodes in different plant species have been described, the transduction cascades involved in the underlying molecular processes have not been yet fully understood. This includes the mode of action of nematode effectors that are putative Avr genes. In addition, during the last years, transcriptomic analysis helped to identify a substantial amount of genes differentially expressed during the incompatible and/or compatible interactions. Some of them are plant susceptibility factors that play important roles in the interactions with cyst and/or root-knot nematodes, providing insights in the mechanisms of infection. Hence, these susceptibility factors are promising candidates to provide additional resistance, as their loss/gain of function impairs nematode success.

Acknowledgements This work was supported by the Spanish Government (grant AGL2013-48787 to C. Escobar, and CSD2007-057 and PCIN-2013-053 to C. Fenoll) and by the Castilla-la Mancha Government (PEII-2014-020-P to CF). Javier Cabrera was supported by a fellowship from the Ministry of Education, Spain.

References

Aditya J, Lewis J, Shirley NJ, Tan H-T, Henderson M, Fincher GB, Burton RA, Mather DE, Tucker MR (2015) The dynamics of cereal cyst nematode infection differ between susceptible and resistant barley cultivars and lead to changes in (1,3;1,4)-beta-glucan levels and *HvCslF* gene transcript abundance. New Phytol 207(1):135–147

Afzal AJ, Srour A, Goil A, Vasudaven S, Liu T, Samudrala R, Dogra N, Kohli P, Malakar A, Lightfoot DA (2013) Homo-dimerization and ligand binding by the leucine-rich repeat domain at *RHG1/RFS2* underlying resistance to two soybean pathogens. BMC Plant Biol 13:43

Albert M (2013) Peptides as triggers of plant defence. J Exp Bot 64(17):5269–5279

Ali MA, Abbas A, Kreil DP, Bohlmann H (2013) Overexpression of the transcription factor *RAP2.6* leads to enhanced callose deposition in syncytia and enhanced resistance against the beet cyst nematode *Heterodera schachtii* in Arabidopsis roots. BMC Plant Biol 13:47

Ali MA, Wieczorek K, Kreil DP, Bohlmann H (2014) The beet cyst nematode *Heterodera schachtii* modulates the expression of *WRKY* transcription factors in syncytia to favour its development in Arabidopsis roots. PLoS One 9(7):e102360

Alkharouf NW, Matthews BF (2004) SGMD: the soybean genomics and microarray database. Nucleic Acids Res 32:D398–D400

Alkharouf NW, Klink VP, Chouikha IB, Beard HS, MacDonald MH, Meyer S, Knap HT, Khan R, Matthews BF (2006) Timecourse microarray analyses reveal global changes in gene expression of susceptible *Glycine max* (soybean) roots during infection by *Heterodera glycines* (soybean cyst nematode). Planta 224(4):838–852

Ammiraju JS, Veremis JC, Huang X, Roberts PA, Kaloshian I (2003) The heat-stable root-knot nematode resistance gene *Mi-9* from *Lycopersicon peruvianum* is localized on the short arm of chromosome 6. Theor Appl Genet 106(3):478–484

Atamian HS, Eulgem T, Kaloshian I (2012) SlWRKY70 is required for *Mi-1*-mediated resistance to aphids and nematodes in tomato. Planta 235(2):299–309

Bagnaresi P, Sala T, Irdani T, Scotto C, Lamontanara A, Beretta M, Rotino GL, Sestili S, Cattivelli L, Sabatini E (2013) Solanum torvum responses to the root-knot nematode Meloidogyne incognita. BMC Genomics 14:540

Barcala M, Garcia A, Cabrera J, Casson S, Lindsey K, Favery B, Garcia-Casado G, Solano R, Fenoll C, Escobar C (2010) Early transcriptomic events in microdissected Arabidopsis nematode-induced giant cells. Plant J 61(4):698–712

Bekal S, Niblack TL, Lambert KN (2003) A chorismate mutase from the soybean cyst nematode *Heterodera glycines* shows polymorphisms that correlate with virulence. Mol Plant Microbe Interact 16(5):439–446

Beneventi MA, da Silva OB, Jr., de Sa ME, Firmino AA, de Amorim RM, Albuquerque EV, da Silva MC, da Silva JP, Campos Mde A, Lopes MJ, Togawa RC, Pappas GJ, Jr., Grossi-de-Sa MF (2013) Transcription profile of soybean-root-knot nematode interaction reveals a key role of phytohormones in the resistance reaction. BMC Genomics 14:322

Bhattarai KK, Li Q, Liu Y, Dinesh-Kumar SP, Kaloshian I (2007) The *MI-1*-mediated pest resistance requires *Hsp90* and *Sgt1*. Plant Physiol 144(1):312–323

Bhattarai KK, Xie QG, Mantelin S, Bishnoi U, Girke T, Navarre DA, Kaloshian I (2008) Tomato susceptibility to root-knot nematodes requires an intact jasmonic acid signaling pathway. Mol Plant Microbe Interact 21(9):1205–1214

Bhattarai KK, Atamian HS, Kaloshian I, Eulgem T (2010) WRKY72-type transcription factors contribute to basal immunity in tomato and Arabidopsis as well as gene-for-gene resistance mediated by the tomato R-gene *Mi-1*. Plant J 63(2):229–240

Bird D, Opperman C, Williamson V (2009) Plant infection by root-knot nematode. In: Berg RH, Taylor C (eds) Plant cell monographs, vol 15. Springer, Berlin, pp 1–13

Bogdanove A (2002) Protein-protein interactions in pathogen recognition by plants. Plant Mol Biol 50(6):981–989

Bohlmann H, Sobczak M (2014) The plant cell wall in the feeding sites of cyst nematodes. Front Plant Sci 5:89

Brown CR, Mojtahedi H, Santo GS, Williamson VM (1997) Effect of the *Mi* gene in tomato on reproductive factors of *Meloidogyne chitwoodi* and *M. hapla*. J Nematol 29(3):416–419

Cai D, Kleine M, Kifle S, Harloff HJ, Sandal NN, Marcker KA, Klein-Lankhorst RM, Salentijn EM, Lange W, Stiekema WJ, Wyss U, Grundler FM, Jung C, Jacobsen E (1997) Positional cloning of a gene for nematode resistance in sugar beet. Science 275(5301):832–834

Caldwell B, Brim C, Ross J (1960) Inheritance of resistance of soybeans to the cyst nematode, *Heterodera glycines*. Agron J 52:635–636

Callahan FE, Jenkins JN, Creech RG, Lawrence GW (1997) Changes in cotton root proteins correlated with resistance to root knot nematode development. J Cotton Sci 1:38–47

Callahan FE, Zhang XD, Ma DP, Jenkins JN, Tucker ML (2004) Comparison of MIC-3 protein accumulation in response to root-knot nematode infection in cotton lines displaying a range of resistance levels. J Cotton Sci 8:186–190

Caromel B, Gebhardt C (2011) Breeding for nematode resistance: use of genomic information. In: Jones J, Gheysen G, Fenoll C (eds) Genomics and molecular genetics of plant-nematode interactions. Springer, Dordrecht, pp 465–492

Chronis D, Chen S, Lu S, Hewezi T, Carpenter SC, Loria R, Baum TJ, Wang X (2013) A ubiquitin carboxyl extension protein secreted from a plant-parasitic nematode *Globodera rostochiensis* is cleaved in planta to promote plant parasitism. Plant J 74(2):185–196

Claverie M, Dirlewanger E, Cosson P, Bosselut N, Lecouls AC, Voisin R, Kleinhentz M, Lafargue B, Caboche M, Chalhoub B, Esmenjaud D (2004) High-resolution mapping and chromosome landing at the root-knot nematode resistance locus *Ma* from Myrobalan plum using a large-insert BAC DNA library. Theor Appl Genet 109(6):1318–1327

Claverie M, Dirlewanger E, Bosselut N, Van Ghelder C, Voisin R, Kleinhentz M, Lafargue B, Abad P, Rosso MN, Chalhoub B, Esmenjaud D (2011) The *Ma* gene for complete-spectrum resistance to *Meloidogyne* species in Prunus is a TNL with a huge repeated C-terminal post-LRR region. Plant Physiol 156(2):779–792

Cook DE, Lee TG, Guo X, Melito S, Wang K, Bayless AM, Wang J, Hughes TJ, Willis DK, Clemente TE, Diers BW, Jiang J, Hudson ME, Bent AF (2012) Copy number variation of multiple genes at *Rhg1* mediates nematode resistance in soybean. Science (New York, NY) 338 (6111):1206–1209

Damiani I, Baldacci-Cresp F, Hopkins J, Andrio E, Balzergue S, Lecomte P, Puppo A, Abad P, Favery B, Hérouart D (2012) Plant genes involved in harbouring symbiotic rhizobia or pathogenic nematodes. New Phytol 194(2):511–522

Das S, Ehlers JD, Close TJ, Roberts PA (2010) Transcriptional profiling of root-knot nematode induced feeding sites in cowpea (*Vigna unguiculata L. Walp.*) using a soybean genome array. BMC Genomics 11:480

Davis E, Haegeman A, Kikuchi T (2011) Degradation of the plant cell wall by nematodes. In: Jones J, Gheysen G, Fenoll C (eds) Genomics and molecular genetics of plant-nematode interactions. Springer, Dordrecht, pp 255–272

de Almeida Engler J, Vieira P, Rodiuc N, de Sa MF G, Engler G (2015) The plant cell cycle machinery: usurped and modulated by plant parasitic nematodes. In: Escobar C, Fenoll C (eds) Plant nematode interactions. A view on compatible interrelationships, vol 73, Advances in botanical research. Elsevier, Oxford, UK, pp 91–118

de Sá ME, Conceição Lopes MJ, de Araujo Campos M, Paiva LV, Dos Santos RM, Beneventi MA, Firmino AA, de Sá MF (2012) Transcriptome analysis of resistant soybean roots infected by Meloidogyne javanica. Genet Mol Biol 35(1(suppl)): 272–282

Djian-Caporalino C, Pijarowski L, Januel A, Lefebvre V, Daubèze A, Palloix A, Dalmasso A, Abad P (1999) Spectrum of resistance to root-knot nematodes and inheritance of heat-stable resistance in pepper (*Capsicum annuum* L.). Theor Appl Genet 99(3-4):496–502

Djian-Caporalino C, Pijarowski L, Fazari A, Samson M, Gaveau L, O'Byrne C, Lefebvre V, Caranta C, Palloix A, Abad P (2001) High-resolution genetic mapping of the pepper (*Capsicum annuum* L.) resistance loci *Me3* and *Me4* conferring heat-stable resistance to root-knot nematodes (*Meloidogyne* spp.). Theor Appl Genet 103(4):592–600

Djian-Caporalino C, Fazari A, Arguel MJ, Vernie T, VandeCasteele C, Faure I, Brunoud G, Pijarowski L, Palloix A, Lefebvre V, Abad P (2007) Root-knot nematode (*Meloidogyne* spp.) *Me* resistance genes in pepper (*Capsicum annuum L.*) are clustered on the P9 chromosome. Theor Appl Genet 114(3):473–486

Doyle EA, Lambert KN (2003) *Meloidogyne javanica* chorismate mutase 1 alters plant cell development. Mol Plant Microbe Interact 16(2):123–131

Dropkin V (1969) The necrotic reaction of tomatoes and other hosts resistant to *Meloidogyne*: reversal by temperature. Phytopathology 59:1632–1637

Dubreuil G, Magliano M, Dubrana MP, Lozano J, Lecomte P, Favery B, Abad P, Rosso MN (2009) Tobacco rattle virus mediates gene silencing in a plant parasitic root-knot nematode. J Exp Bot 60(14):4041–4050

Ernst K, Kumar A, Kriseleit D, Kloos D-U, Phillips MS, Ganal MW (2002) The broad-spectrum potato cyst nematode resistance gene (*Hero*) from tomato is the only member of a large gene family of NBS-LRR genes with an unusual amino acid repeat in the LRR region. Plant J 31 (2):127–136

Escobar C, Brown S, Mitchum M (2011) Transcriptomic and proteomic analysis of the plant response to nematode infection. In: Jones J, Gheysen G, Fenoll C (eds) Genomics and molecular genetics of plant-nematode interactions. Springer, Dordrecht, pp 157–173

Escobar C, Barcala M, Cabrera J, Fenoll C (2015) Overview of root-knot nematodes and giant cells. In: Escobar C, Fenoll C (eds) Plant nematode interactions. A view on compatible interrelationships, vol 73, Advances in botanical research. Elsevier, Oxford, UK, pp 1–32

Esmenjaud D, Minot J, Voisin R, Pinochet J, Salesses G (1994) Inter- and intraspecific resistance variability in *Myrobalan plum*, peach and peach almond rootstocks using 22 root-knot nematode populations. J Am Soc Hortic Sci 119:94–100

Esmenjaud D, Minot JC, Voisin R, Bonnet A, Salesses G (1996) Inheritance of resistance to the root-knot nematode *Meloidogyne arenaria* in Myrobalan plum. Theor Appl Genet 92 (7):873–879

Eulgem T, Somssich IE (2007) Networks of WRKY transcription factors in defense signaling. Curr Opin Plant Biol 10(4):366–371

Fosu-Nyarko J, Jones MGK (2015) Application of biotechnology for nematode control in crop plants. In: Escobar C, Fenoll C (eds) Plant nematode interactions. A view on compatible interrelationships, vol 73, Advances in botanical research. Elsevier, Oxford, UK, pp 339–376

Franco OL, Pereira J, Costa PHA, Rocha TL, Albuquerque EVS, Grossi-de-Sa MF, Carneiro RMDG, Carneiro RG, Mehta A (2010) Methodological evaluation of 2-DE to study root proteomics during nematode infection in cotton and coffee plants. Prep Biochem Biotechnol 40(2):152–163

Gao X, Starr J, Gobel C, Engelberth J, Feussner I, Tumlinson J, Kolomiets M (2008) Maize 9-lipoxygenase *ZmLOX3* controls development, root-specific expression of defense genes, and resistance to root-knot nematodes. Mol Plant Microbe Interact 21(1):98–109

Gardner M, Verma A, Mitchum MG (2015) Emerging roles of cyst nematode effectors in exploiting plant cellular processes. In: Escobar C, Fenoll C (eds) Plant nematode interactions: a view on compatible interrelationships, vol 73, Advances in botanical research. Elsevier, Oxford, UK, pp 259–292

Glazebrook J, Zook M, Mert F, Kagan I, Rogers EE, Crute IR, Holub EB, Hammerschmidt R, Ausubel FM (1997) Phytoalexin-deficient mutants of Arabidopsis reveal that *PAD4* encodes a regulatory factor and that four *PAD* genes contribute to downy mildew resistance. Genetics 146 (1):381–392

Gleason CA, Liu QL, Williamson VM (2008) Silencing a candidate nematode effector gene corresponding to the tomato resistance gene *Mi-1* leads to acquisition of virulence. MPMI 21 (5):576–585

Grant SR, Fisher EJ, Chang JH, Mole BM, Dangl JL (2006) Subterfuge and manipulation: type III effector proteins of phytopathogenic bacteria. Annu Rev Microbiol 60(1):425–449

Hamamouch N, Li C, Hewezi T, Baum TJ, Mitchum MG, Hussey RS, Vodkin LO, Davis EL (2012) The interaction of the novel 30C02 cyst nematode effector protein with a plant beta-1,3-endoglucanase may suppress host defence to promote parasitism. J Exp Bot 63(10):3683–3695

Hammond-Kosack KE, Parker JE (2003) Deciphering plant-pathogen communication: fresh perspectives for molecular resistance breeding. Curr Opin Biotechnol 14(2):177–193

Hare WW (1957) Inheritance of resistance to root-knot nematodes in pepper. Phytopathology 47:455–459

Hewezi T, Howe PJ, Maier TR, Hussey RS, Mitchum MG, Davis EL, Baum TJ (2010) Arabidopsis spermidine synthase is targeted by an effector protein of the cyst nematode *Heterodera schachtii*. Plant Physiol 152(2):968–984

Hogenhout SA, Van der Hoorn RAL, Terauchi R, Kamoun S (2009) Emerging concepts in effector biology of plant-associated organisms. Mol Plant Microbe Interact 22(2):115–122

Hosseini P, Matthews BF (2014) Regulatory interplay between soybean root and soybean cyst nematode during a resistant and susceptible reaction. BMC Plant Biol 14:300

Huang G, Dong R, Allen R, Davis EL, Baum TJ, Hussey RS (2005) Two *chorismate mutase* genes from the root-knot nematode *Meloidogyne incognita*. Mol Plant Pathol 6(1):23–30

Hussey RS, Davis EL (2004) Nematode esophageal glands and plant parasitism. In: Chen ZX, Chen SYA, Dickson DW (eds) Nematology advances and perspectives, volume I. Nematode morphology, physiology and ecology. CABI, Beijing, pp 258–293

Iberkleid I, Vieira P, de Almeida Engler J, Firester K, Spiegel Y, Horowitz SB (2013) Fatty acid- and retinol-binding protein, *Mj-FAR-1* induces tomato host susceptibility to root-knot nematodes. PLoS One 8(5):e64586

Iberkleid I, Ozalvo R, Feldman L, Elbaz M, Patricia B, Horowitz SB (2014) Responses of tomato genotypes to avirulent and *Mi*-virulent *Meloidogyne javanica* isolates occurring in Israel. Phytopathology 104(5):484–496

Ithal N, Recknor J, Nettleton D, Hearne L, Maier T, Baum TJ, Mitchum MG (2007a) Parallel genome-wide expression profiling of host and pathogen during soybean cyst nematode infection of soybean. Mol Plant Microbe Interact 20(3):293–305

Ithal N, Recknor J, Nettleton D, Maier T, Baum TJ, Mitchum MG (2007b) Developmental transcript profiling of cyst nematode feeding cells in soybean roots. Mol Plant Microbe Interact 20(5):510–525

Jablonska B, Ammiraju JS, Bhattarai KK, Mantelin S, Martinez de Ilarduya O, Roberts PA, Kaloshian I (2007) The *Mi-9* gene from *Solanum arcanum* conferring heat-stable resistance to root-knot nematodes is a homolog of *Mi-1*. Plant Physiol 143(2):1044–1054

Jaouannet M, Magliano M, Arguel MJ, Gourgues M, Evangelisti E, Abad P, Rosso MN (2012) The root-knot nematode calreticulin Mi-CRT is a key effector in plant defense suppression. Mol Plant Microbe Interact 26(1):97–105

Jaubert S, Laffaire JB, Abad P, Rosso MN (2002) A polygalacturonase of animal origin isolated from the root-knot nematode *Meloidogyne incognita*. FEBS Lett 522(1-3):109–112

Jaubert S, Milac AL, Petrescu AJ, de Almeida-Engler J, Abad P, Rosso MN (2005) In planta secretion of a calreticulin by migratory and sedentary stages of root-knot nematode. Mol Plant Microbe Interact 18(12):1277–1284

Ji H, Gheysen G, Denil S, Lindsey K, Topping JF, Nahar K, Haegeman A, De Vos WH, Trooskens G, Van Criekinge W, De Meyer T, Kyndt T (2013) Transcriptional analysis through RNA sequencing of giant cells induced by *Meloidogyne graminicola* in rice roots. J Exp Bot 64 (12):3885–3898

Ji H, Gheysen G, Ullah C, Verbeek R, Shang C, De Vleesschauwer D, Hofte M, Kyndt T (2015) The role of thionins in rice defence against root pathogens. Mol Plant Pathol 16(8):870–881

Jones JDG, Dangl JL (2006) The plant immune system. Nature 444(7117):323–329

Jones JT, Furlanetto C, Bakker E, Banks B, Blok V, Chen Q, Phillips M, Prior A (2003) Characterization of chorismate mutase from potato cyst nematode *Globodera pallida*. Mol Plant Pathol 4(1):43–50

Jones JT, Reavy B, Smant G, Prior AE (2004) Glutathione peroxidases of the potato cyst nematode *Globodera Rostochiensis*. Gene 324:47–54

Kaloshian I, Desmond O, Atamian H (2011) Disease resistance-genes and defense responses during incompatible interactions. In: Jones J, Gheysen G, Fenoll C (eds) Genomics and molecular genetics of plant-nematode interactions. Springer, Dordrecht, pp 309–324

Kammerhofer N, Radakovic Z, Regis JM, Dobrev P, Vankova R, Grundler FM, Siddique S, Hofmann J, Wieczorek K (2015) Role of stress-related hormones in plant defence during early infection of the cyst nematode *Heterodera schachtii* in Arabidopsis. New Phytol 207 (3):778–789

Kamoun S (2006) A catalogue of the effector secretome of plant pathogenic oomycetes. Annu Rev Phytopathol 44(4):41–60

Kandoth PK, Ithal N, Recknor J, Maier T, Nettleton D, Baum TJ, Mitchum MG (2011) The soybean *Rhg1* locus for resistance to the soybean cyst nematode *Heterodera glycines* regulates the expression of a large number of stress- and defense-related genes in degenerating feeding cells. Plant Physiol 155(4):1960–1975

Khan RF, Alkharouf N, Beard H, Macdonald M, Chouikha I, Meyer S, Grefenstette J, Knap H, Matthews B (2004) Microarray analysis of gene expression in soybean roots susceptible to the soybean cyst nematode two days post invasion. J Nematol 36(3):241–248

Kiewnick S, Dessimoz M, Franck L (2009) Effects of the *Mi-1* and the *N* root-knot nematode-resistance gene on infection and reproduction of *Meloidogyne enterolobii* on tomato and pepper cultivars. J Nematol 41(2):134–139

Klink VP, Matthews BF (2009) Emerging approaches to broaden resistance of soybean to soybean cyst nematode as supported by gene expression studies. Plant Physiol 151(3):1017–1022

Klink V, Alkharouf N, MacDonald M, Matthews B (2005) Laser capture microdissection (LCM) and expression analyses of *Glycine max* (Soybean) syncytium containing root regions formed by the plant pathogen *Heterodera glycines* (soybean cyst nematode). Plant Mol Biol 59 (6):965–979

Klink V, Overall C, Alkharouf N, MacDonald M, Matthews B (2007a) A time-course comparative microarray analysis of an incompatible and compatible response by *Glycine max* (soybean) to *Heterodera glycines* (soybean cyst nematode) infection. Planta 226(6):1423–1447

Klink VP, Overall CC, Alkharouf NW, MacDonald MH, Matthews BF (2007b) Laser capture microdissection (LCM) and comparative microarray expression analysis of syncytial cells isolated from incompatible and compatible soybean (*Glycine max*) roots infected by the soybean cyst nematode (*Heterodera glycines*). Planta 226(6):1389–1409

Klink VP, Hosseini P, MacDonald MH, Alkharouf NW, Matthews BF (2009a) Population-specific gene expression in the plant pathogenic nematode *Heterodera glycines* exists prior to infection and during the onset of a resistant or susceptible reaction in the roots of the *Glycine max* genotype Peking. BMC Genomics 10:111

Klink VP, Hosseini P, Matsye PD, Alkharouf NW, Matthews BF (2009b) A gene expression analysis of syncytia laser microdissected from the roots of the *Glycine max* (soybean) genotype PI 548402 (Peking) undergoing a resistant reaction after infection by *Heterodera glycines* (soybean cyst nematode). Plant Mol Biol 71(6):525–567

Klink VP, Hosseini P, Matsye PD, Alkharouf NW, Matthews BF (2010a) Syncytium gene expression in *Glycine max* ([PI 88788]) roots undergoing a resistant reaction to the parasitic nematode *Heterodera glycines*. Plant Physiol Biochem 48(2-3):176–193

Klink VP, Overall CC, Alkharouf NW, Macdonald MH, Matthews BF (2010b) Microarray detection call methodology as a means to identify and compare transcripts expressed within syncytial cells from soybean (*Glycine max*) roots undergoing resistant and susceptible reactions to the soybean cyst nematode (*Heterodera glycines*). J Biomed Biotechnol 2010:491217

Klink VP, Hosseini P, Matsye PD, Alkharouf NW, Matthews BF (2011) Differences in gene expression amplitude overlie a conserved transcriptomic program occurring between the rapid and potent localized resistant reaction at the syncytium of the *Glycine max* genotype Peking (PI 548402) as compared to the prolonged and potent resistant reaction of PI 88788. Plant Mol Biol 75(1-2):141–165

Krüger J, Thomas CM, Golstein C, Dixon MS, Smoker M, Tang S, Mulder L, Jones JDG (2002) A tomato cysteine protease required for Cf-2-dependent disease resistance and suppression of autonecrosis. Science 296(5568):744–747

Kyndt T, Denil S, Haegeman A, Trooskens G, Bauters L, Van Criekinge W, De Meyer T, Gheysen G (2012) Transcriptional reprogramming by root knot and migratory nematode infection in rice. New Phytol 196(3):887–900

Lambert KN, Ferrie BJ, Nombela G, Brenner ED, Williamson VM (1999) Identification of genes whose transcripts accumulate rapidly in tomato after root-knot nematode infection. Physiol Mol Plant Pathol 55(6):341–348

Lambert K, Bekal S (2002) Introduction to plant-parasitic nematodes. Plant Health Instructor. doi:10.1094/PHI-I-2002-1218-01

Lecouls A, Salesses G, Minot J, Voisin R, Bonnet A, Esmenjaud D (1997) Spectrum of the *Ma* genes for resistance to *Meloidogyn*e spp. in *Myrobalan plum*. Theor Appl Genet 95:1325–1334

Liu S, Kandoth PK, Warren SD, Yeckel G, Heinz R, Alden J, Yang C, Jamai A, El-Mellouki T, Juvale PS, Hill J, Baum TJ, Cianzio S, Whitham SA, Korkin D, Mitchum MG, Meksem K (2012) A soybean cyst nematode resistance gene points to a new mechanism of plant resistance to pathogens. Nature 492(7428):256–260

Lopes-Caitar VS, de Carvalho MC, Darben LM, Kuwahara MK, Nepomuceno AL, Dias WP, Abdelnoor RV, Marcelino-Guimaraes FC (2013) Genome-wide analysis of the *Hsp20* gene family in soybean: comprehensive sequence, genomic organization and expression profile analysis under abiotic and biotic stresses. BMC Genomics 14:577

Lozano J, Smant G (2011) Survival of plant-parasitic nematodes inside the host. In: Perry RN, Wharton DA (eds) Molecular and physiological basis of nematode survival. CABI, Wallingford, pp 66–85

Lozano-Torres JL, Wilbers RHP, Gawronski P, Boshoven JC, Finkers-Tomczak A, Cordewener JHG, America AHP, Overmars HA, Van't Klooster JW, Baranowski L, Sobczak M, Ilyas M, van der Hoorn RAL, Schots A, de Wit PJGM, Bakker J, Goverse A, Smant G (2012) Dual disease resistance mediated by the immune receptor Cf-2 in tomato requires a common virulence target of a fungus and a nematode. Proc Natl Acad Sci USA 109(25):10119–10124

Macho AP, Zipfel C (2014) Plant PRRs and the activation of innate immune signaling. Mol Cell 54 (2):263–272

Malinovsky FG, Fangel JU, Willats WGT (2014) The role of the cell wall in plant immunity. Front Plant Sci 5:Article 178

Manosalva P, Manohar M, von Reuss SH, Chen S, Koch A, Kaplan F, Choe A, Micikas RJ, Wang X, Kogel KH, Sternberg PW, Williamson VM, Schroeder FC, Klessig DF (2015) Conserved nematode signalling molecules elicit plant defenses and pathogen resistance. Nat Commun 6:7795

Mantelin S, Bhattarai KK, Jhaveri TZ, Kaloshian I (2013) *Mi-1* mediated resistance to *Meloidogyne incognita* in tomato may not rely on ethylene but hormone perception through ETR3 participates in limiting nematode infection in a susceptible host. PLoS One 8(5):e63281

Mantelin S, Thorpe P, Jones JT (2015) Suppression of plant defences by plant-parasitic nematodes. In: Escobar C, Fenoll C (eds) Plant nematode interactions. A view on compatible interrelationships, vol 73, Advances in botanical research. Elsevier, Oxford, UK, pp 325–338

Martinez de Ilarduya O, Moore AE, Kaloshian I (2001) The tomato *Rme1* locus is required for *Mi-1*-mediated resistance to root-knot nematodes and the potato aphid. Plant J 27(5):417–425

Martinez de Ilarduya O, Nombela G, Hwang CF, Williamson VM, Muniz M, Kaloshian I (2004) *Rme1* is necessary for *Mi-1*-mediated resistance and acts early in the resistance pathway. Mol Plant Microbe Interact 17(1):55–61

Matsye PD, Kumar R, Hosseini P, Jones CM, Tremblay A, Alkharouf NW, Matthews BF, Klink VP (2011) Mapping cell fate decisions that occur during soybean defense responses. Plant Mol Biol 77(4-5):513–528

Matsye PD, Lawrence GW, Youssef RM, Kim K-H, Lawrence KS, Matthews BF, Klink VP (2012) The expression of a naturally occurring, truncated allele of an *alpha-SNAP* gene suppresses plant parasitic nematode infection. Plant Mol Biol 80(2):131–155

Matthews BF, Beard H, MacDonald MH, Kabir S, Youssef RM, Hosseini P, Brewer E (2013) Engineered resistance and hypersusceptibility through functional metabolic studies of 100 genes in soybean to its major pathogen, the soybean cyst nematode. Planta 237 (5):1337–1357

Matthews BF, Beard H, Brewer E, Kabir S, MacDonald MH, Youssef RM (2014) Arabidopsis genes, *AtNPR1, AtTGA2* and *AtPR-5*, confer partial resistance to soybean cyst nematode (*Heterodera glycines*) when overexpressed in transgenic soybean roots. BMC Plant Biol 14:96

Mazarei M, Liu W, Al-Ahmad H, Arelli PR, Pantalone VR, Stewart CN Jr (2011) Gene expression profiling of resistant and susceptible soybean lines infected with soybean cyst nematode. Theor Appl Genet 123(7):1193–1206

Melito S, Heuberger AL, Cook D, Diers BW, MacGuidwin AE, Bent AF (2010) A nematode demographics assay in transgenic roots reveals no significant impacts of the *Rhg1* locus LRR-Kinase on soybean cyst nematode resistance. BMC Plant Biol 10:104

Milligan SB, Bodeau J, Yaghoobi J, Kaloshian I, Zabel P, Williamson VM (1998) The root knot nematode resistance gene *Mi* from tomato is a member of the leucine zipper, nucleotide binding, leucine-rich repeat family of plant genes. Plant Cell 10(8):1307–1319

Mitchum MG, Hussey RS, Baum TJ, Wang X, Elling AA, Wubben M, Davis EL (2013) Nematode effector proteins: an emerging paradigm of parasitism. New Phytol 199:879–894

Molinari S, Fanelli E, Leonetti P (2014) Expression of tomato salicylic acid (SA)-responsive pathogenesis-related genes in *Mi-1*-mediated and SA-induced resistance to root-knot nematodes. Mol Plant Pathol 15(3):255–264

Nombela G, Williamson VM, Muñiz M (2003) The root-knot nematode resistance gene *Mi-1.2* of tomato is responsible for resistance against the whitefly *Bemisia tabaci*. Mol Plant Microbe Interact 16(7):645–649

Oliveira JTA, Andrade NC, Martins-Miranda AS, Soares AA, Gondim DMF, Araújo-Filho JH, Freire-Filho FR, Vasconcelos IM (2012) Differential expression of antioxidant enzymes and PR-proteins in compatible and incompatible interactions of cowpea (*Vigna unguiculata*) and the root-knot nematode *Meloidogyne incognita*. Plant Physiol Biochem 51:145–152

Ozalvo R, Cabrera J, Escobar C, Christensen SA, Borrego EJ, Kolomiets MV, Castresana C, Iberkleid I, Brown Horowitz S (2014) Two closely related members of Arabidopsis 13-lipoxygenases (13-LOXs), LOX3 and LOX4, reveal distinct functions in response to plant-parasitic nematode infection. Mol Plant Pathol 15(4):319–332

Paal J, Fau HH, Jost M, Meksem K, Menendez CM, Salamini F, Ballvora A, Gebhardt C (2004) Molecular cloning of the potato *Gro1-4* gene conferring resistance to pathotype Ro1 of the root cyst nematode *Globodera rostochiensis*, based on a candidate gene approach. Plant J 38 (2):285–297

Patel N, Hamamouch N, Li C, Hewezi T, Hussey RS, Baum TJ, Mitchum MG, Davis EL (2010) A nematode effector protein similar to annexins in host plants. J Exp Bot 61(1):235–248

Perry R, Moens M (2011) Introduction to plant-parasitic nematodes; modes of parasitism. In: Jones J, Gheysen G, Fenoll C (eds) Genomics and molecular genetics of plant-nematode interactions. Springer, Dordrecht, pp 3–20

Portillo M, Cabrera J, Lindsey K, Topping J, Andres MF, Emiliozzi M, Oliveros JC, Garcia-Casado G, Solano R, Koltai H, Resnick N, Fenoll C, Escobar C (2013) Distinct and conserved transcriptomic changes during nematode-induced giant cell development in tomato compared with Arabidopsis: a functional role for gene repression. New Phytol 197(4):1276–1290

Postma WJ, Slootweg EJ, Rehman S, Finkers-Tomczak A, Tytgat TO, van Gelderen K, Lozano-Torres JL, Roosien J, Pomp R, van Schaik C, Bakker J, Goverse A, Smant G (2012) The effector SPRYSEC-19 of *Globodera rostochiensis* suppresses CC-NB-LRR-mediated disease resistance in plants. Plant Physiol 160(2):944–954

Postnikova OA, Hult M, Shao J, Skantar A, Nemchinov LG (2015) Transcriptome analysis of resistant and susceptible alfalfa cultivars infected with root-knot nematode *Meloidogyne incognita*. PLoS One 10(2):e0118269

Potenza C, Thomas S, Higgins E, Sengupta-Gopalan C (1996) Early root response to *Meloidogyne incognita* in resistant and susceptible alfalfa cultivars. J Nematol 28:475–484

Prior A, Jones JT, Blok VC, Beauchamp J, McDermott L, Cooper A, Kennedy MW (2001) A surface-associated retinol- and fatty acid-binding protein (Gp-FAR-1) from the potato cyst nematode *Globodera pallida*: lipid binding activities, structural analysis and expression pattern. Biochem J 356(Pt 2):387–394

Puthoff DP, Nettleton D, Rodermel SR, Baum TJ (2003) Arabidopsis gene expression changes during cyst nematode parasitism revealed by statistical analyses of microarray expression profiles. Plant J 33(5):911–921

Puthoff DP, Ehrenfried ML, Vinyard BT, Tucker ML (2007) GeneChip profiling of transcriptional responses to soybean cyst nematode, *Heterodera glycines*, colonization of soybean roots. J Exp Bot 58(12):3407–3418

Robertson L, Robertson WM, Jones JT (1999) Direct analysis of the secretions of the potato cyst nematode *Globodera rostochiensis*. Parasitology 119(Pt 2):167–176

Robertson L, Robertson WM, Sobczak M, Helder J, Tetaud E, Ariyanayagam MR, Ferguson MAJ, Fairlamb A, Jones JT (2000) Cloning, expression and functional characterisation of a peroxiredoxin from the potato cyst nematode *Globodera rostochiensis*. Mol Biochem Parasitol 111(1):41–49

Rossi M, Goggin FL, Milligan SB, Kaloshian I, Ullman DE, Williamson VM (1998) The nematode resistance gene *Mi* of tomato confers resistance against the potato aphid. Proc Natl Acad Sci U S A 95(17):9750–9754

Rosso M-N, Grenier E (2011) Other nematode effectors and evolutionary constraints. In: Jones J, Gheysen G, Fenoll C (eds) Genomics and molecular genetics of plant-nematode interactions. Springer, Dordrecht, pp 287–307

Sacco MA, Koropacka K, Grenier E, Jaubert MJ, Blanchard A, Goverse A, Smant G, Moffett P (2009) The cyst nematode SPRYSEC protein RBP-1 elicits Gpa2- and RanGAP2-dependent plant cell death. PLoS Pathog 5(8):e1000564

Schaff JE, Nielsen DM, Smith CP, Scholl EH, Bird DM (2007) Comprehensive transcriptome profiling in tomato reveals a role for glycosyltransferase in *Mi*-mediated nematode resistance. Plant Physiol 144(2):1079–1092

Seah S, Yaghoobi J, Rossi M, Gleason C, Williamson V (2004) The nematode-resistance gene, *Mi-1*, is associated with an inverted chromosomal segment in susceptible compared to resistant tomato. Theor Appl Genet 108:1635–1642

Sharma P, Jha AB, Dubey RS, Pessarakli M (2012) Reactive oxygen species, oxidative damage, and antioxidative defense mechanism in plants under stressful conditions. J Bot 2012:26

Simonetti E, Alba E, Montes MJ, Delibes A, Lopez-Brana I (2010) Analysis of ascorbate peroxidase genes expressed in resistant and susceptible wheat lines infected by the cereal cyst nematode, *Heterodera avenae*. Plant Cell Rep 29(10):1169–1178

Smant G, Jones J (2011) Suppression of plant defences by nematodes. In: Jones J, Gheysen G, Fenoll C (eds) Genomics and molecular genetics of plant-nematode interactions. Springer, Dordrecht, pp 273–286

Smith PG (1944) Embryo culture of a tomato species hybrid. Proc Am Soc Hortic Sci 44:413–416

Sobczak M, Golinowski W (2011) Cyst nematodes and syncytia. In: Jones J, Gheysen G, Fenoll C (eds) Genomics and molecular genetics of plant-nematode interactions. Springer, Dordrecht, pp 61–82

Sobczak M, Avrova A, Jupowicz J, Phillips MS, Ernst K, Kumar A (2005) Characterization of susceptibility and resistance responses to potato cyst nematode (*Globodera* spp.) infection of

tomato lines in the absence and presence of the broad-spectrum nematode resistance *Hero* gene. Mol Plant Microbe Interact 18(2):158–168
Starr JL, Yang W, Yan Y, Crutcher F, Kolomiets M (2014) Expression of phenylalanine ammonia lyase genes in maize lines differing in susceptibility to *Meloidogyne incognita*. J Nematol 46 (4):360–364
Stergiopoulos I, de Wit PJ (2009) Fungal effector proteins. Annu Rev Phytopathol 47:233–263
Szakasits D, Heinen P, Wieczorek K, Hofmann J, Wagner F, Kreil DP, Sykacek P, Grundler FMW, Bohlmann H (2009) The transcriptome of syncytia induced by the cyst nematode *Heterodera schachtii* in Arabidopsis roots. Plant J 57(5):771–784
Tameling WIL, Nooijen C, Ludwig N, Boter M, Slootweg E, Goverse A, Shirasu K, Joosten MHAJ (2010) RanGAP2 mediates nucleocytoplasmic partitioning of the NB-LRR immune receptor Rx in the Solanaceae, thereby dictating Rx function. Plant Cell 22(12):4176–4194
Tirumalaraju SV, Jain M, Gallo M (2011) Differential gene expression in roots of nematode-resistant and -susceptible peanut (*Arachis hypogaea*) cultivars in response to early stages of peanut root-knot nematode (*Meloidogyne arenaria*) parasitization. J Plant Physiol 168 (5):481–492
Tomczak A, Koropacka K, Smant G, Goverse A, Bakker E (2009) Resistant plant responses. In: Berg RH, Taylor C (eds) Cell biology of plant nematode parasitism, vol 15, Plant cell monographs. Springer, Berlin, pp 83–113
Truong NM, Nguyen C-N, Abad P, Quentin M, Favery B (2015) Function of root knot nematode effectors and their targets in plant parasitism. In: Escobar C, Fenoll C (eds) Plant nematode interactions: a view on compatible interrelationships, vol 73, Advances in botanical research. Elsevier, Oxford, UK, pp 293–324
Uehara T, Sugiyama S, Masuta C (2007) Comparative serial analysis of gene expression of transcript profiles of tomato roots infected with cyst nematode. Plant Mol Biol 63(2):185–194
Uehara T, Sugiyama S, Matsuura H, Arie T, Masuta C (2010) Resistant and susceptible responses in tomato to cyst nematode are differentially regulated by salicylic acid. Plant Cell Physiol 51 (9):1524–1536
Van Damme M, Huibers RP, Elberse J, Van den Ackerveken G (2008) Arabidopsis *DMR6* encodes a putative 2OG-Fe(II) oxygenase that is defense-associated but required for susceptibility to downy mildew. Plant J 54(5):785–793
van der Voort JR, Wolters P, Folkertsma R, Hutten R, van Zandvoort P, Vinke H, Kanyuka K, Bendahmane A, Jacobsen E, Janssen R, Bakker J (1997) Mapping of the cyst nematode resistance locus *Gpa2* in potato using a strategy based on comigrating AFLP markers. Theor Appl Genet 95(5-6):874–880
van der Vossen EA, Kanyuka K, Bendahmane A, Sandbrink H, Baulcombe DC, Bakker J, Stiekema WJ, Klein-Lankhorst RM (2000) Homologues of a single resistance-gene cluster in potato confer resistance to distinct pathogens: a virus and a nematode. Plant J 23(5):567–576
Vanholme B, Kast P, Haegeman A, Jacob J, Grunewald WIM, Gheysen G (2009) Structural and functional investigation of a secreted chorismate mutase from the plant-parasitic nematode *Heterodera schachtii* in the context of related enzymes from diverse origins. Mol Plant Pathol 10(2):189–200
Veremis JC, van Heusden AW, Roberts PA (1999) Mapping a novel heat-stable resistance to *Meloidogyne* in *Lycopersicon peruvianum*. Theor Appl Genet 98(2):274–280
Veronico P, Giannino D, Melillo MT, Leone A, Reyes A, Kennedy MW, Bleve-Zacheo T (2006) A novel lipoxygenase in pea roots. Its function in wounding and biotic stress. Plant Physiol 141 (3):1045–1055
Villeth GR, Carmo LS, Silva LP, Fontes W, Grynberg P, Saraiva M, Brasileiro AC, Carneiro RM, Oliveira JT, Grossi-de-Sa MF, Mehta A (2015) Cowpea-*Meloidogyne incognita* interaction: root proteomic analysis during early stages of nematode infection. Proteomics 15 (10):1746–1759

Wan J, Vuong T, Jiao Y, Joshi T, Zhang H, Xu D, Nguyen HT (2015) Whole-genome gene expression profiling revealed genes and pathways potentially involved in regulating interactions of soybean with cyst nematode (*Heterodera glycines* Ichinohe). BMC Genomics 16:148

Watts V (1947) The use of *Lycopersicon peruvianum* as a source of nematode resistance in tomatoes. Proc Am Soc Hortic Sci 49:233

Williamson VM, Roberts PA (2009) Mechanisms and genetics of resistance. In: Perry RN, Moens M, Starr JL (eds) Root-knot nematodes. CABI, Wallingford, pp 301–325

Williamson VM, Ho JY, Wu FF, Miller N, Kaloshian I (1994) A PCR-based marker tightly linked to the nematode resistance gene, *Mi,* in tomato. Theor Appl Genet 87(7):757–763

Wubben MJ, Callahan FE, Hayes RW, Jenkins JN (2008) Molecular characterization and temporal expression analyses indicate that the *MIC* (*Meloidogyne Induced Cotton*) gene family represents a novel group of root-specific defense-related genes in upland cotton (*Gossypium hirsutum* L.). Planta 228(1):111–123

Wubben M, Callahan F, Velten J, Burke J, Jenkins J (2015) Overexpression of MIC-3 indicates a direct role for the *MIC* gene family in mediating upland cotton (*Gossypium hirsutum*) resistance to root-knot nematode (*Meloidogyne incognita*). Theor Appl Genet 128(2):199–209

Wyss U, Zunke U (1986) Observations on the behaviour of second stage juveniles of *Heterodera schachtii* inside host roots. Revue Nématol 9(2):153–165

Wyss U, Grundler FM, Münch A (1992) The parasitic behaviour of second-stage juveniles of *Meloidogyne incognita* in roots of *Arabidopsis thaliana*. Nematologica 38:98–111

Youssef RM, MacDonald MH, Brewer EP, Bauchan GR, Kim KH, Matthews BF (2013) Ectopic expression of *AtPAD4* broadens resistance of soybean to soybean cyst and root-knot nematodes. BMC Plant Biol 13:67

Zhang Z-Q (2013) Animal biodiversity: an update of classification and diversity in 2013. Zootaxa 3703(1):5–11

Zhang X-D, Callahan FE, Jenkins JN, Ma D-P, Karaca M, Creech RG (2002) A novel root specific gene, *MIC-3,* with increased expression in nematode-resistant cotton (*Gossypium hirsutum* L.) after root-knot nematode infection. Biochem Biophys Acta 1576:214–218

Zhou J, Jia F, Shao S, Zhang H, Li G, Xia X, Zhou Y, Yu J, Shi K (2015) Involvement of nitric oxide in the jasmonate-dependent basal defense against root-knot nematode in tomato plants. Front Plant Sci 6:193

Belowground Defence Strategies Against Migratory Nematodes

Michael G.K. Jones, Sadia Iqbal, and John Fosu-Nyarko

Abstract The biology of migratory plant parasitic nematodes has been less studied than that of the sedentary endoparasites. The damage they cause is less obvious, their presence and number are more difficult to quantify and they are difficult organisms to study. Nevertheless, they are economically serious pests of many crops, from wheat and barley grown in low rainfall areas to horticultural crops (e.g. *Lilium longiflorum*) and tropical crops such as coffee, banana and sugarcane. The most studied migratory nematodes are the root lesion nematodes, *Pratylenchus* spp., the burrowing nematode *Radopholus similis* and the rice root nematode *Hirschmanniella oryzae*. In the life cycle of migratory nematodes apart from the egg, all stages of juveniles and adults are motile and can enter and leave host roots. They do not induce the formation of a permanent feeding site, but feed from individual host cells. They create pathways for entry of other root pathogens, often resulting in lesions, stunted roots, yellowing of leaves and plants showing symptoms of water stress, leading to yield loss and decreased quality of produce. In terms of genetic plant defences, no major genes for resistance to migratory nematodes have been found, and resistance breeding is usually based on QTL analysis and marker-assisted selection to combine the best minor resistance genes. Feeding damage reduces root function, and root damage and necrotic lesions the nematodes cause can then make them leave the root and seek others to parasitise. Infestation induces classical plant defence responses and changes in host metabolism which reflects the damage they cause, although detailed studies are lacking. New genomic resources are becoming available to study migratory endoparasites, and the knowledge gained can contribute to improved understanding of their interactions with hosts. Notably transcriptomes of *Pratylenchus coffeae*, *Pratylenchus thornei*, *Pratylenchus zeae*, *R. similis* and *H. oryzae* and the first genomic sequence, for *P. coffeae*, are now available. From these data, some candidate effector genes required for parasitism have been identified: many effectors similar to those found in sedentary endoparasites are present, with the exception of those thought to be involved in formation of feeding sites induced by the sedentary parasites. Belowground defence, in the form of enhanced resistance to migratory parasites,

M.G.K. Jones (✉) • S. Iqbal • J. Fosu-Nyarko
Western Australian State Agricultural Biotechnology Centre (SABC), School of Veterinary and Life Sciences, Murdoch University, Perth, WA 6150, Australia
e-mail: M.Jones@murdoch.edu.au

C.M.F. Vos, K. Kazan (eds.), *Belowground Defence Strategies in Plants*, Signaling and Communication in Plants, DOI 10.1007/978-3-319-42319-7_11

may also be achieved by transgenic expression of modified cysteine protease inhibitors (cystatins), anti-root invasion peptides and host-induced gene silencing (RNAi) strategies, demonstrating that migratory nematodes are amenable to control by these technologies. New more environmentally friendly nematicides, combined with better biological control agents, can be applied or used in seed coatings in integrated pest management approaches to defend roots from attack by migratory nematodes.

1 Introduction

The health status of roots at the soil–root interface is thought to underlie about 80 % of all problems of plant growth: root infestation with plant parasitic nematodes is a major contributor to these problems. The responses of plant roots to nematode attack depend on the invading nematode and its lifestyle. Feeding and lifestyle strategies used by plant parasitic nematodes vary and can be divided into ectoparasitic, in which the nematodes remain outside the plant and penetrate tissues with only a small portion of their body, and endoparasitic in which nematodes enter plant tissues completely or with a large portion of their body—the latter are subdivided into migratory and sedentary groups, depending on whether all life stages remain motile or whether they induce feeding sites and become sedentary (Dropkin 1989). These parasitic habits are summarised in Table 1.

The sedentary endoparasites which attack plant roots are discussed in chapter ‘Belowground Signalling and Defence in Host–*Pythium* Interactions’: in this chapter the biology and plant defence strategies against migratory parasitic nematodes

Table 1 Parasitic habits of plant nematodes

Ectoparasites	Endoparasites
Nematodes remain outside the plant or there is minor tissue penetration	Nematodes which enter plant tissues mostly or completely
• Surface tissue feeders For example, *Paratylenchus*, *Trichodorus*, *Tylenchorhynchus*	• Migratory Roots, e.g. *Pratylenchus*, *Hirschmanniella*, *Radopholus* Stems and leaves, e.g. *Ditylenchus* Buds and leaves, e.g. *Anguina*, *Aphelenchoides* Trees, e.g. *Bursaphelenchus*, *Rhadinaphelenchus*
• Subsurface feeders E.g. *Belonolaimus*, *Criconemoides*, *Helicotylenchus*, *Hemicycliophora*, *Longidorus*, *Rotylenchulus*, *Scutellonema*, *Xiphinema*	• Sedentary, semi-endoparasites in roots E.g. *Heterodera*, *Rotylenchus*, *Tylenchulus*
	• Sedentary endoparasites, completely within roots, e.g. *Meloidogyne*, *Nacobbus*

are discussed. The focus is on migratory endoparasites, in particular *Pratylenchus* species usually referred to as root lesion nematodes, the burrowing nematode *R. similis* and *Hirschmanniella* species, which include the rice nematode *H. oryzae*. This largely reflects the view that, from an economic point of view, root lesion nematodes are regarded as the third most important group of plant parasitic nematodes after root-knot (*Meloidogyne* spp.) and cyst nematodes (*Heterodera* and *Globodera*), with the burrowing nematode *R. similis* the fourth most important (Jones et al. 2013).

This ranking for economic importance perhaps partially reflects the fact that infestation by the sedentary endoparasites is much easier to recognise than that for the migratory nematodes, since obvious galls or cysts are not present, and the ranking clearly does not hold for all crops and environments. Migratory nematodes are the most damaging nematodes in cereal crops in many areas of dry land agriculture, such as in the Australian wheat belt (Vanstone et al. 2008) and the Pacific Northwest of the USA (Smiley et al. 2014): the increasing practice of no-till agriculture in such regions to preserve topsoil and moisture tends to increase the occurrence of root lesion nematodes. They are also major pests in tropical regions for crops such as sugarcane grown on fine-textured soils (Blair and Stirling 2007) and horticultural crops including coffee and banana (Castillo and Vovlas 2007). In addition, migratory endoparasites such as *Hirschmanniella* spp. are significant pests of rice crops in flooded ecosystems (Bauters et al. 2014; Kyndt et al. 2014).

2 The Biology of Migratory Parasitic Nematodes

Three genera of the Pratylenchidae family are documented as significant pests: these include genera belonging to the subfamilies Pratylenchinae, Hirschmanniellinae and Radopholinae (De Ley and Blaxter 2002; Haegeman et al. 2010). Although many of the root lesion nematodes (*Pratylenchus* species) have been described as economically significant plant pests, of the Radopholinae only *R. similis* is regarded as a major pest, particularly of banana, citrus and black pepper, and of the *Hirschmanniella* species (rice root nematode), *H. oryzae* is the predominant pest (Kyndt et al. 2014).

The number of species of root lesion nematodes (*Pratylenchus* spp.) described so far is between 70 and 89 (Castillo and Vovlas 2007; Subbotin et al. 2008). They are mostly polyphagous, as evidenced by the ability of species such as *P. thornei* and *P. zeae*, isolated, respectively, from the monocots wheat and sugarcane, to be maintained on dicot carrot discs (Tan et al. 2013; Jordaan and De Waele 1988). *Pratylenchus* spp. are migratory, intracellular root endoparasites, and depending on species, host and temperature, their life cycle lasts between 3 and 9 weeks.

A diagrammatic representation of the life cycle of a root lesion nematode is provided in Fig. 1 (from Jones and Fosu-Nyarko 2014), and the life cycles of *R. similis* and *Hirschmanniella* spp. are essentially similar. These migratory nematodes develop within the eggshell to the first stage juvenile (J1), which moults to

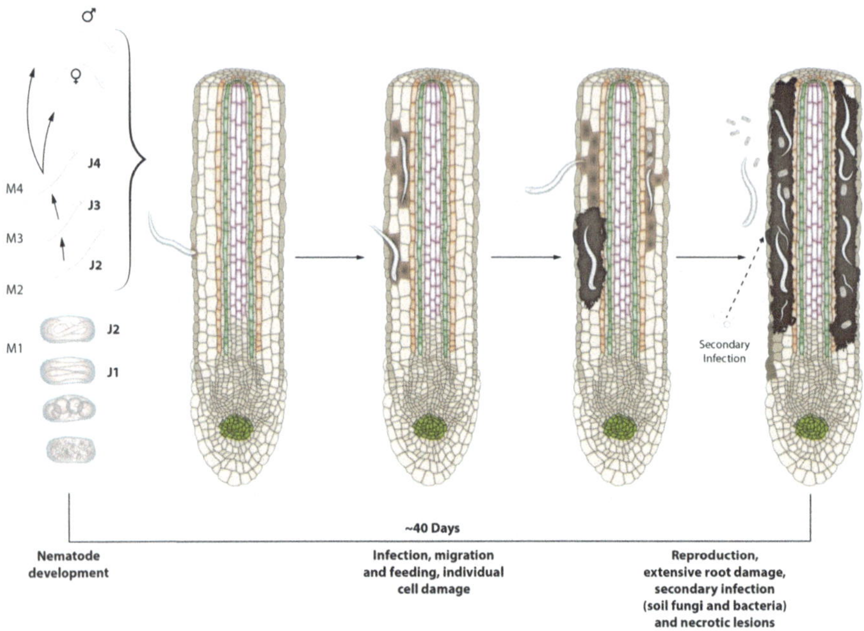

Fig. 1 A diagrammatic representation of the life cycle of *Pratylenchus* (from Jones and Fosu-Nyarko 2014, with permission)

the second-stage juvenile (J2) and then emerges from the eggshell (Fig. 1). However, the difference between migratory and sedentary nematodes is that all subsequent juvenile and adult stages (J2, J3, J4, adults) of the former are worm-like and mobile, and both juvenile and adult stages can enter and leave host plant roots. Some migratory species also infest tuber tissues, and nematodes such as *P. coffeae* and the migratory *Scutellonema bradys* cause major losses when infesting yam tubers in West Africa, in which they continue to multiply in storage. Although these species are migratory endoparasites which usually spend most of their life cycle in host plant roots, they can also be found at the root surface and in nearby soil. Mature females lay eggs both inside infested roots and in nearby soil, and under adverse conditions, these nematodes can survive in soil for several years (Castillo and Vovlas 2007). Reproduction is usually by parthenogenesis, but males occur in some species.

As for other plant parasitic nematodes, root-feeding migratory parasitic nematodes feed by puncturing cells using their hollow mouth stylet. For root lesion nematodes, the J2s tend to feed from the epidermis and root hair cells, but with maturity the nematodes enter roots using their mouth stylet, possibly aided by secretion of plant cell wall-modifying enzymes, and migrate within the root cortex, feeding from the cytoplasm of individual cells, which subsequently die. Dead cells become necrotic, and with additional feeding and tissue damage, typical dark lesions develop in the roots. Development of lesions and further root damage occurs because the nematodes provide entry points for other soil pathogens, such as bacterial (e.g. *Pseudomonas* spp.) and fungal pathogens (e.g. *Fusarium* and *Verticillium* spp.), developing disease complexes which add to the necrosis and root damage (Castillo and Vovlas 2007). The nematodes may leave the roots, particularly from necrotic areas, to feed from new cells or find new host roots. Affected plants are stunted, leaves show early signs of yellowing and roots are short and stubby, with dark lesions. Field infestation is often manifested as patches of poor growth, with more severely affected plants at the centre. Severity is greater under conditions of poor nutrition or water stress.

3 Diagnosing Migratory Nematodes

Understanding the effects of migratory nematodes and finding appropriate strategies for their control first require their identification, and conventional taxonomy based on morphometric measurements is a specialist activity. This has been largely superseded by the development of molecular diagnostic tests, based on differences in ribosomal gene DNA, particularly the Internal Transcribed Spacer (ITS) regions (Al-Banna et al. 2004; Subbotin et al. 2008; Holterman et al. 2009; De Luca et al. 2011; Subbotin et al. 2013), further developed as quantitative polymerase chain reaction (PCR) tests (e.g. Sato et al. 2007; Berry et al. 2008; Yan et al. 2012). Correct identification of the species present is important, because plant resistance to one species does not mean it will be resistant to any other species. For example, wheat cultivars with resistance or tolerance to *P. thornei* are not necessarily resistant or tolerant to *P. neglectus* and vice versa: resistance and tolerance to each species are genetically independent (Smiley and Nicol 2009). A measure of nematode numbers is also important, because overall crop damage reflects the number of nematodes present, and the number of nematodes per gramme of soil at the start of a growing season can be used to predict potential losses and can determine the best cultivar to grow or treatment to apply. The reason why each plant resistance, tolerance or susceptibility may differ when attacked by different root lesion species may be explained partly by differences in the effectors that different nematodes use to enable successful parasitism, and for root lesion nematodes, this is still a developing research topic (see Sect. 5.2).

4 Virus Transmission by Migratory Ectoparasitic Nematodes

It is now well established that many species of migratory ectoparasitic nematodes from the Dorylaimida (*Longidorus*, *Paralongidorus*, *Xiphinema)* and Triplonchida (*Trichodorus*, *Paratrichodorus*), such as the dagger nematodes *Xiphinema index* and *Xiphinema diversicaudatum*, can act as vectors to transmit viruses of the viral genera *Nepovirus* and *Tobravirus*. They acquire and transmit the viruses by feeding on infected and then uninfected roots, either persistently or non-persistently: viruses they transmit include Tobacco ringspot virus (TRSV) and Tobacco rattle virus (TRV). The nepoviruses Grapevine fanleaf virus (GFLV) and Arabis mosaic virus (ArMV) are transmitted in a non-circulative manner and are economically important viruses of vines: precise interactions are required between the components both of the virus and the nematode stylet for virus transmission (Schellenberger et al. 2011). The main defence against virus diseases transmitted by these migratory nematodes is to avoid the introduction of virus-transmitting nematodes using plant biosecurity strategies, if infested to eradicate the nematodes using chemical nematicides or if available to use nematode resistance germplasm or rootstocks.

5 Natural Mechanisms of Plant Resistance to Nematode Attack

Under natural growing conditions, plants are exposed to a range of biotic and abiotic stresses. Among the biotic stresses are various herbivorous organisms feeding on the aboveground and belowground parts of the plant. Belowground attack involves various microorganisms which include nematodes, a diverse and abundant group of multicellular organisms. Plants normally have structural barriers and physiological processes in place that are able to exclude some microbes, parasites and pests from attack or invasion. Conversely, some parasites and pests have evolved mechanisms which aid successful parasitism or infestation of host plants. A compatible parasite–host interaction is when development and reproduction of the parasite are fully supported: the host plant is then referred to as susceptible to infection or infestation. When the development of a parasite is still supported because the host defences do not confer resistance but the parasite grows reasonably well with little apparent damage to the host plant, then the host is tolerant. However, in an incompatible interaction, in which a plant is considered resistant to infection or infestation, its natural, structural, biochemical or physiological defences can prevent invasion, development and/or reproduction of the invading organism. The strategies used by plants to defend themselves against the arsenal of effectors employed by migratory nematodes are discussed in the next sections.

5.1 Root Structure and Barriers to Nematode Infection

For higher plants the root is the main belowground organ and can be invaded by soil-inhabiting migratory parasitic nematodes (although other belowground organs such as tubers can also be attacked). Plants have many natural physical and chemical barriers which can provide protection against pathogens and pests. During root growth in soil, border cells of the root cap become detached (a process termed rhizodeposition) and can secrete antimicrobial proteins, phytoalexins, arabinogalactan proteins and pectins into the extracellular matrix or rhizosphere (Driouich et al. 2013). Border cells or associated extracellular matrix can both attract and repel pathogenic microorganisms. There is ample evidence that *M. incognita* second-stage juveniles (J2) are attracted to and accumulate rapidly around a 1- to 2-mm apical region of pea roots ensheathed by border cells, whereas no such reaction occurs at the root tip of snap bean, indicating possible differences in the perception or response of different plant species to similar root parasites (Zhao et al. 2000). A similar study on the mechanism of resistance to *R. similis* examined the effect that rhizodeposition (root cap cells and exudates) has on infective nematodes: rhizodeposition from both susceptible and resistant cultivars of banana (*Musa acuminata*) attracted nematodes, but the susceptible cultivar appeared to induce temporary quiescence in *R. similis* which lasted for 24 h, whereas nematode quiescence lasted for up to 3 days for the resistant cultivar *Yangambi km5* (Wuyts et al. 2006a). Although these authors concluded that overall there was no indication that rhizodeposition played a part in preformed resistance of *Yangambi km5* against *R. similis*, the relatively longer period of induced quiescence, and cellular responses of border cells to other factors such as aluminium and fungi, suggests that the tightly regulated production of border cells and associated extracellular matrix may play a role in the protection of root tips from some biotic and abiotic stresses (Hawes et al. 2000).

For migratory nematodes or pathogens that reach epidermal cells of the root of host plants, the next physical barrier to overcome is the cell wall. For both monocots and dicots, the plant cell wall is complex: it is composed of polysaccharides, mainly held together by non-covalent bonds, and cell wall proteins. Cellulose constitutes the most abundant polysaccharide and forms the framework to which matrix components are bound. These cellulose microfibrils are composed of associated linear β-1, 4-glucan chains linked by hydrogen bonds, to form an inelastic and insoluble structure. The cellulose microfibrils are embedded in a matrix of non-cellulosic sugar polymers, which include pectins and hemicelluloses, which is further reinforced by structural proteins such as glycoproteins and aromatic compounds (Carpita and Gibeaut 1993; McCann and Roberts 1994). The matrix of primary cell walls of higher plants consists of pectic substances, and the matrix of secondary cell walls are composed of hemicelluloses. Although the overall structures of cell walls of higher plants are similar in both monocots and dicots, there are substantial differences in polysaccharide composition that vary with cell type, cell function, phase of growth and differentiation. Differences in wall composition may well account for some level of resistance/inhibition to invading nematodes (Carpita and McCann 2000). However, the variation in cell wall composition in many instances seems not to present an insurmountable barrier to migratory endoparasitic nematodes, as reflected by the wide host range of many nematodes, encompassing both monocot and dicot plants. With the exception of some migratory ectoparasites, such as dorylaimids with long stylets, which may only use mechanical penetration of host cells, this suggests that successful invasion of host roots reflects strategies that enable invading nematodes to modify cell walls with a range of differences in composition. The latter seems to be a specialty for plant parasitic nematodes in general and migratory endoparasitic nematodes in particular.

5.2 How Migratory Endoparasitic Nematodes Overcome Plant Defences

Many migratory endoparasites have wide host ranges: for this they must have physical attributes, and physiological and evolutionary strategies, that enable them to avoid detection and successfully parasitise many plants. In a compatible interaction, a nematode can breach the barriers presented by cell walls, feed from host cell cytoplasm and suppress host defences. However, in reality, not all available infective juveniles actually succeed in finding and penetrating roots and develop to adults: this suggests that after the initial invasion, host plants may still employ structural, molecular or physiological defences to limit nematode growth and reproduction.

Secretions of the pharyngeal gland cells are thought to play a number of roles. These include suppression of host defences, enabling migration in plant tissues, promotion of nematode feeding (e.g. anticoagulation for migratory endoparasites, formation of feeding tubes for sedentary endoparasites) and digestion of ingested cytoplasm. (Additional functions are proposed for effectors of endoparasites which are involved in processes of host cell modification in the induction of syncytia or giant cells.) The secreted components which are responsible for these activities are generally described as 'effectors'. Here we include cell wall-modifying enzymes as effectors, since they are an important component of the gene products required for plant parasitism and are a unique feature of plant parasitic nematodes.

Study of sedentary endoparasites has been underpinned by the availability of genomic and transcriptomic resources for the bacterial feeding model nematode *Caenorhabditis elegans* and more recently for root-knot and cyst nematodes: similar studies on migratory endoparasites are now emerging. Sequencing of ESTs of *R. similis* and the application of 'next-generation' sequencing technologies to sequence transcriptomes of *H. oryzae* and mixed stages of *P. coffeae*, *P. thornei* and *P. zeae* and more recently the genome of *P. coffeae* now provide the opportunity to identify and characterise effectors that make these migratory nematodes successful parasites (Jacob et al. 2008; Haegeman et al. 2010, 2011; Nicol et al. 2012; Bauters et al. 2014; Fosu-Nyarko et al. 2015; Burke et al. 2015;

Fosu-Nyarko and Jones 2016). Putative effectors of migratory nematodes can now be predicted using software that identifies sequences for proteins likely to be secreted, combined with in situ hybridisation to identify transcripts expressed in gland cells, and sequence similarities and common structural features with effectors already characterised for sedentary endoparasites. Although the focus of nematode-secreted effectors has been on proteins or peptides secreted from the pharyngeal gland cells, other sources of secretions include the chemosensory amphids, the hypodermis, the cuticle, the excretory system and the rectal glands (Truong et al. 2015). For migratory nematodes, little is known about possible secretions from these sources. The current status of potential effectors of migratory nematodes is provided in Table 2.

Probably the best-characterised group of effectors present in plant parasitic nematodes are the cell wall-modifying enzymes. A cocktail of these enzymes (including a range of pectinases, hemicellulases, cellulases and expansins, Wieczorek 2015) appear to be secreted during nematode–host entry and migration and contribute to modifying the structure of host cell walls. Combined with probing with the sclerotised stylet, these enzymes enable nematodes to penetrate and move either intracellularly or intercellularly through root tissues to select appropriate cells to feed from. In situ hybridisation of transcripts and the presence of granules (implying secretory activity) in the subventral gland cells of sedentary endoparasites during migration suggest that these cells are the source of cell wall-modifying enzymes. However, for *Pratylenchus* spp., the subventral glands do not contain obvious granules. Nevertheless, identification of similar transcripts of effectors from recent transcriptomes and genome sequencing data of *Pratylenchus* spp. indicates that they also employ a similar range of cell wall-modifying enzymes to those identified for sedentary endoparasites. Their function is expected to be similar, that is, in hydrolysis of bonds of various polymeric components of primary and secondary cell walls, including pectins, hemicellulose and cellulose (Table 2, Jones and Fosu-Nyarko 2014). Current analysis of available sequences for *R. similis* (7,726 sequences in NCBI) and published reports suggest that this nematode employs only four of the cell wall-modifying enzymes identified for sedentary types; these are beta 1, 4- endoglucanase, xylanase, pectate lyase and cellulose-binding proteins. More work needs to be done to understand how these wall-modifying enzymes function, particularly the role of each in the host–parasite interaction (Jacob et al. 2008; Maier et al. 2013). The transcriptome analysis of *H. oryzae* provides evidence for transcripts putatively encoding a similar repertoire of cell wall-modifying enzymes to that of *Pratylenchus* spp. (Jones and Fosu-Nyarko 2014; Bauters et al. 2014).

In considering the roles of other candidate effectors, the presence of genes encoding proteins secreted by the dorsal glands of plant nematodes further reflects the battle between plants and invading nematodes. In this battle these nematode effectors are responsible for counteracting the effects of plant defences. Such effectors have been characterised better in sedentary nematodes and include proteins suggested to be secreted by nematodes to counter reactive oxygen species (ROS) produced by plants in response to nematode invasion. For example,

Table 2 Nematode effectors of the migratory endoparasites *Pratylenchus* spp., *R. similis* and *H. oryza*

Nematode effector	*Pratylenchus spp.*	*R. similis*	*H. oryzae*	Putative or known function
Cell wall-modifying enzymes				
Endoglucanases	Y	Y	Y	Hydrolysis of beta 1,4-glucan
Pectate lyase	Y	Y	Y	Hydrolysis of alpha 1,4-linkages in pectin
Xylanase	Y	Y	Y	Hydrolysis of xylan
Expansin-like proteins	Y	Not found	Y	Cell wall softening or extension
Endo-1,3-β-glucanase	Y	Not found	Unknown	Hydrolysis of beta 1,3-glucan
Polygalacturonase	Y	Not found	Y	Hydrolysis of alpha 1,4-D-galactosiduronic linkages
Arabinogalactan galactosidase/ arabinase	Y	Not found	Unknown	Hydrolysis of pectin
Cellulose-binding proteins	Y	Y	Unknown	Promote hydrolysis of crystal-line cellulose
β-Mannanase	Not found	Not found	Y	Hydrolysis of -1,4-mannosidic linkages
Poly-α-D-galacturonosidase	Not found	Not found	Y	Hydrolysis of pectic polymers
Protection from host defences				
Thioredoxin	Y	Y	Unknown	Detoxification of ROS
Peroxiredoxin	Y	Not found	Unknown	Detoxification of ROS
Superoxide dismutase	Y	Y	Unknown	Detoxification of ROS
Glutathione-*S*-transferase	Y	Y	Unknown	Detoxification of ROS
Glutathione synthetase	Y	Not found	Unknown	Detoxification of ROS
Glutathione peroxidase	Y	Y	Unknown	Detoxification of ROS
SPRYSEC-RBP-1/ SXP-RAL2	Y	Y	Y	Suppression of host defences
Sec-2/FAR	Y	Y	Unknown	Reduction in host defence response
Transthyretin-like proteins	Y	Y	Unknown	Expressed at parasitic stages, no functional evidence available
Venom allergen-like proteins	Y	Y	Unknown	Suppression of host defences
Targeting regulation and signalling pathways				
Annexin	Y	Y	Unknown	Protection of plant cells against stress

(continued)

Table 2 (continued)

Nematode effector	*Pratylenchus spp.*	*R. similis*	*H. oryzae*	Putative or known function
14-3-3 and 14-3-3b proteins	Y	Y	Unknown	No determined function
SKP-1	Y	Not found	Unknown	Involved in ubiquitination, signal transduction
Ubiquitin extension protein	Y	Y	Unknown	Involved in ubiquitination
Calnexin/ calreticulin/annexin	Y	Y	Unknown	Calcium spiking
Beta-galactoside-binding lectin (galectin)	Y	Y	Unknown	No functional data available for nematodes
Feeding				
Cathepsin L	Y	Y	Unknown	Protein digestion/degradation
Aminopeptidase	Y	Not found	Unknown	Protein digestion/degradation
Initiation and maintenance of feeding site				
C-terminally encoded proteins (CEPs)	Not found	Not found	Unknown	Possibly required for giant cell formation
CLE peptides	Not found	Not found	Unknown	Mimic plant CLEs, no functional evidence available
16D10 CLE-related peptide	Not found	Not found	Unknown	Promotion of giant cell induction
Chorismate mutase	Unclear	Not found	Y	Plant defence suppression, targets SA pathway
19C07 effector	Not found	Not found	Unknown	Modification of auxin influx in syncytium
10A06 effector	Not found	Not found	Unknown	Indirect induction of antioxidant genes in syncytium
7E12 effector	Not found	Not found	Unknown	Promotion of giant cell formation

(Data derived from Jacob et al. 2008; Bauters et al. 2014; Haegeman et al. 2010, 2011; Nicol et al. 2012; Jones and Fosu-Nyarko 2014; Fosu-Nyarko and Jones 2016; Burke et al. 2015)

superoxide dismutase and glutathione peroxidase present at the surface of plant and animal parasitic nematodes have been associated with the role of neutralising oxyradical attack by their host (Waetzig et al. 1999; Robertson et al. 2000; Jones and Fosu-Nyarko 2014). There is also ample evidence that sedentary endoparasites secrete effectors that modulate host cellular functions during establishment and functioning of feeding sites. Some effectors found in root-knot nematodes are involved in the formation of giant cell formation, such as 7E12, CLE peptide and 16D10 CLE-related proteins, whereas others interact with host metabolism to facilitate development of syncytia by cyst nematodes, such as the *Hs19C07*, *Hg30C02* and *10A06* effectors (Huang et al. 2006; Hewezi et al. 2010; Lee et al. 2011; Souza et al. 2011; Hamamouch et al. 2012). Because migratory nematodes do not induce such intricate feeding structures in host tissues, it is not surprising that homologues of the effectors thought to be required for giant cell or syncytium formation have not been identified in migratory nematodes. Nevertheless, in addition to cell wall-modifying enzymes which have now been found in all plant nematodes where there is sufficient molecular data, other common effectors have been identified in secretions and genomes of both sedentary and migratory nematodes. Some are thought to be expressed highly at the parasitic stages (e.g. venom allergen-like proteins, transthyretin-like proteins) or to have roles in other interactions with plant hosts, including targeting and modifying plant signalling pathways (e.g. calreticulin, galectin) (Table 2). Haegeman et al. (2010) suggest a note of caution when extrapolating molecular insights from one group (e.g. *Pratylenchus* spp.) to another (e.g. *Radopholus* spp.) because the taxonomic relationship of *R. similis* and *Pratylenchus* spp. is not firm. Nevertheless, with increasing genomic information on migratory nematodes, our understanding of the function of demonstrated and candidate effectors from specific nematodes will shed more light on how plants defend themselves against migratory nematodes and how in turn the nematodes overcome plant defences.

5.3 Pathogen- and Damage-Associated Molecular Patterns During Nematode Infection

Apart from physical barriers and other basal mechanisms that contribute to resistance to plant pests and pathogens, several defence responses are triggered following root parasitism, including the innate immunity response. Host plants can detect the presence of pathogens using molecules present on the exterior or secreted by the invaders. These molecular signatures, often referred to as pathogen- or microbe-associated molecular patterns (PAMPs or MAMPs), are detected by cell surface receptors or pattern recognition receptors, PRRs. When PRRs of plants survey the apoplast and detect the presence of PAMPs, a PAMP-triggered immunity (PTI) is induced against the invading pathogen (Zipfel 2009). Characteristics of PAMPs and PTI defence against fungi and bacteria have been well studied, and parallels of the process have been drawn for nematode–host interactions. It has been suggested that derivatives of chitin of plant parasitic nematodes may induce PTI, although the nematode cuticle does not contain chitin (Libault et al. 2007). It is however possible that chitin or some of its derivatives may be present in nematode stylets, and on insertion into the plant cell walls, these molecular signatures could be detected by plants, which could lead to responses such as callose deposition which may reduce further invasion by the pathogen (Golinowski et al. 1997). Another facet of PAMP is effector-triggered immunity (ETI), which is specific to strains of a pathogen which secrete unique effectors. As part of the continuing battle between pathogen and host, there is good evidence that fungal plant pathogens and pests can evolve to counteract PAMP-induced plant defences, by selection of mutations of effectors such that they are no longer recognised by the plant or by secreting proteins which prevent PAMP recognition by plant receptors (De Jonge and Thomma 2009). Candidate ETI suppressors or genes linked to possible ETI to nematodes have been reported for sedentary endoparasitic nematodes (Semblat et al. 2001; Sacco et al. 2009; Rehman et al. 2009). For example, the SPRYSEC 19 effector, secreted by the cyst nematode *Globodera rostochiensis*, is known to interact with the leucine-rich repeat domains of receptor proteins in tomato and in doing so possibly suppresses receptor activity (Rehman et al. 2009). At present there is no functional evidence that migratory nematodes secrete such an effector, and for *Pratylenchus* spp., *H. oryzae* and *R. similis* for which transcriptomic and/or genomic sequence data are available, no such specific effector that could trigger ETI has yet been identified (Haegeman et al. 2011; Nicol et al. 2012, Fosu-Nyarko et al. 2015).

Plants also respond to cell damage and stresses that cause mechanical injury to aboveground and belowground parts. This response is mostly against damage-associated molecular pattern (DAMP) molecules released following cellular injury or damage caused by pathogens such as bacteria and fungi (Lotze et al. 2007). Responses to DAMPs are usually systemic and can include the release of redox-sensitive proteins as well as trigger induction of hormone signalling pathways. Movement of migratory nematodes through host roots and the mechanical probing of host cells with the stylet during feeding are likely to cause injury that may elicit such responses from host plants. Generally, plant hormone signalling pathways such as salicylic acid (SA), jasmonic acid (JA) and ethylene (ET) pathways are activated upon infection by many pathogens. While biotrophic pathogens would normally induce the SA pathway, wounding or infection by necrotrophic pathogens often activates the JA and ET pathways (Pieterse and van Loon 1999). It has been suggested that ETI initially activates all three signalling pathways and the plant mobilises resources to support the most effective pathway in combating a particular pathogen (Katagiri and Tsuda 2010). On infection of rice with the migratory nematode *H. oryzae*, JA and ET pathways are activated, while the SA pathway is suppressed, but one week after infection, JA and ET signalling is repressed. Foliar application of JA and ethephon, an exogenous ET, induces systemic defence response in roots against the sedentary endoparasite *Meloidogyne graminicola*, whereas for the migratory endoparasitic *H. oryzae* in rice, all three SA, JA and ET hormonal pathways appear to be essential for defence (Nahar et al. 2011, 2012).

5.4 *Biochemical Responses in Host Plants Following Migratory Nematode Infection*

In response to mechanical damage caused by nematodes, plants produce a range of compounds including ROS. These compounds are toxic to nematodes, but both animal and plant parasitic nematodes are well equipped to metabolise ROS, for example, via the secretion of proteins with antioxidant properties such as peroxiredoxins (Robertson et al. 2000). Production of ROS is associated with a suite of plant defence responses which include activation of signalling pathways and processes which can result in cell wall deposition, synthesis of terpenes, phenolic compounds and nitrogen- and sulphur-containing compounds (Mazid et al. 2011). These responses can be generic and are normally induced locally to eliminate or counteract the invading pathogen but can also be systemic in nature (Bezemer et al. 2004; van Dam 2009). For example, infection of black mustard (*Brassica nigra*) by *P. penetrans* results in increased synthesis of phenolic compounds and glucosinolates in roots, and this innate defence response was also effective in reducing the growth rate of larvae and number of pupae produced by the shoot feeding crucifer insect *Pieris rapae* (L.) (van Dam et al. 2005). The accumulation of isoflavonoid conjugates in roots of alfalfa (*Medicago sativa*) following infection by the stem nematode *Ditylenchus dipsaci* is a classical example of how some plant defence responses are generic and presumptive in nature (Edwards et al. 1995). Transcriptional changes in genes involved in metabolic pathways such as the phenylalanine metabolism, carotenoid biosynthesis and phenylpropanoid biosynthesis following infection by *Pratylenchus* spp. have been associated with induction of plant defence mechanisms (Baldridge et al. 1998; Zhu et al. 2014).

6 Breeding for Resistance to Migratory Nematodes

Some natural genes which confer host resistance to plant parasitic nematodes have been identified in cultivated and wild relatives of crop plants. For sedentary endoparasites, several dominant or semi-dominant resistance genes have been identified, mapped to chromosomal locations or linkage groups, characterised at the molecular level and implemented in a range of economically important crops (Fuller et al. 2008). There has been much less study of genes that confer resistance to migratory nematodes compared to sedentary types, and major dominant genes conferring resistance to migratory species have not yet been found. Not surprisingly, research on mechanisms of host resistance to migratory species has been undertaken mainly in countries and on crops where they cause most damage. For example, for *Pratylenchus* spp., the most detailed work to identify and combine sources of natural resistance to these species has been done with cereals and in most detail on bread wheat (*Triticum aestivum*) and barley (*Hordeum vulgare*) in Australia and the Pacific Northwest of the USA, where infection levels of root lesion nematodes and losses in wheat growing areas are significant (Vanstone et al. 2008; Smiley and Machado 2009; Jones and Fosu-Nyarko 2014).

Eight *Pratylenchus* species are known to attack wheat. In the southern and western wheat belts of Australia, *P. neglectus*, *P. thornei*, *Pratylenchus quasiterioides* (former species *teres*), *P. penetrans*, *P. zeae*, *P. brachyurus* and

P. scribneri are present, with *P. neglectus* the most important (Vanstone et al. 2008), whereas in the northern wheat belt, *P. thornei* and *P. penetrans* cause the most damage (Smiley and Nicol 2009). Genotypes of wheat with different levels of resistance (and tolerance) to specific *Pratylenchus* species have been identified in many breeding programmes using tools for marker-assisted breeding (Table 3). This usually involves large-scale screening of germplasm from wild ancestors or progenitors of crop plant cultivars and mapping of quantitative trait loci (Table 3). A recent marker-assisted selection study for resistance in barley has also identified five QTLs contributing to resistance to *P. neglectus* in barley germplasm (Table 3). However, no major gene conferring resistance to root lesion nematodes has been found, and the mechanisms that underlie resistance to *Pratylenchus* spp. in wheat and barley are not known. Although the identification of QTLs for resistance to migratory

Table 3 Quantitative trait loci of wheat and barley linked to resistance and/or tolerance to *Pratylenchus* species

Nematode species	Major QTLs identified on chromosomes	Plant	References
P. thornei	Examples of QTLs on 2BS, 6DS and 6DL, 6D, 1B, 2B, 3B, 4D, 6D, 7A	Wheat	Thompson et al. (1999) Zwart et al. (2005) Toktay et al. (2006)
	QRlnt.lrc-6D.2, QRlnt.lrc-6D.1 on chromosome 6DL	Wheat	Zwart et al. (2005)
P. neglectus	Examples of QTLs on chromosome 2B, 4DS, 6DS, 7AL		
	QRlnn.lrc-4D.l, QRlnn.lrc-6D.l on chromosome 4DS	Wheat	Zwart et al. (2005)
	Rlnn1 resistance locus on chromosome 7A	Wheat	Williams et al. (2002)
	Pne3H-1, Pne3H-2, Pne5H, Pne6H and Pne7H on Chromosomes 3H, 5H, 6H and 7H	Barley	Sharma et al. (2011)
P. penetrans	Rlnn1 resistance locus on chromosome 7A	Wheat	Williams et al. (2002)
P. neglectus and P. penetrans	Examples of QTLs on chromosome 1B, 2B and 6D	Wheat	Toktay et al. (2006)
	Rlnnp6H resistance on chromosome 6H	Barley	Galal et al. (2014)
P. thornei and P. neglectus	Xbarc 183 on chromosome 6DS	Wheat	Zwart et al. (2005)

endoparasites is an important advance, there is a need for further detailed study to identify new, more effective and durable sources of natural resistance to these nematodes.

7 Resistance to Migratory Nematodes in Tropical Crops

Banana and plantain (*Musa* spp.) constitute the eighth most important staple food crop worldwide. The most damaging migratory nematodes of these crops are the endoparasites *R. similis*, *Pratylenchus goodeyi*, *P. coffeae* and the spiral nematode *Helicotylenchus multicinctus*, together with *Meloidogyne* spp., with combinations of these nematode pests varying with locality (Karakaş 2007; Tripathi et al. 2015). The search for resistance genes against these species, especially against *R. similis*, has largely focussed on *Musa* spp. Using traditional nematode screening methods either by inoculating samples in vitro or in glasshouses or using existing infection at field conditions, many recent *Musa* cultivars have been scored for resistance to nematodes, mainly to *R. similis* but to a lesser extent to *Pratylenchus* spp. and *H. multicinctus* (Elsen et al. 2002; Moens et al. 2005). Among the most well-known nematode-resistant *Musa* spp. are a triploid AAA cultivar, *Yangambi km5*, with high resistance to both *R. similis* and *P. goodeyi*, and the AA diploid *Pisang Jari Buaya*, resistant to *R. similis* (Pinochet and Rowe 1979; Wehunt et al. 1978; Sarah et al. 1993; Price 1994; Fogain and Gowen 1998). Accessions from gene pools of these resistant cultivars have been used as sources of resistance in *Musa* breeding programmes with some success (Pinochet and Rowe 1979; Viaene et al. 2003). In one of the few reports on genetic resistance screening, using 81 banana diploid hybrids, it appeared that resistance to *R. similis* is controlled by two dominant genes, both with additive and interactive effects (Dochez et al. 2009).

Otherwise, investigations on mechanisms of resistance of *Musa* spp. to *R. similis* and *Pratylenchus* spp. have largely focussed on characteristics of root structures and the biochemical responses of resistant and susceptible cultivars on infection. The presence of more preformed phenolic cells in roots of the resistant cultivar *Yangambi km5* suggests that the formation and this type of cell play a role in its defence (Fogain and Gowen 1998). However, resistant cultivar *Pisang Jari Buaya* may have a different resistance mechanism, because it has fewer preformed phenolic cells in roots, but appears to have more cells with lignified walls than cultivars susceptible to *R. similis* (Fogain and Gowen 1998). A possible role of cell wall lignification may also be evident for other resistant and partially resistant *Musa* cultivars, and this suggests that infection by migratory endoparasites may induce lignification and suberisation of endodermal cells, so limiting invasion of the vascular bundle (Collingborn et al. 2000; Valette et al. 1998). Differential accumulation of the secondary metabolites phenylalanine ammonia-lyase, peroxidase and polyphenol oxidase in roots of resistant and susceptible cultivars of banana infected with *R. similis* has been associated with levels of resistance to the nematode pest (Wuyts et al. 2006b).

8 Cultural, Biological and Chemical Control of Migratory Nematodes

8.1 *Rotations with Non-host Crops*

Apart from natural resistance genes or transgenic approaches, the three main approaches used to control plant parasitic nematodes are cultural, biological and chemical. Cultural control relates to developing crop rotation systems which include one or more crop plants which are non-hosts for a particular nematode. The nematode population should then be reduced substantially during the non-host period of the rotation, with the aim of reducing the threshold levels of the damaging nematode to levels below those that result in crop losses. Rotation is more effective if more than one non-host crop species is available in the rotation, and the effectiveness depends on the nematode species and also whether it has an ability to survive for long periods in the absence of a good host. For migratory nematodes with a wide host range, this strategy may not always work well.

In order to study alternative crops suitable for rotations with wheat in the Pacific Northwest of the USA, Smiley et al. (2014) surveyed 30 crop species and cultivars to look for cultivars with reduced reproductive efficiency or as potential non-hosts of *P. neglectus* and *P. thornei*. Poor hosts of both species were identified in chickpea, pea, safflower and sunflower cultivars and some grasses, but more crop cultivars were found to be good hosts for both species: the latter included cultivars of oat, chickpea and lentil. Ten brassica species (canola, mustard, camelina), sudan grass and a sudan grass/sorghum hybrid were good hosts only of *P. neglectus*, and other cultivars of lentil and pea were good hosts for *P. thornei*. The defence mechanisms of these non-host plants to migratory nematodes have not been investigated: such information would contribute to development of resistance to economically important hosts of these damaging nematode pests. Similar studies have been undertaken in Australia, which showed, for example, that densities of *P. neglectus*, but not of *P. thornei*, were likely to be increased after canola (Taylor et al. 2000; Hollaway et al. 2000), although in Australian environments the choices available for alternative cash crops to wheat or barley are relatively limited. The use of non-host crops in rotations to reduce populations of migratory nematodes is a simple approach but needs further study. Smiley et al. (2014) commented that it is likely that reduced efficiency of wheat production is associated with rotations that include multiple crops that are each good hosts of *Pratylenchus* spp., such as now appears to be very likely for some wheat–food legume or wheat–brassica rotations.

8.2 *Biological and Chemical Control of Migratory Nematodes*

A range of nematophagous bacteria and fungi can be found in nematode-suppressive soils, but in the past the success of biological control agents, such as natural predators or pathogens, used to reduce nematode numbers, was limited (Kerry 1997). Biological control was more inconsistent, less effective and slower acting than control normally achieved with chemicals. The use of nematicidal chemicals for nematode control is not always cost effective or environmentally acceptable, especially for broadscale agriculture or for small-scale farms in developing countries. In addition, the phasing out of long-standing chemical nematicides, such as Temik (aldicarb), Mocap (ethoprophos) and Nemacur (fenamiphos), has spurred research to develop more effective and environmentally benign methods of chemical and biological control of plant nematodes. Research by various commercial organisations has led to the development of new seed coating technologies and biocontrol agents which are now commercially available and are much more effective than previous generations of biological control agents. For example, Bayer CropScience now markets VOTiVO, based on *Bacillus firmus* root colonising bacteria which colonises root surfaces and reduces nematode access to root-feeding sites, and Velum (fluopyram), a new class of chemical nematicide which inhibits mitochondrial respiration in nematodes; Syngenta markets AVICTA (abamectin), which has broader anthelmintic and insecticidal properties; and a contact nematicide Nimitz (fluensulfone) has been passed for nematode control for vegetable crops. Other biological control agents such as the entomopathogenic fungus *Paecilomyces* and the parasitic bacterium *Pasteuria penetrans* are also available commercially (the latter was initially developed to control sting nematodes in turfgrass by Pasteuria Bioscience, which was acquired by Syngenta in 2012). Such biological control agents can be included in an integrated pest management approach and are stable enough to be applied as a seed coating, so reducing the chemical load on the field: most are toxic to migratory nematodes. Early protection and establishment of crop seedlings provides a much greater opportunity for a crop to reach its full yield potential.

9 Transgenic Approaches to Migratory Nematode Resistance

Much research has been undertaken to develop transgenic (biotechnological) strategies for nematode control. These include interfering with nematode location of roots, reducing entry into and migration in roots, preventing formation or disturbing the functions of feeding cells of endoparasites and delivery of compounds via plants that interfere with different aspects of nematode life cycles (Fosu-Nyarko and Jones

2015). The focus of the vast majority of such studies has been on sedentary endoparasites.

The earliest transgenic strategies for nematode control were based on plant cystatins, inhibitors of nematode cysteine proteases which interfere with nematode digestion (Urwin et al. 1997; Vain et al. 1998; Samac and Smigocki 2003). The range of available cystatins has been expanded, with reports of effective resistance against the migratory endoparasite *Ditylenchus destructor* (Gao et al. 2011). The focus of these and subsequent experimental work was on cyst and root-knot nematodes.

To find and enter host roots, invading nematodes must respond to root stimuli and physical and chemical gradients in the rhizosphere: these are mediated by chemosensory and mechanosensory neurons. Interference with nematode chemoreceptors can reduce the ability of nematodes to find host roots, and this strategy has been followed by development of peptides that inhibit acetylcholinesterase, which appear to be taken up by chemoreceptor sensillae via retrograde transport along their neurons to cholinergic synapses (Lilley et al. 2011a). Transgenic plants that secreted this peptide from roots driven by a constitutive promoter (CaMV35S) reduced establishment of *Globodera pallida* (Lilley et al. 2004; Liu et al. 2005): the delivery was refined using expression of the peptide driven by a root cap promoter (MDK4-20) (Lilley et al. 2011b).

The two experimental approaches outlined above have been progressed to confined field tests for transgenic plantain (*Musa* spp.) in Uganda, Africa, to control key migratory nematode pests, which include *R. similis*, *H. multilinctus*, *P. coffeae*, *P. goodeyi* and also endoparasitic root-knot nematodes (Tripathi et al. 2015). In this work, an antifeedant cysteine proteinase inhibitor from maize and an anti-root invasion synthetic peptide were expressed either jointly or separately in banana and subjected to nematode challenge. The results focussing on *R. similis* and *H. multicinctus* showed that the best peptide-expressing transgenic line showed improved agronomic performance relative to non-transgenic controls and provided about 99 % nematode resistance at harvest and that the anti-root invasion peptide appeared to be more effective than the cystatin: in plants expressing both genes, the cystatin appeared to contribute little additional resistance (Tripathi et al. 2015). This work demonstrated that expression of cystatins and/or an anti-root invasion peptide can confer resistance to migratory endoparasites as well as sedentary endoparasites and provide a potential new mode of control of nematodes for banana and other tropical crops (e.g. yam, cassava) which are staple foods of small-scale farmers in Central and West Africa.

As further evidence that root lesion nematode infestation can be reduced by a cystatin, expression of a modified rice cystatin (Oc-IDD86) in the flower crop *Lilium longiflorum* also conferred enhanced resistance to *Pratylenchus penetrans*, reducing nematode numbers by about 75 %, resulting in enhanced growth performance (Vieira et al. 2014).

An alternative approach to that described above is generally described as 'host-induced gene silencing' (HIGS) and involves using transgenic plants to deliver a gene silencing (RNAi) signal in the form of dsRNA to silence a vital gene in the nematode when it ingests cell contents (e.g. Lilley et al. 2012; Jones and Fosu-Nyarko 2014). Research in this area on migratory endoparasitic nematodes lagged behind that on sedentary endoparasitic nematodes, partly because of a lack of genomic resources, combined with the fact that migratory nematodes are more difficult to work with than most sedentary endoparasites. However, increasing genomic and transcriptomic data is now becoming available for migratory endoparasitic nematodes, providing a new resource to identify target genes for their control. As discussed above, 'next-generation sequencing' has been used to generate transcriptome data on *P. coffeae*, *P. thornei*, *P. zeae*, *H. oryzae* and *R. similis* (Haegeman et al. 2011; Nicol et al. 2012; Fosu-Nyarko et al. 2015; Bauters et al. 2014), and genomic data for *P. coffeae* is now also available (Burke et al. 2015). These data now enable identification of new gene targets for RNAi-based control of migratory nematodes (Fosu-Nyarko and Jones 2015).

The most common approach to determining what target genes to use for nematode control involves (1) a bioinformatics phase to identify potential target genes, often based on comparative data from the effects of gene knockout in *C. elegans*, or identified effectors required for successful plant parasitism; (2) their cloning and generation of dsRNA to their sequences; (3) in vitro feeding of motile stages with dsRNA, often in the presence of a neurostimulant to make the nematodes take up the external solution, and assessment of the effects of gene knockdown in the nematodes; (4) based on results from in vitro feeding, production of transgenic plants expressing dsRNA to the nematode target gene; and (5) challenge of the transgenic plants with nematodes in glasshouse experiments to quantify the effects on nematode reproduction.

Optimisation of in vitro feeding conditions and treatment with dsRNA of target genes show that *P. coffeae*, *P. thornei* and *P. zeae* are all amenable to a level of control using RNAi (Haegeman et al. 2011; Tan et al. 2013), and this also holds for transgenic plant resistance (Tan 2015). Thus, there is good reason to expect that all the migratory endoparasitic nematodes are equally amenable to control by the RNAi-based HIGS strategy. Such plant-mediated gene silencing traits in nematodes may be transmitted to the next generation and reduce pathogenicity of nematode offspring on non-RNAi plants, which suggests that there can be epigenetic inheritance of the silencing effect (Elling 2015). The level of resistance obtained by HIGS, if expressed as the percentage reduction in the number of nematodes present compared with susceptible controls, is never 100 %, but a percentage reduction in nematode numbers of up to 90 % or more can be obtained, and this will greatly reduce nematode populations over time. There are many reasons why 100 % resistance by this measure is not achieved (Fosu-Nyarko and Jones 2015), but stacking two (or more) different modes of resistance, such as an RNAi trait and an antifeedant peptide or cystatin, might provide the most effective and durable form of transgenic resistance, preferably in a crop cultivar genotype which expresses the best levels of conventional resistance.

10 Conclusions

The losses caused to crops by infestation with migratory nematodes are difficult to quantify accurately, but in many cases they are equal to or more important than losses caused by sedentary endoparasites. The biology of migratory nematodes is becoming better understood, especially with the availability of new genomic resources. In terms of conventional plant breeding, host plant defences can be improved by marker-assisted selection, which is valuable in combining the best QTLs contributing to resistance against major species. There is also clear evidence that migratory nematodes are amenable to various forms of transgenic control, and new integrated approaches to chemical and biological control are also showing success in protecting crop plant roots against migratory nematodes. In many ways understanding of migratory parasitic nematodes and their interactions with host roots is now emerging from biological darkness into the light.

Acknowledgments We thank Murdoch University (MJ, J F-N) and the Australian Government Endeavour Scholarship scheme (SI) for financial support.

References

Al-Banna L, Ploeg AT, Williamson VM, Kaloshian I (2004) Discrimination of six *Pratylenchus* species using PCR and species-specific primers. J Nematol 36:142–146

Baldridge GD, O'Neill NR, Samac DA (1998) Alfalfa (*Medicago sativa* L.) resistance to the root-lesion nematode, *Pratylenchus penetrans*: defense-response gene mRNA and isoflavonoid phytoalexin levels in roots. Plant Mol Biol 38:999–1010

Bauters L, Haegeman A, Kyndt T, Gheysen G (2014) Analysis of the transcriptome of *Hirschmanniella oryzae* to explore potential survival strategies and host-nematode interactions. Mol Plant Pathol 15:352–363

Berry AD, Fargette M, Spaull VW, Morand S, Cadet P (2008) Detection and quantification of root-knot nematode (*Meloidogyne javanica*), lesion nematode (*Pratylenchus zeae*) and dagger nematode (*Xiphinema elongatum*) parasites of sugarcane using real-time PCR. Mol Cell Probes 22:168–176

Bezemer TM, Wagenaar IR, van Dam NM, van der Putten WH, Wackers FL (2004) Above- and below-ground terpenoid aldehyde induction in cotton, *Gossypium herbaceum*, following root and leaf injury. J Chem Ecol 30:53–67

Blair BL, Stirling GR (2007) The role of plant-parasitic nematodes in reducing yield of sugarcane in fine textured soils in Queensland, Australia. Aust J Exp Agric 2007:620–634

Burke, M, Scholl EH, Bird DMcK, Schaff JE, Coleman, S, Crowell R, Diener S, Gordon O, Graham S, Wang X, Windham E, Wright GM and Opperman CH (2015) The plant parasite *Pratylenchus coffeae* carries a minimal nematode genome. Nematology (in press). doi: 10.1163/15685411-00002901

Carpita NC, Gibeaut DM (1993) Structural models of primary cell walls in flowering plants: consistency of molecular structure with the physical properties of the walls during growth. Plant J 3(1):1–30

Carpita N, McCann M (2000) The cell wall. In: Buchanan B et al (eds) Biochemistry and molecular biology of plants. ASPP, Rockville, pp 52–108

Castillo P, Vovlas N (2007) *Pratylenchus (Nematoda: Pratylenchidae):* diagnosis, biology, pathogenicity and management, nematology monographs and perspectives. Brill Leiden, Boston

Collingborn FM, Gowen SR, Mueller-Harvey I (2000) Investigations into the biochemical basis for nematode resistance in roots of three musa cultivars in response to *Radopholus similis* infection. J Agric Food Chem 48:5297–5301

De Jonge R, Thomma BPHJ (2009) Fungal LysM effectors: extinguishers of host immunity? Trends Microbiol 17:151–157

De Ley P, Blaxter M (2002) Systemic position and phylogeny. In: Lee DL (ed) The biology of nematodes. Taylor and Francis, New York, pp 1–30

De Luca F, Reyes A, Troccoli A, Castillo P (2011) Molecular variability and phylogenetic relationships among different species and populations of *Pratylenchus* (Nematoda: *Pratylenchidae*) as inferred from the analysis of the ITS rDNA. Eur J Plant Pathol 130:415–426

Dochez C, Tenkouano A, Ortiz R, Whyte J, De Waele D (2009) Host plant resistance *to Radopholus similis* in a diploid banana hybrid population. Nematology 11:329–335

Driouich A, Follet-Gueye ML, Vicré-Gibouin M, Hawes M (2013) Root border cells and secretions as critical elements in plant host defense. Curr Opin Plant Biol 16(4):489–495

Dropkin VH (1989) Introduction to nematology, 2nd edn. Wiley, New York, p 304

Edwards R, Mizen T, Cook R (1995) Isoflavonoid conjugate accumulation in the roots of lucerne (*Medicago sativa*) seedlings following infection by the stem nematode (*Ditylenchus dipsaci*). Nematologica 41:51–66

Elling AA (2015) RNA interference as a nematode control strategy. In: Proceedings of the 54th annual meeting of the society of nematologists, East Lansing, p 47

Elsen A, Stoffelen R, Thi Tuyet N, Baimey H, Dupré de Boulois H, De Waele D (2002) *In vitro* screening for resistance to *Radopholus similis* in Musa spp. Plant Sci 163:407–416

Fogain R, Gowen SR (1998) Investigations on possible mechanisms of resistance to nematodes in Musa. Euphytica 92(3):375–381

Fosu-Nyarko J, Jones MGK (2015) Application of biotechnology for nematode control in crop plants. Adv Bot Res 73:339–376

Fosu-Nyarko J, Tan JCH, Gill R, Agrez VG, Rao U, Jones MGK (2015) De novo analysis of the transcriptome of *Pratylenchus zeae* to identify transcripts for proteins required for structural integrity, sensation, locomotion and parasitism. Mol Plant Pathol. doi: 10.1111/mpp.12301

Fosu-Nyarko J, Jones MGK (2016) Advances in understanding the molecular mechanisms of root lesion nematode host interactions. Annu Rev Phytopathol 54. doi:10.1146/annurev-phyto-080615-100257

Fuller VL, Lilley CJ, Urwin PE (2008) Nematode resistance. New Phytol 180:27–44

Galal A, Sharma S, Abou-Elwafa SF, Sharma S, Kopisch-Obuch F, Laubach E, Perovic D, Ordon F, Jung C (2014) Comparative QTL analysis of root lesion nematode resistance in barley. Theor Appl Genet 127:1399–1407

Gao S, Yu B, Yuan L, Li Y, Zhai H, He A, Liu WCL (2011) Production of transgenic sweet potato plants resistant to stem nematodes using oryzacystatin-I gene. Sci Hortic 128:408–414

Golinowski W, Sobczak M, Kurek W, Grymaszewska G (1997) The structure of syncytia. In: Fenoll C, Grundler FMW, Ohl S (eds) Cellular and molecular aspects of plant–nematode interactions. Kluwer Academic, Dordrecht, pp 80–97

Haegeman A, Elsen A, de Waele D, Gheysen G (2010) Emerging molecular knowledge on *Radopholus similis*, an important nematode pest of banana. Mol Plant Pathol 11:315–323

Haegeman A, Joseph S, Gheysen G (2011) Analysis of the transcriptome of the root lesion nematode *Pratylenchus coffeae* generated by 454 sequencing technology. Mol Biochem Parasitol 178:7–14

Hamamouch N, Li C, Hewezi T, Baum TJ, Mitchum MG, Hussey RS, Vodkin LO, Davis EL (2012) The interaction of the novel 30C02 cyst nematode effector protein with a plant β-1,3-endoglucanase may suppress host defence to promote parasitism. J Exp Bot 62:3683–3695

Hawes MC, Gunawardena U, Miyasaka S, Zhao X (2000) The role of root border cells in plant defense. Trends Plant Sci 5(3):128–133

Hewezi T, Howe PJ, Maier TR, Hussey RS, Mitchum MG, Davis EL, Baum TJ (2010) Arabidopsis spermidine synthase is targeted by an effector protein of the cyst nematode *Heterodera schachtii*. Plant Physiol 152:968–984

Hollaway GJ, Taylor SP, Eastwood RF, Hunt CH (2000) Effect of field crops on density of *Pratylenchus* in southeastern Australia: Part 2. *P. thornei*. J Nematol 32:600–608

Holterman M, Karssen G, van den Elsen S, van Megen H, Bakker J, Helder J (2009) Small subunit rDNA-based phylogeny of the Tylenchida sheds light on relationships among some high-impact plant-parasitic nematodes and the evolution of plant feeding. Phytopathology 99:227–235

Huang GZ, Dong RH, Allen R, Davis EL, Baum TJ, Hussey RS (2006) A root-knot nematode secretory peptide functions as a ligand for a plant transcription factor. Mol Plant Microbe Interact 19:463–470

Jacob J, Mitreva M, Vanholme B, Gheysen G (2008) Exploring the transcriptome of the burrowing nematode *Radopholus similis*. Mol Genet Genomics 280:1–17

Jones MGK, Fosu-Nyarko J (2014) Molecular biology of root lesion nematodes (*Pratylenchus* spp.) and their interaction with host plants. Ann Appl Biol 164:163–181

Jones JT, Haegeman A, Danchin EGJ, Gaur HS, Helder J, Jones MGK, Kikuchi T, Manzanilla-Lopez R, Palomares-Rius JE, Wesemael WML, Perry RN (2013) Top 10 plant-parasitic nematodes in molecular plant pathology. Mol Plant Pathol 14:946–961

Jordaan EM, De Waele D (1988) Host status of five weed species and their effects on *Pratylenchus zeae* infestation of maize. J Nematol 20(4):620

Karakaş M (2007) Life cycle and mating behavior of *Helicotylenchus multicinctus* (Nematoda: Hoplolaimidae) on excised Musa cavendishii roots. Biologia 62(3):320–322

Katagiri F, Tsuda K (2010) Understanding the plant immune system. Mol Plant Microbe Interact 23:1531–1536

Kerry B (1997) Biological control of nematodes: prospects and opportunities. In: Maqbool MA, Kerry B (eds) Plant nematode problems and their control in the near east region (FAO Plant Production and Protection Paper—144). ISBN: 92-5-103798-1

Kyndt T, Fernandez D, Gheysen G (2014) Plant-parasitic nematode infections in rice: molecular and cellular insights. Annu Rev Phytopathol 52:135–153

Lee C, Chronis D, Kenning C, Peret B, Hewezi T, Davis EL, Baum TJ, Hussey RS, Bennett M, Mitchum MG (2011) The novel cyst nematode effector protein 19C07 interacts with the Arabidopsis auxin influx transporter LAX3 to control feeding site development. Plant Physiol 155:866–880

Libault M, Wan J, Czechowski T, Udvardi M, Stacey G (2007) Identification of 118 Arabidopsis transcription factor and 30 ubiquitin-ligase genes responding to chitin, a plant-defense elicitor. Mol Plant Microbe Interact 20:900–911

Lilley CJ, Urwin PE, Johnston KA, Atkinson HJ (2004) Preferential expression of a plant cystatin at nematode feeding sites confers resistance to Meloidogyne and Globodera sp. Plant Biotech J 2:3–12

Lilley CJ, Davies LJ, Urwin PE (2011a) RNA interference in plant parasitic nematodes: a summary of the current status. Parasitology 139:630–640

Lilley CJ, Wang D, Atkinson HJ, Urwin PE (2011b) Effective delivery of a nematode-repellent peptide using a root-cap-specific promoter. Plant Biotechnol J 9:151–161

Lilley CJ, Davies LJ, Urwin PE (2012) RNA interference in plant parasitic nematodes: a summary of the current status. Parasitology 139:630–640

Liu B, Hibbard JK, Urwin PE, Atkinson HJ (2005) The production of synthetic chemodisruptive peptides in planta disrupts the establishment of cyst nematodes. Plant Biotechnol J 3:487–496

Lotze MT, Zeh HJ, Rubartelli A, Sparvero LJ, Amoscato AA, Washburn NR, De Vera ME, Liang X, Tor M, Billiar T (2007) The grateful dead: damage-associated molecular pattern molecules and reduction/oxidation regulate immunity. Annu Rev Immunol 220:60–81

Maier TR, Hewezi T, Peng J, Baum TJ (2013) Isolation of whole esophageal gland cells from plant-parasitic nematodes for transcriptome analyses and effector identification. Mol Plant Microbe Interact 26:31–35

Mazid M, Khan TA, Firoz M (2011) Role of nitric oxide in regulation of H_2O_2 mediating tolerance of plants to abiotic stress: a synergistic signalling approach. J Stress Physiol Biochem 7:34–74

McCann MC, Roberts K (1994) Changes in cell wall architecture during cell elongation. J Exp Bot 45(Special Issue):1683–1691

Moens T, Araya M, Swennerf R, De Waele D (2005) Screening of Musa cultivars for resistance to *Helicotylenchus multicinctus*, *Meloidogyne incognita*, *Pratylenchus coffeae* and *Radopholus similis*. Aust Plant Pathol 34(3):299–309

Nahar K, Kyndt T, De Vleesschauwer D, Hofte M, Gheysen G (2011) The jasmonate pathway is a key player in systemically induced defense against root knot nematodes in rice. Plant Physiol 157:305–316

Nahar K, Kyndt T, Nzogela YB, Gheysen G (2012) Abscisic acid interacts antagonistically with classical defense pathways in rice–migratory nematode interaction. New Phytol 196:901–913

Nicol P, Gill R, Fosu-Nyarko J, Jones MGK (2012) *de novo* analysis and functional classification of the transcriptome of the root lesion nematode, *Pratylenchus thornei*, after 454 GS FLX sequencing. Int J Parasitol 42:225–237

Pieterse CM, van Loon LC (1999) Salicylic acid-independent plant defence pathways. Trends Plant Sci 4:52–58

Pinochet J, Rowe P (1979) Progress in breeding for resistance to *Radopholus similis* on bananas. Nematropica 9:76–78

Price NS (1994) Alternate cropping in the management of *Radopholus similis* and *Cosmopolites sordidus* two important pests of banana and plantain. Int J Pest Manage 40:237–244

Rehman S, Postma W, Tytgat T, Prins P, Qin L, Overmars H, Vossen J, Spiridon LN, Petrescu AJ, Goverse A, Bakker J, Smant G (2009) A secreted SPRY domain-containing protein (SPRYSEC) from the plant-parasitic nematode *Globodera rostochiensis* interacts with a CC-NB-LRR protein from a susceptible tomato. Mol Plant Microbe Interact 22:330–340

Robertson L, Robertson WM, Sobczak M, Bakker J, Tetaud E, Arinagayayam MR, Ferguson MAJ, Fairlamb AH, Jones JT (2000) Cloning, expression and functional characterisation of a thioredoxin peroxidise from the potato cyst nematode *Globodera rostochiensis*. Mol Biochem Parasitol 111:41–49

Sacco MA, Koropacka K, Grenier E, Jaubert MJ, Blanchard A, Goverse A, Smant G, Moffett P (2009) The cyst nematode SPRYSEC protein RBP-1 elicits Gpa-2 and RanGAP2-dependent plant cell death. PLoS Pathog 5:e1000564

Samac DA, Smigocki AC (2003) Expression of oryzacystatin I and II in alfalfa increases resistance to the root-lesion nematode. Phytopathology 93:799–804

Sarah JL, Sabatini C, Boisseau M (1993) Differences in pathogenicity to banana (Musa Sp., Cv. Poyo) among isolates of *Radopholus similis* from different production areas of the world. Nemtropica 23:75–79

Sato E, Min YY, Shirakashi T, Wada S, Toyota K (2007) Detection of the root-lesion nematode, *Pratylenchus penetrans* (Cobb), in a nematode community using real time PCR. Jpn J Nematol 37:87–92

Schellenberger P, Sauter C, Lorber B, Bron P, Trapani S, Bergdoll M, Marmonier A, Schmitt-Keichinger C, Lemaire O, Demangeat G, Ritzenthaler C (2011) Structural insights into viral determinants of nematode mediated grapevine fanleaf virus transmission. PLoS Pathog 7: e1002034. doi:10.1371/journal.ppat.1002034

Semblat J-P, Rosso M-N, Husser RS, Abad P, Castagnone-Sereno P (2001) Molecular cloning of a cDNA encoding an amphid secreted putative avirulence protein from the root knot nematode *Meloidogyne incognita*. Mol Plant Microbe Interact 14:72–79

Sharma S, Sharma S, Kopisch-Obuch FJ, Keil T, Laubach E, Stein N, Graner A, Jung C (2011) QTL analysis of root-lesion nematode resistance in barley: 1. *Pratylenchus neglectus*. Theor Appl Genet 122:1321–1330

Smiley RW, Machado S (2009) *Pratylenchus neglectus* reduces yield of winter wheat in dryland cropping systems. Plant Dis 93:263–271

Smiley RW, Nicol JM (2009) Nematodes which challenge global wheat production. In: Carver BF (ed) Wheat science and trade. Wiley, Ames, pp 171–187

Smiley RW, Yan GP, Gourlie JA (2014) Selected Pacific Northwest crops as hosts of *Pratylenchus neglectus* and *P. thornei*. Plant Dis 98:1341–1348

Souza Ddos S, de Souza JD, Grossi-de- Sá M, Rocha TL, Fragoso RR, Barbosa AE, de Oliveira GR, Nakasu EY, de Sousa BA, Pires NF et al (2011) Ectopic expression of a *Meloidogyne incognita* dorsal gland protein in tobacco accelerates the formation of the nematode feeding site. Plant Sci 180:276–282

Subbotin SA, Ragsdale EJ, Mullens T, Roberts PA, Mundo-Ocampo M, Baldwin JG (2008) A phylogenetic framework for root lesion nematodes of the genus *Pratylenchus* (Nematoda): evidence from 18S and D2-D3 expansion segments of 28S ribosomal RNA genes and morphological characters. Mol Phylogenet Evol 48:491–505

Subbotin SA, Waeyenberge L, Moens M (2013) Molecular taxonomy and phylogeny. In: Perry RN, Moens M (eds) Plant nematology, 2nd edn. CABI, Wallingford, pp 40–728

Tan J, Jones MGK, Fosu-Nyarko J (2013) Gene silencing in root lesion nematodes (*Pratylenchus* spp.) significantly reduces reproduction in a plant host. Exp Parasitol 133:166–178

Tan JCH (2015) Characterising putative parasitism genes for root lesion nematodes and for use in RNAi studies. PhD thesis, Murdoch University, Perth, Western Australia

Taylor SP, Holloway G, Hunt C (2000) Effect of field crops on population densities of *Pratylenchus neglectus* and *P. thornei* in Southeastern Australia; Part 1: *P. neglectus*. J Nematol 32:591–599

Thompson JP, Brennan PS, Clewett TG, Sheedy JG, Seymour NP (1999) Progress in breeding wheat for tolerance and resistance to root-lesion nematode (*Pratylenchus thornei*). Australas Plant Pathol 28:45–52

Toktay H, McIntyre CL, Nicol JM, Ozkan H, Elekcioglu HI (2006) Identification of common root-lesion nematode (*Pratylenchus thornei* Sher et Allen) loci in bread wheat. Genome 49:1319–1323

Tripathi L, Babirye A, Roderick H, Tripathi JN, Changa C, Urwin PE, Tushemereirwe WK, Coyne D, Atkinson HJ (2015) Field resistance of transgenic plantain to nematodes has potential for future African food security. Sci Rep 5:1–10

Truong NM, Nguyen CN, Abad P, Quentin M, Favery B (2015) Function of root-knot nematode effectors and their targets in plant parasitism. Adv Bot Res 73:293–324

Urwin PE, Lilley CJ, McPherson MJ, Atkinson HJ (1997) Resistance to both cyst and root-knot nematodes conferred by transgenic *Arabidopsis* expressing a modified plant cystatin. Plant J 12:455–461

Vain P, Worland B, Clarke MC, Richard G, Beavis M, Liu H, Kohli A, Leech M, Snape J, Christou P, Atkinson H (1998) Expression of an engineered cysteine proteinase inhibitor (Oryzacystatin-I_D86) for nematode resistance in transgenic rice plants. Theor Appl Genet 96:266–271

Valette C, Andary C, Geiger JP, Sarah JL, Nicole M (1998) Histochemical and cytochemical investigations of phenols in roots of banana infected by the burrowing nematode *Radopholus similis*. Phytopathology 88(11):1141–1148

van Dam NM (2009) Belowground herbivory and plant defenses. Annu Rev Ecol Evol Syst 40:373–392

van Dam NM, Raaijmakers CE, van der Putten WH (2005) Root herbivory reduces growth and survival of the shoot feeding specialists *Pieris rapae* on *Brassica nigra*. Entomol Exp Appl 115:161–170

Vanstone VA, Hollaway GJ, Stirling GR (2008) Managing nematode pests in the southern and western regions of the Australian cereal industry: continuing progress in a challenging environment. Aust Plant Pathol 37:220–234

Viaene N, Duran LF, Mauricio Rivera J, Duenas J, Rowe P, De Waele D (2003) Responses of banana and plantain cultivars, lines and hybrids to the burrowing nematode *Radopholus similis*. Nematology 5:85–95

Vieira P, Wantoch S, Lilley CH, Chitwood DJ, Atkinson HJ, Kamo K (2014) Expression of a cystatin transgene can confer resistance to root lesion nematodes in *Lilium longiflorum* cv. 'Nellie White'. Transgenic Res. doi:10.1007/s11248-014-9848-2

Waetzig GH, Sobczak M, Grundler FMW (1999) Localization of hydrogen peroxide during the defence response of *Arabidopsis thaliana* against the plant-parasitic nematode *Heterodera glycines*. Nematology 1:681–686

Wehunt EJ, Hutchison DJ, Edwards DI (1978) Reaction of banana cultivars to the burrowing nematode (*Radopholus similis*). J Nematol 10(4):368

Wieczorek K (2015) Cell wall alterations in nematode-infected roots. Adv Bot Res 73:61–90

Williams K, Taylor S, Bogacki P, Pallotta M, Bariana H, Wallwork H (2002) Mapping of the root lesion nematode (*Pratylenchus neglectus*) resistance gene Rlnn1 in wheat. Theor Appl Genet 104:874–879

Wuyts N, de Waele D, Swennen R (2006a) Activity of phenylalanine ammonia-lyase, peroxidase and polyphenol oxidase in roots of banana (*Musa acuminata* AAA, cvs Grande Naine and Yangambi km5) before and after infection with *Radopholus similis*. Nematology 8:201–209

Wuyts N, Maung ZTZ, Swennen R, De Waele D (2006b) Banana rhizodeposition: characterization of root border cell production and effects on chemotaxis and motility of the parasitic nematode *Radopholus similis*. Plant and Soil 283:217–228

Yan G, Smiley RW, Okubara PA (2012) Detection and quantification of *Pratylenchus thornei* in DNA extracted from soil using real-time PCR. Phytopathology 102:14–22

Zhao X, Schmitt M, Hawes MC (2000) Species-dependent effects of border cell and root tip exudates on nematode behavior. Phytopathology 90(11):1239–1245

Zhu S, Tang S, Tang Q, Liu T (2014) Genome-wide transcriptional changes of ramie (*Boehmeria nivea* L. Gaud) in response to root-lesion nematode infection. Gene 552:67–74

Zipfel C (2009) Early molecular events in PAMP-triggered immunity. Curr Opin Plant Biol 12:414–420

Zwart RS, Thompson JP, Godwin ID (2005) Identification of quantitative trait loci for resistance to two species of root-lesion nematode (*Pratylenchus thornei* and *P. neglectus*) in wheat. Aust J Agric Res 56:345–352

Part III
Root Responses to Beneficial Micro-organisms

Root Interactions with Nonpathogenic *Fusarium oxysporum*

Hey *Fusarium oxysporum*, What Do You Do in Life When You Do Not Infect a Plant?

Christian Steinberg, Charline Lecomte, Claude Alabouvette, and Véronique Edel-Hermann

Abstract In this review, we tried to present *Fusarium oxysporum* in an ecological context rather than to confine it in the too classic double play of the nonpathogenic fungus that protects the plant against the corresponding *forma specialis*. Moreover, *F. oxysporum* is sometimes one, sometimes the other, and only the fungus can reveal its hidden face, according to it is or not in front of the target plant. Despite the quality and richness of the studies conducted to date, molecular approaches highlight some of the evolutionary mechanisms that explain the polyphyletic nature of this species, but still they do not identify a nonpathogenic *F. oxysporum*.

This soilborne fungus has primarily an intense saprophytic life, and it finds its place in the functioning of the ecosystem of which it actively occupies all compartments, thanks to an impressive metabolic flexibility and a high enzyme potential. This adaptability is exploited by *F. oxysporum* first to get carbon from different organic sources and energy through variable strategies including nitrate dissimilation under severe anaerobic conditions and also to colonize extreme environments, some of which being dramatically anthropized. This adaptability is also exploited by man for bioremediation of polluted sites, for detoxification of xenobiotic compounds including pesticides, and furthermore for industrial and biotechnological processes. The presence of the fungus in water distribution networks of city stresses again the adaptable nature of the fungus, but more precisely, this highlights the presence of clonal populations worldwide and raises the question of the role of man in the transfer of biological resources.

We conclude in a provocative manner by asking if nonpathogenic *F. oxysporum* would not be the all-purpose fungal tool needed to ensure a good soil functioning.

C. Steinberg (✉) • C. Lecomte • V. Edel-Hermann
INRA, UMR1347 Agroécologie, 17 rue Sully, BP 86510, 21000 Dijon, France
e-mail: christian.steinberg@dijon.inra.fr

C. Alabouvette
Agrene, 47 r Constant Pierrot, 21000 Dijon, France

C.M.F. Vos, K. Kazan (eds.), *Belowground Defence Strategies in Plants*, Signaling and Communication in Plants, DOI 10.1007/978-3-319-42319-7_12

1 Introduction

If there are microorganisms, especially soilborne fungi, that fascinate mycologists, plant pathologists, doctors, and microbial ecologists, without forgetting evolutionists, geneticists and taxonomists, *Fusarium* or more precisely *Fusarium oxysporum* is one of those. We should probably also mention in the list that growers and horticulturists are equally concerned with survival, evolution, and activities of *F. oxysporum* Schlecht, but maybe they do not feel the same fascination as the aforementioned corporations. Indeed, *F. oxysporum* is primarily known for its ability to cause disease on a large number of host plants, while the predominant role of this fungus in soils is essentially determined by its saprophytic activity in raw and rhizospheric soils whether they are cultivated or not, by its biochemical activity in anthropic environments, and by its long survival in various environments (Burgess 1981; Bao et al. 2004; Christakopoulos et al. 1991; Holker et al. 1999).

2 To Be or Not to Be a Nonpathogenic *F. oxysporum*?

F. oxysporum is an ascomycete, belonging to the family of Nectriaceae and the order of Hypocreale. This is an asexual fungus whose teleomorph is unknown. Actually, the *F. oxysporum* species complex includes both pathogenic and nonpathogenic populations, the former being split into more than 100 *formae speciales*, each of them being specific of a plant species (Armstrong and Armstrong 1981; Correll 1991; Baayen et al. 2000). This morphological species is now recognized as a species complex because of its high level of phylogenetic diversity (O'Donnell et al. 2009). Phylogenetic analyses also revealed how diverse is the origin of the pathogenicity of most of the various *formae speciales*. Only a few of them such as *F. oxysporum* f. sp. *albedinis*, *ciceri*, and *loti* are monophyletic (Tantaoui et al. 1996; Wunsch et al. 2009; Demers et al. 2014). Therefore, a great effort of research is devoted to characterize the diversity of *formae speciales* of peculiar interest in agriculture (Elias and Schneider 1992; Kistler 1997; Abo et al. 2005; Lievens et al. 2008; Edel-Hermann et al. 2012) and in horticulture (Loffler and Rumine 1991; Lori et al. 2012; Canizares et al. 2015; Lecomte et al. 2016), in order to identify some specific molecular markers allowing to detect and monitor both pathogenic and nonpathogenic populations in the rhizosphere of host plants (Recorbet et al. 2003; Edel-Hermann et al. 2011). However, to date and despite these efforts, it is still not possible to generally discriminate nonpathogenic populations from pathogenic populations except by the fact that a strain is said to be nonpathogenic if it does not cause any symptom on the plant on which it has been inoculated, but even so it is not possible to say whether this strain is definitively nonpathogenic regardless of the plant species. Thus, the very definition of nonpathogen is blurred because it relies on the absence of a trait that can only be expressed by a pathogenic strain in the presence of the host plant on which it is specifically subservient; we talk about compatibility. So, while highlighting that the polyphyletic nature of the origins of the pathogen status has been acquired in the course of evolution by a fungus that originally is not pathogenic, doubt always exists that a strain incapable of causing symptoms on a given plant is not pathogen of a plant species with which the compatibility was not tested. Nevertheless, the notion of risk associated to this doubt is limited and should not be considered as a foil to the positive role that *F. oxysporum* plays in the biological functioning of the soil and can also play in the protection of plants as a biocontrol agent. Indeed, the already mentioned polyphyletic nature of the origin of pathogenicity in most of the *formae speciales* can be explained especially by the presence and mobility of a large number of transposable or repetitive elements, responsible for timely and random mutations in the genomes of pathogenic strains, and by horizontal transfers of chromosomal regions (Daboussi and Langin 1994; Daviere et al. 2001; Ma et al. 2010; Inami et al. 2012; Schmidt et al. 2013). The presence of many transposable elements in *F. oxysporum*, as in other Deuteromycetes, is probably a consequence of the asexual lifestyle of these fungi and the resulting absence of the meiosis process that normally eliminates repetitive elements (Daboussi 1996). In the case of *formae speciales*, it can be assumed that the ongoing compatible interaction of the pathogen with the plant is an additional selection pressure that strengthens the interest for the phytopathogenic fungus to have generators of diversity and adaptation mechanisms to overcome the defence reactions opposed by the plant. In the case of the few nonpathogenic populations that have been studied so far on that point, it seems they harbor much less transposable elements than pathogenic strains (Migheli et al. 1999). Therefore, one could assume a greater genetic stability from a nonpathogenic population than from a pathogenic population. However, this putative genetic stability for a given population is probably compensated by an incredible diversity within the species giving *F. oxysporum* the ability to colonize a huge variety of environments (Edel et al. 2001; Lori et al. 2004; O'Donnell et al. 2004; Sautour et al. 2012). In addition, the host pathogen-plant compatibility is a mark enabling to appreciate the diversity and evolutionary history of a given *forma specialis*. This kind of reference is not available for the nonpathogenic populations, and although nonpathogenic strains are generally used as a control in the analyses of diversity of pathogenic populations, rare phylogenetic studies are dedicated to the evolutionary history of nonpathogenic populations; therefore, it is difficult to comment on their genetic stability (Inami et al. 2014).

3 Is *F. oxysporum* Only a Soilborne Fungus?

The geographic distribution of formae speciales is probably affected by that of host plants and by anthropogenic activities; however, all the various data in the literature for many years issuing from local surveys have shown that *F. oxysporum* occurs primarily in soils in most parts of the world without recourse to pathogenesis (Park 1963; McKenzie and Taylor 1983; Backhouse et al. 2001). Anyway, the terminology "nonpathogenic" is a default appellation regarding the very likely initial saprotrophic status of this species complex, and studies on the ecology of *F. oxysporum* do not discriminate between pathogenic and nonpathogenic populations. So we will do the same. The places over the world where *F. oxysporum* can be found include natural extreme conditions such as saline soil habitats of the hot arid desert environment (Mandeel 2006), tropical dry forests (Bezerra et al. 2013), Arctic circle (Kommedahl et al. 1988), and environments affected by human activities such as industrially polluted sediments (Massaccesi et al. 2002), metal mine wastes (Ortega-Larrocea et al. 2010), biofilms in household appliances such as washing machines (Babic et al. 2015), and water system of hospitals (Anaissie et al. 2001; Steinberg et al. 2015). It is likely that the diversity hosted by *F. oxysporum* explains the adaptation of the fungi to various niches under various soil and climatic conditions, as well as in water and in the air. Their concentration was estimated to vary between 10^2 and 10^4 propagules per gram of soil (Park 1963; Alabouvette et al. 1984; Larkin et al. 1993,) while it is much less (a few propagules per liter) in seawater or springwater (Palmero et al. 2009). It can reach up to 10^3 propagules per mL when accidentally colonizing water pipes (Sautour et al. 2012). Spores of *F. oxysporum* have been found associated with rain dust (0.1–45 propagules per gram of dust) and transported over long distances including overseas (Palmero et al. 2011). Spores of *F. oxysporum* are also found in the air outdoor as well as in air-conditioned indoor environments (Debasmita et al. 2014; Khan et al. 2009). So clearly *F. oxysporum* is a ubiquitous fungus that is able to adapt to many types of environments, although it is more frequently encountered in the soil where its density is important both in cultivated and noncultivated ecosystems. Focus is generally made on the diversity of pathogenic populations to understand the origins of this particular trait (Baayen et al. 2000; Groenewald et al. 2006; Luongo et al. 2015; O'Donnell et al. 1998, 2004). However, many studies have revealed an incredible intraspecific diversity within *F. oxysporum* (Demers et al. 2015; Edel et al. 2001; Edel-Hermann et al. 2015; Laurence et al. 2012; Lori et al. 2004). It is not forbidden to think that this diversity, although it is often assessed by the analysis of noncoding DNA regions, could explain the ability of *F. oxysporum* to colonize such different environments, among which is the rhizosphere of putative host plants. Abundant and more or less specific exudates released by plant roots in the rhizosphere are a main food source for microorganisms and a driving force of their population density and activities. *F. oxysporum* populations are particularly affected by this privileged habitat, and they are actively involved in the colonization of the rhizospheric soil, the rhizoplane, and also root tissues (Fravel et al. 2003; Landa et al. 2001; Ling et al. 2012; Toyota and Kimura 1992). The selective nature of root exudates linked to the genotype of the host plant determines the composition of the populations of *F. oxysporum* associated with the plant (Edel et al. 1997; Demers et al. 2015). Actually, all the strains do not respond in the same way to released exudates, what explains that the abundance ratios between strains of the same population are different from one to another rhizosphere. This difference in ability to use root exudates of a given plant depends on the own characteristics of each strain but is not linked to the pathogenicity or nonpathogenicity of the strains (Steinberg et al. 1999a, b). Consequently, the ability in using efficiently the root exudates determines the issue of the competition for trophic sources between pathogenic and nonpathogenic *F. oxysporum* and therefore the selection for efficient biocontrol agents (Eparvier and Alabouvette 1994; Olivain et al. 2006) (see below). Nonpathogenic strains of *F. oxysporum* can cross the epidermis cells of the root surface, but they are unable to cause disease (Olivain and Alabouvette 1997). They colonize the root cortex of a plant and may establish as endophytes (Belgrove et al. 2011; Demers et al. 2015), but the main point is that this narrow interaction between nonpathogenic *F. oxysporum* and the host-plant results in the so-called priming effect, i.e., the implementation of defence reactions of the plant that slow down their progress and prevent any further invasion by a pathogenic strain (Aime et al. 2013; Benhamou and Garand 2001). Similarly to the absence of preferential selection between pathogenic and nonpathogenic populations of *F. oxysporum* at the root surface of the host plant (Olivain et al. 2006), there is no clear genetic differentiation in the composition of endophyte populations and rhizosphere populations (Demers et al. 2015).

All these interactions in the rhizosphere of the host plant between pathogenic and nonpathogenic populations of *F. oxysporum* reveal protective ability of the latter against the pathogen and invite to consider the use of nonpathogenic strains in biocontrol strategy against *formae speciales* of *F. oxysporum* or other pests (Alabouvette et al. 2009; Vos et al. 2014).

4 Would There Be a New Robin Hood in the Rhizosphere of Plants to Be Protected?

Evidence of a possible role of nonpathogenic *Fusarium* spp. in controlling pathogens resulted from the observation that soils suppressive to *Fusarium* wilt harbored high populations of nonpathogenic *F. oxysporum* and *F. solani* whose involvement in the mechanism of soil suppressiveness was confirmed experimentally (Rouxel et al. 1979). Strains of *F. oxysporum* were much more efficient in establishing suppressiveness in soil than other species of *Fusarium* (Tamietti and Alabouvette 1986). Moreover, there is a great variability among soilborne nonpathogenic strains of *F. oxysporum* for their capacity to protect plants against their specific pathogens (Forsyth et al. 2006; Nel et al. 2006), and some effective strains have not been isolated from soil but from the stem of healthy plants (Ogawa and Komada 1984; Postma and Rattink 1992). In addition, it is well established that a pathogenic strain applied to a non-host plant is able to protect this plant against further infection by its specific *forma specialis*. A review was recently published by Alabouvette et al. (2009) describing the main modes of action of biological control agents in soil and listing a large number of situations in which selected strains of nonpathogenic *F. oxysporum* succeeded or not in protecting the plant against pathogenic *formae speciales*. Since the publication of this review, many other examples of the protective potential of nonpathogenic *F. oxysporum* were also published (Belgrove et al. 2011; Morocko-Bicevska et al. 2014), and it would be tedious to list them all. Actually what is noticeable is the fact that nonpathogenic *F. oxysporum* have been shown to control not only pathogenic *F. oxysporum* but also *Verticillium dahliae* causing wilting of eggplant, pepper, and cotton (Gizi et al. 2011; Veloso and Díaz 2012; Zhang et al. 2015), nematodes causing damage on banana and tomato roots (Paparu et al. 2009; El-Fattah et al. 2007), and insects such as the sucking *Aphis gossypii* and the whitefly *Trialeurodes vaporariorum* affecting tomato (Martinuz et al. 2012; Menjivar et al. 2012). While in the case of *V. dahliae* on eggplant and cotton, volatile organic compounds produced by the strains of *F. oxysporum* control the pathogen; in the case of nematodes, weevils, and insects, the *F. oxysporum* strains are endophyte and elicit the plant defence reactions of the host plants.

It must be admitted that most of the examples cited here and in the review published in 2009 (Alabouvette et al. 2009) correspond to controlled situations that reveal the potential of nonpathogenic strains, but a very limited number of the most powerful strains are licensed, registered, and available in the market with a biocontrol allegation. The protective capacity in *F. oxysporum* is not a simple trait and many genes are likely to be involved. Identifying some traits linked to the protective capacity would help in differentiating pathogenic from protective strains and in screening among soilborne strains to identify potential protective strains. Success of microbiological control requires a sufficient understanding of the modes of action of the antagonist and also of its interactions with the plant, the pathogen, and the rest of the microbiota. All these studies take time, and most of the biocontrol agents other than *F. oxysporum* and already on the market have been studied for more than 20 years before registration. The work already done and the results obtained with nonpathogenic strains of *F. oxysporum* augur an imminent placing on the market of representatives of this species to control pathogens. It is however necessary to be wary of the too rapid interpretation found in recent papers (Schmidt et al. 2013) concerning the results of Ma et al. (2010). Ma et al. showed that under very special laboratory conditions, the nonpathogenic strain Fo47, isolated from the suppressive soil of Châteaurenard (France) and whose protective capability was already proved (Olivain et al. 2004), was likely to integrate by horizontal transfer, a fragment of chromosome 14, bearer genes involved in the pathogenicity of a strain of *F. oxysporum* f. sp. *lycopersici*. Actually, the experimental conditions were such that the likelihood of such a natural realization is zero; the authors simply wanted to show that the horizontal transfer was possible, which is different from likely. We can thus consider as reliable the strains of nonpathogenic *F. oxysporum* to be used in biological control strategies.

5 Dormant or Active Actor of the Biological Functioning of Soils?

5.1 Carbon Utilization

The distribution of *F. oxysporum* in numerous, complex, and varied environments is explained by the enzymatic machinery at its disposal and its ability to modify its metabolism within the constraints of these environments including microaerobic and very-low-oxygen conditions, which gives it this remarkable adaptability and an important role in the biodegradation of the organic matter. *F. oxysporum* produces indeed a large spectrum of extracellular oxidative enzymes of various types including cellulases, laccases, xylanases, lignin-degrading enzymes, and manganese peroxidases (Falcon et al. 1995; Rodriguez et al. 1996; Silva et al. 2009; Zhou et al. 2010; Xiros et al. 2011; Huang et al. 2015). Apart from study cases dedicated to the ability of *F. oxysporum* to metabolize a given C source or to denitrify a nitrogen-containing substrate (Rodriguez et al. 1996; Takaya and Shoun 2000; Ali et al. 2014), there is no global data to quantify the relative importance of the role of *F. oxysporum*, within the fungal community, in the decomposition, reorganization, and mineralization of organic matter in soils and litter. However, its ubiquitous presence and its high abundance mean that the contribution of this fungus in the carbon and nitrogen cycles must be significant. Beyond its ecological role in the saprophytic phase of *F. oxysporum*, this important enzymatic potential is usable in processes for bioproduction and/or biodegradation of natural resources under solid-state fermentation but also in bioremediation process and phytoextraction of heavy-metal under field conditions. For instance, *F. oxysporum* is used to produce ethanol from agricultural sources such as cereal straw, thanks to its ability to combine both the cellulose and hemicellulose degradation system and the capability to ferment hexoses and pentoses to ethanol (Christakopoulos et al. 1989; Ruiz et al. 2007; Anasontzis et al. 2011; Xiros et al. 2011; Ali et al. 2012). Similarly, *F. oxysporum* appears as an efficient biotechnological partner. It is grown in solid-state fermentation process to degrade by-products of the olive oil production or the citrus-processing industry (Sampedro et al. 2007; Mamma et al. 2008).

5.2 Nitrogen Utilization

Nitrogen sources in the environment including soil are variable in nature (organic and mineral) as in structural complexity. It is often difficult to separate the use of nitrogen from that of carbon, but it nevertheless appears that biomass production and secretion of hydrolytic enzymes to use carbon by *F. oxysporum* is strongly impacted by the nitrogen source at its disposal (Da Silva et al. 2001; Escobosa et al. 2009). This phenomenon has been mainly shown in biotechnology processes to solicit the enzyme potential of *F. oxysporum* to degrade a carbon substrate such as lignin or agriculture by-products or to obtain a product of interest (Cheilas et al. 2000; Panagiotou et al. 2003, 2005; Lee et al. 2011). It is also noticeable that, thanks to the incredible flexibility of its metabolism, *F. oxysporum* adapts to moderately up to severe anaerobic conditions by replacing the energy-producing mechanism of O_2 respiration with the reduction of NO_3^- and NO_2^- to N_2O. Denitrification is a dissimilating metabolic mechanism for nitrate and was described in *F. oxysporum* not so long ago (Shoun and Tanimoto 1991). This dissimilatory nitrate reduction allows *F. oxysporum* to regenerate the cofactor NAD(+) during the denitrification process to then efficiently hydrolyze xylose to achieve its anaerobic growth (Panagiotou et al. 2006). *F. oxysporum* could not only denitrify nitrate through the classical sequential reactions of nitrate and nitrite reductases but it can also reduce nitrate to ammonium through ammonia fermentation (Takaya 2002; Takasaki et al. 2004; Zhou et al. 2010). A deep focus has been given to the specific pathways used by this fungus to denitrify nitrate and nitrite to gain energy. It was shown that *F. oxysporum* denitrification activities are localized in the mitochondria and are coupled to the synthesis of ATP (Kobayashi et al. 1996) and that cytochrome P-450, designated as P450nor, was involved in the respiratory nitrite reduction of *F. oxysporum*, while the equivalent NO reductase (NOR) system in bacteria is derived from cytochrome c-oxidase (Shoun and Tanimoto 1991; Takaya and Shoun 2000; Dalber et al. 2005). Recent studies related to the use of nitrogen by *F. oxysporum* help at explaining the role of soilborne fungi in the nitrogen cycle and more specifically in soils (Long et al. 2013; Mothapo et al. 2015). For instance, fungal denitrifiers including *F. oxysporum* generally do not have the gene encoding N_2O reductase (NosZ) as bacteria have and thus are incapable of reducing N_2O to N_2 (Shoun et al. 2012). Many studies dedicated to the fungal release of N_2O as a powerful greenhouse gas contributing both to global warming and ozone depletion underlined the contribution of *F. oxysporum* to this phenomenon (Shoun et al. 2012; Jirout et al. 2013; Chen et al. 2014; Maeda et al. 2015). An equivalent strategy allows *F. oxysporum* to reduce sulfur in anoxic condition to recover energy (still via NADH cofactor) and ensure efficient oxidation of the carbon source and subsequent fungal growth. As for nitrate dissimilation, the anaerobic sulfur reduction by *F. oxysporum* results in the release of a gas, the hydrogen sulfide (H_2S), but in amounts that are less than those noted for N_2O (Abe et al. 2007; Sato et al. 2011). This reveals how the fungus adapts to anaerobic conditions and replaces the energy-producing mechanism of O_2 respiration by a dissimilative strategy. This ability to reduce sulfide in anoxic conditions can confer a competitive advantage to populations of *F. oxysporum* when *Brassica*, rich in sulfur, are ground and incorporated into the soil to reduce densities of primary inoculums of plant pathogenic fungi (Larkin and Griffin 2007).

5.3 Bioremediation

As mentioned above, *F. oxysporum* has also attracted interest for bioremediation of soil and purification of water due to its capability to detoxify and colonize polluted environments. For instance, *F. oxysporum* excretes alkaline substances that increase the pH of the medium around its mycelium, which affects the status of certain minerals. Thus, by issuing chelators produced during its growth in the presence of glutamate, *F. oxysporum* hydrolyzes coal without producing specific enzymes. On the other side, *Trichoderma viride* produces enzymes attacking coal under alkaline conditions; therefore, these fungi combine solubilization of coal and ligninolyse of humic acids, which enables them to colonize mineral soils (Holker et al. 1999). In an iron ore area in Brazil, *F. oxysporum* associated with mycorrhizal fungi facilitates the solubilization of phosphorus, thus facilitating the installation of legumes to ensure revegetation of the soil (Matias et al. 2009). *F. oxysporum* was isolated from industrially polluted effluents highly contaminated with cadmium alone or cadmium and lead. Thanks to its ability to grow in the presence of heavy metals and its associated metabolic activity, *F. oxysporum* may, in aqueous medium, either sequester cadmium in its mycelial biomass (Massaccesi et al. 2002) or turn Pb^{2+} and Cd^{2+} metal ions into the corresponding carbonates that can then be recovered. Besides the removal of toxic heavy-metal ions from water, the crystals thus created have a specific morphology making them exploitable as biominerals for biological and materials sciences (Sanyal et al. 2005). Moreover, the capability of *F. oxysporum* to reduce extracellularly metal ions and in particular silver ions into silver nanoparticles which have an antibacterial effect has been proposed for the production of sterile clothing for hospitals to prevent infection with pathogenic bacteria such as *Staphylococcus aureus*. In this case, the bioremediation of water is ensured by the cyanogenic bacterium *Chromobacterium violaceum* (Duran et al. 2007). It may be admitted that despite the anthropogenic character of mining and the presence of heavy metals at industrial sites, pollutants, although toxic, are natural constituents of the environment that man has concentrated, certainly, but that *F. oxysporum* particularly ubiquitous fungus was confronted to and was able to adapt to their presence, tolerate them, and even exploit them. By cons, it is notable that the enzymatic equipment of *F. oxysporum* makes it capable of degrading synthetic molecules. So *F. oxysporum* was used to degrade and to detoxify a new chemical class of textile dyes called glycoconjugate azo dye and is proposed in the frame of remediation strategies of textile effluents (Porri et al. 2011). The ability of *F. oxysporum* to grow in the presence of arsenic and to volatilize this element present in polluted environments allows considering its exploitation for the bioremediation of As-contaminated soils, sediments, and effluents (Zeng et al. 2010; Feng et al. 2015). As well, the efficiency with which *F. oxysporum* is capable of extracting the iron from asbestos fibers due to a change of its metabolism and thereby reduce its toxicity makes the fungus a potential candidate for the bioremediation of contaminated sites. First, the internalization of asbestos fibers is prevented in *F. oxysporum*, thanks to its rigid cell wall. Then a proteomic analysis revealed an upregulation of two proteins, homologous of already known proteins in *F. graminearum* and *Coccidioides immitis*, and a rerouting of *F. oxysporum* metabolism to the pentose-phosphate pathway to counteract the deleterious consequences of oxidative stress (Chiapello et al. 2010). Indirectly, *F. oxysporum* also contributes to the bioremediation of soils contaminated with zinc and cadmium or mining soils by facilitating the phytoextraction of heavy metals from the soils by plants introduced for that purpose in the areas concerned (Ortega-Larrocea et al. 2010; Zhang et al. 2012).

6 Adaptation to Human Activities

6.1 A Ticket for the Degradation of Xenobiotics?

With a stated goal of protecting crops, chemical control against pests, either weeds, insects, or plant pathogenic microorganisms, results in a spill of more or less complex molecules, most of which being xenobiotic compounds. The accumulation of these molecules can negatively impact human, animal, plant, and microbial populations under increasing pressure. The enzymatic equipment of *F. oxysporum* allows the fungus to degrade pesticides, including organophosphates such as malathion and fenitrothion which are neurotoxic insecticides (Hasan 1999; Peter et al. 2015). According to the initial concentration (400–1000 ppm) and to the availability of additional nutrients (carbon, nitrogen, phosphate), *F. oxysporum* was capable of degrading malathion in less than 8 days up to 3 weeks of incubation. The insecticide chlordecone is a contaminant found in most of the banana plantations in the French West Indies. Microbial communities were severely negatively affected by this organochlorine, but *F. oxysporum* was able to tolerate the presence of the toxic molecules in soil as well as some few other fungal genera belonging to the Ascomycota phylum (Merlin et al. 2013). However, *F. oxysporum* was the only species able to grow on chlordecone as only carbon source in controlled conditions and to dissipate up to 40 % of chlordecone. So also there, the enzyme potential confers to the fungus a ubiquitous adaptability leading to exploit those skills to address the presence of xenobiotic pesticides in soil and water (Pinto et al. 2012).

6.2 A Ticket for the Hospital?

Nosocomial infections are more and more frequently attributed to the presence of *Fusarium* in hospital settings (Girmenia et al. 2000; Anaissie et al. 2001; Dignani and Anaissie 2004; Sautour et al. 2012). The diseases often affect dramatically immunocompromised patients (Nucci and Anaissie 2007) but can also target more specifically and less dramatically contact-lens wearers and patients with infectious keratitis (Jureen et al. 2008). *F. oxysporum* and *F. solani* are the most dominant species involved among the various *Fusarium* species that have been detected so far (Anaissie et al. 2001; O'Donnell et al. 2007; Short et al. 2011; Scheel et al. 2013). An epidemiological investigation conducted over 2 years in hospital and nonhospital buildings in France revealed the existence of homogeneous populations of *F. oxysporum* and *F. dimerum* common to all contaminated hospital sites (Steinberg et al. 2015). The waterborne isolates tolerated higher concentrations of chlorine dioxide used to disinfect the hospital water distribution systems and of copper sulfate released by copper pipes and higher temperatures than did soilborne isolates but did not show any specific resistance to fungicides. These populations are present at very low densities in natural waters, making them difficult to detect, but they are adapted to the specific conditions offered by the complex water systems of public hospitals in France and probably other localities in the world (Steinberg et al. 2015). Molecular analyses on the genetic diversity of populations of *F. oxysporum* in hospitals brought evidence for the recent release of a clonal lineage geographically widespread (O'Donnell et al. 2004).

These studies conducted by doctors, mycologists, taxonomists, and ecologists led the different hospital departments to take measures to reduce the risk of spread of the fungus in the premises, including minimizing the effects of aerosolization to prevent nosocomial infections, what is quite good of course. They especially highlight the impact of man on the evolution of microorganisms and their distribution throughout the world because here are clonal populations of *F. oxysporum* adapted to urban water supply systems that are found in countries from different continents.

7 Conclusion

There is no doubt that nonpathogenic *F. oxysporum* interact firstly with pathogenic *formae speciales* of *F. oxysporum* or other pathogenic fungal species for the use of trophic resources and space in the rhizosphere of host plants and also with the plant, and they elicit defence reactions. These are the reasons why many strains of nonpathogenic *F. oxysporum* are proposed as biocontrol agents to control the infectious activity of pathogens or pests and reduce the severity of the disease even if not so many strains are actually registered and available on the market. Although this biocontrol activity is particularly important, it would be a shame to reduce *F. oxysporum* to a simple role-playing in the rhizosphere of a plant that distributes the game depending on its compatibility with one or the other of the strains. Indeed, only the interaction with the plant discriminates pathogenic strains from nonpathogenic ones. Molecular markers exist for a few number of *formae speciales*, but for most of the others, these markers, if any, are difficult to identify. The reasons are the very high genetic diversity within this species and the polyphyletic origin of the pathogenicity. In return, this diversity is a major asset for *F. oxysporum* that can colonize and exploit all the compartments of the terrestrial ecosystem, even the most unexpected, whether they are extreme in nature or a result of excessive anthropization. Thanks to a diverse enzymatic equipment and a flexible metabolism, *F. oxysporum* is able to adapt to many environmental conditions and above all to actively contribute to the biochemical processes governing the functioning of the niches used by the fungus, whatever they are.

Beyond the biocontrol activity of *F. oxysporum*, the mechanisms of which are beginning to be elucidated, at least partially, the bioremediation of contaminated soils and the detoxification of harmful xenobiotics used in agriculture become particularly attractive, as well as the potential its enzymatic equipment offers for biotechnological processes including food processing. Finally, its ability to reduce nitrates makes *F. oxysporum* the preferred study model to understand the role of fungi in the denitrification process and particularly in their contribution to the production of N_2O and the resulting greenhouse gas. *F. oxysporum*, whether it is pathogenic or nonpathogenic *F. oxysporum*, deserves its qualification as a ubiquitous fungus because actually it is everywhere and it is active throughout. It appears as the multipurpose fungal toolbox that pathologists sometimes ignore but which nevertheless actively contributes to the global functioning of soil.

References

Abe T, Hoshino T, Nakamura A, Takaya N (2007) Anaerobic elemental sulfur reduction by fungus *Fusarium oxysporum*. Biosci Biotechnol Biochem 71:2402–2407

Abo K, Klein KK, Edel-Hermann V, Gautheron N, Traore D, Steinberg C (2005) High genetic diversity among strains of *Fusarium oxysporum* f. sp. *vasinfectum* from cotton in Ivory Coast. Phytopathology 95:1391–1396

Aime S, Alabouvette C, Steinberg C, Olivain C (2013) The endophytic strain *Fusarium oxysporum* Fo47: a good candidate for priming the defense responses in tomato roots. Mol Plant-Microbe Interact 26:918–926

Alabouvette C, Couteaudier Y, Louvet J (1984) Studies on disease suppressiveness of soils—dynamics of the populations of *Fusarium* spp. and *F. oxysporum* f. sp. *melonis* in a wilt-suppressive and a wilt conducive soil. Agronomie 4:729–733

Alabouvette C, Olivain C, Migheli Q, Steinberg C (2009) Microbiological control of soil-borne phytopathogenic fungi with special emphasis on wilt-inducing *Fusarium oxysporum*. New Phytol 184:529–544

Ali S, Khan M, Fagan B, Mullins E, Doohan F (2012) Exploiting the inter-strain divergence of *Fusarium oxysporum* for microbial bioprocessing of lignocellulose to bioethanol. AMB Express 2:16

Ali SS, Khan M, Mullins E, Doohan F (2014) Identification of *Fusarium oxysporum* genes associated with lignocellulose bioconversion competency. Bioenergy Res 7:110–119

Anaissie EJ, Kuchar RT, Rex JH, Francesconi A, Kasai M, Muller FMC, Lozano-Chiu M, Summerbell MC, Dignani MC, Chanock SJ, Walsh TJ (2001) Fusariosis associated with pathogenic *Fusarium* species colonization of a hospital water system: a new paradigm for the epidemiology of opportunistic mold infections. Clin Infect Dis 33:1871–1878

Anasontzis GE, Zerva A, Stathopoulou PM, Haralampidis K, Diallinas G, Karagouni AD, Hatzinikolaou DG (2011) Homologous overexpression of xylanase in *Fusarium oxysporum* increases ethanol productivity during consolidated bioprocessing (CBP) of lignocellulosics. J Biotechnol 152:16–23

Armstrong GM, Armstrong JK (1981) Formae speciales and races of *Fusarium oxysporum* causing wilt diseases. In: Nelson PE et al (eds) *Fusarium*: diseases, biology, and taxonomy. Pennsylvania State University Press, University Park, pp 391–399

Baayen RP, O'Donnell K, Bonants PJM, Cigelnik E, Kroon J, Roebroeck EJA, Waalwijk C (2000) Gene genealogies and AFLP analyses in the *Fusarium oxysporum* complex identify monophyletic and nonmonophyletic formae speciales causing wilt and rot disease. Phytopathology 90:891–900

Babic MN, Zalar L, Zenko B, Schroers HJ, Deroski S, Gunde-Cimerman N (2015) *Candida* and *Fusarium* species known as opportunistic human pathogens from customer-accessible parts of residential washing machines. Fungal Biol 119:95–113

Backhouse D, Burgess LW, Summerell BA (2001) Biogeographie of *Fusarium*. In: Summerell BA et al (eds) *Fusarium*, Paul Nelson Memorial Symposium. The American Phytopathological Society (APS), St Paul, pp 122–137

Bao J, Fravel D, Lazarovits G, Chellemi D, van Berkum P, O'Neill N (2004) Biocontrol genotypes of *Fusarium oxysporum* from tomato fields in Florida. Phytoparasitica 32:9–20

Belgrove A, Steinberg C, Viljoen A (2011) Evaluation of nonpathogenic *Fusarium oxysporum* and *Pseudomonas fluorescens* for panama disease control. Plant Dis 95:951–959

Benhamou N, Garand C (2001) Cytological analysis of defense-related mechanisms induced in pea root tissues in response to colonization by nonpathogenic *Fusarium oxysporum* Fo47. Phytopathology 91:730–740

Bezerra JDP, Santos MGS, Barbosa RN, Svedese VM, Lima DMM, Fernandes MJS, Gomes BS, Paiva LM, Almeida-Cortez JS, Souza-Motta CM (2013) Fungal endophytes from cactus *Cereus jamacaru* in Brazilian tropical dry forest: a first study. Symbiosis 60:53–63

Burgess LW (1981) General ecology of the Fusaria. In: Nelson PE et al (eds) *Fusarium*: diseases, biology, and taxonomy. Pennsylvania State University Press, University Park, pp 225–235

Canizares MC, Gomez-Lama C, Garcia-Pedraja MD, Perez-Artes E (2015) Study of phylogenetic relationships among *Fusarium oxysporum* f. sp. *dianthi* isolates: confirmation of intrarace diversity and development of a practical tool for simple population analyses. Plant Dis 99:780–787

Cheilas T, Stoupis T, Christakopoulos P, Katapodis P, Mamma D, Hatzinikolaou DG, Kekos D, Macris BJ (2000) Hemicellulolytic activity of *Fusarium oxysporum* grown on sugar beet pulp. Production of extracellular arabinanase. Process Biochem 35:557–561

Chen HH, Mothapo NV, Shi W (2014) The significant contribution of fungi to soil N_2O production across diverse ecosystems. Appl Soil Ecol 73:70–77

Chiapello M, Daghino S, Martino E, Perotto S (2010) Cellular response of *Fusarium oxysporum* to crocidolite asbestos as revealed by a combined proteomic approach. J Proteome Res 9:3923–3931

Christakopoulos P, Macris BJ, Kekos D (1989) Direct fermentation of cellulose to ethanol by *Fusarium oxysporum*. Enzyme Microb Technol 11:236–239

Christakopoulos P, Koullas D, Kekos D, Koukios E, Macris B (1991) Direct conversion of straw to ethanol by *Fusarium oxysporum*: effect of cellulose crystallinity. Enzyme Microb Technol 13:272–274

Correll J (1991) The relationship between formae speciales, races, and vegetative compatibility groups in *Fusarium oxysporum*. Phytopathology 81:1061–1064

Da Silva MC, Bertolini MC, Ernandes JR (2001) Biomass production and secretion of hydrolytic enzymes are influenced by the structural complexity of the nitrogen source in *Fusarium oxysporum* and *Aspergillus nidulans*. J Basic Microbiol 41:269–280

Daboussi MJ (1996) Fungal transposable elements: generators of diversity and genetic tools. J Genet 75:325–339

Daboussi MJ, Langin T (1994) Transposable elements in the fungal plant pathogen *Fusarium oxysporum*. Genetica 93:49–59

Dalber A, Shoun H, Ullrich V (2005) Nitric oxide reductase (P450(nor)) from *Fusarium oxysporum*. J Inorg Biochem 99:185–193

Daviere JM, Langin T, Daboussi MJ (2001) Potential role of transposable elements in the rapid reorganization of the *Fusarium oxysporum* genome. Fungal Genet Biol 34:177–192

Debasmita G, Priyanka D, Tanusree C, Naim U, Das AK (2014) Study of aeromycoflora in indoor and outdoor environment of National Library, Kolkata. Int J Plant Anim Environ Sci 4:663–672

Demers JE, Garzón CD, Jiménez-Gasco MDM (2014) Striking genetic similarity between races of *Fusarium oxysporum* f. sp. *ciceris* confirms a monophyletic origin and clonal evolution of the chickpea vascular wilt pathogen. Eur J Plant Pathol 139:303–318

Demers JE, Gugino BK, Jimenez-Gasco MD (2015) Highly diverse endophytic and soil *Fusarium oxysporum* populations associated with field-grown tomato plants. Appl Environ Microbiol 81:81–90

Dignani MC, Anaissie E (2004) Human fusariosis. Clin Microbiol Infect 10:67–75

Duran N, Marcato PD, De Souza GIH, Alves OL, Esposito E (2007) Antibacterial effect of silver nanoparticles produced by fungal process on textile fabrics and their effluent treatment. J Biomed Nanotechnol 3:203–208

Edel V, Steinberg C, Gautheron N, Alabouvette C (1997) Populations of nonpathogenic *Fusarium oxysporum* associated with roots of four plant species compared to soilborne populations. Phytopathology 87:693–697

Edel V, Steinberg C, Gautheron N, Recorbet G, Alabouvette C (2001) Genetic diversity of *Fusarium oxysporum* populations isolated from different soils in France. FEMS Microbiol Ecol 36:61–71

Edel-Hermann V, Aime S, Cordier C, Olivain C, Steinberg C, Alabouvette C (2011) Development of a strainspecific real-time PCR assay for the detection and quantification of the biological control agent Fo47 in root tissues. FEMS Microbiol Lett 322:34–40

Edel-Hermann V, Gautheron N, Steinberg C (2012) Genetic diversity of *Fusarium oxysporum* and related species pathogenic on tomato in Algeria and other Mediterranean countries. Plant Pathol 322:34–40

Edel-Hermann V, Gautheron N, Mounier A, Steinberg C (2015) *Fusarium* diversity in soil using a specific molecular approach and a cultural approach. J Microbiol Methods 111:64–71

El-Fattah A, Dababat A, Sikora RA (2007) Influence of the mutualistic endophyte *Fusarium oxysporum* 162 on *Meloidogyne incognita* attraction and invasion. Nematology 9:771–776

Elias KS, Schneider RW (1992) Genetic diversity within and among races and vegetative compatibility groups of *Fusarium oxysporum* f.sp. *lycopersici* as determined by isozyme analysis. Phytopathology 82:1421–1427

Eparvier A, Alabouvette C (1994) Use of ELISA and GUS-transformed strains to study competition between pathogenic and non-pathogenic *Fusarium oxysporum* for root colonization. Biocontrol Sci Technol 4:35–47

Escobosa ARC, Figueroa JAL, Corona JFG, Wrobel K (2009) Effect of *Fusarium oxysporum* f. sp *lycopersici* on the degradation of humic acid associated with Cu, Pb, and Ni: an in vitro study. Anal Bioanal Chem 394:2267–2276

Falcon MA, Rodriguez A, Carnicero A, Regalado V, Perestelo F, Milstein O, Delafuente G (1995) Isolation of microorganisms with lignin transformation potential from soil of Tenerife Island. Soil Biol Biochem 27:121–126

Feng QF, Su SM, Zeng XB, Zhang YZ, Li LF, Bai LY, Duan R, Lin ZL (2015) Arsenite resistance, accumulation, and volatilization properties of *Trichoderma asperellum* SM-12F1, *Penicillium janthinellum* SM-12F4, and *Fusarium oxysporum* CZ-8F1. Clean: Soil Air Water 43:141–146

Forsyth LM, Smith LJ, Aitken EAB (2006) Identification and characterization of non-pathogenic *Fusarium oxysporum* capable of increasing and decreasing Fusarium wilt severity. Mycol Res 110:929–935

Fravel D, Olivain C, Alabouvette C (2003) *Fusarium oxysporum* and its biocontrol. New Phytol 157:493–502

Girmenia C, Pagano L, Corvatta L, Mele L, Del Favero A, Martino P (2000) The epidemiology of fusariosis in patients with haematological diseases. Br J Haematol 111:272–276

Gizi D, Stringlis IA, Tjamos SE, Paplomatas EJ (2011) Seedling vaccination by stem injecting a conidial suspension of F2, a non-pathogenic *Fusarium oxysporum* strain, suppresses Verticillium wilt of eggplant. Biol Control 58:387–392

Groenewald S, Van den Berg N, Marasas WFO, Viljoen A (2006) The application of high-throughput AFLP's in assessing genetic diversity in *Fusarium oxysporum* f. sp cubense. Mycol Res 110:297–305

Hasan HAH (1999) Fungal utilization of organophosphate pesticides and their degradation by *Aspergillus flavus* and A-sydowii in soil. Folia Microbiol 44:77–84

Holker U, Ludwig S, Scheel T, Hofer M (1999) Mechanisms of coal solubilization by the deuteromycetes *Trichoderma atroviride* and *Fusarium oxysporum*. Appl Microbiol Biotechnol 52:57–59

Huang YH, Busk PK, Lange L (2015) Cellulose and hemicellulose-degrading enzymes in *Fusarium commune* transcriptome and functional characterization of three identified xylanases. Enzyme Microb Technol 73–74:9–19

Inami K, Yoshioka-Akiyama CY, Yamasaki M, Teraoka T, Arie T (2012) A genetic mechanism for emergence of races in *Fusarium oxysporum* f. sp. *lycopersici*: inactivation of avirulence gene AVR1 by transposon insertion. PLoS One 7(8), e44101

Inami K, Kashiwa T, Kawabe M, Onokubo-Okabe A, Ishikawa N, Perez ER, Hozumi T, Caballero LA, De Baldarrago FC, Roco MJ, Madadi KA, Peever TL, Teraoka T, Kodama M, Arie T (2014) The Tomato wilt fungus *Fusarium oxysporum* f. sp *lycopersici* shares common ancestors with nonpathogenic *F. oxysporum* isolated from wild tomatoes in the Peruvian Andes. Microbes Environ 29:200–210

Jirout J, Simek M, Elhottova D (2013) Fungal contribution to nitrous oxide emissions from cattle impacted soils. Chemosphere 90:565–572

Jureen R, Koh TH, Wang G, Chai LYA, Tan AL, Chai T, Wong YW, Wang Y, Tambyah PA, Beuerman R, Tan D (2008) Use of multiple methods for genotyping *Fusarium* during an outbreak of contact lens associated fungal keratitis in Singapore. BMC Infect Dis 8:92

Khan AAH, Karuppayil SM, Manoharachary C, Kunwar IK, Waghray S (2009) Isolation, identification and testing for allergenicity of fungi from air-conditioned indoor environments. Aerobiologia 25:119–123

Kistler HC (1997) Genetic diversity in the plant-pathogenic fungus *Fusarium oxysporum*. Phytopathology 87:474–479

Kobayashi M, Matsuo Y, Takimoto A, Suzuki S, Maruo F, Shoun H (1996) Denitrification, a novel type of respiratory metabolism in fungal mitochondrion. J Biol Chem 271:16263–16267

Kommedahl T, Abbas HK, Burnes PM, Mirocha CJ (1988) Prevalence and toxigenicity of *Fusarium* species from soils of Norway near the Artic-circle. Mycologia 80:790–794

Landa BB, Navas-Cortes JA, Hervas A, Jimenez-Diaz RM (2001) Influence of temperature and inoculum density of *Fusarium oxysporum* f. sp *ciceris* on suppression of fusarium wilt of chickpea by rhizosphere bacteria. Phytopathology 91:807–816

Larkin RP, Griffin TS (2007) Control of soilborne potato diseases using *Brassica* green manures. Crop Prot 26:1067–1077

Larkin RP, Hopkins DL, Martin FN (1993) Effect of successive watermelon plantings on *Fusarium oxysporum* and other microorganisms in soil suppressive and conducive to fusarium wilt of watermelon. Phytopathology 83:1097–1105

Laurence MH, Burgess LW, Summerell BA, Liew ECY (2012) High levels of diversity in *Fusarium oxysporum* from non-cultivated ecosystems in Australia. Fungal Biol 116:289–297

Lecomte C, Edel-Hermann V, Cannesan MA, Gautheron N, Langlois A, Alabouvette C, Robert F, Steinberg C (2016) *Fusarium oxysporum* f. sp. *cyclaminis*, an underestimated genetic diversity. Eur J Plant Pathol 145:421–431

Lee HS, Kang JW, Kim BH, Park SG, Lee C (2011) Statistical optimization of culture conditions for the production of enniatins H, I, and MK1688 by *Fusarium oxysporum* KFCC 11363P. J Biosci Bioeng 111:279–285

Lievens B, Rep M, Thomma BPHJ (2008) Recent developments in the molecular discrimination of *formae speciales* of *Fusarium oxysporum*. Pest Manag Sci 64:781–788

Ling N, Zhang WW, Tan SY, Huang QW, Shen QR (2012) Effect of the nursery application of bioorganic fertilizer on spatial distribution of *Fusarium oxysporum* f. sp *niveum* and its antagonistic bacterium in the rhizosphere of watermelon. Appl Soil Ecol 59:13–19

Loffler HJM, Rumine P (1991) Virulence and vegetative compatibility of Dutch and Italian isolates of *Fusarium oxysporum* f. sp. *lilii*. J Phytopathol 132:12–20

Long A, Heitman J, Tobias C, Philips R, Song B (2013) Co-occurring anammox, denitrification, and codenitrification in agricultural soils. Appl EnvironMicrobiol 79:168–176

Lori G, Edel-Hermann V, Gautheron N, Alabouvette C (2004) Genetic diversity of pathogenic and nonpathogenic populations of *Fusarium oxysporum* isolated from carnation fields in Argentina. Phytopathology 94:661–668

Lori GA, Petiet PM, Malbran I, Mourelos CA, Wright ER, Rivera MC (2012) Fusarium wilt of cyclamen: pathogenicity and vegetative compatibility groups structure of the pathogen in Argentina. Crop Prot 36:43–48

Luongo L, Ferrarini A, Haegi A, Vitale S, Polverari A, Belisario A (2015) Genetic diversity and pathogenicity of *Fusarium oxysporum* f.sp *melonis* races from different areas of Italy. J Phytopathol 163:73–83

Ma LJ, van der Does C, Borkovich KA, Coleman JJ, Daboussi MJ, Di Pietro A, Dufresne M, Freitag M et al (2010) Comparative genomics reveals mobile pathogenicity chromosomes in *Fusarium*. Nature 464:367–373

Maeda K, Spor A, Edel-Hermann V, Heraud C, Breuil MC, Bizouard F, Toyoda S, Yoshida N, Steinberg C, Philippot L (2015) N_2O production, a widespread trait in fungi. Sci Rep 5. doi:10.1038/srep09697

Mamma D, Kourtoglou E, Christakopoulos P (2008) Fungal multienzyme production on industrial by-products of the citrus-processing industry. Bioresour Technol 99:2373–2383

Mandeel QA (2006) Biodiversity of the genus *Fusarium* in saline soil habitats. J Basic Microbiol 46:480–494

Martinuz A, Schouten A, Menjivar RD, Sikora RA (2012) Effectiveness of systemic resistance toward *Aphis gossypii* (Hom, Aphididae) as induced by combined applications of the endophytes *Fusarium oxysporum* Fo162 and *Rhizobium etli* G12. Biol Control 62:7

Massaccesi G, Romero MC, Cazau MC, Bucsinszky AM (2002) Cadmium removal capacities of filamentous soil fungi isolated from industrially polluted sediments, in La Plata (Argentina). World J Microbiol Biotechnol 18:817–820

Matias SR, Pagano MC, Muzzi FC, Oliveira CA, Carneiro AA, Horta SN, Scotti MR (2009) Effect of rhizobia, mycorrhizal fungi and phosphate-solubilizing microorganisms in the rhizosphere of native plants used to recover an iron ore area in Brazil. Eur J Soil Biol 45:259–266

McKenzie F, Taylor GS (1983) *Fusarium* populations in British soils relative to different cropping practices. Trans Br Mycol Soc 80:409–413

Menjivar RD, Cabrera JA, Kranz J, Sikora RA (2012) Induction of metabolite organic compounds by mutualistic endophytic fungi to reduce the greenhouse whitefly *Trialeurodes vaporariorum* (Westwood) infection on tomato. Plant Soil 352:233–241

Merlin C, Devers M, Crouzet O, Heraud C, Steinberg C, Mougin C, Martin-Laurent F (2013) Characterization of chlordecone-tolerant fungal populations isolated from long-term polluted tropical volcanic soil in the French West Indies. Environ Sci Pollut Res 21:4914–4927

Migheli Q, Lauge R, Daviere JM, Gerlinger C, Kaper F, Langin T, Daboussi MJ (1999) Transposition of the autonomous Fot1 element in the filamentous fungus *Fusarium oxysporum*. Genetics 151:1005–1013

Morocko-Bicevska I, Fatehi J, Gerhardson B (2014) Biocontrol of strawberry root rot and petiole blight by use of non-pathogenic *Fusarium* sp strains. In: 7th International strawberry symposium (Vol. 1049, pp. 599–605)

Mothapo N, Chen HH, Cubeta MA, Grossman JM, Fuller F, Shi W (2015) Phylogenetic, taxonomic and functional diversity of fungal denitrifiers and associated N_2O production efficacy. Soil Biol Biochem 83:160–175

Nel B, Steinberg C, Labuschagne N, Viljoen A (2006) The potential of nonpathogenic *Fusarium oxysporum* and other biological control organisms for suppressing fusarium wilt of banana. Plant Pathol 55:217–223

Nucci M, Anaissie E (2007) Fusarium infections in immunocompromised patients. Clin Microbiol Rev 20:695

O'Donnell K, Kistler HC, Cigelnik E, Ploetz RC (1998) Multiple evolutionary origins of the fungus causing Panama disease of banana: concordant evidence from nuclear and mitochondrial gene genealogies. Proc Natl Acad Sci USA 95:2044–2049

O'Donnell K, Sutton DA, Rinaldi MG, Magnon KC, Cox PA, Revankar SG, Sanche S, Geiser DM, Juba JH, van Burik JAH, Padhye A, Anaissie EJ, Francesconi A, Walsh TJ, Robinson JS (2004) Genetic diversity of human pathogenic members of the *Fusarium oxysporum* complex inferred from multilocus DNA sequence data and amplified fragment length polymorphism analyses: evidence for the recent dispersion of a geographically widespread clonal lineage and nosocomial origin. J Clin Microbiol 42:5109–5120

O'Donnell K, Sarver BAJ, Brandt M, Chang DC, Noble-Wang J, Park BJ, Sutton DA, Benjamin L, Lindsley M, Padhye A, Geiser DM, Ward TJ (2007) Phylogenetic diversity and microsphere array-based genotyping of human pathogenic fusaria, including isolates from the multistate contact lens—associated US Keratitis outbreaks of 2005 and 2006. J Clin Microbiol 45:2235–2248

O'Donnell K, Gueidan C, Sink S, Johnston PR, Crous PW, Glenn A, Riley R, Zitomer NC, Colyer P, Waalwijk C, van der Lee T, Moretti A, Kang S, Kim HS, Geiser DM, Juba JH, Baayen RP, Cromey M, Bithell S, Sutton DA, Skovgaard K, Ploetz R, Kistler HC, Elliott M, Davis M, Sarver BAJ (2009) A two-locus DNA sequence database for typing plant and human pathogens within the *Fusarium oxysporum* species complex. Fungal Genet Biol 46:936–948

Ogawa K, Komada H (1984) Biological control of Fusarium wilt of sweet potato by non-pathogenic *Fusarium oxysporum*. Ann Phytopathol Soc Jpn 50:1–9

Olivain C, Alabouvette C (1997) Colonization of tomato root by a non-pathogenic strain of *Fusarium oxysporum*. New Phytol 137:481–494

Olivain C, Alabouvette C, Steinberg C (2004) Production of a mixed inoculum of *Fusarium oxysporum* Fo47 and *Pseudomonas fluorescens* C7 to control Fusarium diseases. Biocontrlol Sci Technol 14:227–238

Olivain C, Humbert C, Nahalkova J, Fatehi J, L'Haridon F, Alabouvette C (2006) Colonization of tomato root by pathogenic and nonpathogenic *Fusarium oxysporum* strains inoculated together and separately into the soil. Appl Environ Microbiol 72:1523–1531

Ortega-Larrocea MD, Xoconostle-Cazares B, Maldonado-Mendoza IE, Carrillo-Gonzalez R, Hernandez-Hernandez J, Garduno MD, Lopez-Meyer M, Gomez-Flores L, Gonzalez-Chavez MDA (2010) Plant and fungal biodiversity from metal mine wastes under remediation at Zimapan, Hidalgo, Mexico. Environ Pollut 158:1922–1931

Palmero D, Iglesias C, de Cara M, Lomas T, Santos M, Tello JC (2009) Species of *Fusarium* isolated from river and sea water of Southeastern Spain and pathogenicity on four plant species. Plant Dis 93:377–385

Palmero D, Rodriguez JM, de Cara M, Camacho F, Iglesias C, Tello JC (2011) Fungal microbiota from rain water and pathogenicity of *Fusarium* species isolated from atmospheric dust and rainfall dust. J Ind Microbiol Biotechnol 38:13–20

Panagiotou G, Kekos D, Macris BJ, Christakopoulos P (2003) Production of cellulolytic and xylanolytic enzymes by *Fusarium oxysporum* grown on corn stover in solid state fermentation. Ind Crops Prod 18:37–45

Panagiotou G, Christakopoulos P, Olsson L (2005) The influence of different cultivation conditions on the metabolome of *Fusarium oxysporum*. J Biotechnol 118:304–315

Panagiotou G, Christakopoulos P, Grotkjaer T, Olsson L (2006) Engineering of the redox imbalance of *Fusarium oxysporum* enables anaerobic growth on xylose. Metabol Eng 8:474–482

Paparu P, Dubois T, Coyne D, Viljoen A (2009) Dual inoculation of *Fusarium oxysporum* endophytes in banana: effect on plant colonization, growth and control of the root burrowing nematode and the banana weevil. Biocontrol Sci Technol 19:639–655

Park D (1963) Presence of *Fusarium oxysporum* in soils. Trans Br Mycol Soc 46:444–448

Peter L, Gajendiran A, Mani D, Nagaraj S, Abraham J (2015) Mineralization of Malathion by *Fusarium oxysporum* strain JASA1 isolated from sugarcane fields. Environ Prog Sustainable Energy 34:112–116

Pinto AP, Serrano C, Pires T, Mestrinho E, Dias L, Teixeira DM, Caldeira AT (2012) Degradation of terbuthylazine, difenoconazole and pendimethalin pesticides by selected fungi cultures. Sci Total Environ 435:402–410

Porri A, Baroncelli R, Guglielminetti L, Sarrocco S, Guazzelli L, Forti M, Catelani G, Valentini G, Bazzichi M, Franceschi M, Vannacci G (2011) *Fusarium oxysporum* degradation and detoxification of a new textile-glycoconjugate azo dye (GAD). Fungal Biol 115:30–37

Postma J, Rattink H (1992) Biological control of fusarium wilt of carnation with a nonpathogenic isolate of *Fusarium oxysporum*. Can J Bot 70:1199–1205

Recorbet G, Steinberg C, Olivain C, Edel V, Trouvelot S, Dumas-Gaudot E, Gianinazzi S, Alabouvette C (2003) Wanted: pathogenesis-related marker molecules for *Fusarium oxysporum*. New Phytol 159:73–92

Rodriguez A, Perestelo F, Carnicero A, Regalado V, Perez R, DelaFuente G, Falcon MA (1996) Degradation of natural lignins and lignocellulosic substrates by soil-inhabiting fungi imperfecti. FEMS Microbiol Ecol 21:213–219

Rouxel F, Alabouvette C, Louvet J (1979) Recherches sur la résistance des sols aux maladies. IV—Mise en évidence du rôle des *Fusarium* autochtones dans la résistance d'un sol à la Fusariose vasculaire du Melon. Annal Phytopathol 11:199–207

Ruiz E, Romero I, Moya M, Sanchez S, Bravo V, Castro E (2007) Sugar fermentation by *Fusarium oxysporum* to produce ethanol. World J Microbiol Biotechnol 23:259–267

Sampedro I, D'Annibale A, Ocampo JA, Stazi SR, Garcia-Romera I (2007) Solid-state cultures of *Fusarium oxysporum* transform aromatic components of olive-mill dry residue and reduce its phytotoxicity. Bioresour Technol 98:3547–3554

Sanyal A, Rautaray D, Bansal V, Ahmad A, Sastry M (2005) Heavy-metal remediation by a fungus as a means of production of lead and cadmium carbonate crystals. Langmuir 21:7220–7224

Sato I, Shimatani K, Fujita K, Abe T, Shimizu M, Fujii T, Hoshino T, Takaya N (2011) Glutathione reductase/glutathione is responsible for cytotoxic elemental sulfur tolerance via polysulfide shuttle in fungi. J Biol Chem 286:20283–20291

Sautour M, Edel-Hermann V, Steinberg C, Sixt N, Laurent J, Dalle F, Aho S, Hartemann P, L'Ollivier C, Goyer M, Bonnin A (2012) *Fusarium* species recovered from the water distribution system of a French university hospital. Int J Hyg Environ Health 215:286–292

Scheel CM, Hurst SF, Barreiros G, Akiti T, Nucci M, Balajee SA (2013) Molecular analyses of *Fusarium* isolates recovered from a cluster of invasive mold infections in a Brazilian hospital. BMC Infect Dis 13:12

Schmidt SM, Houterman PM, Schreiver I, Ma LS, Amyotte S, Chellappan B, Boeren S, Takken FLW, Rep M (2013) MITEs in the promoters of effector genes allow prediction of novel virulence genes in *Fusarium oxysporum*. BMC Genomics 14:21

Short DG, O'Donnell K, Zhang N, Juba JH, Geiser DM (2011) Widespread occurrence of diverse human pathogenic types of the fungus *Fusarium* detected in plumbing drains. J Clin Microbiol 49:4264e4272

Shoun H, Tanimoto T (1991) Denitrification by the fungus *Fusarium oxysporum* and involvement of cytochrome P-450 in the respiratory nitrite reduction. J Biol Chem 266:11078–11082

Shoun H, Fushinobu S, Jiang L, Kim SW, Wakagi T (2012) Fungal denitrification and nitric oxide reductase cytochrome P450nor. Philos Trans R Soc B Biol Sci 367:1186–1194

Silva IS, Grossman M, Durranta LR (2009) Degradation of polycyclic aromatic hydrocarbons (2-7 rings) under microaerobic and very-low-oxygen conditions by soil fungi. Int Biodeterior Biodegrad 63:224–229

Steinberg C, Whipps JM, Wood D, Fenlon J, Alabouvette C (1999a) Mycelial development of *Fusarium oxysporum* in the vicinity of tomato roots. Mycol Res 103:769–778

Steinberg C, Whipps JM, Wood D, Fenlon J, Alabouvette C (1999b) Effects of nutritional sources on growth of one non-pathogenic strain and four strains of *Fusarium oxysporum* pathogenic on tomato. Mycol Res 103:1210–1216

Steinberg C, Laurent J, Edel-Hermann V, Barbezant M, Sixt N, Dalle F, Aho S, Bonnin A, Hartemann P, Sautour M (2015) Adaptation of *Fusarium oxysporum* and *Fusarium dimerum* to the specific aquatic environment provided by the water systems of hospitals. Water Res 76:53–65

Takasaki K, Shoun H, Yamaguchi M, Takeo K, Nakamura A, Hoshino T, Takaya N (2004) Fungal ammonia fermentation, a novel metabolic mechanism that couples the dissimilatory and assimilatory pathways of both nitrate and ethanol—role of acetyl CoA synthetase in anaerobic ATP synthesis. J Biol Chem 279:12414–12420

Takaya N (2002) Dissimilatory nitrate reduction metabolisms and their control in fungi. J Biosci Bioeng 94:506–510

Takaya N, Shoun H (2000) Nitric oxide reduction, the last step in denitrification by *Fusarium oxysporum*, is obligatorily mediated by cytochrome P450nor. Mol Gen Genet 263:342–348

Tamietti G, Alabouvette C (1986) Studies on the disease suppressiveness of soil. 13. Role of non pathogenic *Fusarium oxysporum* in the wilt suppressive mechanisms of a soil from Noirmoutier. Agronomie 6:541–548

Tantaoui A, Ouinten M, Geiger JP, Fernandez D (1996) Characterization of a single clonal lineage of *Fusarium oxysporum* f sp *albedinis* causing Bayoud disease of date palm in Morocco. Phytopathology 86:787–792

Toyota K, Kimura M (1992) Growth of *Fusarium oxysporum f. sp. raphani* in the host rhizosphere. Jpn J Soil Sci Plant Nutr 63:566–570

Veloso J, Díaz J (2012) *Fusarium oxysporum* Fo47 confers protection to pepper plants against *Verticillium dahliae* and *Phytophthora capsici*, and induces the expression of defence genes. Plant Pathol 61:281–288

Vos CM, Yang Y, De Coninck B, Cammue BPA (2014) Fungal (-like) biocontrol organisms in tomato disease control. Biol Control 74:65–81

Wunsch MJ, Baker AH, Kalb DW, Bergstrom GC (2009) Characterization of *Fusarium oxysporum* f. sp *loti Forma Specialis nov.*, a monophyletic pathogen causing vascular wilt of birdsfoot trefoil. Plant Dis 93:58–66

Xiros C, Katapodis P, Christakopoulos P (2011) Factors affecting cellulose and hemicellulose hydrolysis of alkali treated brewers spent grain by *Fusarium oxysporum* enzyme extract. Bioresour Technol 102:1688–1696

Zeng XB, Su SM, Jiang XL, Li LF, Bai LY, Zhang YR (2010) Capability of pentavalent arsenic bioaccumulation and biovolatilization of three fungal strains under laboratory conditions. Clean: Soil Air Water 38:238–241

Zhang XC, Lin L, Chen MY, Zhu ZQ, Yang WD, Chen B, Yang XE, An QL (2012) A nonpathogenic *Fusarium oxysporum* strain enhances phytoextraction of heavy metals by the hyperaccumulator *Sedum alfredii* Hance. J Hazard Mater 229:361–370

Zhang QH, Yang L, Zhang J, Wu MD, Chen WD, Jiang DH, Li GQ (2015) Production of anti-fungal volatiles by non-pathogenic *Fusarium oxysporum* and its efficacy in suppression of Verticillium wilt of cotton. Plant Soil 392:101–114

Zhou ZM, Takaya N, Shoun H (2010) Multi-energy metabolic mechanisms of the fungus *Fusarium oxysporum* in low oxygen environments. Biosci Biotechol Biochem 74:2431–2437

Belowground Defence Strategies in Plants: The Plant–*Trichoderma* Dialogue

Ainhoa Martinez-Medina, Maria J. Pozo, Bruno P.A. Cammue, and Christine M.F. Vos

Abstract *Trichoderma* spp. are cosmopolitan soil fungi that hold great promise as biocontrol organisms. Their biocontrol capacity was initially thought to be based on their direct suppressive effects on plant pathogens, with most strains showing mycoparasitic potential and producing a large variety of enzymes and secondary metabolites. More recently however *Trichoderma* was also recognized as an opportunistic plant root colonizer that can trigger induced systemic resistance (ISR) in the plant, typically leading to a more rapid and robust systemic activation of defences after pathogen attack. As our understanding of the *Trichoderma*–plant interaction advances, it is becoming increasingly clear that *Trichoderma* is initially also perceived by the host plant as a potential invader. *Trichoderma* thus needs to find a way to deal with the plant defence response, either by avoiding or suppressing it, in order to establish a durable interaction with their host. In this chapter, we cover our current knowledge on the initial dialogue between *Trichoderma* and its host, including the defence responses mounted by the host plant and how *Trichoderma*

A. Martinez-Medina
German Centre for Integrative Biodiversity Research (iDiv) Halle-Jena-Leipzig, Institute of Ecology, Friedrich Schiller University Jena, Jena, Germany

M.J. Pozo
Department of Soil Microbiology and Symbiotic Systems, Estación Experimental del Zaidín, Consejo Superior de Investigaciones Científicas, Granada, Spain

B.P.A. Cammue
Centre of Microbial and Plant Genetics, University of Leuven, Leuven, Belgium

Department of Plant Systems Biology, Vlaams Instituut voor Biotechnologie (VIB), Ghent, Belgium

C.M.F. Vos (✉)
Centre of Microbial and Plant Genetics, KU Leuven, Kasteelpark Arenberg 20, 3001 Leuven, Belgium

Department of Plant Systems Biology, Vlaams Instituut voor Biotechnologie (VIB), Technologie park 927, 9052 Ghent, Belgium

Commonwealth Scientific and Industrial Research Organisation (CSIRO) Agriculture, 306 Carmody Road, St Lucia, Brisbane, QLD 4067, Australia

Scientia Terrae Research Institute, Fortsesteenweg 30A, Sint-Katelijne-Waver, Belgium
e-mail: cvo@scientiaterrae.org

C.M.F. Vos, K. Kazan (eds.), *Belowground Defence Strategies in Plants*, Signaling and Communication in Plants, DOI 10.1007/978-3-319-42319-7_13

attempts to circumvent it. Next, we describe how the host plant can benefit from this interaction. *Trichoderma* colonization can indeed prime the host defence, enabling it to react faster and stronger to subsequent pathogen attack. We then conclude with examples of *Trichoderma*-induced resistance and direct antagonism against different types of soil pathogens and pests.

1 Introduction

Trichoderma spp. are among the most commonly isolated saprotrophic fungi in the soil (Harman et al. 2004). They are highly opportunistic and can adapt to a wide range of climatological and ecological conditions. This is illustrated by the fact that *Trichoderma* strains cannot only be found in soils all over the world, but some are also capable of colonizing plant roots, aboveground plant parts, and numerous other substrates such as wood and even other fungi (Druzhinina et al. 2011). While *Trichoderma reesei* strains are widely used in industrial applications due to their prolific production of cellulose- and chitin-degrading enzymes (Seidl et al. 2009), other strains have been of interest for many years due to their biocontrol potential (Lorito et al. 2010). Originally the direct antagonistic potential of *Trichoderma* against plant pathogenic fungi was assumed to be the main explanation for the observed biocontrol effects (Shoresh et al. 2010). A survey of 1100 *Trichoderma* strains found that all strains possessed mycoparasitic potential, thus illustrating the importance of this trait in the genus (Druzhinina et al. 2011). In addition, an overrepresentation of genes encoding cell-wall-degrading enzymes (CWDEs) was found in the genome of three sequenced *Trichoderma* species as compared to other related fungi (Kubicek et al. 2011; Mukherjee et al. 2013). Mycoparasitism is indeed thought to be the ancestral lifestyle of *Trichoderma*, but they also produce an impressive amount of antimicrobial compounds and enzymes (Kubicek et al. 2011). These comprise both volatile and nonvolatile compounds, including pyrones, trichothecenes and terpenoids, as well as non-ribosomal peptides such as peptaibols, which are all able to kill plant pathogens (Mukherjee et al. 2013; Hermosa et al. 2013). More recently, however, it was shown that *Trichoderma* can also protect the host plant against infection even when there is no direct contact between the *Trichoderma* and the pathogen, indicating that *Trichoderma*-mediated biocontrol may also occur through plant-mediated mechanisms (reviewed in Vos et al. 2014).

In addition, specific *Trichoderma* strains can promote plant growth or protect against abiotic stresses (Shoresh et al. 2010; Brotman et al. 2013). The combination of these different modes of action, together with their high reproductive capacity, ability to survive under unfavorable conditions, and high nutrient utilization efficiency, makes *Trichoderma* species highly promising biocontrol organisms (Benitez et al. 2004). As a consequence, several *Trichoderma* strains such as *Trichoderma harzianum* T22 have already been registered as biopesticide or biofertilizer (Lorito et al. 2010), and a better understanding of the mechanisms driving the beneficial effects will probably lead to the selection of additional strains for future agronomic applications.

Despite their numerous beneficial properties for plants, it is now becoming increasingly clear that plants originally perceive *Trichoderma* as an invading microbe and mount a defence response in an attempt to limit root colonization. It is highly interesting to decipher the dialogue that takes place between the plant and the colonizing *Trichoderma*. This knowledge can lead to a better understanding about the thin red line between beneficial and pathogenic plant–microbe interactions and could lead to the discovery and selection of strains with increased rhizosphere competence.

2 Root Immune Signaling During *Trichoderma* Colonization

2.1 Plant Immune Signaling

Immune signaling in plants is initiated upon the recognition of general elicitors, which are broadly conserved within a wide range of microbes, including pathogenic and beneficial ones. These compounds are called microbe-associated molecular patterns (MAMPs or PAMPs in the case of pathogens) and include a diversity of molecules such as flagellin, bacterial lipopolysaccharide, chitin, or peptidoglycans (Gomez-Gomez and Boller 2002; Erbs and Newman 2003; Montesano et al. 2003). MAMPs are recognized by the plant as "nonself molecules" by transmembrane pattern recognition receptors (PRRs) located in the plasma membrane in the plant (reviewed in Boller and Felix 2009), leading to a signal transduction cascade and the activation of MAMP- or PAMP-triggered immunity (MTI or PTI) (Jones and Dangl 2006). The latter response can also be triggered by specific plant components released upon pathogen attack which are recognized by the plant as "nonself activities." Such components are termed damage- or danger-associated molecular patterns (DAMPs) and can include cutin monomers and cellodextrins resulting from the degradation by the pathogen of the plant's cutin and cell wall (Fauth et al. 1998; Aziz et al. 2007). In an ongoing evolutionary arms race, successful pathogens have evolved to minimize host immune stimulation and secrete effector molecules to bypass this first line of defence, by suppressing PTI signaling, thus facilitating colonization and causing effector-triggered susceptibility of the plant to the disease (Jones and Dangl 2006). In turn, plants can perceive these effectors or their modified target proteins and activate immune responses that are quicker, more prolonged, and more robust than those in PTI, resulting in effector-triggered immunity (ETI) (Jones and Dangl 2006; Boller and He 2009). Typically, the activation of PTI/MTI and ETI triggers a series of common early defence-related events including the generation of reactive oxygen species, activation of mitogen-activated protein kinases (MAPKs), extracellular alkalinization, and protein phosphorylation with associated gene regulation that ultimately restricts the growth of the microbial invader (Gimenez-Ibanez and Rathjen 2010). The onset of PTI and

ETI often triggers induced resistance in tissues distal from the site of infection, enhancing the defence-related capacity in still undamaged plant parts (Dempsey and Klessig 2012). This form of pathogen-induced resistance is commonly known as systemic acquired resistance (SAR) (Spoel and Dong 2012) and confers enhanced resistance against a broad spectrum of shoot and root pathogens.

Plant defence responses are in general coordinated by small molecules that act as signal transducers and regulate the production of downstream defence molecules (Ausubel 2005; Jones and Dangl 2006). Among them is the well-established importance of salicylic acid (SA), jasmonic acid (JA), and ethylene (Et) as primary signals in the regulation of the plant immunity (reviewed in Pieterse et al. 2009). Although there are some exceptions, biotrophic pathogens are generally sensitive to the defence responses regulated by SA, while necrotrophs and chewing insects are commonly deterred by defences controlled by JA and ET (reviewed in Pieterse et al. 2009). Further SA- and JA-/ET-regulated pathways often interact in an antagonistic manner, through a complex network of regulatory interactions termed cross talk. It is indeed the hormonal cross talk between different signaling pathways that provides the plant with a powerful capacity to finely regulate its immune response to specific invaders (reviewed in Pieterse et al. 2009).

2.2 *Modulation of Host Immunity by Root Beneficial Microbes*

In nature, plants normally grow in the presence of hundreds of microbial species, including nonpathogenic and beneficial microbes. Well-known examples of beneficial microbes are arbuscular mycorrhizal fungi (AMF) that aid in the uptake of water and minerals (van der Heijden et al. 1998) and rhizobial bacteria that fix atmospheric nitrogen for the plant (Spaink 2000). Although it would seem counterproductive to raise a defence response against beneficial microbes, a growing body of evidence suggests that beneficial microbes in the rhizosphere are initially recognized by the plant as potential invaders, triggering an immune response in the host roots upon MAMP perception (reviewed in Zamioudis and Pieterse 2012). In order to establish a mutualistic plant–microbe relationship, it is therefore essential that the beneficial microbes interfere with the host immune system.

Symbiotic microbes have evolved different strategies to reduce stimulation of the host's immune system and/or to suppress the immune response elicited in the host root after MAMP perception (reviewed in Zamioudis and Pieterse 2012). For example, some previous studies on plant interaction with AMF revealed that during early stages of the interaction, the plant reacts to the presence of AMF by activating some defence-related responses that are subsequently suppressed (reviewed in Jung et al. 2012). These studies suggest that hosts initially treat symbiotic fungi as potential invaders and activate a defence program, which is then countered by the mycorrhizal symbionts. Indeed, the AMF *Rhizophagus irregularis (formerly*

Glomus intraradices) secretes a protein, SP7, in the apoplastic or periarbuscular space during the interaction with the host which acts as effector in order to short-circuit the plant defence program, leading to the accommodation of the fungus within the plant roots (Kloppholz et al. 2011). Similarly, leguminous hosts initially recognize their symbiotic rhizobial partners as a potential threat and incite a defence program which includes the transcriptional activation of defence- and stress-related genes. However, the same cluster of genes is downregulated at later stages of root nodule formation (reviewed in Soto et al. 2009). Several rhizobial MAMPs, including common structural components such as lipopolysaccharides and exopolysaccharides, suppress the immune response in leguminous hosts (Albus et al. 2001; Tellstroem et al. 2007). Collectively, these observations indicate that root endosymbionts have adapted and refined some of their strategies to interact with their hosts, whereas plant hosts may have also evolved to discriminate between friends and foes in the rhizosphere, adapting their perception mechanisms and defence responses to the encountered invader.

2.3 Trichoderma *Colonization Elicits Host Defences*

Although the rhizosphere is among the most common ecological niches for *Trichoderma*, several species seem to have evolved further toward new ecological niches and are able to grow endophytically inside the roots as facultative endophytes (Fig. 1a; Druzhinina et al. 2011). Their plant endophytic behavior is probably a more recent evolution compared to their ancestral mycoparasitic lifestyle (Kubicek et al. 2011). It seems reasonable to assume that *Trichoderma* has been attracted to the plant rhizosphere due to the presence of fungal prey as well as

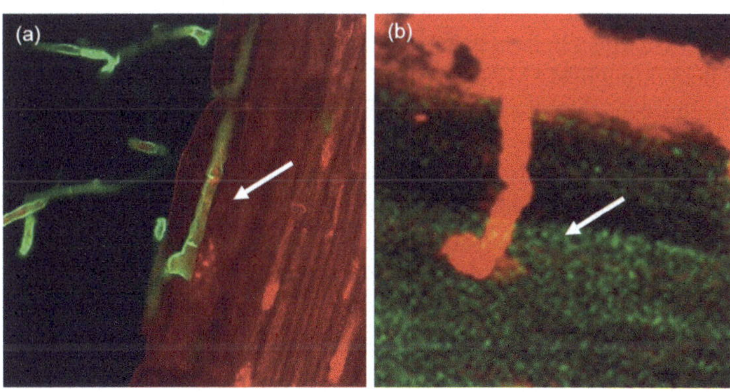

Fig. 1 Root colonization by *Trichoderma*. Confocal laser scanning microscopy images from *Trichoderma*-colonized *Arabidopsis* roots showing (**a**) green fluorescing *Trichoderma harzianum* T78 mycelium (WGA-Alexa Fluor 488) in the root surface and inside cortical cells (indicated by *arrow*) and (**b**) *red* fluorescing T78 hyphae (Texas Red) forming an appressoria-like structure on the *Arabidopsis* root surface (indicated by *arrow*)

nutrients derived from plant roots. Phylogenetic analyses position some of the endophytic *Trichoderma* spp. indeed on a terminal position in their clades, supporting the idea that the ability to endophytically colonize plant tissues is a selected trait acquired more recently in evolution (Druzhinina et al. 2011). The observation that *T. harzianum* was able to use the extraradical hyphae of AMF as a gateway entry into potato roots also fits into this context (De Jaeger et al. 2010).

Development of new technological (*in vivo* confocal microscopy) and molecular (transformed fungi expressing GFP markers) tools has allowed us to provide a clearer description of the sequence of events leading to *Trichoderma* colonization of roots. During root colonization, *Trichoderma* hyphae coil around the roots and form appressoria-like structures on the root surface (Fig. 1b), and after penetrating the root, they grow intercellularly in the epidermis and cortex (Yedidia et al. 1999; Chacon et al. 2007; Velazquez-Robledo et al. 2011; Alonso-Ramirez et al. 2014). Occasionally intracellular growth has been observed too, in which case the colonized cells appeared to remain viable (Chacon et al. 2007). Penetration of the epidermis by *Trichoderma* and subsequent ingress into the outer cortex require the secretion of a battery of cell-wall lytic enzymes and other proteins by the fungal hyphae (Viterbo et al. 2004; Brotman et al. 2008).

Previous studies have evidenced that host recognition of *Trichoderma* MAMPs and/or molecules released during the initial stages of the interaction results in the activation of a quick and often transient defence response, with the concurrent accumulation of defence-related compounds including callose deposition, antimicrobial reactive oxygen species, and phytoalexins (Yedidia et al. 1999, 2000; Chacon et al. 2007; Contreras-Cornejo et al. 2011; Salas-Marina et al. 2011). Indeed *Trichoderma* expresses a collection of MAMPs and elicitors that activate the plant basal immunity upon recognition by the host (Mukherjee et al. 2013). These MAMPs/elicitors include enzymes or peptides, oligosaccharides, and other low-molecular-weight compounds released by the action of specific *Trichoderma* enzymes on fungal and plant cell walls (Woo et al. 2006). Table 1 gives some examples of MAMPs/elicitors that are produced by various *Trichoderma* strains.

Other than through MAMPs, *Trichoderma* colonization can also be detected by the plant through the production of DAMPs. By proteomic, genomic, and transcriptomic approaches, Moran-Diez et al. (2009) characterized the gene *Thpg1* coding an endopolygalacturonase, a plant cell-wall-degrading enzyme required for efficient root colonization by *T. harzianum*. ThPG1 hydrolyzes plant pectin and produces oligogalacturonides that act as DAMPs, activating innate immunity in the host plant (de Lorenzo et al. 2011; Benedetti et al. 2015). *Trichoderma* colonization triggers, therefore, a wide array of plant defence responses during early stages of the asymptomatic colonization of the roots.

Table 1 Examples of MAMPs/elicitors produced by *Trichoderma* spp.

MAMP/ elicitor	Species	Process involved	Plant responses	References
Xylanases	*T. viride* *T. reesei* *T. harzianum*	Degradation of xylans constituting plant cell wall	Hypersensitive response Et biosynthesis Cell death Oxidative burst Defence gene expression	Avni et al. (1994) Yano et al. (1998) Do Vale et al. (2012)
Cellulases	*T. longibrachiatum* *T. harzianum*	Degradation of cellulose constituting plant cell wall	Oxidative burst Induction of Et and SA pathway Induction of peroxidase and chitinase activities	Martinez et al. (2001) Do Vale et al. (2012)
Swollenin protein Swo	*T. reesei*	Cell-wall disruption during saprophytic growth Host root colonization	Expression of chitinase and beta-glucanase genes	Brotman et al. (2008)
Cerato-platanin protein SM1	*T. virens* *T. reesei*	Hyphal growth Conidiation	Oxidative burst Expression of defence genes	Djonovic et al. (2006), Gaderer et al. (2015)
Cerato-platanin protein SM2	*T. virens*	Spore maturation	Induced resistance against *Cochliobolus heterostrophus*	Gaderer et al. (2015)
Cerato-platanin protein Epl-1	*T. harzianum*	Self cell-wall protection and recognition Regulation of mycoparasitism-related gene expression Modulation of mycoparasitic hyphal coiling	Expression of defence-related genes	Gomes et al. (2015)
avr4 and avr9 homologues	*T. atroviride* *T. harzianum* *T. viride*	Avr4 protects *Trichoderma* against plant chitinases	Hypersensitive response Expression of defence genes	Harman et al. (2004), Marra et al. (2006),
18-Mer peptaibols	*T. virens*	*Trichoderma*–host communication	Induction of JA and SA pathway Induction of defence responses against *Pseudomonas syringae*	Viterbo et al. (2007)
Trichokonins	*T. pseudokoningii*	Unknown	Oxidative burst Induction of phenolic compounds	Luo et al. (2010)

2.4 Trichoderma *Counteracts Host Defences to Establish Successful Root Colonization*

Similar to the situation that occurs during mycorrhizal and rhizobial symbiosis, endophytic *Trichoderma* minimize stimulation of the host's immune system, to successfully colonize the roots. Large-scale gene expression profiling studies have revealed that, already a few hours after *Trichoderma* inoculation, a widespread gene transcript reprogramming occurs in the host roots, which is preceded by a transient repression of the plant immune responses, most likely to allow root colonization (Moran-Diez et al. 2012; Brotman et al. 2013). The transcriptional response activated in roots upon *Trichoderma* colonization seems to share some similarities with the two-wave transcriptional reprogramming reported for mycorrhizal and rhizobial symbioses (Liu et al. 2003; Heller et al. 2008; Maunoury et al. 2010). For instance, the upregulation of the WRKY group III transcription factors WRKY41, WRKY53, and WRKY55 induced 24 h after root colonization by *Trichoderma asperelloides* T203 was repressed at 48 h, together with the expression of other defence-related transcripts. Among the downregulated genes by T203 were several genes coding for plant cytochrome P450 monooxygenases (CYP712A2, CYP712A1, CYP93D1, and CYP76G1), which are involved in the synthesis and metabolism of diverse plant defence compounds (Morant et al. 2003; Brotman et al. 2013). In a similar study, Moran-Diez et al. (2012) found that colonization of *Arabidopsis* roots by *T. harzianum* T34 was accompanied by the downregulation of defence-related genes and transcription factors as PR-1 (pathogenesis related 1), FMO1 (flavin monooxygenase 1), WRKY54, and two glutathione transferases. The authors suggested that T203 and T34 can fine-tune the transcriptional regulation of defence-regulated genes in roots to allow colonization (Moran-Diez et al. 2012; Brotman et al. 2013). Similarly, Gupta et al. (2014) recently suggested the ability of *T. asperelloides* to manipulate host nitric oxide (NO) production, which is an important regulator of plant defences. The authors found a weak and transient increase in NO accumulation in *Arabidopsis* roots following *T. asperelloides* inoculation.

Although the molecular basis for the manipulation of plant defences by *Trichoderma* is still lacking, a recent study made it clear that large transcriptional changes also occur in the fungus when coming into contact with plant roots. Moran-Diez et al. (2015) demonstrated the large transcriptional reprogramming occurring in *Trichoderma virens* hyphae when establishing contact with tomato or maize roots. Interestingly, this response seemed to be partially dependent on the host plant species involved in the interaction. In addition, genome-wide screening approaches showed that some filamentous fungi including *Trichoderma* have large numbers of proteins containing LysM motifs (Gruber et al. 2011; Kubicek et al. 2011; Seidl-Seiboth et al. 2013). Some of these proteins are involved in suppressing host defences by sequestering chitin oligosaccharides, which act as elicitors of plant defence responses (Gust et al. 2012). During infection, plant chitinases release chitin oligomers from the fungal cell wall. For example, the LysM proteins Ecp6 and Slp1 from the fungal pathogens *Cladosporium fulvum* and *Magnaporthe grisea*, respectively, bind to the released chitin oligomers, thus preventing recognition of these molecules by the plant, which otherwise would elicit defence responses. The LysM protein TAL6 from *Trichoderma atroviride* also shows an ability to sequester some forms of polymeric chitin. These findings might suggest a role for TAL6 in attenuating plant defences to facilitate root colonization. However, the authors found a more important role of this protein in self-signaling processes during fungal growth rather than fungal–plant interactions (Seidl-Seiboth et al. 2013).

Trichoderma also has the ability to manipulate the phytohormone regulatory network. Salicylic acid (SA) is an important regulator of defence signaling against biotrophic pathogens (Pieterse et al. 2009). Being mutualistic microbes, endophytic *Trichoderma* strains are likely to be sensitive to SA-regulated defence responses, as has also been demonstrated for mycorrhizal (de Roman et al. 2011) and rhizobial (Stacey et al. 2006) symbioses. Indeed, Alonso-Ramirez et al. (2014) reported a negative effect of SA signaling on the intensity of *Trichoderma* colonization. Several studies evidenced the ability of *Trichoderma* to produce substantial amounts of phytohormone-like compounds and/or to induce *de novo* biosynthesis of several phytohormones in their host such as auxins, cytokinins, and gibberellins (Contreras-Cornejo et al. 2009; Sofo et al. 2011; Martinez-Medina et al. 2014). Several of these phytohormones have been demonstrated to negatively cross-communicate with the SA signaling pathway, affecting the outcome of the immune response (reviewed in Pieterse et al. 2009). Furthermore, certain *Trichoderma* strains produce 1-aminocyclopropane-1-caboxylic acid deaminase (ACCD), which degrades the ET precursor ACC, resulting in reduced ET production in the plant (Viterbo et al. 2010; Martinez-Medina et al. 2014). Hence, *Trichoderma* might produce phytohormones or interfere with hormonal plant biosynthesis and signaling, in order to attenuate the relative strength of the defence response via hormonal cross-talk mechanisms.

Besides the attempts to avoid detection and broad-spectrum suppression of the plant innate immunity, it has also been suggested that *Trichoderma* has the capacity to neutralize host defence responses. A proteome analysis identified a protein that is a homologue of Avr4 from *C. fulvum* in *T. harzianum* T22 and *T. atroviride* P1 (Harman et al. 2004). It has been demonstrated that Avr4 protects *Trichoderma viride* against hydrolysis by plant chitinases by binding to chitin present in its cell wall (van den Burg et al. 2006). Taken together, it thus seems that *Trichoderma* has evolved different strategies for reducing stimulation and/or evading the host's immune system, similarly to obligate root symbionts.

2.5 Trichoderma *Affects Root System Architecture*

Activation of the symbiotic (SYM) program in host roots by mycorrhizal fungi or rhizobial bacteria leads to remodeling of root architecture, even before physical contact between both partners (Olah et al. 2005), most likely to promote colonization (Gutjahr and Paszkowski 2013). Although no activation of a SYM program has been described during *Trichoderma* colonization (Lace et al. 2015), some

Trichoderma species can actively influence root system architecture, mainly by enhancing lateral root formation, which could be interpreted as a means of increasing colonization success. Contreras-Cornejo et al. (2009) observed an increased number of lateral roots in *Arabidopsis*, after inoculation with *T. atroviride* or *T. virens*, showing the ability of both species for promoting root branching through an auxin-dependent mechanism. Noticeably, no effects were observed in primary root length. Apart from auxin, also ET- and mitogen-activated protein kinase 6 (MAPK6) signaling seems to be further required for the modulation of root system architecture by *Trichoderma* (Contreras-Cornejo et al. 2015). In addition, the cysteine-rich cell-wall protein QID74 of *T. harzianum* has been described to modify the architecture of cucumber roots, increasing the number and length of secondary roots (Samolski et al. 2012). Furthermore, this remodeling in root architecture can take place even without physical contact between both partners. Hung et al. (2013) found an increase in lateral roots of *Arabidopsis* plants exposed to volatile organic compounds from *T. viride*. This might indicate a reprogramming in host roots occurring even before physical contact. Similarly, 6-pentyl-2H-pyran-2-one (6-PP), a major volatile produced by *Trichoderma*, induced lateral root formation in *Arabidopsis* (Garnica-Vergara et al. 2015). It has been suggested that root responses to 6-PP involve components of auxin transport and signaling and the ET-response modulator EIN2 (Garnica-Vergara et al. 2015). These findings suggest that some *Trichoderma* volatiles may be interpreted by plants as trans-kingdom signals to modulate plant morphogenesis.

2.6 *The Host Plant Regulates* Trichoderma *Colonization*

Although *Trichoderma* seems to be able to manipulate host immunity, the fact that the colonization is limited to the root epidermis and the first layers of cortical cells (Fig. 1a) indicates a feedback system in the plant that controls the colonization. Indeed, *Trichoderma* intercellular growth induces the surrounding plant root cells to deposit cell-wall material and produce phenolic compounds. This plant reaction limits the *Trichoderma* growth inside the root (Yedidia et al. 1999, 2000; Chacon et al. 2007). A similar regulation of colonization is commonly observed during mycorrhizal and rhizobial symbiosis, balancing the cost and benefits of the symbiosis. This phenomenon is termed autoregulation of the symbiosis and prevents excessive colonization over a critical threshold (Vierheilig et al. 2008; Mortier et al. 2012). Although the mechanisms by which the host can control *Trichoderma* colonization are not yet understood, the importance of the hormone SA in controlling *Trichoderma* root colonization was recently reported. By studying the colonization pattern of *T. harzianum* in the *Arabidopsis* SA-impaired mutant *sid2*, Alonso-Ramirez et al. (2014) observed that SA signaling plays an important role in controlling *Trichoderma* colonization, as *T. harzianum* colonization in *sid2* was not restricted to the epidermal and cortical level, but extended into the vascular vessels. This uncontrolled *Trichoderma* invasion had a detrimental effect on plant growth. Similarly, SA signaling seems to have a negative effect on mycorrhizal colonization (de Roman et al. 2011) and on rhizobial infection and nodulation (Van Spronsen et al. 2003). This means that in a well-established *Trichoderma* root endophytic association, plant defence mechanisms are tightly regulated by the two partners to allow maintaining the interaction at mutualistic levels. As a side effect, this regulation may directly impact the plant interaction with other community members (Pieterse et al. 2014). Furthermore, often the effects of *Trichoderma* on host immunity are not restricted to the root, but they are also manifested in aboveground plant tissues rendering the complete plant more resistant to a broad spectrum of plant pathogens (Martinez-Medina et al. 2010, 2013; Mathys et al. 2012).

3 Induced Systemic Resistance Triggered by *Trichoderma*

3.1 The Biocontrol Effect of Trichoderma *Can Be Plant Mediated*

As stated above, protection of *Trichoderma*-colonized plants against diseases has also often been observed in the absence of direct contact between the biocontrol organism and the pathogen. This has been evidenced in several studies investigating the biocontrol effect of *Trichoderma* root colonization against leaf pathogens (Segarra et al. 2009; Martinez-Medina et al. 2013). In these studies it was verified that the *Trichoderma* strain did not colonize the aboveground plant parts, thus indicating that the two microbes were clearly spatially separated. A reduction in disease symptoms using such a setup has been demonstrated for a wide range of pathogens, including bacteria, fungi, and oomycetes (reviewed by Shoresh et al. 2010). But also for soilborne pathogens and nematodes, it has been demonstrated several times that the biocontrol effect caused by *Trichoderma* root colonization was plant mediated and not through direct antagonism. This effect could be demonstrated by using a split-root setup, physically separating the part of the root system inoculated with a *Trichoderma* strain from the root part infected by a pest or pathogen. For example, using *Trichoderma hamatum* T382, Khan et al. (2004) demonstrated that the significant reduction in root and crown rot caused by *Phytophthora capsici* in cucumber was plant mediated. Similarly, *Trichoderma koningiopsis* and *T. harzianum* induced systemic protection in roots against the pathogen *F. oxysporum* (Moreno et al. 2009) and the root-knot nematode *Meloidogyne javanica*, respectively (Selim et al. 2014). Additional evidence is provided by Howell et al. (2000), who showed that despite mutations affecting its capacity for mycoparasitism or antibiotic production, *T. virens* was still able to control *Rhizoctonia solani* infection in cotton. Furthermore, Shoresh et al. (2010) demonstrated that the biocontrol effect on *Pythium ultimum* by *T. harzianum* T22 was not only due to mycoparasitism, as it also required a functional *NPR1* (non-expressor of PR gene 1) in *Arabidopsis*, which is a central transcriptional regulator in the activation of SA-dependent defence responses and a mediator of SA–JA cross talk (Shoresh et al. 2010; Pieterse et al. 2014).

3.2 Induced Systemic Resistance and Priming

In the above examples, the biocontrol effect has occurred because *Trichoderma* has induced systemic resistance (ISR) in its host. The classic definition of ISR is based on our understanding of the systemic plant defence response induced by plant growth-promoting rhizobacteria (PGPR; Van Loon et al. 1998). It is typically described as a systemic response initiated by root colonization of a beneficial microbe, in contrast to SAR which is typically conceived as triggered by local pathogen infection. While SAR is furthermore defined as an SA-dependent defence response leading to the activation of PR genes, ISR is typically JA/ET mediated and does not involve the direct activation of PR genes. However, experimental evidence suggests a considerate overlap between both induced resistance responses (Mathys et al. 2012; Pieterse et al. 2014).

As detailed in the previous section, the initial detection of *Trichoderma* by the plant activates plant signaling pathways and leads to the reprogramming of plant gene expression. These modulations may result in preconditioning of the plant tissues for a more efficient activation of plant defences upon pathogen attack, a phenomenon referred to as priming. Besides by *Trichoderma* root colonization, the plant can be primed by treatment with various other beneficial microbes as well as by pathogen infection, wounding, or treatment with chemicals. Priming preconditions the plant to be in an alert state. Only upon attempted pathogen invasion, this alert state leads to a faster and/or stronger activation of defences in the plant, resulting in an enhanced level of resistance. Defence responses are costly for the plant, which has to seek a balance between investing in growth or defence. In comparison to constitutively activated defences, the principle of priming thus provides a great fitness benefit to the plant (Conrath 2011).

Several reviews have been published on the molecular mechanisms driving defence priming (Balmer et al. 2015; Conrath et al. 2015). Although priming by beneficial microbes is typically JA/ET dependent, the involvement of other signaling pathways such as SA or abscisic acid (ABA) is also becoming evident (Mathys et al. 2012; Martinez-Medina et al. 2013). In addition, the root-specific transcription factor R2R3-type MYB gene MYB72 was identified as crucial for both *Trichoderma*- and rhizobacteria-induced ISR, suggesting that MYB72 is a node of convergence in the ISR signaling pathway triggered by different beneficial microbes (Segarra et al. 2009). Apart from the potentiation of defence-related gene expression, the plant can also be primed for the formation of structural barriers such as callose depositions at pathogen entry sites (Fig. 2; Pieterse et al. 2014).

Below we will discuss the two phases that can be distinguished in the systemic response of the plant to *Trichoderma*, termed the ISR-prime and ISR-boost phase, which refer to the defence status of the plant in the absence or presence of pathogen infection, respectively (Mathys et al. 2012).

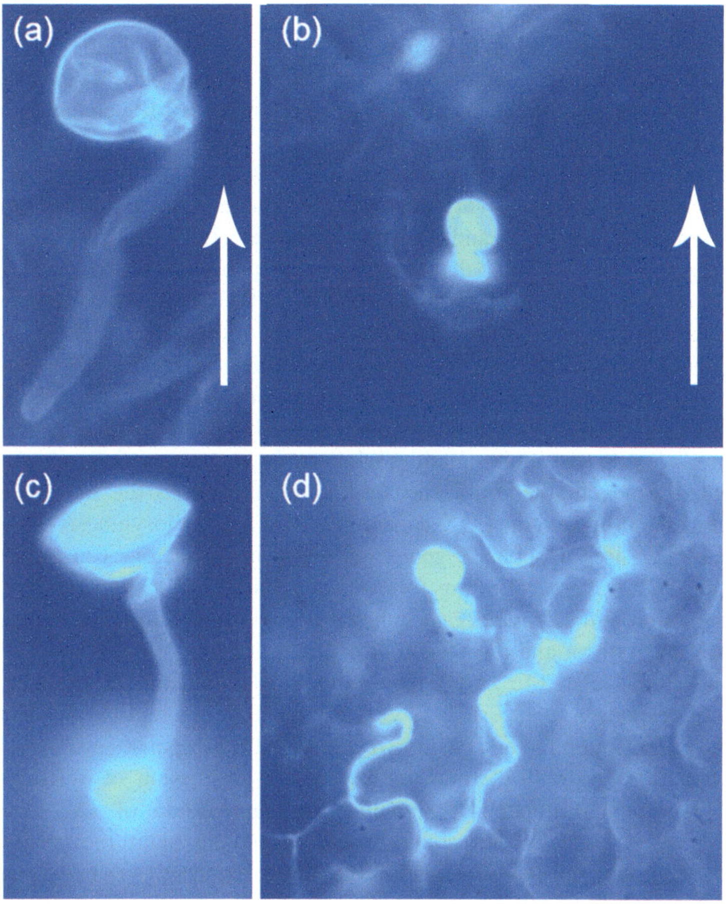

Fig. 2 *Trichoderma* primes for enhanced callose deposition. Microscopic images of the biotrophic pathogen *Hyaloperonospora parasitica* growing on leaves of *Arabidopsis* plants not induced with *Trichoderma harzianum* T78 (**a**, **b**) or induced with T78 (**c**, **d**). In (**c**), the *arrow* shows callose deposition below the appressoria at the end of the germ tube of the pathogen. In (**d**), the *arrow* indicates a hypersensitive respons (HR) in *Trichoderma*–plants

3.3 The Induced Root Response in the Trichoderma*–Plant Interaction: The ISR-Prime Phase*

As detailed in the previous section, various MAMPs can be released by *Trichoderma* and evoke a plant defence response. Several studies have investigated the plant response to *Trichoderma* colonization in great detail, however, mainly focusing on the leaf instead of the root transcriptome (Alfano et al. 2007; Mathys et al. 2012; Moran-Diez et al. 2012; Perazzolli et al. 2012). The changes induced by *Trichoderma* during plant root colonization include in general alterations in the aboveground plant parts in terms of hormone signaling, production of secondary metabolites, and control of ROS damage (reviewed by Vos et al. 2015).

Activation of hormone signaling regulates the defence network of the plant, translating the early signaling events after *Trichoderma* MAMP perception into the activation of effective defence responses. Both JA-/ET- and SA-mediated signal transduction pathways can be activated by *Trichoderma*, but the impact of a particular pathway seems to vary depending on the experimental conditions, the specific *Trichoderma* strains, and the plant species involved. The most comprehensive study so far of the global transcriptome response of *Arabidopsis* roots to *Trichoderma* was performed by Brotman et al. (2013). The authors used microarray analysis revealing extensive reprogramming of the root transcriptome as early as 24 h after the onset of colonization by *T. asperelloides* T203. Enriched functional categories according to GO analysis included response to biotic and abiotic stress, response to different stimuli such as chitin, as well as the biological processes of hormone biosynthesis and signaling. Interestingly, up to 7 % of the total upregulated genes in the roots appeared to be related to the biosynthesis of oxylipins, including several LOX genes, involved in JA biosynthesis. Via qPCR analysis of 137 stress responsive genes and transcription factors, gene modulation in the roots by *T. asperelloides* was followed at 9, 24, and 48 h after colonization. A large proportion of the *Trichoderma*-induced genes appeared to function in JA, ET, and auxin metabolism and response. For example, various WRKY and ERF transcription factors, related to JA/ET regulation and JA signaling, were induced, as well as JA-responsive genes such as vegetative storage protein (VSP). The EIN2 and EIN4 transcription factors, positive regulators of ET responses, were induced as well. The T203 strain furthermore enhanced the expression of WRKY18 and WRKY40, which stimulate JA signaling via suppression of JAZ repressors (Brotman et al. 2013). Other studies focusing on the plant root response to *Trichoderma* colonization also observed the upregulation of LOX1 (lipoxygenase 1) in the roots of *Arabidopsis* after colonization by *T. atroviride* IMI206040, together with the upregulation of PDF1.2a, a marker gene for JA-/ET-mediated signaling in *Arabidopsis* encoding plant defensin 1.2 (Penninckx et al. 1998; Salas-Marina et al. 2011). On the other hand, the SA-inducible PR genes PR1 and PR2 also showed increased expression in *Arabidopsis* roots in the same study. The activation of gibberellin (GA) production in the *Trichoderma* ISR prime was demonstrated by Chowdappa et al. (2013), who reported an increase in GA in tomato roots after application of *T. harzianum* OTPB3.

As is observed for aboveground responses in the plant–*Trichoderma* spp. interaction (Mathys et al. 2012), secondary metabolite production is another important aspect of the response of plant roots to *Trichoderma* in the ISR-prime phase. The phenylpropanoid pathway is a major source for antimicrobial phenolics and SA precursors and is found involved in the ISR-prime response in various studies. In *T. asperelloides*-colonized *Arabidopsis* roots, Brotman et al. (2013) observed increased expression of PAL1 and PAL2, encoding phenylalanine ammonium lyase which is a key enzyme in the first step of the pathway, as well as of 4CL, which encodes 4-coumarate–CoA ligase, involved in the final step of the phenylpropanoid pathway (Fraser and Chapple 2011). The last step in the biosynthesis of the most abundant phytoalexin in *Arabidopsis*, camalexin, is catalyzed by the cytochrome P-450 enzyme CYP71B15 or PAD3 (Ferrari et al. 2003), and upregulation of PAD3 (phytoalexin deficient 3) was found in *Arabidopsis* roots after colonization by various *Trichoderma* strains (Salas-Marina et al. 2011; Brotman et al. 2013). *Trichoderma* colonization also affects the oxidative stress response in plant roots. A peroxidase-encoding gene was observed to be induced in *Arabidopsis* roots colonized by *T. atroviride* (Salas-Marina et al. 2011), and Brotman et al. (2013) also reported the increased expression of genes encoding antioxidant enzymes such as MDAR (encoding a monodehydroascorbate reductase) in the *T. asperelloides*-colonized roots of *Arabidopsis* and cucumber. From the few studies that have focused on the plant root response to *Trichoderma*, it is thus clear that the colonization already leads to elaborate changes in the plant. However, these are still altered and/or magnified when a pathogen enters into the equation, indicating the start of the ISR-boost phase.

3.4 *The Induced Root Response in the* Trichoderma*–Plant–Pathogen Interaction: The ISR-Boost Phase*

The plant-mediated biocontrol effect of *Trichoderma* against soilborne pathogens has been reported in various studies, but in-depth investigations of the three-party *Trichoderma*–plant–pathogen interaction are scarce. This type of studies could, however, provide us with valuable insights into how disease is controlled in *Trichoderma*-treated plants. So far, research on the plant side of this three-party interaction has focused primarily on aboveground plant parts rather than on roots (Mathys et al. 2012; Perazzolli et al. 2012). Gupta et al. (2014) investigated the interaction of *T. asperelloides* and *F. oxysporum* in *Arabidopsis* roots, focusing on the production of NO. Infection of *Arabidopsis* roots by *F. oxysporum* leads to rapid formation of NO, a response that was actively suppressed in *T. asperelloides*-colonized roots and that was linked to transcriptional changes in NO-responsive genes. The induction of defence-associated receptor kinases by *F. oxysporum*, which may be required for disease development, was also reduced by *T. asperelloides* colonization. The authors observed a similar plant response with this particular *Trichoderma* strain against the soilborne pathogens *Verticillium dahliae* and *Pseudomonas syringae* pv. *tomato* DC3000 (Gupta et al. 2014). Martinez-Medina et al. (2010, 2014) investigated the effect of *Trichoderma* strains against *Fusarium* wilt in melon caused by *F. oxysporum* f. sp. *melonis* and found that the biocontrol activity of several *Trichoderma* strains against the pathogen correlated to the induction of ABA and ET and the cytokinin transzeatin riboside in melon shoots, while also attenuating the pathogen-induced responses in the plant. In the three-party interaction *T. harzianum* Tr6–cucumber–*F. oxysporum* f. sp. *radicis-cucumerinum*, Alizadeh et al. (2013) observed a primed expression of the defence-related genes encoding a chitinase, glucanase, and PAL. Similarly, Gallou et al. (2009) investigated the expression of defence-related marker genes in the three-party interaction of *T. harzianum* MUCL29707 and *R. solani* in potato roots. Both organisms were co-inoculated and the expression of defence-related genes was followed from 1 to 7 days after inoculation. A biocontrol effect was observed even though the *T. harzianum* strain did not penetrate the root cells. The authors reported the primed expression of the LOX and GST1 (glutathione-*S*-transferase 1) genes in potato roots, since these genes were highly induced in the three-party interaction but not by the pathogen or the beneficial fungus alone (Gallou et al. 2009). Howell et al. (2000) also observed an ISR effect of *T. virens* against *R. solani* in cotton seedlings, with a strong correlation between the abilities of the strains to induce the biosynthesis of terpenoid phytoalexins and their biocontrol capacity against *R. solani*. Even extracts from the mycelium of *Trichoderma longibrachiatum* have been shown to induce ISR in tobacco seedlings against *Phytophthora parasitica*, which was concomitant with the induced expression of PR1b and PR5c (Chang et al. 1998).

Altogether these studies indicate that *Trichoderma* can indeed protect its host plant by priming it for defence responses upon pathogen attack, although the precise defence responses may vary according to the specific biological interaction. Our knowledge on *Trichoderma*–plant–pathogen tripartite interactions is at this point still fragmentary and especially so when considering root responses and soilborne pathogens, but this is expected to improve in the future with scientific attention increasingly shifting toward the roots (De Coninck et al. 2015).

4 Impact of *Trichoderma* on Other Rhizosphere Organisms

Upon association of their roots with *Trichoderma*, plants can also benefit from the impact that *Trichoderma* can have on other rhizosphere organisms, both beneficial and detrimental. In this section we give some examples on the impact of *Trichoderma* on root pathogens, nematodes, and root-feeding insects, as well as on other beneficial root symbionts.

4.1 Impact of Trichoderma *on Root Pathogens*

Trichoderma species are well known for their direct antagonistic capacity against various microbes, and numerous studies have addressed this topic. Mycoparasitism is thought to be the ancestral lifestyle of the genus, and most strains thus display this capacity (Kubicek et al. 2011). It is a multistep process in which an early recognition stage precedes the actual physical contact. In order to locate its prey, *Trichoderma* constitutively releases low levels of cell-wall-degrading enzymes (CWDEs) such as chitinases, glucanases, and proteases. When cell-wall fragments of a possible target are detected, the fungus grows directionally toward it, while producing higher amounts of CWDEs. *Trichoderma* then attaches to its prey, the mycelium coils around it, and appressoria are formed to penetrate the hyphae (Mukherjee et al. 2013). For example, *T. harzianum* is highly efficient as a mycoparasite and produces a large amount of CWDEs in the presence of *F. oxysporum* cell walls (Lopez-Mondejar et al. 2011). Transformants of *T. virens* overexpressing specific glucanases were more effective in their *in vitro* inhibition of *P. ultimum* and *R. solani*, and the higher enzymatic activity of these strains also correlated with the enhanced protection of cotton seedlings against the same pathogens (Djonovic et al. 2007). Atanasova et al. (2013) reported that the specific mycoparasitic strategies can differ between *Trichoderma* species. The authors performed a comparative transcriptomic study of the hyphal response of strongly mycoparasitic *T. virens* and *T. atroviride* strains, just before physical contact with *R. solani*. While in *T. atroviride* genes encoding secondary metabolites as well as CWDEs were highly expressed, induced genes in the *T. virens* strain were mainly involved in the biosynthesis of gliotoxin. *Trichoderma* can indeed also produce a wide array of antifungal compounds to directly antagonize their rhizosphere competitors (reviewed by Hermosa et al. 2014). The commercially available strain *T. harzianum* T22 produces, for example, an azaphilone, which was shown to inhibit the growth of *P. ultimum*, *Gaeumannomyces graminis*, and *R. solani* (Vinale et al. 2006). Cardoza et al. (2007) partially silenced a gene encoding a key enzyme in the biosynthesis of terpene compounds in *T. harzianum* and demonstrated that the strain had reduced antifungal activity against *R. solani* and *F. oxysporum*. This finding indicates again the importance of such metabolites in the direct antagonistic activity of *Trichoderma*.

4.2 *Impact of* Trichoderma *on Root-Parasitic Nematodes and Root-Feeding Insects*

Many studies have shown the protective effect of *Trichoderma* against infection by root-parasitic nematodes in a range of monocot and dicot hosts, including economically important crops such as tomato (Sharon et al. 2007), potato (El-Shennawy et al. 2012), wheat (Zhang et al. 2014), bean (El-Nagdi and Abd-El-Khair 2014), and eggplant (Bokhari 2009). The majority of these studies focus on the most damaging parasitic nematodes, i.e., the root-knot nematodes *Meloidogyne* and the cyst nematodes *Heterodera* and *Globodera*. Additionally, a few studies have also demonstrated the potential of *Trichoderma* to protect plants against the migratory nematodes *Xiphinema index* (Darago et al. 2013) and *Pratylenchus penetrans* (Miller and Anagnostakis 1977) and the reniform nematode *Rotylenchulus reniformis* (Bokhari 2009). Several *Trichoderma* spp. including *T. harzianum*, *T. asperellum*, *T. longibrachiatum*, *T. viride*, and *T. atroviride* have shown to strongly reduce the population of nematodes in the rhizosphere by affecting egg hatching (Sahebani and Hadavi 2008; Szabo et al. 2013) and/or increasing second-stage juveniles mortality (Sharon et al. 2009; Zhang et al. 2015). Furthermore, *Trichoderma* can also affect nematode root penetration or slow down their further life-stage development in the host plant (Oyekanmi et al. 2007; Affokpon et al. 2011). Apart from the direct impact of *Trichoderma* on plant–nematode interactions, several studies further demonstrated the capability of *Trichoderma* to improve the performance of other biocontrol organisms such as nematode-trapping fungi (Szabo et al. 2012) or *Pseudomonas fluorescens* (Siddiqui and Shaukat 2004). The efficacy of *Trichoderma* to reduce nematode pressure seems to be influenced by the time of *Trichoderma* inoculation, i.e., before, during, or after nematode infection. In general, application of *Trichoderma* before planting or co-inoculation with the nematodes optimizes the plant protection, as a good preestablishment of the fungus in the rhizosphere seems to be important for nematode control (Dababat et al. 2006; Tariq Javeed and Al-Hazmi 2015).

In contrast to the well-known effect on nematodes, to our knowledge, the impact of *Trichoderma* on root-feeding insects has been considered in only a few systems, and the mechanistic basis for their interactions is often unclear. Indeed, despite *Trichoderma* and root herbivores sharing the same ecological niche, the vast majority of research in this area derives from the effects of *Trichoderma* on aboveground, rather than belowground, insect herbivores. This is surprising given that *Trichoderma* might affect root-feeding insects via both direct and indirect plant-mediated effects. In the studies performed by Razinger et al. (2014a, b), an increased mortality of the cabbage maggot (*Delia radicum*) due to the activity of different *Trichoderma* strains was observed in either *in vitro* or soil tests.

4.3 *Impact of* Trichoderma *on Other Beneficial Root Symbionts*

Beneficial organisms that share the rhizosphere can also be influenced by the presence of *Trichoderma*, including plant growth-promoting rhizobacteria and fungi (Bae and Knudsen 2005), nematode-trapping fungi (Maehara and Futai 2000), and even meso- and macrofauna (Maraun et al. 2003). Here we focus mainly on the impact of *Trichoderma* on the plant interaction with other root endophytic organisms that also establish intimate relationships within plant roots. Among them, the establishment of mycorrhizal and rhizobial symbiosis has been shown to be influenced by *Trichoderma*. This is not surprising given the strong impact of *Trichoderma* on the defence response of roots. However, more intriguing are the contrasting effects for the outcome of the interaction, ranging from positive to detrimental effects on the mycorrhizal or rhizobial symbiosis. For example, *T. harzianum* T78 increased the mycorrhization by AMF such as *R. irregularis*, but did not affect the mycorrhization by *Funneliformis mosseae* (formerly known as *Glomus mosseae*) (Martinez-Medina et al. 2009; Martinez-Medina et al. 2011a, b).

In a previous study, on the other hand, a different isolate of *T. harzianum* was found to decrease colonization of soybean roots by *F. mosseae*. The reduction of mycorrhizal colonization by *Trichoderma* has been attributed mainly to the induction of antimicrobial compounds in the plant roots and to its mycoparasitic activity (Wyss et al. 1992). For instance, by using *in vivo* imaging of green fluorescent protein-tagged lines, Lace et al. (2015) observed the strong ability of *T. atroviride* to parasitize *Gigaspora gigantea* and *Gigaspora margarita* hyphae through wall breaking and degradation. It seems that the outcome of the interaction between *Trichoderma* and AMF can be strongly influenced by both the partner's genotypes and the experimental setup.

Some *Trichoderma* species have been shown to affect the interaction between legumes and their rhizobial partners. Likewise, positive, negative, or neutral effects have been observed. For example, a promotion of the rhizobial symbiosis by different isolates of *Trichoderma* was observed in the field (Gupta et al. 2005; Rudresh et al. 2005; Saber et al. 2009). There is less information regarding the impact of *Trichoderma* on plant interaction with *Piriformospora indica*. To our knowledge, only one study performed by Anith et al. (2011) addressed the impact of *Trichoderma* on root colonization by this sebacinal fungus. In their study the authors found an inhibition *of P. indica* by *T. harzianum in vitro* and in root colonization of black pepper. However, inoculation of plants with *P. indica* and subsequently with *T. harzianum* resulted in higher root colonization by *P. indica* and synergistic beneficial effects on plant growth.

5 Conclusions

The great versatility of *Trichoderma* species, displaying a wide array of characteristics, explains why they are the most widespread fungi used for biocontrol purposes. While mycoparasitism is perceived as their ancestral lifestyle, the fact that they can colonize and protect plants remains highly promising and intriguing. The dialogue between a root-colonizing *Trichoderma* and its host plant comprises many stages, which we have tried to address in this chapter. Evidence is accumulating that plants in first instance also react to beneficial root colonizers as if they were potential pathogens. Successful root colonizers such as *Trichoderma* spp. have found a way to escape or suppress this defence response so that a mutualistic relationship can be established to the benefit of both organisms. The plant will be primed for a faster and stronger defence against pathogen attack and can also take advantage of the impact of *Trichoderma* on other microorganisms residing in the rhizosphere.

When compiling recent literature that addresses the different stages in the *Trichoderma*–plant interaction, it becomes apparent how little we actually know about the plant responses to *Trichoderma* taking place in the roots. This seems counterintuitive, since roots are the first points of contact of the plant with its colonizer. When we stop looking at plant roots as a black box, we can start uncovering the initial processes taking place in plant root–microbe interactions. We will not only gain a deeper insight into the plant–*Trichoderma* interaction, which might give us additional clues on the mechanisms of ISR, but also on the differences that separate a pathogen from a beneficial microbe. This knowledge can guide us to improve the use of our current biocontrol arsenal of *Trichoderma* strains and aid in the discovery of new ones for eventual agricultural applications.

Acknowledgments AMM gratefully acknowledges the support of the German Centre for Integrative Biodiversity Research (iDiv) Halle-Jena-Leipzig funded by the German Research Foundation (FZT 118). CV is funded by a Marie Curie International Outgoing Fellowship within the 7th European Community Framework Programme (PIOF-GA-2013-625551).

References

Affokpon A, Coyne DL, Htay CC et al (2011) Biocontrol potential of native *Trichoderma* isolates against root-knot nematodes in West African vegetable production systems. Soil Biol Biochem 43:600–608

Albus U, Baier R, Holst O et al (2001) Suppression of an elicitor-induced oxidative burst reaction in *Medicago sativa* cell cultures by *Sinorhizobium meliloti* lipopolysaccharides. New Phytol 151:597–606

Alfano G, Ivey MLL, Cakir C et al (2007) Systemic modulation of gene expression in tomato by *Trichoderma hamatum* T382. Phytopathology 97:429–437

Alizadeh H, Behboudi K, Agmadzadeh M et al (2013) Induced systemic resistance in cucumber and *Arabidopsis thaliana* by the combination of *Trichoderma harzianum* Tr6 and *Pseudomonas* sp. PS14. Biol Control 65:14–23

Alonso-Ramirez A, Poveda J, Martin I et al (2014) Salicylic acid prevents *Trichoderma harzianum* from entering the vascular system of roots. Mol Plant Pathol 15:823–831

Anith KN, Faseela KM, Archana PA et al (2011) Compatibility of *Piriformospora indica* and *Trichoderma harzianum* as dual inoculants in black pepper (*Piper nigrum* L.). Symbiosis 55:11–17

Atanasova L, Le Crom S, Gruber S et al (2013) Comparative transcriptomics reveals different strategies of *Trichoderma* mycoparasitism. BMC Genomics 14:121

Ausubel FM (2005) Are innate immune signaling pathways in plants and animals conserved? Nat Immunol 6:973–979

Avni A, Bailey BA, Mattoo AK et al (1994) Induction of ethylene biosynthesis in nicotiana-tabacum by a *Trichoderma viride* xylanase is correlated to the accumulation of 1-aminocyclopropane-1-carboxylic acid (acc) synthase and acc oxidase transcripts. Plant Phys 106:1049–1055

Aziz A, Gauthier A, Bezier A et al (2007) Elicitor and resistance-inducing activities of beta-1,4 cellodextrins in grapevine, comparison with beta-1,3 glucans and alpha-1,4 oligogalacturonides. J Exp Bot 58:1463–1472

Bae YS, Knudsen GR (2005) Soil microbial biomass influence on growth and biocontrol efficacy of *Trichoderma harzianum*. Biol Control 32:236–242

Balmer A, Pastor V, Gamir J et al (2015) The prime-ome: towards a holistic approach to priming. Trends Plant Sci 20:443–452

Benedetti M, Pontiggia D, Raggi S et al (2015) Plant immunity triggered by engineered in vivo release of oligogalacturonides, damage-associated molecular patterns. Proc Natl Acad Sci USA 112:5533–5538

Benitez T, Rincon AM, Limon MC et al (2004) Biocontrol mechanisms of *Trichoderma* strains. Int Microbiol 7:249–260

Bokhari FM (2009) Efficacy of some *Trichoderma* species in the control of *Rotylenchulus reniformis* and *Meloidogyne javanica*. Arch Phytopathol Plant Protect 42:361–369

Boller T, Felix G (2009) A renaissance of elicitors: perception of microbe-associated molecular patterns and danger signals by pattern-recognition receptors. Annu Rev Plant Biol 60:379–406

Boller T, He SY (2009) Innate immunity in plants: an arms race between pattern recognition receptors in plants and effectors in microbial pathogens. Science 324:742–744

Brotman Y, Briff E, Viterbo A et al (2008) Role of swollenin, an expansin-like protein from *Trichoderma*, in plant root colonization. Plant Phys 147:779–789

Brotman Y, Landau U, Cuadros-Inostroza A et al (2013) *Trichoderma*-plant root colonization: escaping early plant defense responses and activation of the antioxidant machinery for saline stress tolerance. PLoS Pathog 9:e1003221

Cardoza RE, Hermosa R, Vizcaino JA et al (2007) Partial silencing of a hydroxy-methylglutaryl-CoA reductase-encoding gene in *Trichoderma harzianum* CECT 2413 results in a lower level of resistance to lovastatin and lower antifungal activity. Fungal Genet Biol 44:269–283

Chacon MR, Rodriguez-Galan O, Benitez T et al (2007) Microscopic and transcriptome analyses of early colonization of tomato roots by *Trichoderma harzianum*. Int Microbiol 10:19–27

Chang PFL, Xu Y, Narasimhan ML et al (1998) Induction of pathogen resistance and pathogenesis-related genes in tobacco by a heat-stable *Trichoderma* mycelial extract and plant signal messengers. Physiol Plant 100:341–352

Chowdappa P, Kumar SPM, Lakshmi MJ et al (2013) Growth stimulation and induction of systemic resistance in tomato against early and late blight by *Bacillus subtilis* OTPB1 or *Trichoderma harzianum* OTPB3. Biol Control 65:109–117

Conrath U (2011) Molecular aspects of defense priming. Trends Plant Sci 16:524–531

Conrath U, Beckers GJ, Langenbach CJ et al (2015) Priming for enhanced defense. Annu Rev Phytopathol 53:97–119

Contreras-Cornejo HA, Macías-Rodríguez L, Beltrán-Peña E et al (2011) *Trichoderma*-induced plant immunity likely involves both hormonal-and camalexin-dependent mechanisms in *Arabidopsis thaliana* and confers resistance against necrotrophic fungi *Botrytis cinerea*. Plant Signal Behav 6:1554–1563

Contreras-Cornejo HA, Macías-Rodríguez L, Cortés-Penagos C et al (2009) *Trichoderma virens*, a plant beneficial fungus, enhances biomass production and promotes lateral root growth through an auxin-dependent mechanism in Arabidopsis. Plant Phys 149:1579–1592

Contreras-Cornejo HA, Lopez-Bucio JS, Mendez-Bravo A et al (2015) Mitogen-activated protein kinase 6 and ethylene and auxin signaling pathways are involved in arabidopsis root-system architecture alterations by *Trichoderma atroviride*. Mol Plant-Microbe Interact 28:701–710

Dababat AA, Sikora RA, Hauschild R (2006) Use of *Trichoderma harzianum* and *Trichoderma viride* for the biological control of *Meloidogyne incognita* on tomato. Commun Agric Appl Biol Sci 71:953–961

Darago A, Szabo M, Hracs K et al (2013) In vitro investigations on the biological control of *Xiphinema index* with *Trichoderma* species. Helminthologia 50:132–137

De Coninck B, Timmermans P, Vos C, Cammue B, Kazan K (2015) What lies beneath: below-ground defense strategies in plants. Trends Plant Sci 20:91–101

De Jaeger N, Declerck S, de la Providencia IE (2010) Mycoparasitism of arbuscular mycorrhizal fungi: a pathway for the entry of saprotrophic fungi into roots. FEMS Microbiol Ecol 73:312–322

de Lorenzo G, Brutus A, Savatin DV et al (2011) Engineering plant resistance by constructing chimeric receptors that recognize damage-associated molecular patterns (DAMPs). FEBS Lett 585:1521–1528

de Roman M, Fernandez I, Wyatt T et al (2011) Elicitation of foliar resistance mechanisms transiently impairs root association with arbuscular mycorrhizal fungi. J Ecol 99:36–45

Dempsey DA, Klessig DF (2012) SOS—too many signals for systemic acquired resistance? Trends Plant Sci 17:538–545
Djonovic S, Pozo MJ, Dangott LJ et al (2006) Sm1, a proteinaceous elicitor secreted by the biocontrol fungus *Trichoderma virens* induces plant defense responses and systemic resistance. Mol Plant Microbe Interact 19:838–853
Djonovic S, Vittone G, Mendoza-Herrera A et al (2007) Enhanced biocontrol activity of *Trichoderma virens* transformants constitutively coexpressing β-1,3- and β-1,6-glucanase genes. Mol Plant Pathol 8:469–480
Do Vale LHF, Gomez-Mendoza DP, Kim M-S et al (2012) Secretome analysis of the fungus *Trichoderma harzianum* grown on cellulose. Proteomics 12:2716–2728
Druzhinina IS, Seidl-Seiboth V, Herrera-Estrella A et al (2011) *Trichoderma*: the genomics of opportunistic success. Nat Rev Microbiol 9:749–759
El-Nagdi WMA, Abd-El-Khair H (2014) Biological control of *Meloidogyne incognita* and *Fusarium solani* in dry common bean in the field. Arch Phytopathol Plant Protect 47:388–397
El-Shennawy MZ, Khalifa EZ, Ammar MM et al (2012) Biological control of the disease complex on potato caused by root-knot nematode and Fusarium wilt fungus. Nematol Mediterr 40:169–172
Erbs G, Newman MA (2003) The role of lipopolysaccharides in induction of plant defence responses. Mol Plant Pathol 4:421–425
Fauth M, Schweizer A, Buchala C et al (1998) Cutin monomers and surface wax constituents elicit H_2O_2 in conditioned cucumber hypocotyl segments and enhance the activity of other H_2O_2 elicitors. Plant Physiol 117:1373–1380
Ferrari S, Plotnikova JM, De Lorenzo G et al (2003) Arabidopsis local resistance to *Botrytis cinerea* involves salicylic acid and camalexin and requires EDS4 and PAD2, but not SID2, EDS5 or PAD4. Plant J 35:193–205
Fraser CM, Chapple C (2011) The phenylpropanoid pathway in Arabidopsis. Arabidopsis Book 9: e0152
Gaderer R, Lamdan NL, Frischmann A et al (2015) Sm2, a paralog of the *Trichoderma* cerato-platanin elicitor Sm1, is also highly important for plant protection conferred by the fungal-root interaction of *Trichoderma* with maize. BMC Microbiol 15:2
Gallou A, Cranenbrouck S, Declerck S (2009) *Trichoderma harzianum* elicits defense response genes in roots of potato plantlets challenged by *Rhizoctonia solani*. Eur J Plant Pathol 124:219–230
Garnica-Vergara A, Barrera-Ortiz S, Munoz-Parra E et al (2015) The volatile 6-pentyl-2H-pyran-2-one from *Trichoderma atroviride* regulates *Arabidopsis thaliana* root morphogenesis via auxin signaling and ETHYLENE INSENSITIVE 2 functioning. New Phytol 209:1496–1512
Gimenez-Ibanez S, Rathjen JP (2010) The case for the defense: plants versus *Pseudomonas syringae*. Microbes Infect 12:428–437
Gomes EV, Costa MN, de Paula RG et al (2015) The Cerato-Platanin protein Epl-1 from *Trichoderma harzianum* is involved in mycoparasitism, plant resistance induction and self cell wall protection. Sci Rep 5:17998
Gomez-Gomez L, Boller T (2002) Flagellin perception: a paradigm for innate immunity. Trends Plant Sci 7:251–256
Gruber S, Vaaje-Kolstad G, Matarese F et al (2011) Analysis of subgroup C of fungal chitinases containing chitin-binding and LysM modules in the mycoparasite *Trichoderma atroviride*. Glycobiology 21:122–133
Gupta SB, Thakur KS, Singh A et al (2005) Exploiting *Trichoderma viride* for improving performance of chickpea in wilt complex affected area of Chhattisgarh. Adv Plant Sci 18:609–614
Gupta KJ, Mur LAJ, Brotman Y (2014) *Trichoderma asperelloides* suppresses nitric oxide generation elicited by *Fusarium oxysporum* in *Arabidopsis* roots. Mol Plant Microbe Interact 27:307–314

Gust AA, Willmann R, Desaki Y et al (2012) Plant LysM proteins: modules mediating symbiosis and immunity. Trends Plant Sci 17:495–502

Gutjahr C, Paszkowski U (2013) Multiple control levels of root system remodeling in arbuscular mycorrhizal symbiosis. Front Plant Sci 4:204

Harman GE, Howell CR, Viterbo A et al (2004) *Trichoderma* species—opportunistic, avirulent plant symbionts. Nat Rev Microbiol 2:43–56

Heller G, Adomas A, Li G et al (2008) Transcriptional analysis of *Pinus sylvestris* roots challenged with the ectomycorrhizal fungus *Laccaria bicolor*. BMC Plant Biol 8:19

Hermosa R, Belen Rubio M, Cardoza RE et al (2013) The contribution of *Trichoderma* to balancing the costs of plant growth and defense. Int Microbiol 16:69–80

Hermosa R, Cardoza RE, Belen Rubio M et al (2014) Secondary metabolism and antimicrobial metabolites of *Trichoderma*. In: Gupta VK et al (eds) Biotechnology and biology of *Trichoderma*. Elsevier, Amsterdam, pp 125–137

Howell CR, Hanson LE, Stipanovic RD et al (2000) Induction of terpenoid synthesis in cotton roots and control of *Rhizoctonia solani* by seed treatment with *Trichoderma virens*. Phytopathology 90:248–252

Hung R, Lee S, Bennett JW (2013) *Arabidopsis thaliana* as a model system for testing the effect of *Trichoderma* volatile organic compounds. Fungal Ecol 6:19–26

Jones JDG, Dangl JL (2006) The plant immune system. Nature 444:323–329

Jung SC, Martinez-Medina A, Lopez-Raez JA et al (2012) Mycorrhiza-induced resistance and priming of plant defenses. J Chem Ecol 38:651–664

Khan J, Ooka JJ, Miller SA, Madden LV, Hoitink HAJ (2004) Systemic resistance induced by *Trichoderma hamatum* 382 in cucumber against *Phytophthora* crown rot and leaf blight. Plant Dis 88:280–286

Kloppholz S, Kuhn H, Requena N (2011) A secreted fungal effector of *Glomus intraradices* promotes symbiotic biotrophy. Curr Biol 21:1204–1209

Kubicek CP, Herrera-Estrella A, Seidl-Seiboth V et al (2011) Comparative genome sequence analysis underscores mycoparasitism as the ancestral life style of *Trichoderma*. Genome Biol 12:R40

Lace B, Genre A, Woo S et al (2015) Gate crashing arbuscular mycorrhizas: in vivo imaging shows the extensive colonization of both symbionts by *Trichoderma atroviride*. Environ Microbiol Rep 7:64–77

Liu J, Blaylock LA, Endre G et al (2003) Transcript profiling coupled with spatial expression analyses reveals genes involved in distinct developmental stages of an arbuscular mycorrhizal symbiosis. Plant Cell 15:2106–2123

Lopez-Mondejar R, Ros M, Pascual JA (2011) Mycoparasitism-related genes expression of *Trichoderma harzianum* isolates to evaluate their efficacy as biological control agent. Biol Control 56:59–66

Lorito M, Woo SL, Harman GE et al (2010) Translational research on *Trichoderma*: from 'Omics to the Field'. Annu Rev Phytopathol 48:395–417

Luo Y, Zhang D-D, Dong X-W et al (2010) Antimicrobial peptaibols induce defense responses and systemic resistance in tobacco against tobacco mosaic virus. FEMS Microbiol Lett 313:120–126

Maehara N, Futai K (2000) Population changes of the pinewood nematode, *Bursaphelenchus xylophilus* (Nematoda: Aphelenchoididae), on fungi growing in pine-branch segments. Appl Entomol Zool 35:413–417

Maraun M, Martens H, Migge S et al (2003) Adding to 'the enigma of soil animal diversity': fungal feeders and saprophagous soil invertebrates prefer similar food substrates. Eur J Soil Biol 39:85–95

Marra R, Ambrosino P, Carbone V et al (2006) Study of the three-way interaction between *Trichoderma atroviride*, plant and fungal pathogens by using a proteomic approach. Curr Genet 50:307–321

Martinez C, Blanc F, Le Claire E et al (2001) Salicylic acid and ethylene pathways are differentially activated in melon cotyledons by active or heat-denatured cellulase from *Trichoderma longibrachiatum*. Plant Phys 127:334–344
Martinez-Medina A, Pascual JA, Lloret E et al (2009) Interactions between arbuscular mycorrhizal fungi and *Trichoderma harzianum* and their effects on Fusarium wilt in melon plants grown in seedling nurseries. J Sci Food Agricult 89:1843–1850
Martinez-Medina A, Pascual JA, Perez-Alfocea F et al (2010) *Trichoderma harzianum* and *Glomus intraradices* modify the hormone disruption induced by *Fusarium oxysporum* infection in melon plants. Phytopathology 100:682–688
Martinez-Medina A, Roldan A, Albacete A et al (2011a) The interaction with arbuscular mycorrhizal fungi or *Trichoderma harzianum* alters the shoot hormonal profile in melon plants. Phytochemistry 72:223–229
Martinez-Medina A, Roldan A, Pascual JA (2011b) Interaction between arbuscular mycorrhizal fungi and *Trichoderma harzianum* under conventional and low input fertilization field condition in melon crops: growth response and fusarium wilt biocontrol. Appl Soil Ecol 47:98–105
Martinez-Medina A, Fernandez I, Sanchez-Guzman MJ et al (2013) Deciphering the hormonal signaling network behind the systemic resistance induced by *Trichoderma harzianum* in tomato. Front Plant Sci 4:206
Martinez-Medina A, Alguacil MDM, Pascual JA et al (2014) Phytohormone profiles induced by *Trichoderma* isolates correspond with their biocontrol and plant growth-promoting activity on melon plants. J Chem Ecol 40:804–815
Mathys J, De Cremer K, Timmermans P et al (2012) Genome-wide characterization of ISR induced in *Arabidopsis thaliana* by *Trichoderma hamatum* T382 against *Botrytis cinerea* infection. Front Plant Sci 3:108
Maunoury N, Redondo-Nieto M, Bourcy M et al (2010) Differentiation of symbiotic cells and endosymbionts in *Medicago truncatula* nodulation are coupled to two transcriptome-switches. PLoS One 5:e9519
Miller PM, Anagnostakis S (1977) Suppression of *Pratylenchus penetrans* and *Tylenchorhynchus dubius* by *Trichoderma viride*. J Nematol 9:182–183
Montesano M, Brader G, Palva ET (2003) Pathogen derived elicitors: searching for receptors in plants. Mol Plant Pathol 4:73–79
Moran-Diez E, Hermosa R, Ambrosino P et al (2009) The ThPG1 endopolygalacturonase is required for the *Trichoderma harzianum*-plant beneficial interaction. Mol Plant Microbe Interact 22:1021–1031
Moran-Diez E, Rubio B, Dominguez S et al (2012) Transcriptomic response of *Arabidopsis thaliana* after 24 h incubation with the biocontrol fungus *Trichoderma harzianum*. J Plant Physiol 169:614–620
Moran-Diez ME, Trushina N, Lamdan NL et al (2015) Host-specific transcriptomic pattern of *Trichoderma virens* during interaction with maize or tomato roots. BMC Genomics 16:15
Morant M, Bak S, Moller BL et al (2003) Plant cytochromes P450: tools for pharmacology, plant protection and phytoremediation. Curr Opin Biotechnol 14:151–162
Moreno CA, Castillo F, Gonzalez A et al (2009) Biological and molecular characterization of the response of tomato plants treated with *Trichoderma koningiopsis*. Physiol Mol Plant Pathol 74:111–120
Mortier V, Holsters M, Goormachtig S (2012) Never too many? How legumes control nodule numbers. Plant Cell Environ 35:245–258
Mukherjee PK, Horwitz BA, Herrera-Estrella A et al (2013) *Trichoderma* research in the genome era. Annu Rev Phytopathol 15:105–129
Olah B, Briere C, Becard G et al (2005) Nod factors and a diffusible factor from arbuscular mycorrhizal fungi stimulate lateral root formation in *Medicago truncatula* via the DMI1/DMI2 signaling pathway. Plant J 44:195–207
Oyekanmi EO, Coyne DL, Fagade OE et al (2007) Improving root-knot nematode management on two soybean genotypes through the application of *Bradyrhizobium japonicum*, *Trichoderma*

pseudokoningii and *Glomus mosseae* in full factorial combinations. Crop Protect 26:1006–1012
Penninckx IA, Thomma BP, Buchala A et al (1998) Concomitant activation of jasmonate and ethylene response pathways is required for induction of a plant defensin gene in Arabidopsis. Plant Cell 10:2103–2113
Perazzolli M, Moretto M, Fontana P et al (2012) Downy mildew resistance induced by *Trichoderma harzianum* T39 in susceptible grapevines partially mimics transcriptional changes of resistant genotypes. BMC Genomics 13:660
Pieterse CMJ, Leon-Reyes A, Van der Ent S et al (2009) Networking by small-molecule hormones in plant immunity. Nat Chem Biol 5:308–316
Pieterse CMJ, Zamioudis C, Berendsen RL et al (2014) Induced systemic resistance by beneficial microbes. Annu Rev Phytopathol 52:347–375
Razinger J, Lutz M, Schroers HJ et al (2014a) Direct plantlet inoculation with soil or insect-associated fungi may control cabbage root fly maggots. J Invertebr Pathol 120:59–66
Razinger J, Lutz M, Schroers HJ et al (2014b) Evaluation of insect associated and plant growth promoting fungi in the control of cabbage root flies. J Econ Entomol 107:1348–1354
Rudresh DL, Shivaprakash MK, Prasad RD (2005) Effect of combined application of *Rhizobium*, phosphate solubilizing bacterium and *Trichoderma* spp. on growth, nutrient uptake and yield of chickpea (*Cicer aritenium* L.). Appl Soil Ecol 28:139–146
Saber WIA, El-Hai KMA, Ghoneem KM (2009) Synergistic effect of *Trichoderma* and *Rhizobium* on both biocontrol of chocolate spot disease and induction of nodulation, physiological activities and productivity of *Vicia faba*. Res J Microbiol 4:286–300
Sahebani N, Hadavi N (2008) Biological control of the root-knot nematode *Meloidogyne javanica* by *Trichoderma harzianum*. Soil Biol Biochem 40:2016–2020
Salas-Marina MA, Silva-Flores MA, Uresti-Rivera EE et al (2011) Colonization of *Arabidopsis* roots by *Trichoderma atroviride* promotes growth and enhances systemic disease resistance through jasmonic acid/ethylene and salicylic acid pathways. Eur J Plant Pathol 131:15–26
Samolski I, Rincon AM, Mary Pinzon L et al (2012) The *qid74* gene from *Trichoderma harzianum* has a role in root architecture and plant biofertilization. Microbiology 158:129–138
Segarra G, Van der Ent S, Trillas I, Pieterse CMJ (2009) MYB72, a node of convergence in induced systemic resistance triggered by a fungal and a bacterial beneficial microbe. Plant Biol 11:90–96
Seidl V, Song L, Lindquist E et al (2009) Transcriptomic response of the mycoparasitic fungus *Trichoderma atroviride* to the presence of a fungal prey. BMC Genomics 10:567
Seidl-Seiboth V, Zach S, Frischmann A et al (2013) Spore germination of *Trichoderma atroviride* is inhibited by its LysM protein TAL6. FEBS J 280:1226–1236
Selim ME, Mahdy ME, Sorial ME et al (2014) Biological and chemical dependent systemic resistance and their significance for the control of root-knot nematodes. Nematology 8:1–11
Sharon E, Chet I, Viterbo A et al (2007) Parasitism of *Trichoderma* on *Meloidogyne javanica* and role of the gelatinous matrix. Eur J Plant Pathol 118:247–258
Sharon E, Chet I, Spiegel Y (2009) Improved attachment and parasitism of *Trichoderma* on *Meloidogyne javanica* in vitro. Eur J Plant Pathol 123:291–299
Shoresh M, Harman GE, Mastouri F (2010) Induced systemic resistance and plant responses to fungal biocontrol agents. Annu Rev Phytopathol 48:21–43
Siddiqui IA, Shaukat SS (2004) Trichoderma harzianum enhances the production of nematicidal compounds in vitro and improves biocontrol of *Meloidogyne javanica* by *Pseudomonas fluorescens* in tomato. Lett Appl Microbiol 38:169–175
Sofo A, Scopa A, Manfra M et al (2011) *Trichoderma harzianum* strain T-22 induces changes in phytohormone levels in cherry rootstocks (*Prunus cerasus* × *P. canescens*). Plant Growth Regul 65:421–425
Soto MJ, Dominguez-Ferreras A, Perez-Mendoza D et al (2009) Mutualism versus pathogenesis: the give-and-take in plant-bacteria interactions. Cell Microbiol 11:381–388

Spaink HP (2000) Root nodulation and infection factors produced by rhizobial bacteria. Annu Rev Microbiol 54:257–288
Spoel SH, Dong XN (2012) How do plants achieve immunity? Defence without specialized immune cells. Nat Rev Immunol 12:89–100
Stacey G, McAlvin CB, Kim S-Y et al (2006) Effects of endogenous salicylic acid on nodulation in the model legumes *Lotus japonicus* and *Medicago truncatula*. Plant Phys 141:1473–1481
Szabo M, Csepregi K, Galber M et al (2012) Control plant-parasitic nematodes with *Trichoderma* species and nematode-trapping fungi: the role of chi18-5 and chi18-12 genes in nematode egg-parasitism. Biol Control 63:121–128
Szabo M, Urban P, Viranyi F et al (2013) Comparative gene expression profiles of *Trichoderma harzianum* proteases during in vitro nematode egg-parasitism. Biol Control 67:337–343
Tariq Javeed M, Al-Hazmi AS (2015) Effect of *Trichoderma harzianum* on *Meloidogyne javanica* in tomatoes as influenced by time of the fungus introduction into soil. J Pure Appl Microbiol 9:535–539
Tellstroem V, Usadel B, Thimm O et al (2007) The lipopolysaccharide of *Sinorhizobium meliloti* suppresses defense-associated gene expression in cell cultures of the host plant *Medicago truncatula*. Plant Phys 143:825–837
van den Burg HA, Harrison SJ, Joosten MHAJ et al (2006) *Cladosporium fulvum* Avr4 protects fungal cell walls against hydrolysis by plant chitinases accumulating during infection. Mol Plant Microbe Interact 19:1420–1430
van der Heijden MGA, Klironomos JN, Ursic M et al (1998) Mycorrhizal fungal diversity determines plant biodiversity, ecosystem variability and productivity. Nature 396:69–72
Van Loon LC, Bakker PA, Pieterse CMJ (1998) Systemic resistance induced by rhizosphere bacteria. Annu Rev Phytopathol 36:453–483
van Spronsen PC, Tak T, Rood AMM et al (2003) Salicylic acid inhibits indeterminate-type nodulation but not determinate-type nodulation. Mol Plant Microbe Interact 16:83–91
Velazquez-Robledo R, Contreras-Cornejo HA, Macias-Rodriguez L et al (2011) Role of the 4-phosphopantetheinyl transferase of *Trichoderma virens* in secondary metabolism and induction of plant defense responses. Mol Plant Microbe Interact 24:1459–1471
Vierheilig H, Steinkellner S, Khaosaad T et al (2008) The biocontrol effect of mycorrhization on soilborne fungal pathogens and the autoregulation of the AM symbiosis: one mechanism, two effects? In: Varma A (ed) Mycorrhiza: state of the art, genetics and molecular biology, eco-function, biotechnology, eco-physiology, structure and systematics. Springer, Berlin, pp 307–320
Vinale F, Marra R, Scala F et al (2006) Major secondary metabolites produced by two commercial *Trichoderma* strains active against different phytopathogens. Lett Appl Microbiol 43:143–148
Viterbo A, Harel M, Chet I (2004) Isolation of two aspartyl proteases from *Trichoderma asperellum* expressed during colonization of cucumber roots. FEMS Microbiol Lett 238:151–158
Viterbo A, Wiest A, Brotman Y et al (2007) The 18mer peptaibols from *Trichoderma virens* elicit plant defence responses. Mol Plant Pathol 8:737–746
Viterbo A, Landau U, Kim S et al (2010) Characterization of ACC deaminase from the biocontrol and plant growth-promoting agent *Trichoderma asperellum* T203. FEMS Microbiol Lett 305:42–48
Vos C, Yang Y, De Coninck B, Cammue BPA (2014) Fungal (-like) biocontrol organisms in tomato disease control. Biol Control 74:64–81
Vos CM, De Cremer K, Cammue B et al (2015) The toolbox of *Trichoderma* spp. in the biocontrol of *Botrytis cinerea* disease. Mol Plant Pathol 16:400–412
Woo SL, Scala F, Ruocco M et al (2006) The molecular biology of the interactions between *Trichoderma* spp., phytopathogenic fungi, and plants. Phytopathology 96:181–185
Wyss P, Boller T, Wiemken A (1992) Testing the effect of biological-control agents on the formation of vesicular arbuscular mycorrhiza. Plant Soil 147:159–162

Yano A, Suzuki K, Uchimiya H et al (1998) Induction of hypersensitive cell death by a fungal protein in cultures of tobacco cells. Mol Plant-Microbe Interact 11:115–123

Yedidia I, Benhamou N, Chet I (1999) Induction of defense responses in cucumber plants (*Cucumis sativus* L.) by the biocontrol agent *Trichoderma harzianum*. Appl Environ Microbiol 65:1061–1070

Yedidia I, Benhamou N, Kapulnik Y et al (2000) Induction and accumulation of PR proteins activity during early stages of root colonization by the mycoparasite *Trichoderma harzianum* strain T-203. Plant Physiol Biochem 38:863–873

Zamioudis C, Pieterse CMJ (2012) Modulation of host immunity by beneficial microbes. Mol Plant Microbe Interact 25:139–150

Zhang S, Gan Y, Xu B et al (2014) The parasitic and lethal effects of *Trichoderma longibrachiatum* against *Heterodera avenae*. Biol Control 72:1–8

Zhang S, Gan Y, Xu B (2015) Biocontrol potential of a native species of *Trichoderma longibrachiatum* against *Meloidogyne incognita*. Appl Soil Ecol 94:21–29

Defence Reactions in Roots Elicited by Endofungal Bacteria of the Sebacinalean Symbiosis

Ibrahim Alabid and Karl-Heinz Kogel

Abstract The Alphaproteobacterium Rhizobium radiobacter F4 (*Rr*F4) was originally detected as an endofungal bacterium associated with the endophytic basidiomycete Piriformospora indica that forms a beneficial symbiosis with a wide range of green plants. While attempts to cure *P. indica* from *Rr*F4 repeatedly failed, the bacterium could be isolated and grown in pure culture. In contrast to some other endofungal bacteria, the genome size of *Rr*F4 is not reduced. Instead, it shows a high degree of similarity to the plant pathogenic R. *radiobacter* (formerly: *Agrobacterium tumefaciens*) C58, except vibrant differences in both the tumor-inducing (pTi) and the accessor (pAt) plasmids, which can explain the loss of *Rr*F4's pathogenicity. Similar to its fungal host, *Rr*F4 colonizes plant roots without host preference and forms aggregates of attached cells and dense biofilms at the root surface of maturation zones. *Rr*F4-colonized plants show increased biomass and enhanced resistance against bacterial and fungal leaf pathogens. Resistance mediated by *Rr*F4 is dependent on the plant's jasmonate-based induced systemic resistance (ISR) pathway while the systemic acquired resistance (SAR) pathway is non-operative as shown by genetic analysis. Based on these findings we concluded that *Rr*F4- and *P. indica*-induced pattern of defence gene expression are similar. However, in clear contrast to *P. indica*, but similar to plant growth promoting rhizobacteria (PGPR), *Rr*F4 colonized not only the root outer cortex but spread beyond the endodermis into the stele. Based on our findings *Rr*F4 is an efficient plant growth promoting bacterium.

1 The Sebacinalean Symbiosis

Land plant strategies to protect themselves from invading pathogens and abiotic stress include establishing beneficial associations with soilborne microbes such as plant growth-promoting rhizobacteria and fungi from various taxa (Zamioudis and

I. Alabid • K.-H. Kogel (✉)
Institute of Phytopathology, Justus Liebig University Giessen, Giessen, Germany
e-mail: Karl-Heinz.Kogel@agrar.uni-giessen.de

C.M.F. Vos, K. Kazan (eds.), *Belowground Defence Strategies in Plants*, Signaling and Communication in Plants, DOI 10.1007/978-3-319-42319-7_14

Pieterse 2012). A wide range of monocotyledonous and dicotyledonous plants are associated with higher fungi of the order *Sebacinales* (*Basidiomycota*) to form the sebacinalean symbioses (Selosse et al. 2009; Weiss et al. 2011; Riess et al. 2014). The root endophyte *Piriformospora indica* is a representative fungus of the *Sebacinales*; it was discovered in the Indian Thar Desert in 1996 (Varma et al. 1999) and since then studied in many laboratories (Peškan-Berghöfer et al. 2004; Waller et al. 2005; Pedrotti et al. 2013; for review see Qiang et al. 2012). Unlike true endomycorrhizal fungi, *P. indica* is not an obligate biotroph and thus can be cultured without host plants on synthetic media containing complex and minimal substrates (Deshmukh et al. 2006; Oelmüller et al. 2009; Lahrmann et al. 2013). The fungus asexually produces pear-shaped chlamydospores with 8–25 nuclei (Verma et al. 1998). Having a surprisingly wide range of host plants, *P. indica* and related *Sebacina vermifera* species support plant biomass, along with local and systemic resistance to a variety of microbial pathogens (Waller et al. 2005; Deshmukh et al. 2006; Stein et al. 2008; Glaeser et al. 2016). Several studies also have demonstrated that *P. indica* increases yield as judged from agronomic parameters (Peškan-Berghöfer et al. 2004; Waller et al. 2005; Achatz et al. 2010; Fakhro et al. 2010). *P. indica* has been studied intensively in barley (*Hordeum vulgare*) and *Arabidopsis* (*Arabidopsis thaliana*) although nonhost plants have not been identified yet.

P. indica improves the nutritional status of plants such as tobacco, where around 50 % increase in NADH-dependent nitrate reductase (NR) activity was observed in the roots colonized by the endophyte (Sherameti et al. 2005). Its growth-promoting effect on maize was dependent on a fungal phosphate transporter (*PiPT*) that probably mediates phosphate transport to the host plant (Yadav et al. 2010). Plant growth also could be induced by hyphal wall fragments, suggesting the involvement of receptors at the plant cell surface. Consistent with this, a fungal cell wall extract induced a rapid increase in the root cells' intracellular calcium concentration, suggesting signaling and reprogramming of host cells early in the root colonization (Vadassery et al. 2009).

2 *Piriformospora indica* Elicits Systemic Resistance to Microbial Pathogens

Upon root colonization, *P. indica* induces systemic resistance in leaves of barley and *Arabidopsis* against the respective appropriate powdery mildew fungi *Blumeria graminis* f. sp. *hordei* and *Golovinomyces orontii* (Waller et al. 2005; Stein et al. 2008). In *Arabidopsis*, resistance requires jasmonic acid (JA) signaling and the cytoplasmic activity of non-expressor of pathogenesis-related 1 (NPR1). Such a requirement is indicative of the canonical induced systemic resistance (ISR) defence pathway. Interestingly, operable JA signaling and biosynthesis not only is required for *P. indica*'s potential for enhancing the plant's immune status but also is essential for successful root colonization. Thus, it has been argued that *P. indica* exploits the antagonistic action of the plant defence hormones JA and salicylic acid (SA) to suppress at least part of the SA-mediated plant response as a prerequisite for successful root colonization (Jacobs et al. 2011).

3 Endofungal Bacteria in the Sebacinalean Symbioses

Sharma and coworkers showed that members of the *Sebacinales* fungi regularly undergo complex tripartite interactions involving plants and bacteria of different genera (Sharma et al. 2008). Endofungal bacteria identified in association with the genera *Piriformospora* and *Sebacina* belong to two genera of Gram-negative (*Rhizobium* and *Acinetobacter*) as well as two genera of Gram-positive (*Paenibacillus* and *Rhodococcus*) bacteria. The most comprehensively studied example of a tripartite sebacinalean symbiosis is the association of *P. indica* with the Alphaproteobacterium *Rhizobium radiobacter* *Rr*F4 (syn. *Agrobacterium tumefaciens*). Fluorescence in situ hybridization (FISH) using a *Rhizobium*-specific probe confirmed the endocellular association of *Rr*F4 with fungal hyphae and chlamydospores. *Rr*F4 could be isolated from *P. indica* and propagated in axenic cultures, demonstrating that the bacterium is not entirely dependent on its fungal host. However, attempts to cure *P. indica* of its bacterial associate have failed (Sharma et al. 2008). Antibiotics, which killed or inhibited bacterial growth with high efficacy in in vitro axenic cultures, were virtually ineffective in stably removing the bacteria from the fungus, suggesting an intricate association, and possibly a critical role of *Rr*F4 in fungal survival. Sharma and allies also could show that *Rr*F4 contains a *virD2* gene indicating that a Ti plasmid is present; however, the *isopentenyltransferase (IPT)* gene, which is associated with cytokinin biosynthesis, could not be detected, which explained the nonpathogenic nature of the bacterium. A more detailed comparative analysis of *Rr*F4's genome showed a high degree of similarity to the plant-pathogenic *R. radiobacter* C58 (formerly: *Agrobacterium tumefaciens* C58), except clear differences in both the tumor-inducing (pTi) and the accessor (pAt) plasmids, which can explain the loss of *Rr*F4's pathogenicity (Glaeser et al. 2016).

4 Beneficial Activity of *Rhizobium radiobacter* *Rr*F4

Intriguingly, when roots are inoculated with *Rr*F4, plants reach higher biomasses and develop systemic resistance to microbial pathogens, reminding of *P. indica*'s activity. All plant species tested, including *Arabidopsis* and barley, had increased shoot and root fresh weights in various growth substrates when their roots were initially dip-inoculated with the bacterium. Quantitative PCR analyses showed increased amounts of *Rr*F4 cells over time of infection, demonstrating that the

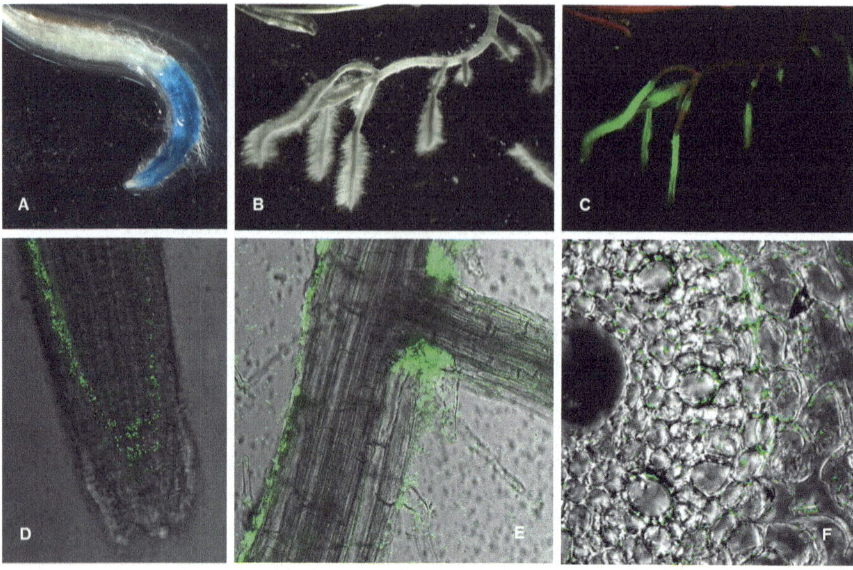

Fig. 1 Colonization of barley and *Arabidopsis* roots by GUS- and GFP-expressing *Rhizobium radiobacter Rr*F4. (**a**) Barley root segment infected with GUS-expressing bacteria at 5 dpi; the root hair zone shows a GUS positive stain. (**b**, **c**) *Arabidopsis* roots uninfected (**b**) and infected (**c**) with GFP-expressing bacteria at 30 dpi. (**d**) *Arabidopsis* root hair zone colonized by GFP-expressing bacteria at 7 dpi with single attached bacteria and aggregates. (**e**) *Rr*F4 forms biofilms and aggregates at the surface of *Arabidopsis* primary root, mostly at the sites of lateral root protrusion at 21 dpi. (**f**) Cross section of barley roots showing colonization of the central cylinder by GFP-tagged bacteria at 21 dpi cells. Images (**b–f**) were taken by confocal laser scanning microscope

bacterium is capable to multiply outside its fungal host in association with roots (Glaeser et al. 2016). Bacteria tagged with beta-glucuronidase (GUS) or green fluorescence protein (GFP) could be visualized to show typical accumulation patterns in root hair zones of primary and lateral roots, mostly at the sites of lateral root protrusion (Fig. 1). Importantly, the cross section of barley roots confirmed colonization of the central cylinder by bacteria at 21 days after inoculation (dpi). Scanning electron microscopy additionally showed that the bacterium forms biofilms and aggregates on the rhizoplane (Fig. 2).

5 *Rhizobium radiobacter Rr*F4 Mediates Disease Resistance via the Induced Systemic Resistance Pathway

*Rr*F4-colonized *Arabidopsis* developed resistance to the plant-pathogenic bacterium *Pseudomonas syringae* pv. *tomato* DC3000 (*Pst*). Similarly, *Rr*F4-treated wheat was protected against the leaf streak disease caused by the bacterium

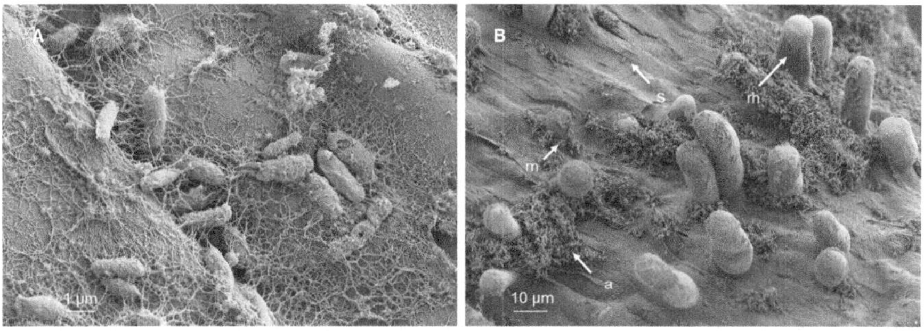

Fig. 2 Colonization of barley primary roots by *Rhizobium radiobacter* *Rr*F4 analyzed by scanning electron microscopy. (**a**) Single bacteria attached to the root surface distal to the tip area. (**b**) Different stages in biofilm formation at the rhizoplane of the root hair zone: single bacteria attached to the rhizoplane (s); microcolonies formed through multiplication of single attached cells (m); larger cell aggregates (a); root hairs (rh) (the figure was taken from Dr. Martin Hardt, Biomedical Research Centre Seltersberg, Justus Liebig University Giessen, Germany)

Xanthomonas translucens pv. *translucens* (*Xtt*; Glaeser et al. 2016). These results show that *Rr*F4, like its host *P. indica*, induces root-initiated systemic resistance against leaf-associated microbial pathogens. In an attempt to define the mechanism of the plants' higher immune status, *Arabidopsis* mutants indicative of both the ISR and the systemic acquired resistance (SAR) pathways were assessed. To this end, *Rr*F4-treated *NahG* plants that overexpress *SA-degrading salicylate hydroxylase* and the mutants *npr1-3* and *ethylene-insensitive 1* (*ein2-1*) displayed systemic resistance against *Pst*, similar to wild-type plants. In contrast, the mutants *jasmonate-resistant1-1* (*jar1-1*), *jasmonate-insensitive1-1* (*jin1-1*), and *npr1-1*, all of which are indicative of the JA pathway, were fully compromised for *Rr*F4-mediated resistance to *Pst*. The data show that JA signaling is required for *Rr*F4-mediated resistance in *Arabidopsis,* while the SA pathway, along with the nuclear localization of NPR1, which is abolished in *npr1-3*, is not. Consistent with this, the JA marker genes *Vegetative storage protein 2* (*VSP2*) and *Plant defensin 1.2* (*PDF1.2*) were strongly induced in *Pst*-infected plants when they were pretreated with *Rr*F4 compared to non-pretreated plants at 24 and 48 h post inoculation (hpi). In contrast, induction of SA-regulated *Pathogenesis-related 1* (*PR1*) and the ethylene (ET)-regulated gene *Ethylene receptor factor 1* (*ERF1*) was not detected in these plants. The data show that *Rr*F4 mediates systemic resistance via the same mechanism as its fungal host *P. indica*, thereby raising the yet unresolved question whether the beneficial activity in the sebacinalean symbiosis may stem at least partly from the bacterium. The data also are consistent with the observation that ISR is commonly accompanied by only weak systemic up- or downregulation of defence genes before challenge inoculation, a phenomenon termed "priming" (Van Wees et al. 1999; Conrath et al. 2006). Priming does not require major metabolic changes in the absence of a challenging pathogen or pest. However, due to previous priming by inducing biotic or abiotic agents, plants can more efficiently activate cellular defence in response to a subsequent challenge inoculation (Conrath et al. 2006). Overall the defence mechanisms mediated by endofungal *Rr*F4 further support the view that rhizobacteria-mediated ISR is dependent on JA and independent of SA signaling (Pieterse et al. 1998; Van Wees et al. 2008), although there are reports that showed convincing evidence for rhizobacteria-induced systemic resistance via the SA pathways (Barriuso et al. 2008; De Vleesschauwer and Höfte 2009). For instance, the root-colonizing *Pseudomonas fluorescens* strain SS101 (*Pf.* SS101) induced resistance in *Arabidopsis* against *Pst* via the SAR pathway (van de Mortel et al. 2012).

JA and its derivatives play an important role in the symbiosis of plants with other higher fungi (Van der Ent et al. 2009). Like in the sebacinalean symbiosis (Schäfer et al. 2009), the hormone also accumulated in the arbuscular mycorrhizal symbiosis (Hause et al. 2002), supporting the idea that JA signaling is a mutual strategy of plants to control colonization by beneficial endophytic fungi. Interestingly, JA is not required for *P. indica*-mediated growth promotion and higher seed yield in *Arabidopsis* (Camehl et al. 2010), suggesting that pathways leading to either immunity or growth can be molecularly separated.

6 Ethylene Signaling in the Sebacinalean Symbiosis

The *Arabidopsis ein2-1* mutant, which is impaired in ET signaling, showed about 29 % less colonization by *Rr*F4 compared with the wild type, suggesting that ET also supports bacterial development in plant roots (Alabid, unpublished data). Consistent with this, ET supported root colonization of *P. indica* in barley and *Arabidopsis* (Khatabi et al. 2012). These authors found increased concentrations of free 1-aminocyclopropane-1-carboxylic acid (ACC) during early colonization stages (60 and 120 hpi) with *P. indica* in barley roots and induction of *1-aminocyclopropane-1-carboxylic acid synthase 1* (*ACS1*) and *ACS8* in *Arabidopsis*. ET-responsive *ERF1* transcripts were elevated both in local and distal *P. indica*-colonized *Arabidopsis* roots at 3 and 5 dpi (Pedrotti et al. 2013). In line with these results, two ESTs encoding *1-aminocyclopropane-1-carboxylic acid oxidase* (*ACC oxidase*), which is involved in ET synthesis by *P. indica*, were induced at early time points of the colonization (3 dpi, Schäfer et al. 2009). Also in agreement with these results, the *Arabidopsis* mutants *constitutive triple response 1* (*ctr1-1*) and *ET over expresser 1* (*eto1-1*), which encode constitutive ET signaling and enhanced ET biosynthesis, respectively, were significantly more colonized with *P. indica* at 14 dpi, while less fungal colonization (about 20 %) was detected in the ET-insensitive mutant *ein2-1* at 3 dpi compared to the wild type (Khatabi et al. 2012).

7 Salicylic Acid Signaling in the Sebacinalean Symbiosis

Several studies have demonstrated negative effects of SA signaling on the rate and intensity of rhizobial infection and nodulation (Martinez-Abarca et al. 1998; Van Spronsen et al. 2003; Stacey et al. 2006). This plant hormone has long been fixed as a key factor of the plant immune system (Feys and Parker 2000; Vlot et al. 2009; Dempsey and Klessig 2012). *Rr*F4, like its host *P. indica*, induced SA-responsive (*PR1b* and *PR10*) genes as well as *Calmodulin binding protein 60-like* (*CBP60*) only at early time points (3 dpi), while they were downregulated later in the symbiosis (Deshmukh and Kogel 2007; Pedrotti et al. 2013; Glaeser et al. 2016). In accordance with this finding are studies that revealed upregulation of defence-related genes during early stages of plant–fungus interaction, while they were downregulated as the symbiosis progressed (Gao et al. 2004; Grunwald et al. 2004; Harrison 2005; Hause and Fester 2005). Remarkably, overexpression of *NPR1* in *Medicago truncatula* suppressed root hair deformation in response to *Sinorhizobium meliloti*, whereas RNAi-mediated *NPR1* knockdown resulted in accelerated root hair curling, suggesting that SA affects this symbiosis through NPR1 (Peleg-Grossman et al. 2009).

8 Gibberellin Signaling in the Sebacinalean Symbiosis

It is common knowledge that beneficial microbes initially are recognized by plants as potential pathogens. As a result, a transient defence response is induced, typically including generation of reactive oxygen species (ROS) and callose deposition (Vos et al. 2014). At a first glance, it seems paradoxical that *P. indica* can successfully colonize so many plants while at the same time inducing local and systemic resistance to challenger root pathogens. An explanation for this has only recently been found. As a prerequisite of successful colonization, *P. indica* actively suppresses part of the plant's immune system by interfering with phytohormone signaling (Schäfer et al. 2009: Jacobs et al. 2011). A global transcriptome analysis of *P. indica*-colonized barley roots showed that the fungus suppressed SA-mediated defence and additionally altered gibberellin (GA) metabolism. Plants that were impaired in GA synthesis and perception could hardly be colonized by *P. indica* (Schäfer et al. 2009). During root colonization the fungus induced genes involved in the C-methyl-D-erythritol 4-phosphate (MEP) pathway as well as genes immediately lying downstream of this pathway. For example, the gene encoding a putative geranylgeranyl diphosphate synthase (GGPS), which catalyzes the conversion of isopentenyl pyrophosphate (IPP) and dimethylallyl diphosphate (DMAPP) into geranylgeranyl diphosphate (GGDP), was induced at 3 and 7 dpi. In addition, downstream of GGPS, two *Kaurene synthase* genes [*ent-KS1a* and *ent-KS-like 4* (*ent-KSL4*)] were differentially regulated in *P. indica*-colonized barley roots at 1, 3, and 7 dpi. The terpene cyclases Copalyl diphosphate synthase (CPS) and

Kaurene synthases play key roles in GA biosynthesis (Otomo et al. 2004). Consistent with these findings, the GA biosynthesis mutant *ga1-6* showed lower degrees of colonization, suggesting that *P. indica* recruits GA signaling to help root cell colonization (Jacobs et al. 2011). Importantly, *1-Deoxy-D-xylulose 5-phosphate synthase* (*DXS*) as well as *ent-KS1a* and *ent- KSL4* also were induced in barley roots colonized by *Rr*F4 (Glaeser et al. 2016). This finding is in line with previous results showing that GA-regulated gene *Exp-PT1* (*Phosphatidylinositol N-acetylglucosaminyltransferase subunit P-related*), which is suppressed by GA (Zentella et al. 2007), was downregulated in *P. indica*-colonized *Arabidopsis* roots at 7 dpi (Jacobs et al. 2011).

9 Auxin Signaling in the Sebacinalean Symbiosis

P. indica is able to produce the plant growth-promoting auxin indoleacetic acid (IAA) in liquid culture (Sirrenberg et al. 2007; Hilbert et al. 2012). The induction of auxin signaling is an active strategy employed by pathogens to establish plant–microbe compatibility (Navarro et al. 2006; Wang et al. 2007). For instance, exogenous application of auxin increased the susceptibility to microbial colonization via manipulation of root defence (Hilbert et al. 2013). However, *P. indica* produced only low amounts of auxins, and the expression of auxin-regulated genes was not altered in colonized *Arabidopsis* roots (Vadassery et al. 2008). Thus, these authors claimed that the auxin levels had little or no effect on *P. indica*-mediated growth promotion. Instead, *P. indica* produced relatively high levels of cytokinins, and the cytokinin levels in colonized *Arabidopsis* roots were higher than in non-colonized roots (Vadassery et al. 2008). Since auxin inhibits cytokinin biosynthesis, both hormones can interact to control plant development (Nordstrom et al. 2004), and the observed root growth promotion induced by both *P. indica* and *Rr*F4, respectively, may be due to changes in the auxin-to-cytokinin ratio.

Regardless of the significance of auxin produced in the sebacinalean symbiosis, it is interesting to note that *Rr*F4 can produce IAA in the presence of tryptophan (Sharma et al. 2008). Thus, it remains unclear whether the fungus itself, the bacterium, or even both partners contribute to production (or induction) of that and probably all other phytohormones. This question is part of the key issue about what is the critical contribution of either microbial partner to the beneficial effects in the sebacinalean symbiosis. Moreover, this question must be extended in gaining clarity on which partner regulates defence reactions in the plant. To answer either question, more attempts are required to cure *Piriformospora indica* from *Rhizobium radiobacter Rr*F4.

References

Achatz B, von Ruden S, Andrade D, Neumann E, Pons-Kuhnemann J, Kogel KH, Franken P, Waller F (2010) Root colonization by *Piriformospora indica* enhances grain yield in barley under diverse nutrient regimes by accelerating plant development. Plant Soil 333:59–70

Barriuso J, Solano BR, Gutiérrez Mañero FJ (2008) Protection against pathogen and salt stress by four plant growth-promoting rhizobacteria isolated from *Pinus* sp. on *Arabidopsis thaliana*. Phytopathology 98:666–672

Camehl I, Sheramethi I, Venus Y, Bethke G, Varma A, Lee J, Oelmüller R (2010) Ethylene signalling and ethylene-targeted transcription factors are required to balance beneficial and nonbeneficial traits in the symbiosis between the endophytic fungus *Piriformospora indica* and *Arabidopsis thaliana*. New Phytol 185:1062–1073

Conrath U, Beckers GJ, Flors V, Garcia-Agustin P, Jakab G, Mauch F, Newman MA, Pieterse CM, Poinssot B, Pozo MJ, Pugin A, Schaffrath U, Ton J, Wendehenne D, Zimmerli L, Mauch-Mani B (2006) Priming: getting ready for battle. Mol Plant Microbe Interact 19:1062–1071

De Vleesschauwer D, Höfte M (2009) Rhizobacteria-induced systemic resistance. In: van Loon LC (ed) Plant innate immunity. Academic/Elsevier Science, London, pp 223–281

Dempsey DA, Klessig DF (2012) SOS-too many signals for systemic acquired resistance? Trend Plant Sci 17:538–545

Deshmukh SD, Kogel KH (2007) *Piriformospora indica* protects barley from root rot caused by *Fusarium graminearum*. J Plant Dis Prot 114:263–268

Deshmukh SD, Hückelhoven R, Schäfer P, Imani J, Sharma M, Weiss M et al (2006) The root endophytic fungus *Piriformospora indica* requires host cell death for proliferation during mutualistic symbiosis with barley. Proc Natl Acad Sci USA 103:18450–18457

Fakhro A, Andrade-Linares DR, von Bargen S, Bandte M, Buttner C, Grosch R, Schwarz D, Franken P (2010) Impact of *Piriformospora indica* on tomato growth and on interaction with fungal and viral pathogens. Mycorrhiza 20:191–200

Feys BJ, Parker JE (2000) Interplay of signalling pathways in plant disease resistance. Trends Genet 16:449–455

Gao LL, Knogge W, Delp G, Smith FA, Smith SE (2004) Expression patterns of defense-related genes in different types of arbuscular mycorrhizal development in wild-type and mycorrhiza-defective mutant tomato. Mol Plant Microbe Interact 17:1103–1113

Glaeser SP, Imani J, Alabid I, Guo HJ, Kämpfer P, Hardt M, Blom J, Rothballer M, Hartmann A, Kogel KH (2016) Non-pathogenic *Rhizobium radiobacter* F4 deploys plant beneficial activity independent of its host *Piriformospora indica*. ISME J 10:871–884

Grunwald U, Nyamsuren O, Tamasloukht M, Lapopin L, Becker A, Mann P, Gianinazzi-Pearson-V, Krajinski F, Franken P (2004) Identification of mycorrhiza-regulated genes with arbuscular development-related expression profile. Plant Mol Biol 55:553–566

Harrison MJ (2005) Signaling in the arbuscular mycorrhizal symbiosis. Annu Rev Microbiol 59:19–42

Hause B, Fester T (2005) Molecular and cell biology of arbuscular mycorrhizal symbiosis. Planta 221:184–196

Hause B, Maier W, Miersch O, Kramell R, Strack D (2002) Induction of jasmonate biosynthesis in arbuscular mycorrhizal barley roots. Plant Physiol 130:1213–1220

Hilbert M, Voll ML, Ding Y, Hofmann J, Sharma M, Zuccaro A (2012) Indole derivative production by the root endophyte *Piriformospora indica* is not required for growth promotion but for biotrophic colonization of barley roots. New Phytol 196:520–534

Hilbert M, Nostadt R, Zuccaro A (2013) Exogenous auxin affects the oxidative burst in barley roots colonized by *Piriformospora indica*. Plant Signal Behav 8(4), e23572

Jacobs S, Zechmann B, Molitor A, Trujillo M, Petutschnig E, Lipka V, Kogel KH, Schäfer P (2011) Broad-spectrum suppression of innate immunity is required for colonization of *Arabidopsis* roots by the fungus *Piriformospora indica*. Plant Physiol 156:726–740

Khatabi B, Molitor A, Lindermayr C, Pfiffi S, Durner J, von Wettstein D, Kogel KH, Schäfer P (2012) Ethylene supports colonization of plant roots by the mutualistic fungus *Piriformospora indica*. PLoS One 7:e35502

Lahrmann U, Ding Y, Banhara A, Rath M, Hajirezaei MR, Döhlemann S, von Wirén N, Parniske M, Zuccaro A (2013) Host-related metabolic cues affect colonization strategies of a root endophyte. Proc Natl Acad Sci USA 110:13965–13970

Martinez-Abarca F, Herrera-Cervera JA, Bueno P, Sanjuan J, Bisseling T, Olivares J (1998) Involvement of salicylic acid in the establishment of the *Rhizobium meliloti*-Alfalfa symbiosis. Mol Plant Microbe Interact 11:153–155

Navarro L, Dunoyer P, Jay F, Arnold B, Dharmasiri N, Estelle M, Voinnet O, Jones JD (2006) A plant miRNA contributes to antibacterial resistance by repressing auxin signaling. Science 312:436–439

Nordstrom A, Tarkowski P, Tarkowska D, Norbaek R, Astot C, Dolezal K, Sandberg G (2004) Auxin regulation of cytokinin biosynthesis in Arabidopsis thaliana: a factor of potential importance for auxin-cytokinin-regulated development. Proc Natl Acad Sci USA 101:8039–8044

Oelmüller R, Sherameti I, Tripathi S, Varma A (2009) *Piriformospora indica*, a cultivable root endophyte with multiple biotechnological applications. Symbiosis 49:1–17

Otomo K, Kanno Y, Motegi A, Kenmoku H, Yamane H, Mitsuhashi W, Oikawa H, Toshima H, Itoh H, Matsuoka M, Sassa T, Toyomasu T (2004) Diterpene cyclases responsible for the biosynthesis of phytoalexins, momilactones A, B, and oryzalexins A–F in rice. Biosci Biotechnol Biochem 68:2001–2006

Pedrotti L, Mueller MJ, Waller F (2013) *Piriformospora indica* root colonization triggers local and systemic root responses and inhibits secondary colonization of distal roots. PLoS One 8:e69352

Peleg-Grossman S, Golani Y, Kaye Y, Melamed-Book N, Levine A (2009) NPR1 protein regulates pathogenic and symbiotic interactions between *Rhizobium* and legumes and non-legumes. PLoS One 4:e8399

Peškan-Berghöfer T, Shahollari B, Giong PH, Hehl S, Markert C, Blanke V, Kost G, Varma A, Oelmüller R (2004) Association of *Piriformospora indica* with *Arabidopsis thaliana* roots represents a novel system to study beneficial plant-microbe interactions and involves early plant protein modifications in the endoplasmic reticulum and at the plasma membrane. Physiol Plant 122:465–477

Pieterse CM, van Wees SC, van Pelt JA, Knoester M, Laan R, Gerrits H, Weisbeek PJ, van Loon LC (1998) A novel signaling pathway controlling induced systemic resistance in *Arabidopsis*. Plant Cell 10:1571–1580

Qiang X, Weiss M, Kogel KH, Schäfer P (2012) *Piriformospora indica*-a mutualistic basidiomycete with an exceptionally large plant host range. Mol Plant Pathol 13:508–518

Riess K, Oberwinkler F, Bauer R, Garnica S (2014) Communities of endophytic sebacinales associated with roots of herbaceous plants in agricultural and grassland ecosystems are dominated by *Serendipita herbamans* sp. nov. PLoS One 9:e94676

Schäfer P, Pfiffi S, Voll LM, Zajic D, Chandler PM, Waller F, Scholz U, Pons-Kühnemann J, Sonnewald S, Sonnewald U, Kogel KH (2009) Manipulation of plant innate immunity and gibberellin as factor of compatibility in the mutualistic association of barley roots with *Piriformospora indica*. Plant J 59:461–474

Selosse MA, Dubois MP, Alvarez N (2009) Do Sebacinales commonly associate with plant roots as endophytes? Mycol Res 113:1062–1069

Sharma M, Schmid M, Rothballer M, Hause G, Zuccaro A, Imani J, Kämpfer P, Domann E, Schafer P, Hartmann A, Kogel KH (2008) Detection and identification of bacteria intimately associated with fungi of the order Sebacinales. Cell Microbiol 10:2235–2246

Sherameti I, Shahollari B, Venus Y, Altschmied L, Varma A, Oelmüller R (2005) The endophytic fungus *Piriformospora indica* stimulates the expression of nitrate reductase and the starch-degrading enzyme glucan-water dikinase in tobacco and *Arabidopsis* roots through a homeodomain transcription factor that binds to a conserved motif in their promoter. J Biol Chem 280:26241–26247

Sirrenberg A, Goebel C, Grondc S, Czempinskic N, Ratzinger A, Karlovsky P et al (2007) *Piriformospora indica* affects plant growth by auxin production. Physiol Plant 131:581–589
Stacey G, McAlvin CB, Kim SY, Olivares J, Soto MJ (2006) Effects of endogenous salicylic acid on nodulation in the model legumes *Lotus japonicus* and *Medicago truncatula*. Plant Physiol 141:1473–1481
Stein E, Molitor A, Kogel KH, Waller F (2008) Systemic resistance in Arabidopsis conferred by the mycorrhizal fungus *Piriformospora indica* requires jasmonic acid signaling and the cytoplasmic function of NPR1. Plant Cell Physiol 49:1747–1751
Vadassery J, Ritter C, Venus Y, Camehl I, Varma A, Shahollari B, Novák O, Strnad M, Ludwig-Müller J, Oelmüller R (2008) The role of auxins and cytokinins in the mutualistic interaction between Arabidopsis and *Piriformospora indica*. Mol Plant Microbe Interact 21:1371–1383
Vadassery J, Ranf S, Drzewiecki C, Mithöfer A, Mazars C, Scheel D, Lee J, Oelmüller R (2009) A cell wall extract from the endophytic fungus *Piriformospora indica* promotes growth of *Arabidopsis* seedlings and induces intracellular calcium elevation in roots. Plant J 59:193–206
van de Mortel J, de Vos R, Dekkers E, Pineda A, Guillod L, Bouwmeester K, van Loon J, Dicke M, Raaijmakers J (2012) Metabolic and transcriptomic changes induced in *Arabidopsis* by the rhizobacterium *Pseudomonas fluorescens* SS101. Plant Physiol 160:2173–2188
Van der Ent S, Van Wees SCM, Pieterse CMJ (2009) Jasmonate signaling in plant interactions with resistance-inducing beneficial microbes. Phytochemistry 70:1581–1588
Van Spronsen PC, Tak T, Rood AMM, van Brussel AAN, Kijne JW, Boot KJM (2003) Salicylic acid inhibits indeterminate-type nodulation but not determinate-type nodulation. Mol Plant Microbe Interact 16:83–91
Van Wees SC, Luijendijk M, Smoorenburg I, Van Loon LC, Pieterse CM (1999) Rhizobacteria-mediated induced systemic resistance (ISR) in *Arabidopsis* is not associated with a direct effect on expression of known defense-related genes but stimulates the expression of the jasmonate-inducible gene Atvsp upon challenge. Plant Mol Biol 41:537–549
Van Wees SC, Van der Ent S, Pieterse CM (2008) Plant immune responses triggered by beneficial microbes. Curr Opin Plant Biol 11:443–448
Varma A, Verma S, Sudah SN, Franken P (1999) *Piriformospora indica*, a cultivable plant growth-promoting root endophyte. Appl Environ Microbiol 65:2741–2744
Verma S, Varma A, Rexer KH, Hassel A, Kost G, Sarabhoy A (1998) *Piriformospora indica*, gen. et sp. nov., a new root-colonizing fungus. Mycology 90:896–903
Vlot AC, Dempsey DA, Klessig DF (2009) Salicylic acid, a multifaceted hormone to combat disease. Annu Rev Plant Pathol 47:177–206
Vos C, Yang Y, Deconinck B, Cammue BPA (2014) Fungal (-like) biocontrol organisms in tomato disease control. Biol Control 74:65–81
Waller F, Achatz B, Baltruschat H, Fodor J, Becker K, Fischer M, Heier T, Hückelhoven R, Neumann C, von Wettstein D, Franken P, Kogel KH (2005) The endophytic fungus *Piriformospora indica* reprograms barley to salt-stress tolerance, disease resistance, and higher yield. Proc Natl Acad Sci USA 102:13386–13391
Wang D, Pajerowska-Mukhtar K, Culler AH, Dong X (2007) Salicylic acid inhibits pathogen growth in plants through repression of the auxin signaling pathway. Curr Biol 17:1784–1790
Weiss M, Selosse MA, Rexer KH, Urban A, Oberwinkler F (2011) Sebacinales everywhere: previously overlooked ubiquitous fungal endophytes. PLoS One 108:1003–1010
Yadav VM, Kumar DK, Deep H, Kumar R, Sharma T, Tripathi N, Tuteja A, Saxena K, Johri AK (2010) A phosphate transporter from the root endophytic fungus *Piriformospora indica* plays a role in phosphate transport to the host plant. J Biol Chem 285:26532–26544
Zamioudis C, Pieterse CMJ (2012) Modulation of host immunity by beneficial microbes. Mol Plant Microbe Interact 25:139–150
Zentella R, Zhang ZL, Park M, Thomas SG, Endo A, Murase K, Fleet CM, Jikumaru Y, Nambara E, Kamiya Y et al (2007) Global analysis of DELLA direct targets in early gibberellin signaling in Arabidopsis. Plant Cell 19:3037–3057

Mitigating Abiotic Stresses in Crop Plants by Arbuscular Mycorrhizal Fungi

Katia Plouznikoff, Stéphane Declerck, and Maryline Calonne-Salmon

Abstract Abiotic stresses [i.e., salinity, drought, high temperatures, and pollutants such as trace elements (TEs) and/or petroleum, crude oil, and PAHs] have detrimental effects on plant growth, fitness, and yield. They can cause significant production losses at a time when food needs are constantly increasing. The development of tolerant/resistant crops and innovative/alternative methods to alleviate abiotic stresses have thus become of major concern in our societies. One promising strategy is the use of arbuscular mycorrhizal fungi (AMF) that form symbiotic associations with the vast majority of agricultural and horticultural important crops.

Here we summarized the impact of abiotic stresses on the AMF life cycle and physiology. If these organisms are usually affected by abiotic stresses, they are also frequently reported to improve growth and tolerance of plants under these conditions. The mechanisms most often described concern (1) improved plant nutrition; (2) accumulation and use of sugars, polyamines, abscisic acid (ABA), and lipids; (3) tolerance to induced oxidative stress; (4) modification in plant physiology; and (5) root and fungal chelation and inactivation of pollutants. The association of crops with AMF thus offers interesting perspectives to increase/maintain crop production under stressed environmental conditions.

1 Introduction

Earth is expected to be inhabited by some 9000 million people by 2050, and a recent report by the FAO estimates that farmers will have to produce 70 % more to meet the needs of this population (FAO 2009). Within the same period, the global temperature is projected to increase by 2.5 °C with major impacts on plant growing conditions, on the emergence of new pests and diseases and on an increase in water scarcity and desertification. The challenges that agriculture has to face to feed the future population are thus becoming more and more pressing. Their fulfillment will require wide-ranging solutions, including improved crop varieties with higher

---

K. Plouznikoff • S. Declerck (✉) • M. Calonne-Salmon
Earth and Life Institute, Applied Microbiology, Mycology, Université Catholique de Louvain, Croix du Sud, 2 box L7.05.06, 1348, Louvain-la-Neuve, Belgium
e-mail: stephan.declerck@uclouvain.be

C.M.F. Vos, K. Kazan (eds.), *Belowground Defence Strategies in Plants*, Signaling and Communication in Plants, DOI 10.1007/978-3-319-42319-7_15

yield; increased N, P, and water use efficiency; ecologically sustainable management practices; the converting of marginal lands into productive areas and restoration of degraded areas (Lal 2000); and optimal use of agricultural inputs without increasing negative environmental impacts associated with agriculture. There is thus a need for new flexible crop varieties that can resist abiotic and biotic stress factors without putting unacceptable pressure on scarce land and water resources.

Multiple abiotic stresses defined as *outside (non-living) factors which can cause harmful effects to plants, such as soil conditions, drought and, extreme temperatures* (Newell-McGloughlin and Burke 2014) affect negatively plant growth, development, and crop productivity. Soils, earth's nonrenewable resources, can be deeply affected and degraded by abiotic stress factors, and 2015 has been chosen by the FAO as the international year of soils (http://www.fao.org/soils-2015/en/) in order to promote their protection. Managing plant environmental stresses is the foundation of sustainable agriculture (http://www.fao.org/emergencies/emergency-types/drought/en/).

Salinity, drought, and high temperature have become serious problems in many regions, not only because of a higher risk for public health and the environment but also because of negative effects on the yield. These abiotic stresses are also common to many agricultural areas around the globe and severely affect plant productivity. For instance, in the USA between 1980 and 2004, drought stress caused some $20 billion in damages (Mittler 2006). Finally, the contamination of agricultural soils with trace elements (TEs) and organic pollutants may represent under some specific conditions a threat to agricultural soil. The loading of ecosystems with TEs can be due to excessive fertilizer and pesticide use, irrigation, atmospheric deposition, and pollution by waste materials. The risks caused by polyaromatic hydrocarbons (PAHs) are anecdotic, although a recent study reported pollution in agricultural soil and vegetables from Tianjin (China), a site close to an urban district and irrigated with wastewater (Tao et al. 2004). Interestingly, the current concern in the application of biochar in agricultural soil may warn about the risks caused by PAHs. Indeed, biochar contains PAHs at various levels and its high sorptive capacity may facilitate the persistence of PAHs in soil.

The usage of crop varieties with improved root architecture associated with a beneficial rhizosphere microbiome to boost yield and protect plants against biotic and abiotic stresses is likely to prove crucial for increasing future agricultural production. Among the microorganisms, arbuscular mycorrhizal fungi (AMF) are of particular interest. They live at the interface between plant and soil; help plants fend off disease; stimulate growth; crowd out space that would be taken up by pathogens; promote resistance to drought, salinity, TEs, etc.; or influence crop yield by more efficient acquisition of nutrients. Thus, an increasing number of scientists and farmers alike think their exploitation and valorization represent the next revolution in agriculture.

Here we review the impact of abiotic stresses (salinity, drought, high temperatures, TEs, and PAHs) on the symbionts and symbiotic association in the first part and examine the role of AMF in mitigating these stresses in crop plants and depict the mechanisms involved in the second part.

2 Impact of Abiotic Stresses on AMF

2.1 *Environmental Abiotic Stresses*

2.1.1 Soil Salinity

Salinity is one of the major problems affecting soil in the world. Salinization results from natural factors or anthropogenic activities. It results in the degradation of soils and, in some cases, is responsible for irreparable losses to their productive capacity, with great extensions of arable land becoming sterile (Maganhotto de Souza Silva and Francisconi Fay 2012). Irrigation with groundwater, irrational use of easily soluble fertilizers, and poor drainage conditions are the main causes for salinity in agroecosystems (Copeman et al. 1996; Al-Karaki 2000; Priyadharsini and Muthukumar 2015; Singh 2015). Estimates by FAO indicate that of the 250 million hectares of irrigated land in the world, approximately 50 % already show salinization and soil saturation problems. More than 250 million hectares of irrigated land are damaged by salt, and 1.5 million hectares are taken out of production each year as a result of high salinity levels (Bot et al. 2000; Munns and Tester 2008; Priyadharsini and Muthukumar 2015). In India, about 8.1 million hectares are salinized, of which 3.1 million in coastal regions (Yadav et al. 1983; Tripathi et al. 2007). In Europe, salt excess affects 3.8 million hectares, mainly in the Mediterranean countries (EEA 1995), and this tendency is increasing, mainly in Spain, Hungary, and Greece (de Paz et al. 2004; Jones et al. 2012). In Nordic countries, the use of salt to remove ice from highways in winter produces localized salinization phenomena (Jones et al. 2012). Moreover, salinization is expected to increase in the future due to the increasing temperatures and decrease in rainfall worldwide (Maganhotto de Souza Silva and Francisconi Fay 2012).

Salinity is one of the cosmopolitan threats to crop production worldwide (Munns and Tester 2008; Priyadharsini and Muthukumar 2015). It affects seed germination, plant growth and vigor, and thus crop productivity (Giri et al. 2003; Mathur et al. 2007; Munns and Tester 2008; Carillo et al. 2011). Salt deposition in the soil results in hyperionic and hyperosmotic stresses in organisms (Evelin et al. 2013), which may limit the growth of organisms such as plants and fungi due to specific ion toxicity or osmotic stress. The stressed plants present nutritional disorders, oxidative stress, alteration of metabolic processes, membrane disorganization, reduction of cell division and expansion, and genotoxicity (Hasegawa et al. 2000; Munns 2002; Zhu 2007). However, the importance of these factors is relative to species and concentration of ions involved (Brownell and Schneider 1985).

AMF have been repeatedly mentioned in saline environments (Khan 1974; Allen and Cunningham 1983; Pond et al. 1984; Rozema et al. 1986; Ho 1987; Juniper and Abbott 1993). Species such as *Funneliformis geosporum* or belonging to the *Rhizophagus irregularis* clade have been reported in roots of halophytes (Bothe 2012). The dominant salt marsh grass *Puccinellia* sp. showed variable degrees of

AMF colonization, with many specimens recorded without any traces of AMF colonization (Hildebrandt et al. 2001; Landwehr et al. 2002). Conversely, salt aster (*Aster tripolium*) was reported to be strongly dependent on AMF (Mason 1928; Boullard 1959). Recently, Yamato et al. (2008) identified two different *Glomus* spp. in salt coastal vegetation on Okinawa Island (Japan), and a phylogenetic analysis showed that one AMF is closely related to *R. irregularis*. The authors also reported that colonization rates of this AMF were not reduced when cultivated in pots in the presence of 200 mM of NaCl.

Under controlled laboratory conditions, most studies did not separate the direct effects of salinity on AMF from plant-mediated effects. Indeed, spore germination is the sole stage that can be studied independently of the symbiotic association with a plant (Daniels and Graham 1976; Hepper 1979; Daniels and Trappe 1980; Elias and Safir 1987; Gianinazzi-Pearson et al. 1989; Juniper and Abbott 1993, 2006). The available literature indicated that concentrations of NaCl from 4.30×10^{-2} M to 2.14×10^{-1} M inhibited spore germination (Hirrel 1981; Estaun 1990, 1991; Juniper and Abbott 1991, 1993, 2006; Al-Karaki 2000). This impact seemed to be fungistatic since Hirrel (1981) and Koske (1981) reported a "germination recovery" following incubation of the stressed spores in the absence of salt.

Salt was also reported to impact the growth of hyphae and germ tube. For instance, the germ tube of *F. mosseae* was inhibited by NaCl and mannitol below 0.75 MPa osmotic potential in vitro (Estaun 1990). In some AMF, the germ tube growth is stimulated by the proximity of a root (Mosse and Hepper 1975) and by root exudates (Graham 1972). This stimulation could be altered in saline conditions since exudation is greatly influenced by soil chemistry and soil moisture (Rovira 1969). This impact on spore germination and germ tube growth and morphology could thus negatively affect the association process and AMF survival as suggested by Calonne et al. (2010).

Most studies reported a decrease in root colonization under salt stress conditions at electric conductivity ranging from 1.4 dS/m (control condition), 4.7 dS/m (moderate), and 7.4 dS/m (high) (Poss et al. 1985; Pfeiffer and Bloss 1988; Al-Karaki 2000; Tian et al. 2004; Kaya et al. 2009; Hajiboland et al. 2010; He and Huang 2013). Curiously, at 150 mM NaCl, inhibition appeared more pronounced during the early stages of root colonization than in the late stages (McMillen et al. 1998). Beltrano et al. (2013) showed that under the high salt condition (200 mM NaCl), root colonization was reduced by 28 % compared to control. Arbuscule and vesicle abundance decreased by 75 % with 200 mM NaCl. Viability of hyphae, expressed by SDH activity, was reduced over 50 % at 200 mM.

A few studies also demonstrated an increase in AMF colonization under salt stress conditions under salinity levels ranging from 7.3 to 92.0 dS/m (Aliasgharzadeh et al. 2001; Porcel et al. 2012). This contrasting observation may be related to various factors among which the AMF species (Daei et al. 2009). For instance, Ruiz-Lozano and Azcón (2000) compared the growth and colonization of two AMF species, one isolated from a saline soil (*Glomus* sp.) and one isolated from a nonsaline soil (*G. deserticola*), under increasing salt concentrations (0.25, 0.50, or 0.75 g NaCl/kg dry soil). In the presence of low salt levels, root colonization by the *Glomus* sp. was lower than that of *G. deserticola*. In contrast, at the highest salt levels, both species showed a similar percentage of colonization (Ruiz-Lozano and Azcón 1995). Furthermore, the authors observed that *G. deserticola* was more efficient to establish common mycelial networks and colonize neighbor plants than the AMF isolated from the saline soil (Ruiz-Lozano and Azcón 1995). In a study conducted by Jahromi et al. (2008) under in vitro conditions, the spore production of *R. irregularis* was drastically reduced under NaCl conditions. These results invite researchers to search for salt-tolerant AMF species and to test if these AMF isolates maintain colonization capacity and symbiosis efficiency under saline conditions (Evelin et al. 2009).

2.1.2 Drought

Salt and drought stresses share some common properties and generally result in impaired key physiological functions in living organisms such as fungi (Daffonchio et al. 2015). One component of salinity is hyperosmotic stress, resulting in a water deficit that is comparable with a drought-induced water deficit (Daffonchio et al. 2015). Under salty or drought conditions, water moves from plant cells to the soil solution, and as a result, cells shrink and ultimately collapse and die (Brady and Weil 2008).

In drought-stressed soils, Egerton-Warburton et al. (2007) and Querejeta et al. (2009) reported a dominance of *Glomus*/*Rhizophagus* species in AMF communities in soils of xeric habitats, while species belonging to the genus *Scutellospora*, *Gigaspora*, or *Acaulospora* were less abundant. Moreover, they demonstrated that under drought stress conditions, the proportion of *Gigaspora* species within the AMF tended to decrease, whereas the proportion of *Glomus* species tended to increase proportionally (Querejeta et al. 2009). Some AMF species were isolated from Arabian arid regions and deserts: *F. mosseae*, *Claroideoglomus etunicatum*, *R. fasciculatus*, and *G. aggregatum*, *Diversispora aurantia*, *D. omaniana*, *S. africanum*, and undescribed *Paraglomus* species (Dhar et al. 2015; Symanczik et al. 2015).

Up to date, only a few studies have investigated the impact of different water regimes on the AMF life cycle and ecology. This may be related to technical difficulties to observe the mycorrhizal development in drought stress experiments. Spore germination was increased (Douds and Schenck 1991), decreased (Tommerup 1984; Estaun 1990; Douds and Schenck 1991), or unaffected (Douds and Schenck 1991) by soil drying. These differences were partially related to the AMF species considered (Augé 2001).

Similarly, root colonization was investigated under drought stress conditions, in laboratory as well as in the field. As previously reviewed by Augé (2001), a large number of studies showed that drought only affected root colonization in about half of the reports examined, and curiously the level of root colonization was increased rather than decreased. These surprising results could be explained by the differences in host plant and AMF species studied, the origin of fungi, and the culture conditions (controlled or field conditions). For instance, root colonization in lettuce changed under different water regimes: under dry conditions, the level of colonization decreased in roots inoculated by *F. mosseae* or *P. occultum* but remained constant when inoculated by *G. deserticola* or *C. etunicatum* (Ruiz-Lozano et al. 1995). Aroca et al. (2008) observed that lettuce plants cultivated under well-watered conditions showed 45–50 % of mycorrhizal root length by *R. irregularis* BEG121, while in contrast, plants subjected to drought showed 62–70 % of root colonization. Recently, it has been shown that the AMF species composition in hyperarid plain of Oman changed with water regimes (Symanczik et al. 2015). Under well-watered and drying cycle conditions, it was dominated by *D. omaniana*, while under drought stress conditions, *S. africanum* and *Paraglomus* spp. were more abundant (Symanczik et al. 2015). Again, a decline in AMF colonization was observed in greenhouse (Manoharan et al. 2010; Zou et al. 2015) and field experiments (Ryan and Ash 1996; Al-Karaki et al. 2004) under drought stress. However, Augé (2001) reported that in field situations, chronic drought periods may promote more extensive colonization. Upon 43 flowering plants examined on a fallow agricultural site in Germany, 40 were heavily colonized by AMF in a low soil moisture habitat, and 29 were heavily colonized on a comparable but high soil moisture habitat (Kühn 1991; Kühn et al. 1991). Plant colonization was shown to vary between seasons (Clark et al. 2009), with the highest values observed during fall and the lowest during summer drought periods (Apple et al. 2005) and according to wet and dry years (Querejeta et al. 2009).

In more detail, changes in water availability due to changes in precipitation were shown to influence the abundance of arbuscules and vesicles (Martínez-García et al. 2012; Symanczik et al. 2015). These authors noticed that the abundance of vesicles and arbuscules drastically decreased in plants subjected to rainfall reduction. Martínez-García et al. (2012) concluded that when precipitation changed seasonally but annual precipitation remained the same, a shrub species endemic from the most arid systems in SE Spain had mycorrhiza with a highest production of arbuscules. However, when there was a reduction in annual precipitation, the number of vesicles decreased suggesting less investment in the production and maintenance of these structures because they are storage structures rather than transfer structures. To summarize, they suggested that increased drought conditions consequently to climate change in the region of study may enhance arbuscule production to favor water transfer as long as drought intensity does not affect growth of internal fungal structures (Martínez-García et al. 2012).

The production of extraradical mycelium was also reported to be impacted by drought conditions. For instance, the production of extraradical mycelium under drought conditions was the highest in soils colonized by native species of *F. mosseae* and *R. irregularis* as compared to exotic species (Marulanda et al. 2007). Symanczik et al. (2015) suggested that a community of native AMF species can buffer against different water regimes, as reflected by the constant production of extraradical mycelium in well-watered, drying cycles and drought-stressed regimes. From these studies, it appeared that native AMF species are better adapted to drought stress than introduced species.

2.1.3 High Temperatures

Global climate is predicted to change dramatically over the next century (Houghton et al. 2001) becoming a major threat to agriculture (Maya and Matsubara 2013). Temperature is one of the main factors that regulate the growth and productivity of plants (Allakhverdiev et al. 2008). High temperatures cause a series of morphological, physiological, and biochemical changes in plants via their effects on photochemical and biochemical reactions, as well as on photosynthetic pigments (Wahid et al. 2007; Zhu et al. 2011a).

It is generally admitted that low temperature impacts the AMF development (Smith and Bowen 1979; Baon et al. 1994; Zhang et al. 1995; Gavito et al. 2003; Heinemeyer and Fitter 2004; Liu et al. 2004a, b; Hawkes et al. 2008; Wu and Zou 2010; Zhu et al. 2010a, b; Latef and Chaoxing 2011a; Karasawa et al. 2012; Chen et al. 2013, 2014; Liu et al. 2013; Barrett et al. 2014). Most of these studies demonstrated a decrease in root colonization, extraradical length, and spore production in the presence of temperature under 18 °C, whatever the origin (warm or cold soil) of the AMF. However, species diversity remains high as shown by Gai et al. (2012). These authors identified 52 AMF species in cold elevated areas in Tibet mountains. This included ten species belonging to *Acaulospora*; 18 to *Glomus*; five to *Funneliformis*; three to *Ambispora* and *Gigaspora*; two to *Scutellospora*, *Rhizophagus*, *Claroideoglomus*, *Sclerocystis*, and *Pacispora*; and one to *Diversispora*, *Archaeospora*, and *Paraglomus*. The dominant species were *G. aggregatum*, *F. geosporum*, and *R. clarus* (Gai et al. 2012).

High temperatures may affect differently the development of AMF. However, only one study to our knowledge reported on the effect of high temperatures on spore germination (Schenck et al. 1975). These authors demonstrated that the germination of spores of *Racocetra coralloidea* and *Fuscutata heterogama* decreased when cultivated in Petri dishes incubated above 34 °C.

Most studies reported that warm soil conditions differentially alter AMF activity. Root colonization generally decreased under temperatures exceeding 30 °C (Bowen 1987; Martin and Stutz 2004), and soil temperatures above 40 °C were generally lethal to AMF (Bendavid-Val et al. 1997; Martin and Stutz 2004). However, the degree to which temperature affected the AMF varied with species (Schenck and Smith 1982) and their origin. With *F. mosseae*, root colonization reached its maximum at 24–25 °C in pots heated in a water bath (Schenck and Smith 1982) or when the pots were stored in growth chambers at temperature of 25 °C (Wu 2011) and decrease at higher temperatures. For other species (i.e., *C. claroideum*, *R. clarus*, *A. laevis*, *S. pellucida*, *Endogone macrocarpa*, *Endogone* sp.), root colonization only declined at soil temperatures above 30 °C (Schenck and Schroder 1974; Schenck and Smith 1982) or when pots were stored in growth chambers at temperature of 30 °C (Raju et al. 1990). One work focusing on

R. irregularis also showed an optimal colonization temperature when pots were stored in growth chambers at temperature of 30 °C and was severely reduced when the chamber reached temperatures between 32.1 and 38 °C (Martin and Stutz 2004). Temperatures above 35 °C generally induced important decrease in root colonization. Nevertheless, some species seemed not impacted and showed increase in root colonization. Indeed, *Racocetra gregaria*, *Gi. margarita*, *G. ambisporum*, and *R. irregularis* had their maximum percentage of root colonization when soil was heated at 36 °C (Schenck and Smith 1982; Smith and Roncadori 1986). Martin and Stutz (2004) studied two *Glomus*/*Rhizophagus* species (*R. irregularis* and *Glomus* sp. AZ112), the last isolated in Arizona where high temperatures are usually recorded. They demonstrated that in the presence of high temperatures in the growth chamber (between 32.1 and 38 °C), root colonization and abundance of arbuscules decreased with *R. irregularis*, whereas it increased with *Glomus* sp. AZ112. In other studies, no difference in *C. etunicatum* and *R. fasciculatum* root colonization was noticed in a range of temperatures of 25–40 °C in the growth chamber (Zhu et al. 2011a; Maya and Matsubara 2013).

A number of studies also reported the impact of temperature on the development of the extraradical mycelium. For instance, the hyphal length of *R. irregularis* isolated from Quebec increased at temperatures between 18 and 30 °C under in vitro culture conditions. To the contrary, *G. cerebriforme* also isolated from Quebec showed a decline in extraradical growth at temperatures above 24 °C (Gavito et al. 2005).

High temperatures (above 30 °C) affected sporulation of *C. claroideum*, *S. pellucida*, *Racocetra gregaria*, and *R. clarus*, while *F. mosseae*, *Gi. decipiens*, and *A. laevis* spore productions were impacted in soils heated at 32–36 °C (Schenck and Smith 1982) or incubated in Petri dishes (Costa et al. 2013). Sporulation of *Endogone* sp. was optimal in soils heated at 35 °C but was inhibited at 41 °C, in parallel to plant senescence (Schenck and Schroder 1974).

2.2 *Anthropogenic Abiotic Stresses*

2.2.1 Trace Elements (TEs)

Soil contaminations by TEs, originating from anthropogenic activities, are of great concern worldwide because of their persistence and toxicity for humans and the environment (Wong et al. 2002; Huang et al. 2006). Unfortunately, estimations of accumulated TEs in soils are scarce. One global estimation mentioned that TEs are the major pollutants found in European soils (Panagos et al. 2013). They are released into the environment via various anthropogenic activities such as mining, energy and fuel production, electroplating, wastewater sludge treatment, and agriculture (Abioye 2011). The study of Nicholson et al. (2003) conducted in England and Wales demonstrated that atmospheric deposition, livestock manures, sewage sludge, inorganic fertilizers and lime, agrochemicals, irrigation water, and industrial by-product "wastes" and composts accounted for the principal sources of TE in agricultural soils. In China, Huang et al. (2007a, b) noticed that the increase of Cd and Hg in agricultural soils was attributed to the long-term use of agrochemicals. Indeed, many P fertilizers and pesticides contain TEs, including Cd and Cu (Lugon-Moulin et al. 2006; Nziguheba and Smolders 2008; Kabata-Pendias 2011). Moreover, TE atmospheric deposition induced by traffic at the vicinity of roads and highways was demonstrated all over the world (Albasel and Cottenie 1985; Kelly et al. 1996; Pagotto et al. 2001; Turer et al. 2001; Sezgin et al. 2004; Viard et al. 2004; Saeedi et al. 2009).

Several studies reported a high diversity of AMF in TE-contaminated areas (Del Val et al. 1999a; Vallino et al. 2006; Long et al. 2010; Zarei et al. 2010; Hassan et al. 2011; Schneider et al. 2013; Krishnamoorthy et al. 2015; Yang et al. 2015) with *F. mosseae* and *R. irregularis* as dominant species (Zarei et al. 2010; Hassan et al. 2011; Krishnamoorthy et al. 2015). However, the species' richness and diversity were reported to decrease from un- or low-contaminated sites to highly contaminated sites (Del Val et al. 1999a; Hassan et al. 2011; Yang et al. 2015).

Even some authors observed the absence of change in mycorrhizal colonization in the presence of TEs (Andrade et al. 2010); more generally, these pollutants have been reported to affect spore germination, root colonization, extraradical mycelium development, and sporulation of AMF (Weissenhorn et al. 1993, 1994, 1995; Del Val et al. 1999b; Regvar et al. 2001; Shalaby 2003; Pawlowska and Charvat 2004; González-Guerrero et al. 2005; Zarei et al. 2008, 2010; Cornejo et al. 2013; Kelkar and Bhalerao 2013; Abdelmoneim et al. 2014; Gavito et al. 2014; Spagnoletti et al. 2014; Spagnoletti and Lavado 2015). The main cause was attributed to a fungitoxic effect, resulting in a certain inability of AMF to colonize the root system and/or to propagate in the rhizosphere. The AMF can also be indirectly affected by TEs in soils. Indeed, carbohydrate concentrations in plant tissues can be modified by Cd toxicity (Seregin and Ivanov 2001) and may thus indirectly affect arbuscule abundance (Repetto et al. 2003). However, AMF isolated from sludge-polluted sites showed higher tolerance to TEs in comparison to isolates from unpolluted soils (Gildon and Tinker 1983; Weissenhorn et al. 1993, 1995; Díaz et al. 1996; Del Val et al. 1999b). For instance, Del Val et al. (1999b) compared four AMF isolates colonizing *Sorghum bicolor* and *Allium porrum* for their tolerance to heavy metals. They noticed that *Glomus* sp. isolated from a nonpolluted soil was the most sensitive AMF, while *C. claroideum* 7 isolated from a contaminated soil was the more tolerant. In non-contaminated soils, *F. mosseae* and *Glomus* spp. were the most effective in terms of root colonization as compared to the two *C. claroideum* species, whereas *C. claroideum* 7 was slightly more efficient in the polluted soil as compared to the other isolates (*Glomus* sp., *F. mosseae*, and *C. claroideum* 2) from nonpolluted soils (Del Val et al. 1999b). A similar observation was made by Weissenhorn et al. (1995). These authors noticed that root colonization in a polluted soil was higher with a strain of *F. mosseae* isolated from a polluted soil than with a strain isolated from a nonpolluted soil.

The impact of TEs on the AMF life cycle was often attributed to the induction of an oxidative stress in AMF. This was suggested by the increase in malondialdehyde

(MDA, a lipid peroxidation biomarker) in the extraradical mycelium (González-Guerrero et al. 2007). González-Guerrero et al. (2009) described the antioxidant mechanisms involved in the resistance to oxidative stress caused by TEs. Some genes involved in reactive oxygen species (ROS) homeostasis have been identified and characterized in AMF: two CuZn-SODs (CuZn-superoxide dismutases), ten genes putatively encoding GSTs (glutathione *S*-transferases), one Grx (glutaredoxin), one gene encoding a protein involved in vitamin B6 biosynthesis, and three MTs (metallothioneins) (Ouziad et al. 2005; Waschke et al. 2006; Benabdellah et al. 2009a, b; González-Guerrero et al. 2010a). Other strategies possibly contributing to TE tolerance appear to be involved as well, which is indicated by the significantly enhanced expression of an MT and a Zn transporter gene, particularly under Cu stress (Hildebrandt et al. 2007).

2.2.2 Petroleum and Polycyclic Aromatic Hydrocarbons

PAHs originate mainly from pyrolysis of organic matter and fossil fuel. Anthropogenic sources of PAHs are car traffic, industries, waste incinerators, and domestic heating via both atmospheric transport and local activities (Manoli and Samara 1999; Blanchard et al. 2004).

Mineral oil and PAHs are frequently encountered in polluted soils. They contribute jointly to nearly 35 % of pollutants found in soils in Europe (Panagos et al. 2013). Most of the studies were conducted near industrial sites (Bewley et al. 1989; Ellis et al. 1991; Mueller et al. 1991; Erickson et al. 1993; Juhasz 1998; Leyval and Binet 1998; Bispo et al. 1999; Bogan et al. 1999; Joner et al. 2002, 2006; Joner and Leyval 2003; Nadal et al. 2004; Potin et al. 2004; Biache et al. 2008; Rezek et al. 2008; Kacálková and Tlustoš 2011), whereas agricultural fields and rural sites were less investigated (Wild and Jones 1995; Trapido 1999; Nam et al. 2003; Cai et al. 2007; Zuo et al. 2007; Maliszewska-Kordybach et al. 2008). PAH deposition in agricultural fields close to roads and airports has, however, been demonstrated (Tuháčková et al. 2001; Crépineau et al. 2003; Crépineau-Ducoulombier and Rychen 2003). Similarly, accumulation of PAHs in rural sites and agricultural soils may result from atmospheric transport over long distances (Halsall et al. 2001). In some rural sites in Poland, concentration of PAHs between 100 and 395 μg/kg of soil, with a maximum of 7264 μg/kg soil, have been reported (Maliszewska-Kordybach et al. 2008).

Wastewater and sewage sludge could also pose a problem of PAH contamination of agricultural soils. Indeed, sewage sludge applied as fertilizer may increase content of PAHs in soils (Cai et al. 2008). Consequently, there is an increasing concern about the accumulation of organic contaminants in plant-soil systems amended with sewage sludge (Cai et al. 2008). These authors demonstrated that most of the PAHs recovered in radish resulted from a soil-to-root transfer and translocation. Tao et al. (2006) further demonstrated the accumulation of PAHs in rice grown in soils contaminated by wastewater irrigation. These authors measured a higher PAH accumulation in roots than in shoots. Moreover, grains and internodes accumulated much lower amounts of PAHs than leaves, hulls, or ear axes.

Another potential source of PAH in agricultural soils is the biochar amendment. Biochar is a solid by-product obtained by the pyrolysis or gasification of biomass in a low or zero oxygen environment (Yargicoglu and Reddy 2014). Biochar is considered much more effective than other organic matter to make nutrients available to plants and increasing crop yields. However, biochar usually contains phytotoxic and potential carcinogenic compounds such as PAHs and heavy metals (Zheng et al. 2010; Hale et al. 2012; Hilber et al. 2012; Keiluweit et al. 2012; Kloss et al. 2012; Yargicoglu and Reddy 2014; Yargicoglu et al. 2015). It is worth to notice that the biochar feedstock, as well as the temperature and duration of pyrolysis, can make a significant difference to the final concentration of PAHs (Hale et al. 2012; Quilliam et al. 2013; Yargicoglu and Reddy 2014). Recently, Quilliam et al. (2013) showed that biochar can reduce the degradation of PAHs in agricultural soils, which could increase the concentration of soil PAHs in the short term but also affect the long-term persistence of PAHs in the environment. Impact of long-term biochar application on soil beneficial microorganisms should thus be investigated in the future.

AMF species diversity is searched in petroleum-contaminated soils for less than 10 years (Huang et al. 2007a, b; Hassan et al. 2014; Iffis et al. 2014; de la Providencia et al. 2015). For instance, in a sedimentation basin dedicated to the storage of petroleum-hydrocarbon wastes from a former petrochemical plant, 21 AMF taxa were detected belonging to *Claroideoglomus, Diversispora, Rhizophagus*, and *Paraglomus* (de la Providencia et al. 2015). In highly polluted soils in Canada, *R. irregularis* dominated the relative abundance of AMF species, whereas it was dominated by *Claroideoglomeraceae* species in low-contaminated ones (Hassan et al. 2014; Iffis et al. 2014). *F. constrictum* and *F. mosseae* were predominant species in a petroleum-contaminated soil in China (Huang et al. 2007a, b).

PAHs and diesel have been reported to impact the AMF life cycle (Leyval and Binet 1998; Gaspar et al. 2002; Liu et al. 2004a, b; Rabie 2004, 2005; Kirk et al. 2005; Alarcón et al. 2006; Verdin et al. 2006; Franco-Ramírez et al. 2007; Debiane et al. 2008, 2009, 2011; Tang et al. 2009; Aranda et al. 2013; Calonne et al. 2014a; Driai et al. 2015). Only the works of Kirk et al. (2005) showed no differences on spore germination of *R. irregularis* and *G. aggregatum* under in vitro conditions with increased concentrations of petroleum or diesel [0.5, 1, 2, and 3 % (vol/vol)] in the absence or presence of roots. Curiously, in the presence of plant roots, a decrease in germ tube length between AMF in contaminated and non-contaminated medium with 0.5 % of diesel was marked. The authors suspected that the signals released by the plant roots were perhaps not reaching the AMF in diesel-contaminated substrate (Kirk et al. 2005).

PAHs were demonstrated to induce an oxidative stress in AMF. This was evidenced by the increase in MDA content and the perturbations in unsaturated fatty acid quantities in the extraradical mycelium of *R. irregularis* (Debiane et al. 2011). Lipid metabolism is one of the major metabolisms in AMF since the fungus can be qualified as "oleaginous" (Gaspar et al. 1994). A number of studies have reported on the impact of PAHs on the AMF metabolism, as reviewed in Dalpé et al. (2012). PAHs affected the sterol biosynthesis pathway via a decrease of [1-^{14}C]acetate incorporation in sterol precursors (i.e., 4α-methylsterols) (Calonne et al. 2014b, c). This finding is in agreement with Debiane et al. (2011), which reported a decrease in the 4-demethylsterol content in *R. irregularis* exposed to PAHs. Secondly, when the fungus developed in the presence of benzo[a]pyrene (B[a]P, a high molecular weight PAH), whereas the phosphatidylcholine (the major phospholipid in AMF) quantity decreased, the [1-^{14}C]acetate incorporation in this phospholipid increased. These data could indicate that the AMF promotes the phosphatidylcholine biosynthesis probably in order to regenerate this phospholipid altered by the PAH (Debiane et al. 2011; Calonne et al. 2014c). Storage lipids are also affected by PAHs. Indeed, despite the biosynthesis activation of triacylglycerols (TAGs) in the presence of B[a]P, their quantity decreased in the extraradical phase of the fungus when *R. irregularis* was exposed to this pollutant. This suggested the involvement of the fungal TAG metabolism to cope with B[a]P toxicity by (1) providing carbon skeletons and energy necessary for membrane regeneration and/or for B[a]P translocation and degradation or (2) activating the phosphatidic acid and hexose metabolisms which may be involved in cellular stress defence (Calonne et al. 2014c).

3 AMF-Mediated Plant Protection Mechanisms Against Abiotic Stresses

3.1 Preamble

In the last decade, most studies on abiotic stress factors described a better growth of AMF-colonized crops as compared to non-colonized controls (see Table 1), suggesting an increased tolerance of mycorrhizal plants to abiotic stresses (Dodd and Ruiz-Lozano 2012).

This was reported under salt stress for onions (Hirrel 1981), lettuce (Ruiz-Lozano et al. 1996; Ruiz-Lozano and Azcón 2000), tomato (Al-Karaki 2000), maize (Feng et al. 2002), and pearl millet (Borde et al. 2011) and drought stress for tomato (Beltrano et al. 2013; Latef and Chaoxing 2011b), wheat (Zhou et al. 2015), strawberry (Boyer et al. 2015), and sunflower foxtail millet (Gong et al. 2015). A better growth was also recorded in maize, trifoliate orange, and cyclamen associated with AMF under heat stress (Zhu et al. 2010a, b; Wu 2011; Maya and Matsubara 2013). Finally, in the presence of TEs and petroleum, mineral nutrition was improved in the presence of AMF and resulted in an increased plant growth (Andrade et al. 2004) (see Table 1).

Several mechanisms were suggested by Rivera-Becerril et al. (2005) and Smith and Read (2008):

Table 1 Impact of arbuscular mycorrhizal fungi (AMF) on abiotic stress alleviation in crops reported over the last decade

Stress	AMF	Host plant	Growth conditions	Improved parameters: Growth promotion	Nutrition	Antioxidant system	Osmotic balance	Photosynthesis	Remarks	References
Salt										
0.8 % NaCl in the substrate (8 g/L NaCl)	Mixture of *F. geosporum* and *R. intraradices* (INVAM 167)	*Lycopersicon esculentum Mill.* var. Tamina	Pot culture greenhouse	=			+		Roots −*LePIP1*, −*LeTIP*, =*LePIP2*, =*LeNHX1*, =*LeNHX2* Leaf +*LePIP2*, +*LeTIP*, =*LeNHX1*, =*LeNHX2*, =*LePIP1*	Ouziad et al. (2006)
0.5 and 1 % NaCl solution	*F. mosseae*	*Lycopersicon esculentum* cv. Zhongzha No. 9	Pot culture greenhouse	+					Leakage value of macromolecules =(50 mM) and −(100 mM), +SOD, APX, POD, =CAT, −MDA	He et al. (2007)
50, 100, 150, and 200 mM NaCl	*C. etunicatum* (Sh21)	*Glycine max*	Pot culture greenhouse	+	+P (shoot), +K (shoot), +Ca, +Zn (50 mM), =Zn (100–200 mM)		+		−Proline, −Na	Sharifi et al. (2007)
1.5 and 3 g NaCl/L	Mixture of F. *mosseae*, *R. intraradices*/*irregularis*, *F. coronatum*	*Lactuca sativa* L.	Pot culture greenhouse	+	+K (leaves), −K (roots), +P (shoots)			+	+Leaf area, −Na, −Cl	Zuccarini (2007)
4, 6, and 8 dS/m	*F. mosseae*	*Cajanus cajan* (L.)	Pot culture greenhouse			+			+Leghemoglobin, +SOD, APX, CAT, POX, GR, −MDA	Garg and Manchanda. (2008)
50 mM NaCl	*R. irregularis* (MUCL43194)	*Lactuca sativa* L. cv. Romana	Pot culture glasshouse	+		+	+		+RWC (shoot), −proline, −ABA −*Lsp5cs*, −*LsLea*, −*Lsnced*	Jahromi et al. (2008)
100 mM NaCl									+RWC (shoot), −proline, −ABA =Lsp5cs, −LsLea, =Lsnced	

(continued)

Table 1 (continued)

Stress	AMF	Host plant	Growth conditions	Improved parameters: Growth promotion	Nutrition	Antioxidant system	Osmotic balance	Photosynthesis	Remarks	References
0.5, 1.0, 1.5, and 2.0 g NaCl/kg dry substrate	*F. mosseae*	*Zea mays*	Pot culture greenhouse	=			+	+	−Intercellular O_2, +chlorophyll, +photosynthetic rate, +Fv/Fm, +ϕPSII, +qP, +RWC (leaves), +stomatal conductance	Sheng et al. (2008)
EC = 0.860 mS/cm and 5.6 mS/cm	*R. intraradices*	*Ocimum basilicum* L.	Pot culture greenhouse	+				+	−Na, −Cl, +Fv/Fm	Zuccarini and Okurowska (2008)
75 ppm NaCl	*R. intraradices*/ *irregularis*	*Capsicum annuum* L.	Pot culture growth chamber	+	+P, K, Ca (shoot) −P, −Ca, +K (root)				−Na	Turkmen et al. (2008)
	Gi. margarita				+P, −K, −Ca (shoot) −P, −K, −Ca (root)					
50 and 100 mM NaCl	*R. clarus*	*Capsicum annum*	Pot culture glasshouse	+	+P (leaf)		+	+	−Proline, membrane permeability	Kaya et al. (2009)
75 mM NaCl	*Gi. rosea* BEG9	*Cucumis sativus* L.	Pot culture growth chamber	+				+	+Number of leaves, +leaf area, +root length, root surface area and tip number, +photosynthetic performance index, +Fv/Fm	Gamalero et al. (2010)
EC = 5 dS/m and 10 dS/m	*R. intraradices*	*Solanum lycopersicum* L. cultivars Behta and Piazar	Pot culture greenhouse	+	+P, Ca, K +Ca/Na, +K/Na	+	+	+	+Electron transport rate, +APX, CAT, POD, SOD, proteins, −H2O2, MDA, +proline +stomatal conductance to water vapor, +transpiration, +Fv/Fm, F/Fo, PSII +photosynthetic rate + activity of ROS scavenging enzymes, $-H_2O_2$, −lipid peroxidation, +proline, +stomatal conductance, +protection photochemical processes of PSII	Hajiboland et al. (2010)

0.5 and 1.0 % NaCl solution	*F. mosseae*	*Lycopersicon esculentum* L. cultivar Zhongzha No. 9	Pot culture greenhouse	+		+			+SOD, POD, ASA-POD, =CAT, $-O_2$, −MDA, +leaf area	Huang et al. (2010)
		Lycopersicon esculentum L.							+SOD, POD, ASA-POD, CAT $+O_2$ production rate, −MDA	
2 and 4 g NaCl/kg soil	*F. mosseae*	*Lactuca sativa* L. cv. Tafalla	Mesocosm	+					+Na (soil), +glomalin-related soil protein +% Aggregate stability	Kohler et al. (2010)
100 mM NaCl	Mixture of *R. intraradices*/ *irregularis*, *F. mosseae*, *G. aggregatum*, *R. clarus*, *G. monosporum*, *G. deserticola*, *P. brasilianum*, *C. etunicatum*, *Gi. margarita*	*Lycopersicon esculentum* L.	Pot culture greenhouse	+				+		Demir et al. (2011)
42 mM NaCl		Grafted *Lycopersicum esculentum* Gökçe F1, Maxifort, and Beaufort		+						Oztekin et al. (2013)
30 and 60 mM NaCl	*R. irregularis* (MUCL43194)	*Fragaria* × *ananassa*	Pot culture greenhouse	+					+Root growth, −fine root length, +medium root length	Fan et al. (2011, 2012)
50 and 100 mM NaCl	*F. mosseae*	*Lycopersicon esculentum* L. cv. Zhongzha No. 105	Pot culture greenhouse	+	+P, =K	+		+	+Fruit fresh weight and fruit yield, +SOD, CAT, POD, APX, −MDA, −Na, +leaf area	Latef and Chaoxing (2011b)
1, 2, 4, and 8 mM NaCl	*F. mosseae*, *R. intraradices*	*Capsicum annuum*	Pot culture greenhouse		+P	+	+	+	*F. mosseae:* +total carotenoid content but with *R. intraradices*, +SOD, +APX, +CAT, −MDA, +RWC	Cekic et al. (2012)
50, 100, and 200 mM NaCl	*R. intraradices* (GA4)	*Capsicum annuum* L.	Pot culture greenhouse	+	+K, +P		+		+Leaves, cell membrane stability, leaf area, −Na, proline	Beltrano et al. (2013)

(continued)

Table 1 (continued)

Stress	AMF	Host plant	Growth conditions	Improved parameters: Growth promotion	Nutrition	Antioxidant system	Osmotic balance	Photosynthesis	Remarks	References
66 and 100 mM NaCl	*R. intraradices*, *C. etunicatum*, *S. constrictum*	*Zea mays* L.	Pot culture glasshouse	+	+K		+		*+Expression of ZmAKT2,ZmSOS1, ZmSKOR (homeostasis regulation)*, −proline (roots), −Na, −Cl	Estrada et al. (2013a)
66 mM NaCl	*R. intraradices*, *C. etunicatum*, *F. constrictum* (native from saline soils), *R. intraradices* (isolate EEZ 58)			=Ri collect, Sc (SDW), +Ri, Ce (SDW)	+K+ (shoots), =all except +*C. etunicatum* (roots)			+	−Cl- Ri collect, Ce, =Cl- Ri, Sc (shoots), +Cl- all (roots)	Estrada et al. (2013b)
100 mM NaCl				−Ri collect, Sc (SDW), +Ri, Ce (SDW)	+K			+	−Cl- all, except =Sc (shoot), +Cl- all (roots)	
50, 100, 200 mM NaCl	*R. intraradices* (CMCC Wep319)	*Trigonella foenum-graecum*	Pot culture under natural conditions			+	+	+	−Plasma membrane detachment, +cytoplasm vesicles, −nucleus chromatin condensation, +chloroplasts structure, −number thylakoids, +plastoglobules −Putrescine, +spermine and spermidine, +total sugars −Proline, +glycine betaine, +αtocopherol, mitochondria conservation	Evelin et al. (2013)
EC = 0.9, 4.2, and 7.1 mS/cm	*F. mosseae*	*Lycopersicum esculentum* cv. Zhongzha No. 9	Pot culture greenhouse	+			+		−Transcription vacuolar Na^+/H^+ antiporter gene (*LeNHX1*), +leaf area, +stem sap flow, +transpiration, +root water uptake capacities −Na^+ and Cl^- content	He and Huang (2013)

EC = 4.7, and 9.4 dS/m	Mixture of *Glomus* spp.	*Triticum aestivum* L. cultivar Giza 168	Pot culture greenhouse	+	+N, P, K, Fe, Zn, Cu	+			+Putrescine, −spermine and spermidine (Giza168), −Na	Talaat, and Shawky (2013)
		Triticum aestivum L. cultivar Sids1							−Putrescine, +spermine and spermidine (Sids1), −Na	
		Triticum aestivum L. cultivars Sids 1 and Giza 168		+	+N, K, nitrates			+	+Stomatal conductance, +net CO_2 assimilation rate, +free amino acids, $-H_2O_2$, −lipid peroxidation, +carbohydrates, proteins, RWC, soluble sugars, +proline, glycine betaine	Talaat, and Shawky (2014)
200 mM NaCl	*F. caledonium* (DAOM 193528) *F. mosseae* (DAOM 194475) *R. irregularis* (MUCL 43194) (individual and mixture)	*Fragaria* × *ananassa* Duch.	Pot culture greenhouse	+					+Fruit quality	Sinclair et al. (2013, 2014)
EC = 4.19 dS/m	*R. iranicus* var. tenuihypharum sp. nova	*Lactuca sativa* L.	Pot culture greenhouse	+	+N, Ca, K		+	+	Urban-treated wastewater, −Na, +stomatal conductance, +WUE	Vicente-Sánchez et al. (2014)
2 and 4 g NaCl/kg soil	*F. mosseae*	*Phaseolus vulgaris* L.	Pot culture greenhouse	+	+N, P, K		+		+Proline, =soluble sugars, −Na, +shoot water content	Younesi and Moradi (2014a)
EC = 6 and 12 dS/m	*F. mosseae*	*Glycine max*	Pot culture greenhouse	+		+			+Nodulation with rhizobacteria, +SOD, CAT, POX, GR, =APX	Younesi and Moradi (2014b)
50 and 100 mM NaCl	Mixture of *F. mosseae*, *R. intraradices*, *C. etunicatum*	*Vicia faba*	Pot culture greenhouse	+	+P, +Ca	+		+	+Acid and alkaline phosphatase, -MDA, +SOD,CAT,PX,APX, +nodule activity, +polyamines (putrescine, spermidine, spermine)	Abeer et al. (2014)
50 and 150 mM NaCl		*Solanum lycopersicum* L. var. Castle rock		+		+	+		+Proline, −MDA, +membrane stability index	Abeer et al. (2015)
25 mM NaCl	*C. etunicatum*	*Zea mays*	Pot culture greenhouse	+	+N, P, K, Ca, Mg, =Na				−Proline	Lee et al. (2015)

(continued)

Table 1 (continued)

Stress	AMF	Host plant	Growth conditions	Improved parameters: Growth promotion	Nutrition	Antioxidant system	Osmotic balance	Photosynthesis	Remarks	References
Drought										
Water irrigation EC = 2.4 dS/m soil EC = 4.4 dS/m	*F. mosseae*	*Lycopersicon esculentum* Mill. Cv. Marriha	Pot culture greenhouse	−SDW, +RDW	+P, K, Zn, Cu, Fe				+Yield, +fruit fresh yield, −Na (shoot)	Al-Karaki (2006)
0.50, 0.75, 1.00, and 1.25 mm irrigation water	*R. intraradices* (# TNAU 120-02)	*Lycopersicum esculentum*	Field experiment	+(moderate drought), =(severe drought)	+P		+		+Leaf relative water content, +WUE	Subramanian et al. (2006)
75 % of field capacity during 10 days	*R. irregularis* BEG 121	*Lactuca sativa* L. cv. Romana	Pot culture greenhouse	+			+		+ABA level regulation, +root hydraulic conductivity, −transpiration rate	Aroca et al. (2008)
Soil water content reduced to 12 % of well watered (34.5 % of WHC)	*R. intraradices/irregularis* (BEG 110)	*Ipomea batatas*	Pot culture glasshouse	+	+P, +Zn, +Cu		=		=Evapotranspiration	Neumann et al. (2009)
40 % and 20 % volume of well-watered regime	*R. intraradices/irregularis*	*Oryza sativa*	Pot culture growth chamber	=		+	+	+	$-H_2O_2$, −lipid peroxides, =stomatal conductance, =proline, +glutathione (shoot)	Ruiz-Sánchez et al. (2010)
10 % of weight	*F. mosseae*	*Fragaria*	Pot culture greenhouse			+	+	+	−MDA content, −plasma membrane conductivity, $+H^+$ ATPase activity, +CAT, +APX, +leaf pigment, −chlorophyll decomposition, +free proline, +soluble protein, +transport speed, +soluble sugars, +osmoregulation	Yin et al. (2010)
35–45 % of the relative substrate moisture content (63 % field capacity)	*F. mosseae*, *G. versiforme*, *R. intraradices/irregularis*	*Cucumis melo* L.	Pot culture greenhouse	+ (*F. mosseae* >*G. versiforme* >*R. intraradices/irregularis*)		+	+		+SOD, +GR, −POD, +CAT, +WUE, +soluble sugars	Huang et al. (2011)
50 % reduced water for 4 weeks and 25 % 2 weeks more	*R. intraradices/irregularis*	*Oryza sativa* L.	Pot culture greenhouse	+		+		+	+Ascorbate, +oxidative damage to lipids, +stomatal conductance −Shoot water potential, proline (shoot)	Ruíz-Sánchez et al. (2011)

55 % of relative soil water content, 4 weeks	*C. etunicatum*	*Zea mays* L.	Pot culture greenhouse	=SDW, =RDW		+	+	+	−MDA, +SOD, +CAT, +SOD activity (roots) +POD (leaves and roots), −relative permeability of leaf membrane, +leaf proline, −root proline	Zhu et al. (2011b)
									$-CO_2$ concentration, +transpiration rate, +stomatal conductance, +RWC, +WUE, +Net photosynthetic rate, +chlorophyll concentration, +maximal fluorescence, +maximum quantum PSII, +potential photochemical efficiency, −primary fluorescence	Zhu et al. (2012)
Cyclic drought at 2/3 and 1/2 of FC	Mixture of *R. intraradices* and *F. mosseae*	*Batavia Rubia Munguía* L. *sativa* L. var. Capitata	Pot culture greenhouse	+		+		+	+Carotenoids, +anthocyanins, +phenolics	Baslam and Goicoechea (2012)
Available water capacity <45 %	Mixture of *R. irregularis* (BEG140) *F. mosseae* (BEG95) *C. etunicatum* (BEG92) *C. claroideum* (BEG96) *G. microaggregatum* (BEG56) *F. geosporum* (BEG199)	*Capsicum annuum* L. 'Slávy' F1	Field experiment in trays				+	+	+transpiration rate, +gs, +WUE, +photosynthetic rate	Jezdinský et al. (2012)
−0.6 and −1.2 MPa	*C. etunicatum*, *Endogone versiformis*, *R. intraradices*	*Cicer arietinum* L.	Pot culture greenhouse	+		+		+	+Chlorophyll a (leaf), +POD, POX, −CAT, −lipid peroxidation, +leaf soluble protein	Sohrabi et al. (2012)
Irrigating after 60 and 80 % water depletion	*F. mosseae*, *G. hoi*	*Helianthus annuus* cv. Alestar	Field experiment	+	+P, +N (leaves and seeds)				+Dry matter, +seeds' weight and size, +oil yields	Gholamhoseini et al. (2013)
7 weeks, 33 % field capacity	*F. mosseae*	*Zea mays* L.	Pot culture greenhouse	+	+P	+	+		−Proline, +soluble proteins	Abdelmoneim et al. (2014)

(continued)

Table 1 (continued)

Stress	AMF	Host plant	Growth conditions	Improved parameters					Remarks	References
				Growth promotion	Nutrition	Antioxidant system	Osmotic balance	Photosynthesis		
Two levels of irrigation regimes (normal and deficit)	Autochthonous strains	*Sorghum bicolor* (L.) *Moench*	Field experiment	+Grain yield	+P					Afshar et al. (2014)
4 days (short-term drought) 12 days (sustained drought)	*R. intraradices* (strain EEZ 58)	*Zea mays* L. Potro	Pot culture greenhouse	+			+		+Sap flow rate, +root hydraulic conductance *Short term*: +*ZmTIP1;2*, +*ZmPip1;4*, *ZmPIP2;4*, *ZmNIP1;2*, *ZmTIP2;3*, +*ZmPIP2;2*, *ZmTIP1;1*, +*ZmPIP1;2*, *ZmPIP1;6*, *ZmSIP2;1* *Long term:* −*ZmPIP1;1*, *ZmPIP1;3*, *ZmNIP2;2*, −*ZmTIP1;2*, −*ZmPIP2;2*, *ZmTIP1;1*, =*ZmPip1;4*, *ZmPIP2;4*, *ZmNIP1;2*, *ZmTIP2;3*, =*ZmPIP1;2*, *ZmPIP1;6*, *ZmSIP2;1*, −*ZmNIP1;1*, *ZMTIP4;1*, *ZmTIP4;2*	Bárzana et al. (2014)
35 % field water capacity	*R. intraradices* (BCG AH01)	*Hordeum vulgare* L. cv. Pallas	Pot culture growth chamber	+	+P		+	+	+Leaf water potential, +transpiration rate, +stomatal conductance, +WUE	Li et al. (2014)
50 % of water holding capacity, for 8 weeks	Mixture of *F. constrictum* (EEZ198) *D. aurantium* (EEZ199) *Archaespora trappei* (EEZ200) *G. versiforme* (EEZ201) *P. occultum* (EEZ202)	*Zea mays*	Pot culture greenhouse	=	=N, =C, +P, =K, Mg, Ca, B, Fe, Br	+	+	=	Shoot: +root hydraulic conductivity, −proline, +GH, +*ZmPIP1;6*, +*ZmPIP2;5*, −*ZmPIP2;4*, −*ZmPIP2;6* Root: −MDA, $-H_2O_2$, −proline, −GH, =*ZmPip1;1*, =*ZmPIP1;3*, +*ZmPIP2;3*, +*ZmPIP2;4*	Armada et al. (2015)

60 and 70 % of water lost by evapotranspiration	*F. mosseae* (BEG25) *F. geosporum* (BEG11) (individual and mixture)	*Fragaria* × *ananassa*	Pot culture greenhouse	+			+	-	+WUE, −chlorophyll content/leaf area	Boyer et al. (2015)
−0.68 MPa	*R. intraradices*/*irregularis* (BCG AH01)	*Setaria italica* L.	Pot culture greenhouse	+		+			+SOD, +CAT, +POD, +GR, −MDA, $-H_2O_2$, $-O_2$	Gong et al. (2015)
7 % of volumetric soil moisture	*F. constrictum*, *Glomus* sp., *G. aggregatum* (individual and mixture)	*Glycine max*	Pot culture greenhouse	+SDW, −RDW	+P, +N	+	+	=	−R/S ratio, −MDA, =soluble sugars, +proline (*Glomus* sp., *G. aggregatum*), +shoot water content (*F. constrictum*, *Glomus* sp.) −Shoot water content (*G. aggregatum* and mixture)	Grümberg et al. (2015)
60 % (0.14 Mpa) and 40 % (0.38 MPa) of field capacity	*R. intraradices* (BGC BJ09)	*Zea mays* L.	Pot culture greenhouse	+	+P (shoot), +N, +Mg		+		−WUE, +leaf moisture % of fresh weight	Zhao et al. (2015)
40 % field capacity	*R. intraradices*/*irregularis* (Mix MYKE PRO SP3)	*Triticum aestivum* var. Superb	Pot culture greenhouse	+	+N, P (stem, grain), +ΔC13 (leaves)		+		+Relative water content, +WUE, +leaf area	Zhang et al. (2015)
40 % SRWC, restoration at 80 % soil moisture until harvest maturity	Mixture of *R. irregularis* (BEG140) *F. mosseae* (BEG95) *F. geosporum*, *C. claroideum*	*Triticum aestivum* L. cultivar "Vinjett" / *Triticum aestivum* L. cultivar 1110	Pot culture outdoor	+	−N (grains), +N (roots)					Zhou et al. (2015)
Temperature										
35 and 40 °C	*C. etunicatum*	*Zea mays* genotype Zhengdan 958	Pot culture greenhouse	+		+	+ (Leaf) = (Roots)		Root =MDA, −membrane relative permeability, proline +Soluble sugars, SOD, CAT, POD Leaf −MDA, membrane relative permeability, proline =Soluble sugars, +SOD, CAT, POD	Zhu et al. (2010a)
				=			+	+	+Photosynthetic rate, transpiration rate,	Zhu et al. (2011a)

(continued)

Table 1 (continued)

Stress	AMF	Host plant	Growth conditions	Improved parameters: Growth promotion	Nutrition	Antioxidant system	Osmotic balance	Photosynthesis	Remarks	References
									stomatal conductance, water relative content, water use efficiency −Intracellular CO_2 concentration +Fv/Fm, Fv/Fo, =Fo, −Fm	
Trace elements and metalloids										
Al (12, 24, 48 ppm)	*R. irregularis* (MUCL43194)	Transformed *Daucus carota* root clone DC1	In vitro	=(12, 24 ppm) +(48 ppm)						Gavito et al. (2014)
Pb (1, 4, 10, 100 ppm)				=(1, 4 ppm) +(10 ppm) −(100 ppm)						
As (pots watered every alternate day with 1, 2, 5, 10 mg/L)	*F. mosseae* (UK115)	*Lens culinaris* L. cv. Titore	Pot culture greenhouse	+	+P					Sadeque Ahmed et al. (2006)
Soil contaminated with As (+other TEs) (24, 185, 287 mg/kg soil)	*F. caledonium* (90036)	*Zea mays* L. cv. ChengHai-1	Pot culture greenhouse	=(low and medium As) +(high As)	=P (roots) +P (shoot at high As)				Root and shoot P content increased in the presence of high concentration of As	Bai et al. (2008)
	Mixture including *Glomus* spp. and *Acaulospora* spp.				−P (low As in roots) +P (high As in roots) +P (shoot at medium and high As)					
As (75 and 150 mg/kg soil)	*F. mosseae* BEG 167	*Zea mays* L. cv. Nongda 108	Pot culture glasshouse	=	=					Wang et al. (2008)
	Acaulospora morrowiae (BEG194)			−(root dry wt) =(shoot)	+P (shoot) =P (root)					
As (0, 2.08, and 4.16 mg/kg)	*R. irregularis* (MUCL43194)	*Hordeum vulgare* L. cv. Golden Promise	Pot culture glasshouse	−	−P (shoot) =P (root)					Christophersen et al. (2009)
As (2.5 mg/kg soil)				−(shoot) =(roots)	=P				+*HvPht1;8* −*HvPht1;1*, *HvPht1;2* AMF provides P to the host plant and protect it against As uptake	
As (40, 80 mg/kg soil)	*F. geosporum*, *F. mosseae*, *G. versiforme* (individual and mixture)	*Oryza sativa* L.	Pot culture	+(shoot and roots, especially at 80 mg/kg) =(grains)	+P (especially grain)					Chan et al. (2013)

As (250 mg/kg soil)	*R. intraradices/ irregularis* (AEGIS®)	*Chicorium endivia* cv. Natacha	Pot culture greenhouse	=	+P concentration in leaves =(Roots)					Pigna et al. (2014)
As (25, 50 mg/kg soil)	*R. irregularis* (vch 0011)	Soybean NIDERA 4990	Pot culture greenhouse	+						Spagnoletti and Lavado (2015)
Cd (20 μmol/L)	*R. irregularis*	*Helianthus annuus L.*		+	+P (shoot)			+	−Guaiacol peroxidase	Andrade et al. (2008)
Cd (25, 50 mg/kg) + Pb (500, 800 mg/kg soil)	*F. mosseae*	*Cajanus cajan* (L.) *Millsp.*	Pot culture	+		+			−MDA, electrolyte leakage +SOD, CAT POD, GR	Garg and Aggarwal (2011, 2012)
Cd (25, 50 mg/kg soil)	*F. mosseae*	*Cajanus cajan* (L.) Millsp.	Pot culture greenhouse	+		+	+	+	+Phytochelatins	Garg and Chandel (2012)
Cd (0.3, 0.6, 0.9 mM)	*F. mosseae*	*Triticum aestivum* cv. Sardari39	Pot culture growth chamber	+(0.3 mm) =(0.6 and 0.9 mM)				+	=Fv/Fm +Performance index	Shahabivand et al. (2012)
Cd (0.1, 0.5 mM)	*F. mosseae*	*Capsicum annuum* L. cv. Zhongjiao 105	Pot culture	+	+P	+		+	−MDA, =SOD, −POD, +APX +Total sugar	Latef (2013)
Cd (10 and 20 mg/kg)	*R. irregularis, F. mosseae*	*Zea mays* L.	Pot culture greenhouse	+ (shoot)	+P	+			−CAT, POD, SOD =Polyphenol oxidase, proline	Aghababaei et al. (2014) Aghababaei and Raiesi (2015)
		Helianthus annuus L.		=	+P	+			−CAT, POD, SOD =Polyphenol oxidase, proline	
Cd (0.02, 0.20 mM)	*R. irregularis* (BGC USA05) *F. constrictum* (BGC USA02) *F. mosseae* (BGC NM04A)	*Zea mays* L.	Pot culture growth chamber	=(Shoot and roots) +(*G. constrictum* and F. *mosseae* at 0.02 mM						Liu et al. (2014)
Cd (0.1, 0.5, 1.0 mg Cd/L)	*F. mosseae, A. leavis*	*Zea mays* L.	Pot culture greenhouse	+				+	−Proline	Abdelmoneim et al. (2014)
Cu (0. 5, 1.0, 1.5 mg Cu/L)				+						
Cd (100 μM)	Mixture of *F. mosseae, R. intraradices, C. etunicatum*	*Helianthus annuus* L., cultivar, Sakha 53	Pot culture controlled conditions	+	+P	+		+	−MDA, H_2O_2, proline, relative membrane permeability, total phenols +C18:1, C18:2, C18:3 −SOD, POD, +GR, GPX +Acid phosphatases, = alkaline phosphatases	Abd-Allah et al. (2015)

(continued)

Table 1 (continued)

Stress	AMF	Host plant	Growth conditions	Improved parameters: Growth promotion	Nutrition	Antioxidant system	Osmotic balance	Photosynthesis	Remarks	References
Zn (100, 300, and 900 mg/kg soil)	*C. etunicatum* (IAC-42)	*Canavalia ensiformis* (L.) D.C.	Pot culture greenhouse	+	+P, Ca, Mg, Cu =N, K, Mn	+			−MDA (100 mg/kg), =MDA (300, 900 mg/kg) +CAT (100, 300 mg/kg), −CAT (900 mg/kg) +APX (300 mg/kg), =APX (100 and 900 mg/kg) +GR −Amino acids, proline	Andrade et al. (2009)
Cu (50, 150, and 450 mg/dm^3 soil)				+		=			=MDA, APX, SOD, −CAT, GR, +proline, phytochelatins, =free amino acids +GSH (50, 150 mg/dm^3 soil), =GSH (450 mg/dm^3 soil)	Andrade et al. (2010)
Cu (1.5, 3.5, 5.5, 7.5 mM)	*R. irregularis* *C. etunicatum*	*Lycopersicon esculentum*	Pot culture growth chamber	+		+		+	+Soluble sugars =APX, GPX (+*R. irregularis* in shoots), =(roots)	Malekzadeh et al. (2012)
Cu (150, 300, 450 mg/kg soil)	Mixture of autochthonous Glomeromycota	*Helianthus annuus* L.	Pot culture greenhouse	+(shoot) =(root)					Glomalin production superior for *G. claroideum* than autochthonous AMF	Meier et al. (2012)
	C. claroideum			=(shoot) but + at 450 mg/kg) +(roots)						
Ni (20–40 μg/g sand)	*C. etunicatum* (SFONL and SBH 56)	*Sorghum vulgare*	Pot culture greenhouse	+	−P					Amir et al. (2013)
Pb (500, 1000 mg/kg soil)	*F. mosseae* *R. irregularis*	*Zea mays* L.	Pot culture greenhouse	+	+P				+Glomalin	Vahideh et al. (2013)
Zn (0.065, 25, 65 mg/kg soil)	Inoculum from Zn-contaminated soil	*Solanum lycopersicum* L. CV. 76R	Pot culture growth chamber	+	+P					Cavagnaro et al. (2010)
Zn (25, 50, 75 mg/kg soil)				+						Watts-Williams and Cavagnaro (2012)
Zn (2, 20, 50 mg/kg soil)				−						Watts-Williams et al. (2015)

Soil from Pb-polluted waste disposal site	*R. irregularis* BEG75 (from nonpolluted soil	*Zea mays* L. cv. TATO-260	Pot culture greenhouse	+	=(Shoot) =(Roots)				The M plants inoculated with BEG75 had the lowest P concentrations in their shoots, whereas the plants inoculated with Pb isolates AMF showed the highest concentrations	Sudová and Vosátka (2007)
	R. irregularis (PH5-OS) (from Pb-polluted substrate)				=P (shoot) =P (slightly AM roots) +P (highly AM roots)					
	R. irregularis (PH5-IS) (from Pb-polluted substrate, maintained for 45 months in nonpolluted substrate)				=(Shoot) =(Slightly AM roots) +(Highly AM roots)					
Soil contaminated with Cu (848.3 mg/kg) + Zn (4466 mg/kg) + Pb (1141 mg/kg) + Cd (7.0 mg/kg soil)	*F. caledonium* (90036)	*Zea mays* L.	Pot culture greenhouse	+					+Phosphatase and urease activities in soil	Wang et al. (2006)
				=	+P (roots) =P (shoot)					Wang et al. (2007)
	Mixture of *Gi. margarita* (ZJ37), *Gi. decipiens* (ZJ38), *S. gilmori* ZJ39, *Acaulospora* spp., and *Glomus* spp.			=					+Phosphatase and urease activities in soil	Wang et al. (2006)
				=	+P (roots) =P (shoot)					Wang et al. (2007)
C1 polymetallic soil (As, Cr, Mn, and Ni)	*R. intraradices*/*irregularis*	*Helianthus annuus* L.	Pot culture growth chamber	=	+	+		+	−MDA, SOD, POD	Păun et al. (2015)
C2 polymetallic soil (As, Cr, Mn, and Ni)				+	+	+		+		
PAHs, petroleum, diesel, coal										
ANT (6.25, 12.5, 25, and 50 mg/L)	*R. irregularis* (MUCL43194)	Transformed *Cichorium intybus* roots	In vitro	=		+			−MDA, =SOD, −8-OhdG (50 mg/L) No DNA fragmentation in M and NM roots	Debiane et al. (2008)
ANT (30 and 140 mg/L)				+					+POD	Verdin et al. (2006)
B[a]P (35, 70, 140, 280 μM)				+		+			−MDA, 8-OHdG, +SOD −POD (35 and 140 μM)	Debiane et al. (2009)

(continued)

Table 1 (continued)

Stress	AMF	Host plant	Growth conditions	Improved parameters					Remarks	References
				Growth promotion	Nutrition	Antioxidant system	Osmotic balance	Photosynthesis		
ANT, PHE	*R. irregularis*, *G. versiforme*	*Allium porrum* L. cv. Musselburgh	Pot culture greenhouse	+	+ (N and P)					Liu and Dalpé (2009)
ANT, PHE, dBT (60, 120, 240 μM)	*R. custos*	Transformed *Daucus carota* L.	In vitro	+(ANT and PHE) =(dBT)					ANT did not impacted NM and M root dry weight, whereas PHE and dBT did	Aranda et al. (2013)
Diesel (0.05, 0.1, 0.25, 0.5, 1 % of M medium)	*R. irregularis* (MUCL43194)	Transformed *Cichorium intybus* roots	In vitro	+(In the presence of low diesel content: 0.05, 0.1, and 0.25 %)		+			=MDA −SOD +POD	Driai et al. (2015)
Diesel (2, 6, 10, 15 g/kg soil)	*G. constrictum*	*Zea mays* cultivars Yuyu 22	Pot culture	+		+	+	+	*G. constrictum* was isolated in a petroleum-polluted soil −Proline, MDA (10 and 15 g/kg) +SOD, CAT, POD	Tang et al. (2009)
Petroleum (5, 10 g/kg soil)	*R. intraradices/irregularis*	*Avena sativa* cv. Baiyan No. 7	Pot culture greenhouse	+		+	+		−MDA, proline +SOD, CAT, POD +Ureases, sucrases, dehydrogenases	Xun et al. (2015)
Petroleum (3.5 g/kg soil)				+		+			+SOD, urease, dehydrogenase =CAT, POD	Dong et al. (2014)
Crude oil (2, 4, 8 % soil)	*F. mosseae*	*Phaseolus vulgaris* L.	Pot culture screen house	+				+		Nwoko (2014)
Coal mine	*G. aggregatum* (BGC HK02D), *R. intraradices* (BGC BJ09), *F. mosseae* (BGC XJ01)	*Zea mays* L.	Pot culture	+	+(N, P, K, C)					Guo et al. (2014)

Each sign +, =, and − indicates an increase, an equal, or a decrease effect of AMF on host plant as compared to non-mycorrhized (NM) controls, respectively

- Improved mineral nutrition
- Qualitative and quantitative changes of sugars, polyamines, and lipids
- Increased tolerance to oxidative stress
- Modifications in plant physiology (photosynthetic activity, osmotic balance, etc.)
- Root and fungal chelation and inactivation/exclusion of pollutants

3.2 *Improved Mineral Nutrition*

In the literature from the last 10 years, most studies on the impact of abiotic stresses on nutrition of AMF-colonized plants were focused on P, N, and K and to a lesser extent Ca, Mg, Cu, Fe, or S (see Table 1).

The principal mechanisms explaining the increased nutrient content in M plants as compared to NM plants were related to the capacity of extraradical hyphae to extend far beyond the root depletion zone and to solubilize mineral nutrients by the release of organic acids and enzymes. A faster movement of nutrients, mainly P and N, into the hyphae as well as a stimulation of plant P transporters was also noticed. Finally, an increased nodule formation in leguminous plants was observed in the presence of AMF (Bolan 1991; Smith and Read 2008; Christophersen et al. 2009).

3.2.1 Phosphorus

Salinity, drought, high temperatures, TEs, and PAHs have been reported to impact P uptake and accumulation as well as P use efficiency in NM plants. In the presence of AMF, these effects were less marked (Table 1).

Root colonization was mentioned as a key aspect in P accumulation under stress conditions. For instance, in a soil contaminated with lead, Sudová and Vosátka (2007) observed that maize plants heavily colonized by AMF had higher P contents than plants poorly colonized. One reason could be related to the expression of specific P gene transporters in AMF-colonized plants. Similarly, Christophersen et al. (2009) observed in *Hordeum vulgare* L. colonized by *R. irregularis* MUCL 43194 and grown in the presence of As at 2.5 mg/kg soil the downregulation of two plant P transporter genes in M roots as compared to NM roots, while the specific mycorrhizal P transporter gene *HvPht1;8* was upregulated. This indicates that the mycorrhizal P transport pathway is the preferred. By this specific mechanism, the AMF protects its host against the root absorption of As, which can enter the roots via P transporters (Christophersen et al. 2009). Indeed, because As(V) is a P analogue, it is effectively transported across the plasma membrane of plants via P transporters, apparently competing with P (Asher and Reay 1979; Meharg et al. 1994; Christophersen et al. 2009).

3.2.2 Nitrogen

Salt reduces the uptake of N by competition of chloride with nitrate at the level of membrane transporters (Talaat and Shawky 2013). Interestingly, in AMF-colonized crops, several studies mentioned an increased accumulation of N under salt stress conditions (see Table 1). Colonization of leguminous plant by AMF can increase the number of nodules and thus the N content (Giri and Mukerji 2004; Rabie and Almadini 2005; Garg and Manchanda 2008). For instance, Giri and Mukerji (2004) reported a strong effect of AMF inoculation on nodule formation in *Sesbania aegyptiaca* and *Sesbania grandiflora* under salt stress. These authors observed higher leghemoglobin content and nitrogenase activity in M plants.

Several studies focusing on the impact of TEs and diesel on N accumulation reported marked differences between M and NM plants (see Table 1). For instance, a higher accumulation of N was observed in leguminous plants (*Glycine max* L., *Sesbania rostrata*, *Sesbania cannabina*) colonized by *G. macrocarpum* and *F. mosseae* in a TE multi-contaminated soil (Andrade et al. 2004; Lin et al. 2007), while in the presence of diesel (Hernández-Ortega et al. 2012), an equal content of N was reported in M and NM plants. As explained by Hernández-Ortega et al. (2012), the equal content of N in M and NM diesel-stressed roots could be explained by a reduction in nitrate reductase activity in M roots. Indeed, M plants could reduce their direct N uptake because of an increase in inorganic N transfer to the host by the AMF.

3.2.3 Na^+, Cl^-, and Other Minerals

The role of AMF in Na^+ accumulation in plants remains controversial. Some studies reported an increased Na^+ uptake and concentration in shoots of AMF-colonized plants under high saline soils (Allen and Cunningham 1983; Evelin et al. 2009), while others, in most cases, noticed a decrease (Sharifi et al. 2007; Zuccarini and Okurowska 2008; Talaat and Shawky 2011). AMF decreased the Na:Ca ratio and increased the K:Na ratio in *Vicia faba* plants and pepper seedlings thus reducing Na^+ toxicity effects (Rabie and Almadini 2005; Turkmen et al. 2008).

Differences in Cl^- uptake and accumulation in M plants were reported (Evelin et al. 2009). The increase in Cl^- uptake and accumulation could be related to the carbon drain imposed by mycorrhizal hyphae on plants, which enhances the translocation of highly mobile anions like Cl^- from the soil (Buwalda et al. 1983; Graham and Syvertsen 1984).

In saline soils, Talaat and Shawky (2013) noticed a substantial reduction of K accumulation in wheat tissues. They attributed this observation to the competition between Na^+ and K^+ at the level of absorption sites (Epstein and Rains 1987). However, in the presence of AMF, these authors demonstrated that K^+ uptake was significantly increased in wheat, highlighting the regulation of the expression of K^+/Na^+ pumps and their increased activity.

Mycorrhizal plants have shown greater absorption of Mg under salt stress (Wu et al. 2010), while Cantrell and Linderman (2001) reported increased Ca^{2+} uptake in M lettuce (Table 1).

3.3 Qualitative and Quantitative Changes of Sugars, PAs, Abscisic Acid, and Lipids

3.3.1 Sugars

The accumulation of soluble sugars or carbohydrates in AMF-colonized plants has been proposed as a defence mechanism against salt, drought, or Cd and Cu (Porcel and Ruiz-Lozano 2004; Liu et al. 2011; Sheng et al. 2011; Malekzadeh et al. 2012; Talaat and Shawky 2013; Latef 2013). This accumulation, as compared to NM plants, may result from an increased plant photosynthesis (Sheng et al. 2011; Ruiz-Lozano et al. 2012a, b) or growth (Wu and Xia 2006).

Under heavy saline or drought conditions, the structure and function of the PSII reaction center may be damaged and the electron transport in photosynthetic apparatus disrupted (Baker 2008), while these impacts are less marked in AMF-colonized plants (Zhu et al. 2012). In TE-contaminated soils, the alleviated effect of Cu on chlorophyll content and carbohydrate metabolism was explained by a reduced concentration of this pollutant in the shoot (Malekzadeh et al. 2012).

3.3.2 Polyamines

Polyamines (PAs) are aliphatic nitrogen compounds involved in a wide range of regulatory processes such as plant growth promotion, cell division, DNA replication, and cell differentiation (Evans and Malmberg 1989; Groppa and Benavides 2008). They play a specific role in preventing photooxidative damage (Løvaas 1997). The involvement of PAs in abiotic stress response has been proved. An accumulation of putrescine, spermidine, and spermine (three major PAs) and betaine was reported (Groppa and Benavides 2008; Lingua et al. 2008). However, their role in stress alleviation remains to be elucidated (Alcázar et al. 2006).

Under saline conditions, PAs have been proposed as candidates for the regulation of root development (Sannazzaro et al. 2007). Indeed, in AMF-colonized plants, a higher content of total free PAs and glycine betaine was noticed and thus resulted in improved root growth as compared to NM plants (Al-Garni 2006; Sannazzaro et al. 2007). In a recent study, Talaat and Shawky (2013) reported an increase in putrescine associated with low contents of spermidine and spermine in one genotype of wheat colonized by an AMF, while the second wheat genotype showed a decrease in putrescine and increase in spermidine and spermine. In both cases, mycorrhizal symbiosis protected these genotypes (especially the first one) against salinity. They concluded that modulation of PA pool can be one of the mechanisms used by AMF to improve wheat adaptation to saline soils. Although salt stress strongly promotes diamine oxidase activity and PA oxidation (Xing et al. 2007), resulting in ROS accumulation, Talaat and Shawky (2013) reported a reduced activity of diamine oxidase and PA oxidase in salt-stressed M wheat thus reducing oxidative damage.

In the presence of Zn, free and conjugated putrescines decreased and increased, respectively, in NM poplar grown with the TE (Lingua et al. 2008). This demonstrates a Zn-induced stress in this plant. On the other hand, both free and conjugated putrescine concentrations reached values in M-stressed plants which were identical to those obtained for NM poplar grown in the absence of Zn. This suggests that the metal toxicity is mitigated by the presence of the AMF (Lingua et al. 2008).

The molecular basis behind the role of PAs in alleviating stress is still unclear. However, there is evidence that they can act at several metabolic levels, as antioxidant scavengers, and facilitate metal ion compartmentation (Bors et al. 1989; Sharma and Dietz 2006; Lingua et al. 2008).

3.3.3 Proline

Proline can act in the antioxidant system, regulating redox potentials and acting as a hydroxyl radical scavenger and as a mean of reducing acidity in the cell (Prasad and Saradhi 1995; Sharma and Dietz 2006; Zhu et al. 2010a; Ruiz-Lozano et al. 2012a, b). In the osmotic balance, accumulation of proline has been reported to increase plant osmoprotection and to protect macromolecules against denaturation (Kishor et al. 1995, 2005).

Several studies have described a higher proline concentration in M plants as compared to NM plants at different salinity, drought, TEs, or petroleum levels (Hare et al. 1999; Herrera-Rodríguez et al. 2007; Sharifi et al. 2007; Tang et al. 2009; Talaat and Shawky 2011; Xun et al. 2015). This indicates a role in mediating osmotic adjustment by lowering water potential in stressed plants (Sharma and Dietz 2006; Ashraf and Foolad 2007). In a recent study conducted on lettuce grown under drought stress conditions, Ruíz-Lozano et al. (2011) demonstrated that NM plants accumulated more proline in their shoots than M plants, while the reverse was observed in roots.

Conversely, in other studies (Wang et al. 2004; Rabie and Almadini 2005; Alarcón et al. 2008; Jahromi et al. 2008; Andrade et al. 2010; Tang et al. 2009; Zhu et al. 2010a; Sheng et al. 2011; Xun et al. 2015), NM plants were reported to accumulate more proline than M plants under salt, drought, high temperatures, TEs, or petroleum. In this way, proline can only be regarded as a stress indicator suggesting that a less stressed plant accumulates less proline in cells.

3.3.4 Abscisic Acid

Abscisic acid (ABA) is a key hormone in several physiological processes. It affects ion transport in guard cells and influences stomatal conductance and aperture in response to changing water availability and thus plant turgescence (Roelfsema et al. 2004).

A number of studies have reported that AMF plant colonization can alter ABA levels in the host plant under abiotic stresses (Duan et al. 1996; Ludwig-Müller 2000; Estrada-Luna and Davies 2003). Higher foliar water status was associated with lower xylem sap ABA concentrations in M plants (Duan et al. 1996). Reduced ABA content in leaves may be a strategy of AMF plants to improve water relations under drought stress (Barker and Tagu 2000; Ruiz-Lozano 2003; Hause et al. 2007).

3.3.5 Lipids

In a recent study, Debiane et al. (2012) suggested that root colonization by AMF may decrease lipid peroxidation in plants under PAH stress. A decrease of polyunsaturated fatty acids (C18:1, C18:2, and C18:3) was noticed in NM roots, while it remained unchanged in M roots grown in the presence of benzo[a]pyrene (Debiane et al. 2012) or was superior in M sunflower as compared to NM ones grown in the presence of Cd (Abd-Allah et al. 2015). Moreover, modification of root sterol composition was hypothesized to help avoiding translocation of PAHs in root tissues and consequently protect the host against these toxicants (Debiane et al. 2012). A quantitative modification of fatty acids, especially unsaturated fatty acids (C18:1, C18:2, and C18:3) in leaves and the ratio saturated/unsaturated fatty acids in roots, was also observed in M *Miscanthus* × *giganteus* plants (Firmin et al. 2015). These modifications could be considered as a restoration of the membrane optimal lipid properties. In addition, protein expression of an annexin was increased in M plants (Repetto et al. 2003; Aloui et al. 2009). The putative major function of this protein in Golgi-mediated secretion and maturation of newly synthetized cell membrane and wall materials (Repetto et al. 2003) suggests an increase in membrane lipid production in M plants.

3.4 Increased Tolerance to Oxidative Stress

Numerous studies have focused on environmental factors inducing an oxidative stress and the production of ROS that could interact with polyunsaturated fatty acids to generate malondialdehyde (MDA) or with DNA and proteins and cause cell damage or death (Gill and Tuteja 2010). The control of oxidant levels is achieved by antioxidative systems composed of nonenzymatic (e.g., ascorbate, glutathione, polyphenols, tocopherol, vitamins C, E, B6) and enzymatic [e.g., superoxide dismutase (SOD), peroxidase (POD), catalase (CAT), and ascorbate peroxidase (APX)] ROS scavengers (Schützendübel and Polle 2002; Ferrol et al. 2009).

AMF have been reported to protect plants against abiotic-induced oxidative stresses (Ouziad et al. 2005; Hildebrandt et al. 2007; Andrade et al. 2009) caused by salt, drought, high temperatures, or pollutants. This was demonstrated by reduced accumulation of MDA (see Table 1), mainly in roots, and decreased genomic alteration (Zhu et al. 2010a; Wu 2011; Latef and Chaoxing 2011a, b, 2014; Latef 2013; Firmin et al. 2015). Concomitantly, AMF contribute to enhance antioxidant enzymatic and nonenzymatic scavenging systems (SOD, CAT, POD, APX) in plants grown in the presence of excess salt, high temperatures, TEs, petroleum, or PAHs (see Table 1).

SOD activity in AMF was reported about 20 years ago, but genes involved in SOD expression and regulation are still to be identified. Indeed, *F. mosseae* possesses a CuZn-SOD activity, and mycorrhizal clover roots exhibit two additional SOD isoforms as compared to NM roots: a mycCuZn-SOD, specific for the mycorrhizal association, and a Mn-SOD in nodules (Palma et al. 1993). Ruiz-Lozano et al. (2001) observed a marked increase in the expression of the *Mn-sod II* gene in mycorrhizal lettuce plants under drought stress conditions. This overexpression was correlated to an enhanced tolerance of plants to drought. These authors suggested that mycorrhizal protection against oxidative stress caused by drought may be an important mechanism of protection. The gene encoding a CuZn-SOD has been identified in *Gi. margarita* by Lanfranco et al. (2005).

More recently, a *GintSOD1* gene encoding a functional protein that scavenges ROS was identified in *R. irregularis* by González-Guerrero et al. (2010a). The upregulation of *GintSOD1* transcripts in the *R. irregularis* fungal mycelia treated with Cu indicated that the gene product might be involved in the detoxification of ROS. Salinity also induced an upregulation of this gene, providing evidence for a role of *GintSOD1* in the fungal response to the induced oxidative stress (Estrada et al. 2013a).

The involvement of nonenzymatic antioxidant systems such as GSSG/GSH (glutathione and its oxidized form) was also reported in AMF-colonized *Miscanthus* × *giganteus* protection against oxidative stress (Firmin et al. 2015). Indeed, GSH is considered as a major scavenger of ROS and a precursor of phytochelatins which chelate metals. Nonenzymatic mechanisms induced in AMF-colonized plants under abiotic stresses also include compounds able to scavenge directly several ROS, such as ascorbic acid (AsA), glutathione (GSH), α-tocopherol, polyphenols, or flavonoids (Wu et al. 2006a, b; Huang et al. 2008; Wu and Zou 2009; Matsubara 2010; Wu et al. 2010; Scheibe and Beck 2011; Abbaspour et al. 2012; Ruiz-Lozano et al. 2012a, b; Maya and Matsubara 2013). Recently, Aloui et al. (2012) described that *R. irregularis* colonization of *M. truncatula* roots alleviates cadmium stress via the accumulation of isoflavonoids and their derivates, reinforcing the hypothesis that AMF colonization buffered the effect of TE in plant roots.

The improved tolerance of M poplar clones to TEs was mainly associated with a reduced expression of antioxidant genes, both in roots and in leaves (Pallara et al. 2013). In a comparative proteomic approach, Aloui et al. (2009) provided evidence for *R. intraradices*-dependent down-accumulation of Cd stress-plant responsive proteins and concomitant up-accumulation of mycorrhiza-related proteins putatively involved in reducing Cd oxidative toxicity in M plants. Up-accumulated proteins included a cyclophilin, a guanine nucleotide-binding protein, an ubiquitin carboxyl-terminal hydrolase, a thiazole biosynthetic enzyme, an annexin, a glutathione *S*-transferase (GST)-like protein, and an S-adenosylmethionine synthase (Aloui et al. 2009).

3.5 Modifications in Plant Physiology

3.5.1 Osmotic Adjustment/Gas Exchange

Many authors have reported that plants inoculated with AMF are more resistant to drought conditions (Ruiz-Lozano 2003; Allen 2007; Ruiz-Lozano et al. 2012b; see Table 1). This is mostly related to the capacity of extraradical hyphae to reach smaller pores inaccessible to root hairs (Smith and Read 2008). Increased root or plant hydraulic conductivity, adjustment of osmotic balance, and composition of carbohydrates in the presence of AMF are similarly involved in plant resistance to water shortage (Ruiz-Lozano 2003; Augé 2004; Evelin et al. 2009; Zhu et al. 2010b, 2011a). K^+ and Cl^-, glycine betaine, and carbohydrates such as sucrose, pinitol, and mannitol mainly participate in osmotic adjustments (Ruiz-Lozano et al. 2012a, b). Zhu et al. (2010b, 2011a) further demonstrated that AMF-colonized maize plants had higher leaf relative water content and a better water use efficiency as compared with NM plants stressed by heat. This was probably related to an improved water absorption capacity by *C. etunicatum* and to a lesser content of proline in leaf of M maize plants.

In parallel, AMF interfere in plant water uptake via the production of glycoproteins, such as glomalin, which shapes the soil structure through the formation of microaggregates retaining water (Rillig et al. 2002). In addition, hyphae maintain liquid continuity in the substrate and limit the loss of soil hydraulic conductivity caused by air gaps (Allen 2007; Smith et al. 2010; Ruiz-Lozano et al. 2012b).

Finally, aquaporins, which are key proteins involved in water transport (Javot et al. 2003; Katsuhara et al. 2008; Chaumont and Tyerman 2014; Bárzana et al. 2015), have been reported to be regulated by AMF (Aroca et al. 2007, 2008; Ruiz-Lozano et al. 2009; Ruiz-Lozano and Aroca 2010; Bárzana et al. 2014, 2015). Consequently, M plants regulate better the transcellular water flow and cellular water content (Javot and Maurel 2002; Marjanović et al. 2005; Lee et al. 2010; Ruiz-Lozano et al. 2012b; Bárzana et al. 2014, 2015). Nevertheless, the effects of the AM symbiosis on aquaporin genes depend on the severity of drought stress imposed, on the plant species, and on the specific aquaporin gene considered (Aroca et al. 2007; Ruiz-Lozano and Aroca 2010; Bárzana et al. 2014, 2015). Ouziad et al. (2006) showed that after continuous salt treatment in M *Lycopersicon esculentum*, AMF significantly reduced the mRNA transcripts of *LePIP1* and *LeTIP* but not of *LePIP2* in non-treated controls and salt-stressed roots. Therefore, regulation of PIP and TIP aquaporins was expected to be a key player in the regulation of plant water transport by AM symbiosis.

The role of aquaporins in AM symbiosis was suggested to be more complex than simply regulating plant water status (Maurel and Plassard 2011). As reviewed by Bárzana et al. (2014) and Srivastava et al. (2014), they can participate in glycerol, nitrogen, metalloids, and H_2O_2 transport.

To date, a few studies were conducted on AMF aquaporins (Aroca et al. 2009; Li et al. 2013a, b; Bárzana et al. 2015) located in the extraradical mycelium and in the periarbuscular membrane (Li et al. 2013a). The *GintAQP1* expression was upregulated in the extraradical structures when only a fraction of the mycelium developed in the presence of NaCl (Aroca et al. 2009). Recently, an upregulation of *GintAQP1* gene was observed at 75 mM NaCl in an AMF from a collection, whereas this was not the case in a fungus isolated from a salt-contaminated soil. In contrast, at the highest salinity level (150 mM NaCl), the upregulation was found only in the salt-isolated fungus. Thus, this AMF was able to induce the expression of this aquaporin gene when salt in the culture substrate reached high levels (Bárzana et al. 2015). In the presence of salt stress, Aroca et al. (2009) and Bárzana et al. (2015) found some evidences supporting the idea that fungal aquaporins could compensate the downregulation of host plant aquaporins caused by osmotic stress. Furthermore, under drought stress, aquaporin expression in arbuscule-enriched cortical cells and extraradical mycelia of maize roots were also enhanced significantly, as demonstrated in the presence of polyethylene glycol (Li et al. 2013a, b).

Mycorrhizal plants were found to exhibit a higher stomatal conductance thereby increasing transpiration (Duan et al. 1996; Ruiz-Lozano et al. 1996; Dell'Amico et al. 2002; Jahromi et al. 2008; Sheng et al. 2008). The gas exchange capacity thus increases in M plants (Graham and Syvertsen 1984). As explained by Zhu et al. (2011a), the AM symbiosis provides a high gas exchange capacity by decreasing stomatal resistances and by increasing CO_2 assimilation and transpiration fluxes, as they demonstrated in maize plants submitted to high-temperature stress.

3.5.2 Relative Permeability and Electrolyte Leakage

Electrolyte leakage is a measure of ion leakage caused by membrane damage. Enhancement of membrane lipid peroxidation also causes an increase in membrane permeability, exosmosis of electrolytes, and finally injury to the cell membrane system (Zhu et al. 2010b). Under abiotic stresses, AMF-colonized plants maintain a higher electrolyte concentration and a lower membrane permeability than NM plants by preserving the integrity and stability of the membrane (Feng et al. 2002; Garg and Manchanda 2008; Kaya et al. 2009). This was demonstrated by decreases in MDA production and electrolyte leakage (Zhu et al. 2010b; Garg and Aggarwal 2012; Abd-Allah et al. 2015).

3.5.3 Photosynthesis

Under saline conditions, chlorophyll content of AMF-colonized plants was higher as compared to controls (Giri and Mukerji 2004; Sannazzaro et al. 2006; Zuccarini 2007; Colla et al. 2008; Sheng et al. 2008; see Table 1). This suggested that salt interfered less with chlorophyll synthesis in M plants (Giri and Mukerji 2004), leading to a photosynthetic activity (estimated by chlorophyll content) even superior to the nonstressed NM plants (Feng et al. 2002; Giri et al. 2003; Zuccarini 2007; Colla et al. 2008; Kaya et al. 2009; Hajiboland et al. 2010; Latef and Chaoxing 2011a).

High temperatures, Cd, and petroleum were also reported to impact photosynthesis. However, their effects were less pronounced in M plants than in NM ones (see Table 1). The increase in chlorophyll content may be related to an improved transfer of Mg^{2+} by AMF (Giri et al. 2003; Latef 2013) or a lesser pollutant translocation from soil to roots and aerial part (Malekzadeh et al. 2012).

Improvement of photosynthetic activity, structure, and function of photosynthetic apparatus, photosynthetic index, and PSII reactions has been reported in mycorrhizal plants growing under abiotic stress as compared to NM plants (Sheng et al. 2008; Zuccarini and Okurowska 2008; Hajiboland et al. 2010; Zhu et al. 2011a; Shahabivand et al. 2012). Mycorrhiza-inoculated plants also showed higher non-photochemical quenching as compared to NM plants, which can occur as a result of processes that protect the leaves from light-induced damage (Maxwell and Johnson 2000; Sheng et al. 2008). AM symbiosis also triggers the regulation of energy bifurcation between photochemical and non-photochemical events (Sheng et al. 2008).

The Fv/Fm ratio is a chlorophyll fluorescence measuring parameter that expresses the maximum efficiency of PSII (Lazár 2003). Under abiotic stresses, this ratio is generally higher in M plants as compared to NM plants (see Table 1).

3.6 Root and Fungal Chelation and Inactivation/Exclusion of Pollutants

3.6.1 Plant Intracellular Chelation and Inactivation Is Increased in Mycorrhizal Plants

The role of AMF in the accumulation or exclusion of TEs is mixed. Indeed, a higher (Andrade et al. 2008; Punamiya et al. 2010; Ali et al. 2015), equal (Kelkar and Bhalerao 2013; Aghababaei et al. 2014; Caporale et al. 2014; Pigna et al. 2014), or lower (Aloui et al. 2009; Christophersen et al. 2009; Zhang et al. 2009; Garg and Aggarwal 2011; Liu et al. 2011; Aghababaei and Raiesi 2015) TE concentration was noticed in AMF-colonized plants as compared to NM plants. These contrasting results are more than likely related to the association between the fungus and the plant. Indeed, recent results demonstrated that *S. constrictum* enhanced Cd phytostabilization, whereas *F. mosseae* reduced Cd uptake in maize (Liu et al. 2014). These authors suggested that the mechanisms involved in the TE uptake differ among fungi. Similar results were obtained by Rivera-Becerril et al. (2002), Andrade et al. (2005), and Marques et al. (2006). Redon et al. (2009) and Orłowska et al. (2012) also observed that the origin of AMF (isolated from a polluted or nonpolluted soil) could influence the root and shoot TE accumulation. The site of TE accumulation also differs between AMF-colonized and controls plants. For instance, a lesser accumulation of Cd was observed in shoots of M plant, whereas an equal or increased accumulation was measured in M and NM roots (Huang et al. 2006; Redon et al. 2009; Latef 2013; Aghababaei et al. 2014). The TE considered and its concentration also affected the accumulation. For instance, Shahabivand et al. (2012) observed a lower accumulation of Cd in mycorrhizal roots at 0.3 mM as compared to NM ones, while the accumulation was identical between M and NM roots at concentrations 0.6 and 0.9 mM. It is also important to remind that a lower TE concentration found in M plants may be a consequence of the dilution effect caused by a higher biomass of these plants (Plenchette et al. 1983).

It has been proposed that a shift in root-to-shoot biomass partitioning allowed plants to reduce the incidence of TE-induced stress in photosynthetic organs, a process referred to as allocation plasticity (Audet and Charest 2008; Aloui et al. 2011).

A higher Cu-sorption capacity was observed in the cell walls of M roots compared to NM roots, which could be correlated with a significant increase in uronic acids (Zhang et al. 2009). To avoid free metals in the cell cytosol, cytosolic chelators may induce metal chelation. The best-known chelators are metallothioneins (MTs) (González-Guerrero et al. 2009) and phytochelatins (PCs), involved in the cellular detoxification mechanism by forming stable metal-PC complexes (Garg and Aggarwal 2011). The presence of AMF has been reported to induce MT and PC genes in plant grown in the presence of TEs. This confirmed the important role of the fungal symbiont in the regulation of genes involved in TE chelation (Cicatelli et al. 2010; Pallara et al. 2013). However, in tomato grown in the presence of high concentration of Zn or Cd, Northern blot analysis and qRT-PCR showed an equal expression of *Lemt1*, *Lemt3*, *Lemt4* (encoding MT), *Nramp2* (probably encoding a Zn transporter), and *LePcs1* in any conditions tested (the presence of heavy metal or mycorrhizal association). On the other hand, *Lemt2* and *Nramp1* and *Nramp3* expressions were downregulated upon mycorrhizal colonization under heavy metal stress (Ouziad et al. 2005). The decrease of the transcript formation could be explained by a lower concentration of heavy metal inside the plant cells (Ouziad et al. 2005). According to Rivera-Becerril et al. (2005), whereas the expression of $PsMT_A$ did not differ between M and NM pea plants in the presence of heavy metal, the expression of *hgsh2* (encoding a homoglutathione synthetase, precursor of homophytochelatines) gene was significantly enhanced in AMF-colonized pea roots. This suggested a possible role of the homoglutathione pathway in the plant tolerance to Cd, which was enhanced by mycorrhizal colonization (Rivera-Becerril et al. 2005). This also indicated that Cd chelation pathways does not contribute significantly to metal tolerance strategies operating in the AM symbiosis and argues for an alternative action of the symbiosis at the molecular level (Rivera-Becerril et al. 2005). An increased protein expression of vacuolar H^+-ATP synthase was quantified in *F. mosseae*-colonized pea roots, suggesting a better vacuolar compartmentalization of TEs (Repetto et al. 2003).

PAHs were reported to be stored in lipid bodies of transformed chicory roots cultivated in vitro (Verdin et al. 2006). These authors also demonstrated a lower anthracene accumulation in M roots as compared to non-colonized roots, demonstrating a protective role of AMF in decreasing the pollutant accumulation in the host plant. However, divergent results on PAH accumulation in M and NM plants grown in pot culture were also reported (Binet et al. 2000; Rabie 2005; Verdin et al. 2006; Gao et al. 2011; Wu et al. 2011; Yu et al. 2011). As stated above, this could be explained by the fungal symbiont, plant host, culture conditions, soil properties, and PAHs studied.

3.6.2 TEs' Fungal Intracellular Binding and Inactivation

Ultrastructural localization of TE in AMF demonstrated that these pollutants are accumulated in all fungal structures, but mainly in the fungal wall and vacuole (González-Guerrero et al. 2008). Vesicles within roots have been shown to store more TE than extraradical hyphae (Weiersbye et al. 1999; Orłowska et al. 2008). Three transporters were identified so far: *GintZnT1* (encoding a putative Zn transporter), *GintABC1* (encoding an ABC transporter), and a P-type ATPase (González-Guerrero et al. 2005, 2010b). The cytosolic chelators were identified as organic acids, amino acids, glutathione, and MTs (Lanfranco et al. 2002; González-Guerrero et al. 2007, 2009). Three MTs were identified in AMF (*GrosMT1*, *GmarMT1*, and *GintMT1* in *Gi. rosea*, *Gi. margarita*, and *R. irregularis*, respectively) (Stommel et al. 2001; Lanfranco et al. 2002; González-Guerrero et al. 2007). Nevertheless, these molecules seemed to be more involved in oxidative stress alleviation than in metal homeostasis, as previously thought (González-Guerrero et al. 2007).

Concerning PAHs, only one study demonstrated the accumulation of anthracene in AMF extraradical hyphae and spores (Verdin et al. 2006). These authors demonstrated that the pollutant is accumulated in fungal lipid bodies. However, the exact mechanism of transport from the soil to the AMF lipid bodies remains unknown. All these data indicate that AMF operate an intracellular compartmentalization in order to protect themselves against the negative damage caused by pollutants (Ferrol et al. 2009). Moreover, as reported by Aloui et al. (2009), Cd stress alleviation in M plants grown in contaminated soils is mainly attributed to reduced heavy metal translocation from soil to roots and roots to shoots likely due to Cd immobilization by the extraradical mycelium and intraradical hyphae of AMF, respectively (Joner and Leyval 1997; Joner et al. 2000; Gonzalez-Chavez et al. 2002).

3.6.3 Fungal Extracellular and Cell Wall-Binding Immobilization/Chelation and Inactivation of Pollutants

AMF exude organic acids such as citric, malic, and oxalic acids and amino acids into the rhizosphere, to increase the mobility of metal ions or immobilize and detoxify them through precipitation and complexation (Saraswat and Rai 2011). Glomalin has been reported to stabilize soil. Its concentration in soil however depends on the plant and associated AMF (Rillig et al. 2002). This glycoprotein produced by AMF has been postulated to play a role not only in soil aggregation (Wright and Upadhyaya 1998) but also in Cu, Cd, Zn, As, and Pb sequestration and inactivation in soil (Gonzalez-Chavez et al. 2002; 2004; Cornejo et al. 2008; Ferrol et al. 2009; Amir et al. 2014). Glomalin is also partly located at the AMF wall (Purin and Rillig 2008), which is responsive for 50 % of metal retained (Joner et al. 2000; González-Guerrero et al. 2008). The fungal cell wall also has a high content of chitin with potential metal-binding sites, such as hydroxyls, carboxyls, or amino acids (Strandberg et al. 1981; González-Guerrero et al. 2009). Decreasing the TEs' plant availability plants in soils could be a protective effect conferred by the AMF to its host.

4 Conclusion

Abiotic stresses (i.e., salinity, drought, high temperatures, TEs, and hydrocarbons) are major threats to agriculture, impacting crop yield. Global warming and its cohort of effects (e.g., water scarcity, emergence of new pests, and diseases), combined to an alarming increase of the world population, are major challenges that agriculture has to face in the coming decades. Improved crop varieties, the converting of marginal lands into productive areas, the modifications in management practices, and optimal use of agricultural inputs are among the solutions often considered. In addition, the rhizosphere microbiome and, more precisely, the AMF are increasingly considered since they have been widely reported to increase plant tolerance to several biotic and abiotic stresses. Their application is encouraged by the green wave emerging in the context of sustainable development.

AMF are obligate root symbionts that can develop in disturbed environments and affect plant development in many ways. Under the abiotic stress conditions mentioned above, we noticed that AMF generally improve plant mineral nutrition, especially phosphorus. They induce a better balance of soluble carbohydrate, polyamine, ABA, and lipid content known to be involved in stress alleviation. Oxidative stress mitigation is also frequently reported in AMF-colonized plants as well as pollutant compartmentalization and inactivation.

These observations support the role of AMF in the alleviation of abiotic stresses. The understanding of plant/AMF relationships has increased significantly in the last decade, and although physiological plant parameters affected by AMF under abiotic stress conditions have been well described in the literature, molecular mechanisms behind these effects need further attention.

Only an insignificant fraction of AMF species were isolated from abiotic-stressed soils and their potential investigated. The combination of species adapted to stress environment with, for instance, new crop varieties that can resist abiotic stress factors may represent a novel strategy under agriculture constraints. Indeed, stress alleviation remains fungus, host, and stress level specific. In parallel, the developments of adequate inocula adapted to field applications or agricultural practices favoring local AMF populations are major challenges in the coming years to consider these root symbionts as key players for plant productivity under a changing world.

Acknowledgments The authors thank the European Commission via the ROOTOPOWER consortium [FP7-KBBE-2011-5] (contract #289365) for funding Katia Plouznikoff and the Marie Curie Actions—Intra-European Fellowships (IEF) [FP7-PEOPLE-2013-IEF]—for funding Maryline Calonne-Salmon (under the Grant Agreement No. 623425).

References

Abbaspour H, Saeidi-Sar S, Afshari H et al (2012) Tolerance of mycorrhiza infected pistachio (*Pistacia vera* L.) seedling to drought stress under glasshouse conditions. J Plant Physiol 169: 704–709

Abd-Allah EF, Abeer H, AlQaraki AA et al (2015) Alleviation of adverse impact of cadmium stress in sunflower (*Helianthus annuus* L.) by arbuscular mycorrhizal fungi. Pak J Bot 47: 785–795

Abdelmoneim TS, Moussa TA, Almaghrabi OA et al (2014) Investigation the effect of arbuscular mycorrhizal fungi on the tolerance of maize plant to heavy metals stress. Life Sci J 11:255–263

Abeer H, Abd-Allah EF, Alqarawi AA et al (2014) Alleviation of adverse impact of salinity on faba bean (*Vicia faba* L.) by arbuscular mycorrhizal fungi. Pak J Bot 46:2003–2013

Abeer H, Abd-Allah EF, Alqarawi AA et al (2015) Arbuscular mycorrhizal fungi mitigates NaCl induced adverse effects on *Solanum Lycopersicum* L. Pak J Bot 47:327–340

Abioye OP (2011) Biological remediation of hydrocarbon and heavy metals contaminated soil. In: Pascucci S (ed) Soil contamination. InTech, Rijeka. doi:10.5772/24938

Afshar RK, Jovini MA, Chaichi MR et al (2014) Grain sorghum response to arbuscular mycorrhiza and phosphorus fertilizer under deficit irrigation. Agron J 106:1212–1218

Aghababaei F, Raiesi F, Hosseinpur A (2014) The significant contribution of mycorrhizal fungi and earthworms to maize protection and phytoremediation in Cd-polluted soils. Pedobiologia 57:223–233

Aghababaei F, Raiesi F (2015) Mycorrhizal fungi and earthworms reduce antioxidant enzyme activities in maize and sunflower plants grown in Cd-polluted soils. Soil Biol Biochem 86: 87–97

Al-Garni SS (2006) Increasing NaCl-salt tolerance of a halophytic plant *Phragmites australis* by mycorrhizal symbiosis. Am Eurasian J Agric Environ Sci 1:119–126

Al-Karaki GN (2000) Growth of mycorrhizal tomato and mineral acquisition under salt stress. Mycorrhiza 10:51–54

Al-Karaki GN (2006) Nursery inoculation of tomato with arbuscular mycorrhizal fungi and subsequent performance under irrigation with saline water. Sci Hortic 109:1–7

Al-Karaki GN, McMichael B, Zak J (2004) Field response of wheat to arbuscular mycorrhizal fungi and drought stress. Mycorrhiza 14:263–269

Alarcón A, Delgadillo-Martinez J, Franco-Ramírez A et al (2006) Influence of two polycyclic aromatic hydrocarbons on spore germination, and phytoremediation potential of *Gigaspora margarita-Echynochloa polystachya* symbiosis in benzo[a]pyrene-polluted substrate. Rev Int Contam Ambie 22:39–47

Alarcón A, Davies FT Jr, Autenrieth RL et al (2008) Arbuscular mycorrhiza and petroleum-degrading microorganisms enhance phytoremediation of petroleum-contaminated soil. Int J Phytorem 10:251–263

Albasel N, Cottenie A (1985) Heavy metal contamination near major highways, industrial and urban areas in Belgian grassland. Water Air Soil Pollut 24:103–109

Alcázar R, Marco F, Cuevas JC et al (2006) Involvement of polyamines in plant response to abiotic stress. Biotechnol Lett 28:1867–1876

Ali N, Masood S, Mukhtar T et al (2015) Differential effects of cadmium and chromium on growth, photosynthetic activity, and metal uptake of *Linum usitatissimum* in association with *Glomus intraradices*. Environ Monit Assess 187:1–11

Aliasgharzadeh N, Rastin SN, Towfighi H et al (2001) Occurrence of arbuscular mycorrhizal fungi in saline soils of the Tabriz Plain of Iran in relation to some physical and chemical properties of soil. Mycorrhiza 11:119–122

Allakhverdiev SI, Kreslavski VD, Klimov VV et al (2008) Heat stress: an overview of molecular responses in photosynthesis. Photosynth Res 98:541–550

Allen EB, Cunningham GL (1983) Effects of vesicular-arbuscular mycorrhizae on *Distichlis spicata* under three salinity levels. New Phytol 93:227–236

Allen MF (2007) Mycorrhizal fungi: highways for water and nutrients in arid soils. Vadose Zone J 6:291–297

Aloui A, Recorbet G, Gollotte A et al (2009) On the mechanisms of cadmium stress alleviation in *Medicago truncatula* by arbuscular mycorrhizal symbiosis: a root proteomic study. Proteomics 9:420–433

Aloui A, Recorbet G, Robert F et al (2011) Arbuscular mycorrhizal symbiosis elicits shoot proteome changes that are modified during cadmium stress alleviation in *Medicago truncatula*. BMC Plant Biol 11:75

Aloui A, Dumas-Gaudot E, Daher Z et al (2012) Influence of arbuscular mycorrhizal colonisation on cadmium induced *Medicago truncatula* root isoflavonoid accumulation. Plant Physiol Biochem 60:233–239

Amir H, Lagrange A, Hassaïne N et al (2013) Arbuscular mycorrhizal fungi from New Caledonian ultramafic soils improve tolerance to nickel of endemic plant species. Mycorrhiza 23:585–595

Amir H, Jourand P, Cavaloc Y et al (2014) Role of mycorrhizal fungi in the alleviation of heavy metal toxicity in plants. In: Solaiman ZM, Abbott LK, Varma A (eds) Mycorrhizal fungi: use in sustainable agriculture and land restoration. Springer, New York, pp 241–258

Andrade SAL, Abreu CA, De Abreu MF et al (2004) Influence of lead additions on arbuscular mycorrhiza and Rhizobium symbioses under soybean plants. Appl Soil Ecol 26:123–131

Andrade SAL, Jorge RA, Silveira APD (2005) Cadmium effect on the association of jackbean (*Canavalia ensiformis*) and arbuscular mycorrhizal fungi. Sci Agric 62:389–394

Andrade SAL, da Silveira APD, Jorge RA et al (2008) Cadmium accumulation in sunflower plants influenced by arbuscular mycorrhiza. Int J Phytorem 10:1–13

Andrade SAL, Gratão PL, Schiavinato MA et al (2009) Zn uptake, physiological response and stress attenuation in mycorrhizal jack bean growing in soil with increasing Zn concentrations. Chemosphere 75:1363–1370

Andrade SAL, Gratão PL, Azevedo RA et al (2010) Biochemical and physiological changes in jack bean under mycorrhizal symbiosis growing in soil with increasing Cu concentrations. Environ Exp Bot 68:198–207

Apple ME, Thee CI, Smith-Longozo VL et al (2005) Arbuscular mycorrhizal colonization of *Larrea tridentata* and *Ambrosia dumosa* roots varies with precipitation and season in the Mojave Desert. Symbiosis 39:131–135

Aranda E, Scervino JM, Godoy P et al (2013) Role of arbuscular mycorrhizal fungus *Rhizophagus custos* in the dissipation of PAHs under root-organ culture conditions. Environ Pollut 181: 182–189

Armada E, Azcón R, López-Castillo OM et al (2015) Autochthonous arbuscular mycorrhizal fungi and *Bacillus thuringiensis* from a degraded Mediterranean area can be used to improve physiological traits and performance of a plant of agronomic interest under drought conditions. Plant Physiol Biochem 90:64–74

Aroca R, Porcel R, Ruiz-Lozano JM (2007) How does arbuscular mycorrhizal symbiosis regulate root hydraulic properties and plasma membrane aquaporins in *Phaseolus vulgaris* under drought, cold or salinity stresses? New Phytol 173:808–816

Aroca R, Vernieri P, Ruiz-Lozano JM (2008) Mycorrhizal and non-mycorrhizal *Lactuca sativa* plants exhibit contrasting responses to exogenous ABA during drought stress and recovery. J Exp Bot 59:2029–2041

Aroca R, Bago A, Sutka M et al (2009) Expression analysis of the first arbuscular mycorrhizal fungi aquaporin described reveals concerted gene expression between salt-stressed and nonstressed mycelium. Mol Plant-Microbe Interact 22:1169–1178

Asher CJ, Reay PF (1979) Arsenic uptake by barley seedlings. Funct Plant Biol 6:459–466

Ashraf M, Foolad M (2007) Roles of glycine betaine and proline in improving plant abiotic stress resistance. Environ Exp Bot 59:206–216

Audet P, Charest C (2008) Allocation plasticity and plant–metal partitioning: meta-analytical perspectives in phytoremediation. Environ Pollut 156:290–296

Augé RM (2001) Water relations, drought and vesicular-arbuscular mycorrhizal symbiosis. Mycorrhiza 11:3–42

Augé RM (2004) Arbuscular mycorrhizae and soil/plant water relations. Can J Soil Sci 84: 373–381

Bai J, Lin X, Yin R et al (2008) The influence of arbuscular mycorrhizal fungi on As and P uptake by maize (*Zea mays* L.) from As-contaminated soils. Appl Soil Ecol 38:137–145

Baker NR (2008) Chlorophyll fluorescence: a probe of photosynthesis *in vivo*. Annu Rev Plant Biol 59:89–113

Baon JB, Smith SE, Alston AM (1994) Phosphorus uptake and growth of barley as affected by soil temperature and mycorrhizal infection. J Plant Nutr 17:479–492

Barker SJ, Tagu D (2000) The roles of auxins and cytokinins in mycorrhizal symbioses. J Plant Growth Regul 19:144–154

Barrett G, Campbell CD, Hodge A (2014) The direct response of the external mycelium of arbuscular mycorrhizal fungi to temperature and the implications for nutrient transfer. Soil Biol Biochem 78:109–117

Bárzana G, Aroca R, Bienert GP et al (2014) New insights into the regulation of aquaporins by the arbuscular mycorrhizal symbiosis in maize plants under drought stress and possible implications for plant performance. Mol Plant-Microbe Interact 27:349–363

Bárzana G, Aroca R, Ruiz-Lozano JM (2015) Localized and non-localized effects of arbuscular mycorrhizal symbiosis on accumulation of osmolytes and aquaporins and on antioxidant systems in maize plants subjected to total or partial root drying. Plant Cell Environ 38: 1613–1627

Baslam M, Goicoechea N (2012) Water deficit improved the capacity of arbuscular mycorrhizal fungi (AMF) for inducing the accumulation of antioxidant compounds in lettuce leaves. Mycorrhiza 22:347–359

Beltrano J, Ruscitti M, Arango MC et al (2013) Effects of arbuscular mycorrhiza inoculation on plant growth, biological and physiological parameters and mineral nutrition in pepper grown under different salinity and p levels. J Soil Sci Plant Nutr 13:123–141

Benabdellah K, Merlos MÁ, Azcón-Aguilar C et al (2009a) *GintGRX1*, the first characterized glomeromycotan glutaredoxin, is a multifunctional enzyme that responds to oxidative stress. Fungal Genet Biol 46:94–103

Benabdellah K, Azcón-Aguilar C, Valderas A et al (2009b) *GintPDX1* encodes a protein involved in vitamin B6 biosynthesis that is up-regulated by oxidative stress in the arbuscular mycorrhizal fungus *Glomus intraradices*. New Phytol 184:682–693

Bendavid-Val R, Rabinowitch HD, Katan J et al (1997) Viability of VA-mycorrhizal fungi following soil solarization and fumigation. Plant Soil 195:185–193

Bewley R, Ellis B, Theile P et al (1989) Microbial cleanup of contaminated soils. Chem Ind 23: 778–783

Biache C, Mansuy-Huault L, Faure P et al (2008) Effects of thermal desorption on the composition of two coking plant soils: impact on solvent extractable organic compounds and metal bioavailability. Environ Pollut 156:671–677

Binet P, Portal JM, Leyval C (2000) Fate of polycyclic aromatic hydrocarbons (PAH) in the rhizosphere and mycorrhizosphere of ryegrass. Plant Soil 227:207–213

Bispo A, Jourdain MJ, Jauzein M (1999) Toxicity and genotoxicity of industrial soils polluted by polycyclic aromatic hydrocarbons (PAHs). Org Geochem 30:947–952

Blanchard M, Teil MJ, Ollivon D et al (2004) Polycyclic aromatic hydrocarbons and polychlorobiphenyls in wastewaters and sewage sludges from the Paris area (France). Environ Res 95: 184–197

Bogan BW, Lamar RT, Burgos WD et al (1999) Extent of humification of anthracene, fluoranthene, and benzo[α]pyrene by *Pleurotus ostreatus* during growth in PAH-contaminated soils. Lett Appl Microbiol 28:250–254

Bolan NS (1991) A critical review on the role of mycorrhizal fungi in the uptake of phosphorus by plants. Plant Soil 134:189–207

Borde M, Dudhane M, Jite P (2011) Growth photosynthetic activity and antioxidant responses of mycorrhizal and non-mycorrhizal bajra (*Pennisetum glaucum*) crop under salinity stress condition. Crop Protect 30:265–271

Bors W, Langebartels C, Michel C et al (1989) Polyamines as radical scavengers and protectants against ozone damage. Phytochemistry 28:1589–1595

Bot A, Nachtergaele F, Young A (2000) Land resource potential and constraints at regional and country levels. World soil resources report. FAO, vol 90, p 122

Bothe H (2012) Arbuscular mycorrhiza and salt tolerance of plants. Symbiosis 58:7–16

Boullard B (1959) Relations entre la photopériode et l'abondance des mycorrhizes chez *Aster tripolium* L. Bull Soc Bot Fr 106:131–134

Bowen GD (1987) The biology and physiology of infection and its development. In: Safir GR (ed) Ecophysiology of VA mycorrhizal plants. CRC, Boca Raton, pp 27–57

Boyer LR, Brain P, Xu XM et al (2015) Inoculation of drought-stressed strawberry with a mixed inoculum of two arbuscular mycorrhizal fungi: effects on population dynamics of fungal species in roots and consequential plant tolerance to water deficiency. Mycorrhiza 25: 215–227

Brady NC, Weil RR (2008) The nature and properties of soils, 14th edn. Prentice Hall, Upper Saddle River

Brownell KH, Schneider RW (1985) Roles of matric and osmotic components of water potential and their interaction with temperature in the growth of *Fusarium oxysporum* in synthetic media and soil. Phytopathology 75:53–57

Buwalda JG, Stribley DP, Tinker PB (1983) Increased uptake of bromide and chloride by plants infected with vesicular arbuscular mycorrhizas. New Phytol 93:217–225

Cai QY, Mo CH, Li YH et al (2007) Occurrence and assessment of polycyclic aromatic hydrocarbons in soils from vegetable fields of the Pearl River Delta, South China. Chemosphere 68: 159–168

Cai QY, Mo CH, Wu QT et al (2008) The status of soil contamination by semivolatile organic chemicals (SVOCs) in China: a review. Sci Total Environ 389:209–224

Calonne M, Lounès-Hadj Sahraoui A, Grandmougin-Ferjani A et al (2010) Propiconazole toxicity on the non-target organism, the arbuscular mycorrhizal fungus, *Glomus irregulare*. In: Carisse O (ed) Fungicides. InTech, Rijeka. doi:10.5772/10482

Calonne M, Fontaine J, Tisserant B et al (2014a) Polyaromatic hydrocarbons impair phosphorus transport by the arbuscular mycorrhizal fungus *Rhizophagus irregularis*. Chemosphere 104: 97–104

Calonne M, Fontaine J, Debiane D et al (2014b) Impact of anthracene on the arbuscular mycorrhizal fungus lipid metabolism 1. Botany 92:173–178

Calonne M, Fontaine J, Debiane D et al (2014c) The arbuscular mycorrhizal *Rhizophagus irregularis* activates storage lipid biosynthesis to cope with the benzo[a]pyrene oxidative stress. Phytochemistry 97:30–37

Cantrell IC, Linderman RG (2001) Preinoculation of lettuce and onion with VA mycorrhizal fungi reduces deleterious effects of soil salinity. Plant Soil 233:269–281

Caporale AG, Sarkar D, Datta R et al (2014) Effect of arbuscular mycorrhizal fungi (*Glomus* spp.) on growth and arsenic uptake of vetiver grass (*Chrysopogon zizanioides* L.) from contaminated soil and water systems. J Soil Sci Plant Nutr 14:955–972

Carillo P, Annunziata MG, Pontecorvo G et al (2011) Salinity stress and salt tolerance. In: Shanker AK, Venkateswarlu B (eds) Abiotic stress plants-mechanisms and adaptations. InTech, Rijeka. doi:10.5772/22331

Cavagnaro TR, Dickson S, Smith FA (2010) Arbuscular mycorrhizas modify plant responses to soil zinc addition. Plant Soil 329:307–313

Cekic FO, Ünyayar S, Ortaş İ (2012) Effects of arbuscular mycorrhizal inoculation on biochemical parameters in *Capsicum annuum* grown under long term salt stress. Turk J Bot 36:63–72

Chan WF, Li H, Wu FY et al (2013) Arsenic uptake in upland rice inoculated with a combination or single arbuscular mycorrhizal fungi. J Hazard Mater 262:1116–1122

Chaumont F, Tyerman SD (2014) Aquaporins: highly regulated channels controlling plant water relations. Plant Physiol 164:1600–1618

Chen S, Jin W, Liu A et al (2013) Arbuscular mycorrhizal fungi (AMF) increase growth and secondary metabolism in cucumber subjected to low temperature stress. Sci Hortic 160: 222–229

Chen X, Song F, Liu F et al (2014) Effect of different arbuscular mycorrhizal fungi on growth and physiology of maize at ambient and low temperature regimes. Sci World J 2014(Article ID 956141):7. doi:10.1155/2014/956141

Christophersen HM, Smith FA, Smith SE (2009) Arbuscular mycorrhizal colonization reduces arsenate uptake in barley via downregulation of transporters in the direct epidermal phosphate uptake pathway. New Phytol 184:962–974

Cicatelli A, Lingua G, Todeschini V et al (2010) Arbuscular mycorrhizal fungi restore normal growth in a white poplar clone grown on heavy metal-contaminated soil, and this is associated with upregulation of foliar metallothionein and polyamine biosynthetic gene expression. Ann Bot 106:791–802

Clark JS, Campbell JH, Grizzle H et al (2009) Soil microbial community response to drought and precipitation variability in the Chihuahuan Desert. Microb Ecol 57:248–260

Colla G, Rouphael Y, Cardarelli M et al (2008) Alleviation of salt stress by arbuscular mycorrhizal in zucchini plants grown at low and high phosphorus concentration. Biol Fertil Soils 44: 501–509

Copeman RH, Martin CA, Stutz JC (1996) Tomato growth in response to salinity and mycorrhizal fungi from saline or nonsaline soils. Hortic Sci 31:341–344

Cornejo P, Meier S, Borie G et al (2008) Glomalin-related soil protein in a Mediterranean ecosystem affected by a copper smelter and its contribution to Cu and Zn sequestration. Sci Total Environ 406:154–160

Cornejo P, Pérez-Tienda J, Meier S et al (2013) Copper compartmentalization in spores as a survival strategy of arbuscular mycorrhizal fungi in Cu-polluted environments. Soil Biol Biochem 57: 925–928

Costa FA, Haddad LSAM, Kasuya MCM et al (2013) *In vitro* culture of *Gigaspora decipiens* and *Glomus clarum* in transformed roots of carrot: the influence of temperature and pH. Acta Sci Agron 35:315–323

Crépineau C, Rychen G, Feidt C et al (2003) Contamination of pastures by polycyclic aromatic hydrocarbons (PAHs) in the vicinity of a highway. J Agric Food Chem 51:4841–4845

Crépineau-Ducoulombier C, Rychen G (2003) Assessment of soil and grass polycyclic aromatic hydrocarbon (PAH) contamination levels in agricultural fields located near a motorway and an airport. Agronomie 23:345–348

Daei G, Ardekani MR, Rejali F et al (2009) Alleviation of salinity stress on wheat yield, yield components, and nutrient uptake using arbuscular mycorrhizal fungi under field conditions. J Plant Physiol 166:617–625

Daffonchio D, Hirt H, Berg G (2015) Plant-microbe interactions and water management in arid and saline soils. In: Lugtenberg B (ed) Principles of plant-microbe interactions – microbes for sustainable agriculture. Springer, Leiden

Dalpé Y, Trépanier M, Lounès-Hadj Sahraoui A et al (2012) Lipids of mycorrhizas. In: Hock B (ed) Fungal associations, vol 9. Springer, Berlin, pp 137–169

Daniels BA, Graham SO (1976) Effects of nutrition and soil extracts on germination of *Glomus mosseae* spores. Mycologia 68:108–116

Daniels BA, Trappe JM (1980) Factors affecting spore germination of the vesicular-arbuscular mycorrhizal fungus *Glomus epigaeus*. Mycologia 72:457–471

De la Providencia YE, Stefani FO, Labridy M et al (2015) Arbuscular mycorrhizal fungal diversity associated with *Eleocharis obtusa* and *Panicum capillare* growing in an extreme petroleum hydrocarbon-polluted sedimentation basin. FEMS Microbiol Lett 362:fnv081. doi:10.1093/femsle/fnv081

de Paz JM, Visconti F, Zapata R et al (2004) Integration of two simple models in a geographical information system to evaluate salinization risk in irrigated land of the Valencian Community, Spain. Soil Use Manage 20:333–342

Debiane D, Garçon G, Verdin A et al (2008) *In vitro* evaluation of the oxidative stress and genotoxic potentials of anthracene on mycorrhizal chicory roots. Environ Exp Bot 64:120–127

Debiane D, Garçon G, Verdin A et al (2009) Mycorrhization alleviates benzo[a]pyrene-induced oxidative stress in an *in vitro* chicory root model. Phytochemistry 70:1421–1427

Debiane D, Calonne M, Fontaine J et al (2011) Lipid content disturbance in the arbuscular mycorrhizal, *Glomus irregulare* grown in monoxenic conditions under PAHs pollution. Fungal Biol 115:782–792

Debiane D, Calonne M, Fontaine J et al (2012) Benzo[a]pyrene induced lipid changes in the monoxenic arbuscular mycorrhizal chicory roots. J Hazard Mater 209:18–26

Del Val C, Barea JM, Azcon-Aguilar C (1999a) Diversity of arbuscular mycorrhizal fungus populations in heavy-metal-contaminated soils. Appl Environ Microbiol 65:718–723

Del Val C, Barea JM, Azcón-Aguilar C (1999b) Assessing the tolerance to heavy metals of arbuscular mycorrhizal fungi isolated from sewage sludge-contaminated soils. Appl Soil Ecol 11:261–269

Dell'Amico J, Torrecillas A, Rodriguez P et al (2002) Responses of tomato plants associated with the arbuscular mycorrhizal fungus *Glomus clarum* during drought and recovery. J Agric Sci 138:387–393

Demir K, Başak H, Okay FY et al (2011) The effect of endo-mycorrhiza (VAM) treatment on growth of tomato seedling grown under saline conditions. Afr J Agric Res 6:3326–3332

Dhar PP, Al-Qarawi AA, Mridha MA (2015) Arbuscular mycorrhizal fungal association in Asteraceae plants growing in the arid lands of Saudi Arabia. J Arid Land 7:676–686

Díaz G, Azcón-Aguilar C, Honrubia M (1996) Influence of arbuscular mycorrhizae on heavy metal (Zn and Pb) uptake and growth of *Lygeum spartum* and *Anthyllis cytisoides*. Plant Soil 180:241–249

Dodd IC, Ruiz-Lozano JM (2012) Microbial enhancement of crop resource use efficiency. Curr Opin Biotechnol 23:236–242

Dong R, Gu L, Guo C et al (2014) Effect of PGPR *Serratia marcescens* BC-3 and AMF *Glomus intraradices* on phytoremediation of petroleum contaminated soil. Ecotoxicology 23:674–680
Douds DD, Schenck NC (1991) Germination and hyphal growth of VAM fungi during and after storage in soil at five matric potentials. Soil Biol Biochem 23:177–183
Driai S, Verdin A, Laruelle F et al (2015) Is the arbuscular mycorrhizal fungus *Rhizophagus irregularis* able to fulfil its life cycle in the presence of diesel pollution? Int Biodeter Biodegr 105:58–65
Duan X, Neuman DS, Reiber JM et al (1996) Mycorrhizal influence on hydraulic and hormonal factors implicated in the control of stomatal conductance during drought. J Exp Bot 47: 1541–1550
EEA (1995) Soil. In: Europe's Environment: the Dobris Assessment. European Environment Agency, Copenhagen
Egerton-Warburton LM, Querejeta JI, Allen MF (2007) Common mycorrhizal networks provide a potential pathway for the transfer of hydraulically lifted water between plants. J Exp Bot 58:1473–1483
Elias KS, Safir GR (1987) Hyphal elongation of *Glomus fasciculatus* in response to root exudates. Appl Environ Microbiol 53:1928–1933
Ellis B, Harold P, Kronberg H (1991) Bioremediation of a creosote contaminated site. Environ Technol 12:447–459
Epstein E, Rains DW (1987) Advances in salt tolerance. In: Gabelman WH, Loughman BC (eds) Genetic aspects of plant mineral nutrition. Proceedings of the 2nd international symposium on genetic aspects of plant mineral nutrition, organized by the University of Wisconsin, Madison, June 16–20, 1985. Martinus Nijhoff, Dordrecht, pp 113–126
Erickson DC, Loehr RC, Neuhauser EF (1993) PAH loss during bioremediation of manufactured gas plant site soils. Water Res 27:911–919
Estaun MV (1990) Effect of sodium chloride and mannitol on germination and hyphal growth of the vesicular-arbuscular mycorrhizal fungus *Glomus mosseae*. Agric Ecosyst Environ 29: 123–129
Estaun MV (1991) Effect of NaC1 and mannitol on the germination of two isolates of the vesicular-arbuscular mycorrhizal fungus *Glomus mosseae*. In: Abstracts of the 3rd European Symposium on Mycorrhizas, University of Sheffield, Sheffield
Estrada B, Aroca R, Barea JM et al (2013a) Native arbuscular mycorrhizal fungi isolated from a saline habitat improved maize antioxidant systems and plant tolerance to salinity. Plant Sci 201:42–51
Estrada B, Aroca R, Maathuis FJ et al (2013b) Arbuscular mycorrhizal fungi native from a Mediterranean saline area enhance maize tolerance to salinity through improved ion homeostasis. Plant Cell Environ 36:1771–1782
Estrada-Luna AA, Davies FT (2003) Arbuscular mycorrhizal fungi influence water relations, gas exchange, abscisic acid and growth of micropropagated chile ancho pepper (*Capsicum annuum*) plantlets during acclimatization and post-acclimatization. J Plant Physiol 160: 1073–1083
Evans PT, Malmberg RL (1989) Do polyamines have roles in plant development? Annu Rev Plant Biol 40:235–269
Evelin H, Kapoor R, Giri B (2009) Arbuscular mycorrhizal fungi in alleviation of salt stress: a review. Ann Bot 104:1263–1280
Evelin H, Giri B, Kapoor R (2013) Ultrastructural evidence for AMF mediated salt stress mitigation in *Trigonella foenum-graecum*. Mycorrhiza 23:71–86
FAO (2009) How to feed the world in 2050? Report of the expert meeting on how to feed the world in 2050, 24–26 June 2009. FAO Headquarters, Rome
Fan L, Dalpé Y, Fang C et al (2011) Influence of arbuscular mycorrhizae on biomass and root morphology of selected strawberry cultivars under salt stress. Botany 89:397–403
Fan L, Fang C, Dubé C et al (2012) Arbuscular mycorrhiza alleviates salinity stress of strawberry cultivars under salinity condition. Acta Hortic 926:491–496

Feng G, Zhang F, Li X et al (2002) Improved tolerance of maize plants to salt stress by arbuscular mycorrhiza is related to higher accumulation of soluble sugars in roots. Mycorrhiza 12: 185–190

Ferrol N, Gonzalez-Guerrero M, Valderas A et al (2009) Survival strategies of arbuscular mycorrhizal fungi in Cu-polluted environments. Phytochem Rev 8:551–559

Firmin S, Labidi S, Fontaine J et al (2015) Arbuscular mycorrhizal fungal inoculation protects *Miscanthus* × *giganteus* against trace element toxicity in a highly metal-contaminated site. Sci Total Environ 527:91–99

Franco-Ramírez A, Ferrera-Cerrato R, Varela-Fregoso L et al (2007) Arbuscular mycorrhizal fungi in chronically petroleum-contaminated soils in Mexico and the effects of petroleum hydrocarbons on spore germination. J Basic Microbiol 47:378–383

Gai JP, Tian H, Yang FY et al (2012) Arbuscular mycorrhizal fungal diversity along a Tibetan elevation gradient. Pedobiologia 55:145–151

Gamalero E, Berta G, Massa N et al (2010) Interactions between *Pseudomonas putida* UW4 and *Gigaspora rosea* BEG9 and their consequences for the growth of cucumber under salt-stress conditions. J Appl Microbiol 108:236–245

Gao Y, Li Q, Ling W et al (2011) Arbuscular mycorrhizal phytoremediation of soils contaminated with phenanthrene and pyrene. J Hazard Mater 185:703–709

Garg N, Manchanda G (2008) Effect of arbuscular mycorrhizal inoculation on salt-induced nodule senescence in *Cajanus cajan* (pigeonpea). J Plant Growth Regul 27:115–124

Garg N, Aggarwal N (2011) Effects of interactions between cadmium and lead on growth, nitrogen fixation, phytochelatin, and glutathione production in mycorrhizal *Cajanus cajan* (L.) Millsp. J Plant Growth Regul 30:286–300

Garg N, Aggarwal N (2012) Effect of mycorrhizal inoculations on heavy metal uptake and stress alleviation of *Cajanus cajan* (L.) Millsp. genotypes grown in cadmium and lead contaminated soils. J Plant Growth Regul 66:9–26

Garg N, Chandel S (2012) Role of arbuscular mycorrhizal (AM) fungi on growth, cadmium uptake, osmolyte, and phytochelatin synthesis in *Cajanus cajan* (L.) Millsp. under NaCl and Cd stresses. J Plant Growth Regul 31:292–308

Gaspar ML, Pollero RJ, Cabello MN (1994) Triacylglycerol consumption during spore germination of vesicular-arbuscular mycorrhizal fungi. J Am Oil Chem Soc 71:449–452

Gaspar ML, Cabello MN, Cazau M et al (2002) Effect of phenanthrene and *Rhodotorula glutinis* on arbuscular mycorrhizal fungus colonization of maize roots. Mycorrhiza 12:55–59

Gavito ME, Schweiger P, Jakobsen I (2003) P uptake by arbuscular mycorrhizal hyphae: effect of soil temperature and atmospheric CO_2 enrichment. Glob Chang Biol 9:106–116

Gavito ME, Olsson PA, Rouhier H et al (2005) Temperature constraints on the growth and functioning of root organ cultures with arbuscular mycorrhizal fungi. New Phytol 168:179–188

Gavito ME, Abud YC, Sántiz YM et al (2014) Effects of aluminum and lead on the development of *Rhizophagus irregularis* and roots in root cultures. Environ Eng Manage J 13:2357–2361

Gianinazzi-Pearson V, Branzanti B, Gianinazzi S (1989) *In vitro* enhancement of spore germination and early hyphal growth of a vesicular–arbuscular mycorrhizal fungus by host root exudates and plant flavonoids. Symbiosis 7:243–255

Gildon A, Tinker PB (1983) Interactions of vesicular-arbuscular mycorrhizal infection and heavy metals in plants. New Phytol 95:247–261

Gill SS, Tuteja N (2010) Reactive oxygen species and antioxidant machinery in abiotic stress tolerance in crop plants. Plant Physiol Biochem 48:909–930

Giri B, Mukerji KG (2004) Mycorrhizal inoculant alleviates salt stress in *Sesbania aegyptiaca* and *Sesbania grandiflora* under field conditions: evidence for reduced sodium and improved magnesium uptake. Mycorrhiza 14:307–312

Giri B, Kapoor R, Mukerji KG (2003) Influence of arbuscular mycorrhizal fungi and salinity on growth, biomass, and mineral nutrition of *Acacia auriculiformis*. Biol Fertil Soils 38:170–175

Gholamhoseini M, Ghalavand A, Dolatabadian A et al (2013) Effects of arbuscular mycorrhizal inoculation on growth, yield, nutrient uptake and irrigation water productivity of sunflowers grown under drought stress. Agric Water Manag 117:106–114

Gong M, You X, Zhang Q (2015) Effects of *Glomus intraradices* on the growth and reactive oxygen metabolism of foxtail millet under drought. Ann Microbiol 65:595–602

Gonzalez-Chavez MC, D'haen J, Vangronsveld J et al (2002) Copper sorption and accumulation by the extraradical mycelium of different *Glomus* spp. (arbuscular mycorrhizal fungi) isolated from the same polluted soil. Plant Soil 240:287–297

Gonzalez-Chavez MC, Carrillo-Gonzalez R, Wright SF et al (2004) The role of glomalin, a protein produced by arbuscular mycorrhizal fungi, in sequestering potentially toxic elements. Environ Pollut 130:317–323

González-Guerrero M, Azcón-Aguilar C, Mooney M et al (2005) Characterization of a *Glomus intraradices* gene encoding a putative Zn transporter of the cation diffusion facilitator family. Fungal Genet Biol 42:130–140

González-Guerrero M, Cano C, Azcón-Aguilar C et al (2007) *GintMT1* encodes a functional metallothionein in *Glomus intraradices* that responds to oxidative stress. Mycorrhiza 17: 327–335

González-Guerrero M, Melville LH, Ferrol N et al (2008) Ultrastructural localization of heavy metals in the extraradical mycelium and spores of the arbuscular mycorrhizal fungus *Glomus intraradices*. Can J Microbiol 54:103–110

González-Guerrero M, Benabdellah K, Ferrol N et al (2009) Mechanisms underlying heavy metal tolerance in arbuscular mycorrhizas. In: Azcon-Aguilar C, Barea JM, Gionanizzi S, Gioaninazzi-Pearson V (eds) Mycorrhizas-functional processes and ecological impact. Springer, Berlin, pp 107–122

González-Guerrero M, Oger E, Benabdellah K et al (2010a) Characterization of a CuZn superoxide dismutase gene in the arbuscular mycorrhizal fungus *Glomus intraradices*. Curr Genet 56:265–274

González-Guerrero M, Benabdellah K, Valderas A et al (2010b) *GintABC1* encodes a putative ABC transporter of the MRP subfamily induced by Cu, Cd, and oxidative stress in *Glomus intraradices*. Mycorrhiza 20:137–146

Graham JH (1972) Effect of citrus root exudates on germination of chlamydospores of the vesicular-arbuscular mycorrhizal fungus *Glomus epigaeum*. Mycologia 74:831–835

Graham JH, Syvertsen JP (1984) Influence of vesicular–arbuscular mycorrhiza on the hydraulic conductivity of roots of two citrus rootstocks. New Phytol 97:277–284

Groppa MD, Benavides MP (2008) Polyamines and abiotic stress: recent advances. Amino Acids 34:35–45

Grümberg BC, Urcelay C, Shroeder MA et al (2015) The role of inoculum identity in drought stress mitigation by arbuscular mycorrhizal fungi in soybean. Biol Fertil Soils 51:1–10

Guo W, Zhao R, Fu R et al (2014) Contribution of arbuscular mycorrhizal fungi to the development of maize (*Zea mays* L.) grown in three types of coal mine spoils. Environ Sci Pollut Res 21:3592–3603

Hajiboland R, Aliasgharzadeh N, Laiegh SF et al (2010) Colonization with arbuscular mycorrhizal fungi improves salinity tolerance of tomato (*Solanum lycopersicum* L.) plants. Plant Soil 331: 313–327

Hale SE, Lehmann J, Rutherford D et al (2012) Quantifying the total and bioavailable polycyclic aromatic hydrocarbons and dioxins in biochars. Environ Sci Technol 46:2830–2838

Halsall CJ, Sweetman AJ, Barrie LA et al (2001) Modelling the behaviour of PAHs during atmospheric transport from the UK to the Arctic. Atmos Environ 35:255–267

Hare PD, Cress WA, Van Staden J (1999) Proline synthesis and degradation: a model system for elucidating stress-related signal transduction. J Exp Bot 50:413–434

Hasegawa PM, Bressan RA, Zhu JK et al (2000) Plant cellular and molecular responses to high salinity. Annu Rev Plant Biol 51:463–499

Hassan SED, Boon E, St-Arnaud M et al (2011) Molecular biodiversity of arbuscular mycorrhizal fungi in trace metal-polluted soils. Mol Ecol 20:3469–3483

Hassan SED, Bell TH, Stefani FO et al (2014) Contrasting the community structure of arbuscular mycorrhizal fungi from hydrocarbon-contaminated and uncontaminated soils following willow (*Salix* spp. L.) planting. PLoS One 9:e102838

Hause B, Mrosk C, Isayenkov S et al (2007) Jasmonates in arbuscular mycorrhizal interactions. Phytochemistry 68:101–110

Hawkes CV, Hartley IP, Ineson P et al (2008) Soil temperature affects carbon allocation within arbuscular mycorrhizal networks and carbon transport from plant to fungus. Glob Chang Biol 14:1181–1190

He Z, Huang Z (2013) Expression analysis of *LeNHX1* gene in mycorrhizal tomato under salt stress. J Microbiol 51:100–104

He Z, He C, Zhang Z et al (2007) Changes of antioxidative enzymes and cell membrane osmosis in tomato colonized by arbuscular mycorrhizae under NaCl stress. Colloids Surf B Biointerfaces 59:128–133

Heinemeyer A, Fitter AH (2004) Impact of temperature on the arbuscular mycorrhizal (AM) symbiosis: growth responses of the host plant and its AM fungal partner. J Exp Bot 55:525–534

Hepper CM (1979) Germination and growth of *Glomus caledonius* spores: the effects of inhibitors and nutrients. Soil Biol Biochem 11:269–277

Hernández-Ortega HA, Alarcón A, Ferrera-Cerrato R et al (2012) Arbuscular mycorrhizal fungi on growth, nutrient status, and total antioxidant activity of *Melilotus albus* during phytoremediation of a diesel-contaminated substrate. J Environ Manage 95:S319–S324

Herrera-Rodríguez MB, Pérez-Vicente R, Maldonado JM (2007) Expression of asparagine synthetase genes in sunflower (*Helianthus annuus*) under various environmental stresses. Plant Physiol Biochem 45:33–38

Hilber I, Blum F, Leifeld J et al (2012) Quantitative determination of PAHs in biochar: a prerequisite to ensure its quality and safe application. J Agr Food Chem 60:3042–3050

Hildebrandt U, Regvar M, Bothe H (2007) Arbuscular mycorrhiza and heavy metal tolerance. Phytochemistry 68:139–146

Hildebrandt U, Janetta K, Ouziad F et al (2001) Arbuscular mycorrhizal colonization of halophytes in Central European salt marshes. Mycorrhiza 10:175–183

Hirrel MC (1981) The effect of sodium and chloride salts on the germination of *Gigaspora margarita*. Mycologia 73:610–617

Ho I (1987) Vesicular-arbuscular mycorrhizae of halophytic grasses in the Alvard desert of Oregon Northwest. Science 61:148–151

Houghton JT, Ding YDJG, Griggs DJ, Noguer M, van der Linden PJ, Dai X, Johnson CA (eds) (2001) Climate change 2001: the scientific basis. Contribution of working group I to the third assessment report of the intergovernmental panel on climate change. Published for the Intergovernmental Panel on Climate Change. Cambridge University Press, Cambridge

Huang H, Zhang S, Chen BD et al (2006) Uptake of atrazine and cadmium from soil by maize (*Zea mays* L) in association with the arbuscular mycorrhizal fungus *Glomus etunicatum*. J Agr Food Chem 54:9377–9382

Huang JC, Tang M, Niu ZC et al (2007a) Arbuscular mycorrhizal fungi in petroleum-contaminated soil in Suining area of Sichuan Province Chinese. J Ecol 9:014

Huang LL, Yang C, Zhao Y (2008) Antioxidant defenses of mycorrhizal fungus infection against SO_2-induced oxidative stress in *Avena nuda* seedlings. Bull Environ Contam Toxicol 81: 440–444

Huang SS, Liao QL, Hua M et al (2007b) Survey of heavy metal pollution and assessment of agricultural soil in Yangzhong district, Jiangsu Province, China. Chemosphere 67:2148–2155

Huang Z, He CX, He ZQ et al (2010) The effects of arbuscular mycorrhizal fungi on reactive oxyradical scavenging system of tomato under salt tolerance. Agric Sci China 9:1150–1159

Huang Z, Zou Z, He C et al (2011) Physiological and photosynthetic responses of melon (*Cucumis melo* L) seedlings to three *Glomus* species under water deficit. Plant Soil 339:391–399

Iffis B, St-Arnaud M, Hijri M (2014) Bacteria associated with arbuscular mycorrhizal fungi within roots of plants growing in a soil highly contaminated with aliphatic and aromatic petroleum hydrocarbons. FEMS Microbiol Lett 358:44–54

Jahromi F, Aroca R, Porcel R et al (2008) Influence of salinity on the *in vitro* development of *Glomus intraradices* and on the *in vivo* physiological and molecular responses of mycorrhizal lettuce plants. Microb Ecol 55:45–53

Javot H, Maurel C (2002) The role of aquaporins in root water uptake. Ann Bot 90:301–313

Javot H, Lauvergeat V, Santoni V et al (2003) Role of a single aquaporin isoform in root water uptake. Plant Cell 15:509–522

Jezdinský A, Vojtíšková J, Slezák K et al (2012) Effect of drought stress and *Glomus* inoculation on selected physiological processes of sweet pepper (*Capsicum annuum* L cv 'Slávy'). Acta Univ Agric Silvic Mendel Brun 60:69–76

Joner EJ, Leyval C (1997) Uptake of ^{109}Cd by roots and hyphae of a *Glomus mosseae/Trifolium subterraneum* mycorrhiza from soil amended with high and low concentrations of cadmium. New Phytol 135:353–360

Joner EJ, Leyval C (2003) Rhizosphere gradients of polycyclic aromatic hydrocarbon (PAH) dissipation in two industrial soils and the impact of arbuscular mycorrhiza. Environ Sci Technol 37:2371–2375

Joner EJ, Briones R, Leyval C (2000) Metal-binding capacity of arbuscular mycorrhizal mycelium. Plant Soil 226:227–234

Joner EJ, Corgie SC, Amellal N et al (2002) Nutritional constraints to degradation of polycyclic aromatic hydrocarbons in a simulated rhizosphere. Soil Biol Biochem 34:859–864

Joner EJ, Leyval C, Colpaert JV (2006) Ectomycorrhizas impede phytoremediation of polycyclic aromatic hydrocarbons (PAHs) both within and beyond the rhizosphere. Environ Pollut 142: 34–38

Jones A, Panagos P, Barcelo S et al (eds) (2012) The state of Soil in Europe – A contribution of the JRC to the European Environment Agency's Environment State and Outlook Report – SOER 2010

Juhasz A (1998) Microbial degradation of high molecular weight polycyclic aromatic hydrocarbons. Doctoral Dissertation, Victoria University of Technology

Juniper S, Abbott LK (1991) The effect of salinity on spore germination and hyphal extension of some VA mycorrhizal fungi. Abstracts of the 3rd European Symposium on Mycorrhizas. University of Sheffield, Sheffield

Juniper S, Abbott LK (1993) Vesicular-arbuscular mycorrhizas and soil salinity. Mycorrhiza 4: 45–57

Juniper S, Abbott LK (2006) Soil salinity delays germination and limits growth of hyphae from propagules of arbuscular mycorrhizal fungi. Mycorrhiza 16:371–379

Kabata-Pendias A (ed) (2011) Trace elements in soils and plants, 4th edn. CRC Press, Boca Raton

Kacálková L, Tlustoš P (2011) The uptake of persistent organic pollutants by plants. Open Life Sci 6:223–235

Karasawa T, Hodge A, Fitter AH (2012) Growth, respiration and nutrient acquisition by the arbuscular mycorrhizal fungus *Glomus mosseae* and its host plant *Plantago lanceolata* in cooled soil. Plant Cell Environ 35:819–828

Katsuhara M, Hanba YT, Shiratake K et al (2008) Expanding roles of plant aquaporins in plasma membranes and cell organelles. Funct Plant Biol 35:1–14

Kaya C, Ashraf M, Sonmez O et al (2009) The influence of arbuscular mycorrhizal colonisation on key growth parameters and fruit yield of pepper plants grown at high salinity. Sci Hortic 121: 1–6

Keiluweit M, Kleber M, Sparrow MA et al (2012) Solvent-extractable polycyclic aromatic hydrocarbons in biochar: influence of pyrolysis temperature and feedstock. Environ Sci Technol 46:9333–9341

Kelkar TS, Bhalerao SA (2013) Beneficiary effect of arbuscular mycorrhiza to *Trigonella foenum-graceum* in contaminated soil by heavy metal. Res J Recent Sci 2:29–32

Kelly J, Thornton I, Simpson PR (1996) Urban geochemistry: a study of the influence of anthropogenic activity on the heavy metal content of soils in traditionally industrial and non-industrial areas of Britain. Appl Geochem 11:363–370

Khan AG (1974) The occurrence of mycorrhizas in halophytes, hydrophytes and xerophytes, and of endogone spores in adjacent soils. J Gen Microbiol 81:7–14

Kirk JL, Moutoglis P, Klironomos J et al (2005) Toxicity of diesel fuel to germination, growth and colonization of *Glomus intraradices* in soil and *in vitro* transformed carrot root cultures. Plant Soil 270:23–30

Kishor PK, Hong Z, Miao GH et al (1995) Overexpression of [delta]-pyrroline-5-carboxylate synthetase increases proline production and confers osmotolerance in transgenic plants. Plant Physiol 108:1387–1394

Kishor PK, Sangam S, Amrutha RN et al (2005) Regulation of proline biosynthesis, degradation, uptake and transport in higher plants: its implications in plant growth and abiotic stress tolerance. Curr Sci 88:424–438

Kloss S, Zehetner F, Dellantonio A et al (2012) Characterization of slow pyrolysis biochars: effects of feedstocks and pyrolysis temperature on biochar properties. J Environ Qual 41: 990–1000

Kohler J, Caravaca F, Roldán A (2010) An AM fungus and a PGPR intensify the adverse effects of salinity on the stability of rhizosphere soil aggregates of *Lactuca sativa*. Soil Biol Biochem 42: 429–434

Koske RE (1981) *Gigaspora gigantea*; observations on spore germination of a VA mycorrhizal fungus. Mycologia 73:289–300

Krishnamoorthy R, Kim CG, Subramanian P et al (2015) Arbuscular mycorrhizal fungi community structure, abundance and species richness changes in soil by different levels of heavy metal and metalloid concentration. PLoS One 10:e0128784

Kühn KD (1991) Distribution of vesicular-arbuscular mycorrhizal fungi on a fallow agriculture site. II Wet habitat. Angew Bot 65:187–203

Kühn KD, Weber HC, Dehne HW et al (1991) Distribution of vesicular-arbuscular mycorrhizal fungi on a fallow agriculture site. I: Dry habitat. Angew Bot 65:169–185

Lal R (2000) Soil management in the developing countries. Soil Sci 165:57–72

Landwehr M, Hildebrandt U, Wilde P et al (2002) The arbuscular mycorrhizal fungus *Glomus geosporum* in European saline, sodic and gypsum soils. Mycorrhiza 12:199–211

Lanfranco L, Bolchi A, Ros EC et al (2002) Differential expression of a metallothionein gene during the presymbiotic versus the symbiotic phase of an arbuscular mycorrhizal fungus. Plant Physiol 130:58–67

Lanfranco L, Novero M, Bonfante P (2005) The mycorrhizal fungus *Gigaspora margarita* possesses a CuZn superoxide dismutase that is up-regulated during symbiosis with legume hosts. Plant Physiol 137:1319–1330

Latef AAHA (2013) Growth and some physiological activities of pepper (*Capsicum annuum* L) in response to cadmium stress and mycorrhizal symbiosis. J Agr Sci Technol 15:1437–1448

Latef AAHA, Chaoxing H (2011a) Arbuscular mycorrhizal influence on growth, photosynthetic pigments, osmotic adjustment and oxidative stress in tomato plants subjected to low temperature stress. Acta Physiol Plant 33:1217–1225

Latef AAHA, Chaoxing H (2011b) Effect of arbuscular mycorrhizal fungi on growth, mineral nutrition, antioxidant enzymes activity and fruit yield of tomato grown under salinity stress. Sci Hortic 127:228–233

Latef AAHA, Chaoxing H (2014) Does inoculation with *Glomus mosseae* improve salt tolerance in pepper plants? J Plant Growth Regul 33:644–653

Lazár D (2003) Chlorophyll a fluorescence rise induced by high light illumination of dark-adapted plant tissue studied by means of a model of photosystem II and considering photosystem II heterogeneity. J Theor Biol 220:469–503

Lee SH, Calvo-Polanco M, Chung GC et al (2010) Cell water flow properties in root cortex of ectomycorrhizal (*Pinus banksiana*) seedlings. Plant Cell Environ 33:769–780

Lee Y, Krishnamoorthy R, Selvakumar G et al (2015) Alleviation of salt stress in maize plant by co-inoculation of arbuscular mycorrhizal fungi and *Methylobacterium oryzae* CBMB20. J Korean Soc Appl Bi 58:533–540

Leyval C, Binet P (1998) Effect of polyaromatic hydrocarbons in soil on arbuscular mycorrhizal plants. J Environ Qual 27:402–407

Li T, Hu YJ, Hao ZP et al (2013a) First cloning and characterization of two functional aquaporin genes from an arbuscular mycorrhizal fungus *Glomus intraradices*. New Phytol 197:617–630

Li T, Hu YJ, Hao ZP et al (2013b) Aquaporin genes *GintAQPF1* and *GintAQPF2* from *Glomus intraradices* contribute to plant drought tolerance. Plant Signal Behav 8:e24030

Li T, Lin G, Zhang X et al (2014) Relative importance of an arbuscular mycorrhizal fungus (*Rhizophagus intraradices*) and root hairs in plant drought tolerance. Mycorrhiza 24:595–602

Lin AJ, Zhang XH, Wong MH et al (2007) Increase of multi-metal tolerance of three leguminous plants by arbuscular mycorrhizal fungi colonization. Environ Geochem Health 29:473–481

Lingua G, Franchin C, Todeschini V et al (2008) Arbuscular mycorrhizal fungi differentially affect the response to high zinc concentrations of two registered poplar clones. Environ Pollut 153: 137–147

Liu A, Wang B, Hamel C (2004a) Arbuscular mycorrhiza colonization and development at sub-optimal root zone temperature. Mycorrhiza 14:93–101

Liu A, Dalpé Y (2009) Reduction in soil polycyclic aromatic hydrocarbons by arbuscular mycorrhizal leek plants. Int J Phytorem 11:39–52

Liu H, Tan Y, Nell M et al (2014) Arbuscular mycorrhizal fungal colonization of *Glycyrrhiza glabra* roots enhances plant biomass, phosphorus uptake and concentration of root secondary metabolites. J Arid Land 6:186–194

Liu L, Li Y, Tang J et al (2011) Plant coexistence can enhance phytoextraction of cadmium by tobacco (*Nicotiana tabacum* L) in contaminated soil. J Environ Sci 23:453–460

Liu SL, Luo YM, Cao ZH et al (2004b) Degradation of benzo[a]pyrene in soil with arbuscular mycorrhizal alfalfa. Environ Geochem Health 26:285–293

Liu ZL, Li YJ, Hou HY et al (2013) Differences in the arbuscular mycorrhizal fungi-improved rice resistance to low temperature at two N levels: aspects of N and C metabolism on the plant side. Plant Physiol Biochem 71:87–95

Long LK, Yao Q, Guo J et al (2010) Molecular community analysis of arbuscular mycorrhizal fungi associated with five selected plant species from heavy metal polluted soils. Eur J Soil Biol 46:288–294

Løvaas E (1997) Antioxidative and metal-chelating effects of polyamines. Adv Pharma 38: 119–149

Ludwig-Müller J (2000) Hormonal balance in plants during colonization by mycorrhizal fungi. In: Douds DD, Kapulnik Y (eds) Arbuscular mycorrhizas: physiology and function. Kluwer, Dordrecht, pp 263–285

Lugon-Moulin N, Ryan L, Donini P et al (2006) Cadmium content of phosphate fertilizers used for tobacco production. Agron Sustain Dev 26:151

Maganhotto de Souza Silva CM, Francisconi Fay E (2012) Effect of salinity on soil microorganisms. In: Hernandez Soriano MC (ed) Soil health and land use management. InTech, Rijeka. doi:10.5772/28613

Malekzadeh E, Alikhani HA, Savaghebi-Firoozabadi GR et al (2012) Bioremediation of cadmium-contaminated soil through cultivation of maize inoculated with plant growth–promoting rhizobacteria. Bioremed J 16:204–211

Maliszewska-Kordybach B, Smreczak B, Klimkowicz-Pawlas A et al (2008) Monitoring of the total content of polycyclic aromatic hydrocarbons (PAHs) in arable soils in Poland. Chemosphere 73:1284–1291

Manoharan PT, Shanmugaiah V, Balasubramanian N et al (2010) Influence of AM fungi on the growth and physiological status of *Erythrina variegata* Linn grown under different water stress conditions. Eur J Soil Biol 46:151–156

Manoli E, Samara C (1999) Polycyclic aromatic hydrocarbons in natural waters: sources, occurrence and analysis. Trends Anal Chem 18:417–428
Marjanović Ž, Uwe N, Hampp R (2005) Mycorrhiza formation enhances adaptive response of hybrid poplar to drought. Ann NY Acad Sci 1048:496–499
Marques AP, Oliveira RS, Rangel AO et al (2006) Zinc accumulation in *Solanum nigrum* is enhanced by different arbuscular mycorrhizal fungi. Chemosphere 65:1256–1263
Martin CA, Stutz JC (2004) Interactive effects of temperature and arbuscular mycorrhizal fungi on growth, P uptake and root respiration of *Capsicum annuum* L. Mycorrhiza 14:241–244
Martínez-García LB, de Dios MJ, Pugnaire FI (2012) Impacts of changing rainfall patterns on mycorrhizal status of a shrub from arid environments. Eur J Soil Biol 50:64–67
Marulanda A, Porcel R, Barea JM et al (2007) Drought tolerance and antioxidant activities in lavender plants colonized by native drought-tolerant or drought-sensitive *Glomus* species. Microb Ecol 54:543–552
Mason E (1928) Note on the presence of mycorrhizae in the roots of salt-marsh plants. New Phytol 27:193–195
Mathur N, Singh J, Bohra S et al (2007) Arbuscular mycorrhizal status of medicinal halophytes in saline areas of Indian Thar desert. Int J Soil Sci 2:119–127
Matsubara Y (2010) High temperature stress tolerance and the changes in antioxidative ability in mycorrhizal strawberry plants. In: Sampson NA (ed) Horticulture in the 21st century. Nova Science, New York, pp 179–192
Maurel C, Plassard C (2011) Aquaporins: for more than water at the plant–fungus interface? New Phytol 190:815–817
Maxwell K, Johnson GN (2000) Chlorophyll fluorescence – a practical guide. J Exp Bot 51: 659–668
Maya MA, Matsubara YI (2013) Influence of arbuscular mycorrhiza on the growth and antioxidative activity in cyclamen under heat stress. Mycorrhiza 23:381–390
McMillen BG, Juniper S, Abbott LK (1998) Inhibition of hyphal growth of a vesicular-arbuscular mycorrhizal fungus in soil containing sodium chloride limits the spread of infection from spores. Soil Biol Biochem 30:1639–1646
Meharg AA, Bailey J, Breadmore K et al (1994) Biomass allocation, phosphorus nutrition and vesicular-arbuscular mycorrhizal infection in clones of Yorkshire Fog, *Holcus lanatus* L (Poaceae) that differ in their phosphate uptake kinetics and tolerance to arsenate. Plant Soil 160:11–20
Meier S, Borie F, Curaqueo G et al (2012) Effects of arbuscular mycorrhizal inoculation on metallophyte and agricultural plants growing at increasing copper levels. Appl Soil Ecol 61: 280–287
Mittler R (2006) Abiotic stress, the field environment and stress combination. Trends Plant Sci 11: 15–19
Mosse B, Hepper C (1975) Vesicular-arbuscular mycorrhizal infections in root organ cultures. Physiol Plant Pathol 5:215–223
Mueller JG, Middaugh DP, Lantz SE et al (1991) Biodegradation of creosote and pentachlorophenol in contaminated groundwater: chemical and biological assessment. Appl Environ Microbiol 57:1277–1285
Munns R (2002) Comparative physiology of salt and water stress. Plant Cell Environ 25:239–250
Munns R, Tester M (2008) Mechanisms of salinity tolerance. Annu Rev Plant Biol 59:651–681
Nadal M, Schuhmacher M, Domingo JL (2004) Levels of PAHs in soil and vegetation samples from Tarragona County, Spain. Environ Pollut 132:1–11
Nam JJ, Song BH, Eom KC et al (2003) Distribution of polycyclic aromatic hydrocarbons in agricultural soils in South Korea. Chemosphere 50:1281–1289
Neumann E, Schmid B, Römheld V et al (2009) Extraradical development and contribution to plant performance of an arbuscular mycorrhizal symbiosis exposed to complete or partial rootzone drying. Mycorrhiza 20:13–23

Newell-McGloughlin N, Burke J (2014) Regulatory challenges to commercializing the products of ag biotech. In: Van Alfen NK (ed) Encyclopedia of agriculture and food system, vol 5. Elsevier/Academic, Amsterdam, pp 21–40

Nicholson FA, Smith SR, Alloway BJ et al (2003) An inventory of heavy metals inputs to agricultural soils in England and Wales. Sci Total Environ 311:205–219

Nwoko CO (2014) Effect of arbuscular mycorrhizal (AM) fungi on the physiological performance of *Phaseolus vulgaris* grown under crude oil contaminated soil. J Geosci Environ Protect 2:9

Nziguheba G, Smolders E (2008) Inputs of trace elements in agricultural soils via phosphate fertilizers in European countries. Sci Total Environ 390:53–57

Orłowska E, Mesjasz-Przybyłowicz J, Przybyłowicz W et al (2008) Nuclear microprobe studies of elemental distribution in mycorrhizal and non-mycorrhizal roots of Ni-hyperaccumulator Berkheya coddii. X-Ray Spectro 37:129–132

Orłowska E, Godzik B, Turnau K (2012) Effect of different arbuscular mycorrhizal fungal isolates on growth and arsenic accumulation in *Plantago lanceolata* L. Environ Pollut 168:121–130

Ouziad F, Hildebrandt U, Schmelzer E et al (2005) Differential gene expressions in arbuscular mycorrhizal-colonized tomato grown under heavy metal stress. J Plant Physiol 162:634–649

Ouziad F, Wilde P, Schmelzer E et al (2006) Analysis of expression of aquaporins and Na+/H+ transporters in tomato colonized by arbuscular mycorrhizal fungi and affected by salt stress. Environ Exp Bot 57:177–186

Oztekin GB, Tuzel Y, Tuzel IH (2013) Does mycorrhiza improve salinity tolerance in grafted plants? Sci Hortic 149:55–60

Pagotto C, Remy N, Legret M et al (2001) Heavy metal pollution of road dust and roadside soil near a major rural highway. Environ Technol 22:307–319

Pallara G, Todeschini V, Lingua G et al (2013) Transcript analysis of stress defence genes in a white poplar clone inoculated with the arbuscular mycorrhizal fungus *Glomus mosseae* and grown on a polluted soil. Plant Physiol Biochem 63:131–139

Palma JM, Longa M, Río LA et al (1993) Superoxide dismutase in vesicular arbuscular mycorrhizal red clover plants. Physiol Plant 87:77–83

Panagos P, Van Liedekerke M, Yigini Y et al (2013) Contaminated sites in Europe: review of the current situation based on data collected through a European network. J Environ Public Health 2013(Article ID158764):11. doi:10.1155/2013/158764

Păun A, Neagoe A, Păun M et al (2015) Heavy metal-induced differential responses to oxidative stress and protection by mycorrhization in sunflowers grown in lab and field scales. Pol J Environ Stud 24(3)

Pawlowska TE, Charvat I (2004) Heavy-metal stress and developmental patterns of arbuscular mycorrhizal fungi. Appl Environ Microbiol 70:6643–6649

Pfeiffer CM, Bloss HE (1988) Growth and nutrition of guayule (*Parthenium argentatum*) in a saline soil as influenced by vesicular–arbuscular mycorrhiza and phosphorus fertilization. New Phytol 108:315–321

Pigna M, Caporale AG, Cartes P et al (2014) Effects of arbuscular mycorrhizal inoculation and phosphorus fertilization on the growth of escarole (*Cichorium endivia* L) in an arsenic polluted soil. J Soil Sci Plant Nutr 14:199–208

Plenchette C, Fortin JA, Furlan V (1983) Growth responses of several plant species to mycorrhizae in a soil of moderate P-fertility. Plant Soil 70:199–209

Pond EC, Menge JA, Jarrell WM (1984) Improved growth of tomato in salinized soil by vesicular-arbuscular mycorrhizal fungi collected from saline soils. Mycologia 76:74–84

Porcel R, Ruiz-Lozano JM (2004) Arbuscular mycorrhizal influence on leaf water potential, solute accumulation, and oxidative stress in soybean plants subjected to drought stress. J Exp Bot 55:1743–1750

Porcel R, Aroca R, Ruiz-Lozano JM (2012) Salinity stress alleviation using arbuscular mycorrhizal fungi. A review. Agron Sustain Dev 32:181–200

Poss JA, Pond E, Menge JA et al (1985) Effect of salinity on mycorrhizal onion and tomato in soil with and without additional phosphate. Plant Soil 88:307–319

Potin O, Veignie E, Rafin C (2004) Biodegradation of polycyclic aromatic hydrocarbons (PAHs) by *Cladosporium sphaerospermum* isolated from an aged PAH contaminated soil. FEMS Microbiol Ecol 51:71–78

Prasad KVSK, Saradhi PP (1995) Effect of zinc on free radical and proline in *Brassica juncea* and *Cajanus cajan*. Phytochemistry 39:45–47

Priyadharsini P, Muthukumar T (2015) Insight into the role of arbuscular mycorrhizal fungi in sustainable agriculture, Part I. In: Thangavel P, Sridevi G (eds) Environmental sustainability: role of green technologies. Springer India, New Delhi, pp 3–37

Punamiya P, Datta R, Sarkar D et al (2010) Symbiotic role of *Glomus mosseae* in phytoextraction of lead in vetiver grass (*Chrysopogon zizanioides* (L)). J Hazard Mater 177:465–474

Purin S, Rillig MC (2008) Parasitism of arbuscular mycorrhizal fungi: reviewing the evidence. FEMS Microbiol Lett 279:8–14

Querejeta J, Egerton-Warburton LM, Allen MF (2009) Topographic position modulates the mycorrhizal response of oak trees to interannual rainfall variability. Ecology 90:649–662

Quilliam RS, Rangecroft S, Emmett BA et al (2013) Is biochar a source or sink for polycyclic aromatic hydrocarbon (PAH) compounds in agricultural soils? GCB Bioenergy 5:96–103

Rabie GH (2004) Using wheat-mungbean plant system and arbuscular mycorrhiza to enhance *in-situ* bioremediation. J Food Agric Environ 2:381–390

Rabie GH (2005) Role of arbuscular mycorrhizal fungi in phytoremediation of soil rhizosphere spiked with poly aromatic hydrocarbons. Mycobiology 33:41–50

Rabie GH, Almadini AM (2005) Role of bioinoculants in development of salt-tolerance of *Vicia faba* plants under salinity stress. Afr J Biotechnol 4:210–222

Raju PS, Clark RB, Ellis JR et al (1990) Effects of species of VA-mycorrhizal fungi on growth and mineral uptake of sorghum at different temperatures. Plant Soil 121:165–170

Redon PO, Béguiristain T, Leyval C (2009) Differential effects of AM fungal isolates on *Medicago truncatula* growth and metal uptake in a multimetallic (Cd, Zn, Pb) contaminated agricultural soil. Mycorrhiza 19:187–195

Regvar M, Groznik N, Goljevšček K et al (2001) Diversity of arbuscular mycorrhizal fungi form various disturbed ecosystems in Slovenia. Acta Biol Slovenia 44:27–34

Repetto O, Bestel-Corre G, Dumas-Gaudot E et al (2003) Targeted proteomics to identify cadmium-induced protein modifications in *Glomus mosseae*-inoculated pea roots. New Phytol 157:555–567

Rezek J, in der Wiesche C, Mackova M et al (2008) The effect of ryegrass (*Lolium perenne*) on decrease of PAH content in long term contaminated soil. Chemosphere 70:1603–1608

Rillig MC, Wright SF, Eviner VT (2002) The role of arbuscular mycorrhizal fungi and glomalin in soil aggregation: comparing effects of five plant species. Plant Soil 238:325–333

Rivera-Becerril F, Calantzis C, Turnau K et al (2002) Cadmium accumulation and buffering of cadmium-induced stress by arbuscular mycorrhiza in three *Pisum sativum* L genotypes. J Exp Bot 53:1177–1185

Rivera-Becerril F, van Tuinen D, Martin-Laurent F et al (2005) Molecular changes in *Pisum sativum* L roots during arbuscular mycorrhiza buffering of cadmium stress. Mycorrhiza 16: 51–60

Roelfsema MRG, Levchenko V, Hedrich R (2004) ABA depolarizes guard cells in intact plants, through a transient activation of R and S type anion channels. Plant J 37:578–588

Rovira AD (1969) Plant root exudates. Bot Rev 35:35–57

Rozema J, Arp W, Van Diggelen J et al (1986) Occurrence and ecological significance of vesicular-arbuscular mycorrhiza in the salt marsh environment. Acta Bot Neerl 35:45

Ruiz-Lozano JM (2003) Arbuscular mycorrhizal symbiosis and alleviation of osmotic stress. New perspectives for molecular studies. Mycorrhiza 13:309–317

Ruiz-Lozano JM, Azcón R (1995) Hyphal contribution to water uptake in mycorrhizal plants as affected by the fungal species and water status. Physiol Plant 95:472–478

Ruiz-Lozano JM, Azcón R (2000) Symbiotic efficiency and infectivity of an autochthonous arbuscular mycorrhizal *Glomus* sp from saline soils and *Glomus deserticola* under salinity. Mycorrhiza 10:137–143

Ruiz-Lozano JM, Aroca R (2010) Modulation of aquaporin genes by the arbuscular mycorrhizal symbiosis in relation to osmotic stress tolerance. In: Sechback J, Grube M (eds) Symbioses and stress. Springer, Berlin, pp 357–374

Ruiz-Lozano JM, Azcón R, Gomez M (1995) Effects of arbuscular-mycorrhizal *Glomus* species on drought tolerance: physiological and nutritional plant responses. Appl Environ Microbiol 61:456–460

Ruiz-Lozano JM, Azcon R, Gomez M (1996) Alleviation of salt stress by arbuscular mycorrhizal *Glomus* species in *Lactuca sativa* plants. Physiol Plant 98:767–772

Ruiz-Lozano JM, Collados C, Barea JM et al (2001) Arbuscular mycorrhizal symbiosis can alleviate drought-induced nodule senescence in soybean plants. New Phytol 151:493–502

Ruiz-Lozano JM, del Mar AM, Bárzana G et al (2009) Exogenous ABA accentuates the differences in root hydraulic properties between mycorrhizal and non mycorrhizal maize plants through regulation of PIP aquaporins. Plant Mol Biol 70:565–579

Ruíz-Lozano JM, del Carmen PM, Aroca R et al (2011) The application of a treated sugar beet waste residue to soil modifies the responses of mycorrhizal and non mycorrhizal lettuce plants to drought stress. Plant Soil 346:153–166

Ruiz-Lozano JM, Porcel R, Azcón C et al (2012a) Regulation by arbuscular mycorrhizae of the integrated physiological response to salinity in plants: new challenges in physiological and molecular studies. J Exp Bot 63:4033–4044

Ruiz-Lozano JM, Porcel R, Bárzana G et al (2012b) Contribution of arbuscular mycorrhizal symbiosis to plant drought tolerance: state of the art. In: Aroca R (ed) Plant responses to drought stress. Springer, Berlin, pp 335–362

Ruiz-Sánchez M, Aroca R, Muñoz Y, Polón R, Ruiz-Lozano JM (2010) The arbuscular mycorrhizal symbiosis enhances the photosynthetic efficiency and the antioxidative response of rice plants subjected to drought stress. J Plant Physiol 167:862–869

Ruíz-Sánchez M, Armada E, Muñoz Y et al (2011) Azospirillum and arbuscular mycorrhizal colonization enhance rice growth and physiological traits under well-watered and drought conditions. J Plant Physiol 168:1031–1037

Ryan MH, Ash JE (1996) Colonisation of wheat in southern New South Wales by vesicular-arbuscular mycorrhizal fungi is significantly reduced by drought. Anim Prod Sci 36:563–569

Sadeque Ahmed FR, Killham K, Alexander I (2006) Influences of arbuscular mycorrhizal fungus *Glomus mosseae* on growth and nutrition of lentil irrigated with arsenic contaminated water. Plant Soil 283:33–41

Saeedi M, Hosseinzadeh M, Jamshidi A et al (2009) Assessment of heavy metals contamination and leaching characteristics in highway side soils. Iran Environ Monit Assess 151:231–241

Sannazzaro AI, Ruiz OA, Albertó EO et al (2006) Alleviation of salt stress in *Lotus glaber* by *Glomus intraradices*. Plant Soil 285:279–287

Sannazzaro AI, Echeverría M, Albertó EO et al (2007) Modulation of polyamine balance in *Lotus glaber* by salinity and arbuscular mycorrhiza. Plant Physiol Biochem 45:39–46

Saraswat S, Rai JPN (2011) Prospective application of *Leucaena leucocephala* for phytoextraction of Cd and Zn and nitrogen fixation in metal polluted soils. Int J Phytorem 13:271–288

Scheibe R, Beck E (2011) Drought, desiccation, and oxidative stress. In: Lüttge U, Beck E, Bartels D (eds) Plant desiccation tolerance. Springer, Berlin, pp 209–231

Schenck NC, Schroder VN (1974) Temperature response of *Endogone* mycorrhiza on soybean roots. Mycologia 66:600–605

Schenck NC, Smith GS (1982) Responses of six species of vesicular-arbuscular mycorrhizal fungi and their effects on soybean at four soil temperatures. New Phytol 92:193–201

Schenck NC, Graham SO, Green NE (1975) Temperature and light effect on contamination and spore germination of vesicular-arbuscular mycorrhizal fungi. Mycologia 67:1189–1192

Schneider J, Stürmer SL, Guilherme LRG et al (2013) Arbuscular mycorrhizal fungi in arsenic-contaminated areas in Brazil. J Hazard Mater 262:1105–1115

Schützendübel A, Polle A (2002) Plant responses to abiotic stresses: heavy metal-induced oxidative stress and protection by mycorrhization. J Exp Bot 53:1351–1365

Seregin IV, Ivanov VB (2001) Physiological aspects of cadmium and lead toxic effects on higher plants. Russ J Plant Physiol 48:523–544
Sezgin N, Ozcan HK, Demir G et al (2004) Determination of heavy metal concentrations in street dusts in Istanbul E-5 highway. Environ Int 29:979–985
Shahabivand S, Maivan HZ, Goltapeh EM et al (2012) The effects of root endophyte and arbuscular mycorrhizal fungi on growth and cadmium accumulation in wheat under cadmium toxicity. Plant Physiol Biochem 60:53–58
Shalaby AM (2003) Responses of arbuscular mycorrhizal fungal spores isolated from heavy metal-polluted and unpolluted soil to Zn, Cd, Pb and their interactions *in vitro*. Pak J Biol Sci 6: 1416–1422
Sharifi M, Ghorbanli M, Ebrahimzadeh H (2007) Improved growth of salinity-stressed soybean after inoculation with salt pre-treated mycorrhizal fungi. J Plant Physiol 164:1144–1151
Sharma SS, Dietz KJ (2006) The significance of amino acids and amino acid-derived molecules in plant responses and adaptation to heavy metal stress. J Exp Bot 57:711–726
Sheng M, Tang M, Chen H et al (2008) Influence of arbuscular mycorrhizae on photosynthesis and water status of maize plants under salt stress. Mycorrhiza 18:287–296
Sheng M, Tang M, Zhang F et al (2011) Influence of arbuscular mycorrhiza on organic solutes in maize leaves under salt stress. Mycorrhiza 21:423–430
Sinclair G, Charest C, Dalpé Y et al (2013) Influence of arbuscular mycorrhizal fungi and a root endophyte on the biomass and root morphology of selected strawberry cultivars under salt conditions. Can J Plant Sci 93:997–999
Sinclair G, Charest C, Dalpé Y et al (2014) Influence of colonization by arbuscular mycorrhizal fungi on three strawberry cultivars under salty conditions. Agric Food Sci 23:146–158
Singh A (2015) Soil salinization and waterlogging: a threat to environment and agricultural sustainability. Ecol Indic 57:128–130
Smith GS, Roncadori RW (1986) Responses of three vesicular–arbuscular mycorrhizal fungi at four soil temperatures and their effects on cotton growth. New Phytol 104:89–95
Smith SE, Bowen GD (1979) Soil temperature, mycorrhizal infection and nodulation of *Medicago truncatula* and *Trifolium subterraneum*. Soil Biol Biochem 11:469–473
Smith SE, Read DJ (eds) (2008) Mycorrhizal symbiosis, 3rd edn. Academic, New York
Smith SE, Facelli E, Pope S et al (2010) Plant performance in stressful environments: interpreting new and established knowledge of the roles of arbuscular mycorrhizas. Plant Soil 326:3–20
Sohrabi Y, Heidari G, Weisany W et al (2012) Changes of antioxidative enzymes, lipid peroxidation and chlorophyll content in chickpea types colonized by different *Glomus* species under drought stress. Symbiosis 56:5–18
Spagnoletti F, Lavado RS (2015) The arbuscular mycorrhiza *Rhizophagus intraradices* reduces the negative effects of arsenic on soybean plants. Agronomy 5:188–199
Spagnoletti FN et al (2014) The *in-vitro* and *in-vivo* influence of arsenic on arbuscular mycorrhizal fungi. In: Litter MA, Nicolli HB, Meichtry M et al (eds) One century of the discovery of arsenicosis in Latin America (1914–2014) As 2014: Proceedings of the 5th international congress on arsenic in the environment, Buenos Aires, Argentina, May 11–16, 2014. CRC Press, Boca Raton, pp 375–377
Srivastava AK, Penna S, Nguyen DV et al (2014) Multifaceted roles of aquaporins as molecular conduits in plant responses to abiotic stresses. Crit Rev Biotechnol 8:1–10
Stommel M, Mann P, Franken P (2001) EST-library construction using spore RNA of the arbuscular mycorrhizal fungus *Gigaspora rosea*. Mycorrhiza 10:281–285
Strandberg GW, Shumate SE, Parrott JR (1981) Microbial cells as biosorbents for heavy metals: accumulation of uranium by *Saccharomyces cerevisiae* and *Pseudomonas aeruginosa*. Appl Environ Microbiol 41:237–245
Subramanian KS, Santhanakrishnan P, Balasubramanian P (2006) Responses of field grown tomato plants to arbuscular mycorrhizal fungal colonization under varying intensities of drought stress. Sci Hortic 107:245–253

Sudová R, Vosátka M (2007) Differences in the effects of three arbuscular mycorrhizal fungal strains on P and Pb accumulation by maize plants. Plant Soil 296:77–83
Symanczik S, Courty P-E, Boller T et al (2015) Impact of water regimes on an experimental community of four desert arbuscular mycorrhizal fungal (AMF) species, as affected by the introduction of a non-native AMF species. Mycorrhiza 25:639–647
Talaat NB, Shawky BT (2011) Influence of arbuscular mycorrhizae on yield, nutrients, organic solutes, and antioxidant enzymes of two wheat cultivars under salt stress. J Plant Nutr Soil Sci 174:283–291
Talaat NB, Shawky BT (2013) Modulation of nutrient acquisition and polyamine pool in salt-stressed wheat (*Triticum aestivum* L) plants inoculated with arbuscular mycorrhizal fungi. Acta Physiol Plant 35:2601–2610
Talaat NB, Shawky BT (2014) Protective effects of arbuscular mycorrhizal fungi on wheat (*Triticum aestivum* L) plants exposed to salinity. Environ Exp Bot 98:20–31
Tang M, Chen H, Huang JC et al (2009) AM fungi effects on the growth and physiology of *Zea mays* seedlings under diesel stress. Soil Biol Biochem 41:936–940
Tao S, Cui YH, Xu FL et al (2004) Polycyclic aromatic hydrocarbons (PAHs) in agricultural soil and vegetables from Tianjin. Sci Total Environ 320:11–24
Tao S, Jiao XC, Chen SH et al (2006) Accumulation and distribution of polycyclic aromatic hydrocarbons in rice (Oryza sativa). Environ Pollut 140:406–415
Tian CY, Feng G, Li XL et al (2004) Different effects of arbuscular mycorrhizal fungal isolates from saline or non-saline soil on salinity tolerance of plants. Appl Soil Ecol 26:143–148
Tommerup IC (1984) Effect of soil water potential on spore germination by vesicular-arbuscular mycorrhizal fungi. Trans Br Mycol Soc 83:193–202
Trapido M (1999) Polycyclic aromatic hydrocarbons in Estonian soil: contamination and profiles. Environ Pollut 105:67–74
Tripathi S, Chakraborty A, Chakrabarti K et al (2007) Enzyme activities and microbial biomass in coastal soils of India. Soil Biol Biochem 39:2840–2848
Tuháčková J, Cajthaml T, Novak K et al (2001) Hydrocarbon deposition and soil microflora as affected by highway traffic. Environ Pollut 113:255–262
Turer D, Maynard JB, Sansalone JJ (2001) Heavy metal contamination in soils of urban highways comparison between runoff and soil concentrations at Cincinnati, Ohio. Water Air Soil Pollut 132:293–314
Turkmen O, Sensoy S, Demir S et al (2008) Effects of two different AMF species on growth and nutrient content of pepper seedlings grown under moderate salt stress. Afr J Biotechnol 7: 392–396
Vahideh SZ, Nasser A, Shahin O (2013) Glomalin production by two Glomeral fungi in symbiosis with corn plant under different Pb levels. Int J Agric Res Rev 3:854–863
Vallino M, Massa N, Lumini E et al (2006) Assessment of arbuscular mycorrhizal fungal diversity in roots of *Solidago gigantea* growing in a polluted soil in Northern Italy. Environ Microbiol 8:971–983
Verdin A, Lounès-Hadj Sahraoui A, Fontaine J et al (2006) Effects of anthracene on development of an arbuscular mycorrhizal fungus and contribution of the symbiotic association to pollutant dissipation. Mycorrhiza 16:397–405
Viard B, Pihan F, Promeyrat S et al (2004) Integrated assessment of heavy metal (Pb, Zn, Cd) highway pollution: bioaccumulation in soil, Graminaceae and land snails. Chemosphere 55: 1349–1359
Vicente-Sánchez J, Nicolás E, Pedrero F et al (2014) Arbuscular mycorrhizal symbiosis alleviates detrimental effects of saline reclaimed water in lettuce plants. Mycorrhiza 24:339–348
Wahid A, Gelani S, Ashraf M et al (2007) Heat tolerance in plants: an overview. Environ Exp Bot 61:199–223
Wang FY, Liu RJ, Lin XG et al (2004) Arbuscular mycorrhizal status of wild plants in saline-alkaline soils of the Yellow River Delta. Mycorrhiza 14:133–137

Wang FY, Lin XG, Yin R et al (2006) Effects of arbuscular mycorrhizal inoculation on the growth of *Elsholtzia splendens* and *Zea mays* and the activities of phosphatase and urease in a multi-metal-contaminated soil under unsterilized conditions. Appl Soil Ecol 31:110–119

Wang FY, Lin XG, Yin R (2007) Inoculation with arbuscular mycorrhizal fungus *Acaulospora mellea* decreases Cu phytoextraction by maize from Cu-contaminated soil. Pedobiologia 51: 99–109

Wang ZH, Zhang JL, Christie P et al (2008) Influence of inoculation with *Glomus mosseae* or *Acaulospora morrowiae* on arsenic uptake and translocation by maize. Plant Soil 311:235–244

Waschke A, Sieh D, Tamasloukht M et al (2006) Identification of heavy metal-induced genes encoding glutathione S-transferases in the arbuscular mycorrhizal fungus *Glomus intraradices*. Mycorrhiza 17:1–10

Watts-Williams SJ, Cavagnaro TR (2012) Arbuscular mycorrhizas modify tomato responses to soil zinc and phosphorus addition. Biol Fertil Soils 48:285–294

Watts-Williams SJ, Smith FA, McLaughlin MJ et al (2015) How important is the mycorrhizal pathway for plant Zn uptake? Plant Soil 390:157–166

Weiersbye IM, Straker CJ, Przybylowicz WJ (1999) Micro-PIXE mapping of elemental distribution in arbuscular mycorrhizal roots of the grass, Cynodon dactylon, from gold and uranium mine tailings. Nucl Instrum Meth B 158:335–343

Weissenhorn I, Leyval C, Berthelin J (1993) Cd-tolerant arbuscular mycorrhizal (AM) fungi from heavy-metal polluted soils. Plant Soil 157:247–256

Weissenhorn I, Glashoff A, Leyval C et al (1994) Differential tolerance to Cd and Zn of arbuscular mycorrhizal (AM) fungal spores isolated from heavy metal-polluted and unpolluted soils. Plant Soil 167:189–196

Weissenhorn I, Leyval C, Belgy G et al (1995) Arbuscular mycorrhizal contribution to heavy metal uptake by maize (*Zea mays* L) in pot culture with contaminated soil. Mycorrhiza 5:245–251

Wild SR, Jones KC (1995) Polynuclear aromatic hydrocarbons in the United Kingdom environment: a preliminary source inventory and budget. Environ Pollut 88:91–108

Wong SC, Li XD, Zhang G et al (2002) Heavy metals in agricultural soils of the Pearl River Delta, South China. Environ Pollut 119:33–44

Wright SF, Upadhyaya A (1998) A survey of soils for aggregate stability and glomalin, a glycoprotein produced by hyphae of arbuscular mycorrhizal fungi. Plant Soil 198:97–107

Wu FY, Yu XZ, Wu SC et al (2011) Phenanthrene and pyrene uptake by arbuscular mycorrhizal maize and their dissipation in soil. J Hazard Mater 187:341–347

Wu QS (2011) Mycorrhizal efficacy of trifoliate orange seedlings on alleviating temperature stress. Plant Soil Environ 10:459–464

Wu QS, Xia RX (2006) Arbuscular mycorrhizal fungi influence growth, osmotic adjustment and photosynthesis of citrus under well-watered and water stress conditions. J Plant Physiol 163: 417–425

Wu QS, Xia RX, Zou YN (2006a) Reactive oxygen metabolism in mycorrhizal and non-mycorrhizal citrus (*Poncirus trifoliata*) seedlings subjected to water stress. J Plant Physiol 163:1101–1110

Wu Q, Xia R, Hu Z (2006b) Effect of arbuscular mycorrhiza on the drought tolerance of *Poncirus trifoliata* seedlings. Front Forest China 1:100–104

Wu QS, Zou YN (2009) Mycorrhiza has a direct effect on reactive oxygen metabolism of drought-stressed citrus. Plant Soil Environ 55:436–442

Wu QS, Zou YN (2010) Beneficial roles of arbuscular mycorrhizas in citrus seedlings at temperature stress. Sci Hortic 125:289–293

Wu QS, Zou YN, Liu W et al (2010) Alleviation of salt stress in citrus seedlings inoculated with mycorrhiza: changes in leaf antioxidant defense systems. Plant Soil Environ 56:470–475

Xing SG, Jun YB, Hau ZW et al (2007) Higher accumulation of γ-aminobutyric acid induced by salt stress through stimulating the activity of diamine oxidases in *Glycine max* (L) Merr roots. Plant Physiol Biochem 45:560–566

Xun F, Xie B, Liu S et al (2015) Effect of plant growth-promoting bacteria (PGPR) and arbuscular mycorrhizal fungi (AMF) inoculation on oats in saline-alkali soil contaminated by petroleum to enhance phytoremediation. Environ Sci Pollut Res 22:598–608

Yadav JSP, Bandyopadhyay AK, Bandyopadhyay BK (1983) Extent of coastal saline soils of India. J Indian Soc Coast Agric Res 1:1–6

Yamato M, Ikeda S, Iwase K (2008) Community of arbuscular mycorrhizal fungi in a coastal vegetation on Okinawa island and effect of the isolated fungi on growth of sorghum under salt-treated conditions. Mycorrhiza 18:241–249

Yang Y, Song Y, Scheller HV et al (2015) Community structure of arbuscular mycorrhizal fungi associated with *Robinia pseudoacacia* in uncontaminated and heavy metal contaminated soils. Soil Biol Biochem 86:146–158

Yargicoglu EN, Reddy KR (2014) Evaluation of PAH and metal contents of different biochars for use in climate change mitigation systems. In: Proceedings of international conference on sustainable infrastructure (ICSI), Long Beach, CA, 6–8 Nov 2014

Yargicoglu EN, Sadasivam BY, Reddy KR et al (2015) Physical and chemical characterization of waste wood derived biochars. Waste Manage 36:256–268

Yin B, Wang Y, Liu P et al (2010) Effects of vesicular-arbuscular mycorrhiza on the protective system in strawberry leaves under drought stress. Front Agric China 4:165–169

Younesi O, Moradi A (2014a) Effects of plant growth-promoting rhizobacterium (PGPR) and arbuscular mycorrhizal fungus (AMF) on antioxidant enzyme activities in salt-stressed bean (*Phaseolus Vulgaris* L). Agriculture 60:10–21

Younesi O, Moradi A (2014b) The effects of arbuscular mycorrhizal fungi inoculation on reactive oxyradical scavenging system of soybean (*Glycine max*) nodules under salt stress condition. Agric Conspect Sci 78:321–326

Yu XZ, Wu SC, Wu FY et al (2011) Enhanced dissipation of PAHs from soil using mycorrhizal ryegrass and PAH-degrading bacteria. J Hazard Mater 186:1206–1217

Zarei M, Saleh-Rastin N, Jouzani GS et al (2008) Arbuscular mycorrhizal abundance in contaminated soils around a zinc and lead deposit. Eur J Soil Biol 44:381–391

Zarei M, Hempel S, Wubet T et al (2010) Molecular diversity of arbuscular mycorrhizal fungi in relation to soil chemical properties and heavy metal contamination. Environ Pollut 158: 2757–2765

Zhang B, Chang SX, Anyia AO (2015) Mycorrhizal inoculation and nitrogen fertilization affect the physiology and growth of spring wheat under two contrasting water regimes. Plant Soil 398:47–57

Zhang F, Hamel C, Kianmehr H et al (1995) Root-zone temperature and soybean [*Glycine max* (L) Merr] vesicular-arbuscular mycorrhizae: development and interactions with the nitrogen fixing symbiosis. Environ Exp Bot 35:287–298

Zhang XH, Lin AJ, Gao YL et al (2009) Arbuscular mycorrhizal colonisation increases copper binding capacity of root cell walls of *Oryza sativa* L and reduces copper uptake. Soil Biol Biochem 41:930–935

Zhao R, Guo W, Bi N et al (2015) Arbuscular mycorrhizal fungi affect the growth, nutrient uptake and water status of maize (*Zea mays* L) grown in two types of coal mine spoils under drought stress. Appl Soil Ecol 88:41–49

Zheng W, Sharma BK, Rajagopalan N (2010) Using biochar as a soil amendment for sustainable agriculture. Final Project Report, Illinois Department of Agriculture Sustainable Agriculture Grant Program; SA 09–37

Zhou Q, Ravnskov S, Jiang D et al (2015) Changes in carbon and nitrogen allocation, growth and grain yield induced by arbuscular mycorrhizal fungi in wheat (*Triticum aestivum* L) subjected to a period of water deficit. Plant Growth Regul 75:751–760

Zhu JK (ed) (2007) Plant salt stress eLS. John Wiley & Sons, Chichester. doi:10.1002/9780470015902.a0001300.pub2

Zhu XC, Song FB, Xu H (2010a) Influence of arbuscular mycorrhiza on lipid peroxidation and antioxidant enzyme activity of maize plants under temperature stress. Mycorrhiza 20:325–332

Zhu XC, Song FB, Xu HW (2010b) Arbuscular mycorrhizae improves low temperature stress in maize via alterations in host water status and photosynthesis. Plant Soil 331:129–137

Zhu XC, Song FB, Liu SQ et al (2011a) Effects of arbuscular mycorrhizal fungus on photosynthesis and water status of maize under high temperature stress. Plant Soil 346:189–199

Zhu XC, Song FB, Liu SQ (2011b) Arbuscular mycorrhiza impacts on drought stress of maize plants by lipid peroxidation, proline content and activity of antioxidant system. J Food Agric Environ 9:583–587

Zhu XC, Song FB, Liu SQ, Liu TD, Zhou X (2012) Arbuscular mycorrhizae improves photosynthesis and water status of *Zea mays* L under drought stress. Plant Soil Environ 58:186–191

Zou YN, Srivastava AK, Ni QD et al (2015) Disruption of mycorrhizal extraradical mycelium and changes in leaf water status and soil aggregate stability in rootbox-grown trifoliate orange. Front Microbiol 6:203

Zuccarini P (2007) Mycorrhizal infection ameliorates chlorophyll content and nutrient uptake of lettuce exposed to saline irrigation. Plant Soil Environ 53:283–289

Zuccarini P, Okurowska P (2008) Effects of mycorrhizal colonization and fertilization on growth and photosynthesis of sweet basil under salt stress. J Plant Nutrit 31:497–513

Zuo Q, Duan YH, Yang Y et al (2007) Source apportionment of polycyclic aromatic hydrocarbons in surface soil in Tianjin, China. Environ Pollut 147:303–310

Index

C.M.F. Vos, K. Kazan (eds.), *Belowground Defence Strategies in Plants*, Signaling and Communication in Plants, DOI 10.1007/978-3-319-42319-7

B

C

The manufacturer's authorised representative in the EU is Springer Nature Customer Service Centre GmbH, Europaplatz 3, 69115 Heidelberg, Germany. If you have any concerns regarding our products, please contact ProductSafety@springernature.com

Printed and bound by CPI Group (UK) Ltd, Croydon, CR0 4YY
23/09/2026
02221733-0008

Nakamura Y, Itakura J, Matsuzaki I (1995) Influence of the earthworm *Pheretima hilgendorfi* (Megascolecidae) on *Plasmodiophora brassicae* clubroot galls of cabbage seedlings in pot. Edaphologia 54:39–41
Narisawa K, Tokumasu S, Hashiba T (1998) Suppression of clubroot formation in Chinese cabbage by the root endophytic fungus, *Heteroconium chaetospira*. Plant Pathol 47:206–210
Narisawa K, Ohiki KT, Hashiba T (2000) Suppression of clubroot and Verticillium yellows in Chinese cabbage in the field by the root endophytic fungus, *Heteroconium chaetospira*. Plant Pathol 49:141–146
Narisawa K, Shimura M, Usuki F, Fukuhara S, Hashiba T (2005) Effects of pathogen density, soil moisture, and soil pH on biological control of clubroot in Chinese cabbage by *Heteroconium chaetospira*. Plant Dis 89:285–290
Niwa R, Nomura Y, Osaki M, Ezawa T (2008) Suppression of clubroot disease under neutral pH caused by inhibition of spore germination of *Plasmodiophora brassicae* in the rhizosphere. Plant Pathol 57:445–452
Päsold S, Ludwig-Müller J (2013) Clubroot (*Plasmodiophora brassicae*) formation in Arabidopsis is reduced after treatment with prohexadione-calcium, an inhibitor for oxoglutaric acid dependent dioxygenases. Plant Pathol 62:1357–1365
Päsold S, Siegel I, Seidel C, Ludwig-Müller J (2010) Flavonoid accumulation in *Arabidopsis thaliana* root galls caused by the obligate biotrophic pathogen *Plasmodiophora brassicae*. Mol Plant Pathol 11:545–562
Peng G, Hwang SF, McDonald MR, Lahlali R, Gossen BD, Hynes RH, Boyetchko SM (2010) Management of clubroot (*Plasmodiophora brassicae*) with microbial and synthetic fungicides. Phytopathology 100:S99
Peng G, McGregor L, Lahlali R, Gossen BD, Hwang SF, Adhikari KK, Strelkov SE, McDonald MR (2011) Potential biological control of clubroot on canola and crucifer vegetable crops. Plant Pathol 60:566–574
Peng G, Lahlali R, Hwang S-F, Pageau D, Hynes RK, McDonald MR, Gossen BD, Strelkov SE (2014) Crop rotation, cultivar resistance, and fungicides/biofungicides for managing clubroot (*Plasmodiophora brassicae*) on canola. Can J Plant Pathol 36(S1):99–112
Rademacher W (2000) Growth retardants: effects on gibberellin biosynthesis and other metabolic pathways. Annu Rev Plant Physiol Plant Biol 51:501–531
Rennie DC, Holtz MD, Turkington TK, Leboldus JM, Hwang S-F, Howard RJ, Strelkov SE (2015) Movement of *Plasmodiophora brassicae* resting spores in windblown dust. Can J Plant Pathol. doi:10.1080/07060661.2015.1036362
Robak J (1994) Crop rotation effect on clubroot disease decrease. Acta Hortic 371:223–226
Sakai R, Sato R, Niki H, Sakamura S (1970) Biological activity of phyllosinol, a phytotoxic compound isolated from a culture filtrate of *Phyllosticta* sp. Plant Cell Physiol 11:907–920
Schuller A, Kehr J, Ludwig-Müller J (2014) Laser microdissection coupled to transcriptional profiling of Arabidopsis roots inoculated by *Plasmodiophora brassicae* indicates a role for brassinosteroids in clubroot formation. Plant Cell Physiol 55:392–411
Schwelm A, Fogelqvist J, Knaust A, Jülke S, Lilja T, Bonilla-Rosso G, Karlsson M, Shevchenko A, Choi SR, Dhandapani V, Kim HG, Park JY, Lim YP, Ludwig-Müller J, Dixelius C (2015) The genome sequence of the plant pathogenic Rhizarian *Plasmodiophora brassicae* reveals insights in its biotrophic life cycle and ancestry of chitin synthases. Sci Rep. doi:10.1038/srep11153
Sharma K, Gossen BD, McDonald MR (2011) Effect of temperature on primary infection by *Plasmodiophora brassicae* and initiation of clubroot symptoms. Plant Pathol 60:830–838
Siemens J, Nagel M, Ludwig-Müller J, Sacristán MD (2002) The interaction of *Plasmodiophora brassicae* and *Arabidopsis thaliana*: parameters for disease quantification and screening of mutant lines. J Phytopathol 150:592–605
Siemens J, Keller I, Sarx J, Kunz S, Schuller A, Nagel W, Schmülling T, Parniske M, Ludwig-Müller J (2006) Transcriptome analysis of *Arabidopsis* clubroots indicate a key role for cytokinins in disease development. Mol Plant Microbe Interact 19:480–494

Kühne S, Burth U, Marx P (2006) Biologischer Pflanzenschutz im Freiland. Eugen Ulmer, Stuttgart

Lahlali R, Peng G (2014) Suppression of clubroot by *Clonostachys rosea* via antibiosis and induced host resistance. Plant Pathol 63:447–455

Lahlali R, Peng G, McGregor L, Gossen BD, Hwang SF, McDonald MR (2011) Mechanisms of the biofungicide Serenade (*Bacillus subtilis* GST713) in suppressing clubroot. Biocontrol Sci Technol 21:1351–1362

Lahlali R, Peng G, Gossen BD, McGregor L, Yu FQ, Hynes RK, Hwang S-F, McDonald MR, Boyetchko SM (2013) Evidence that the biofungicide Serenade (*Bacillus subtilis*) suppresses clubroot on canola via antibiosis and induced host resistance. Phytopathology 103:245–254

Lahlali R, McGregor L, Song T, Gossen BD, Narisawa K, Peng G (2014) *Heteroconium chaetospira* induces resistance to clubroot via upregulation of host genes involved in jasmonic acid, ethylene, and auxin biosynthesis. PLoS One 9(4):e94144

Lecourieux D, Ranjeva R, Pugin A (2006) Calcium in plant defence-signalling pathways. New Phytol 171:249–269

Lee SO, Choi GJ, Choi YH et al (2008) Isolation and characterization of endophytic actinomycetes from Chinese cabbage roots as antagonists to *Plasmodiophora brassicae*. J Microbiol Biotechnol 18:1741–1746

Lovelock DA, Donald CE, Conlan XA, Cahill DM (2013) Salicylic acid suppression of clubroot in broccoli (*Brassicae oleracea* var. *italica*) caused by the obligate biotroph *Plasmodiophora brassicae*. Australas Plant Pathol 42:141–153

Ludwig-Müller J (2009) Plant defence—what can we learn from clubroots? Australas Plant Pathol 38:318–324

Ludwig-Müller J, Kasperczyk N, Schubert B, Hilgenberg W (1995) Identification of salicylic acid in Chinese cabbage and its possible role during root infection with *Plasmodiophora brassicae*. Curr Top Phytochem 14:39–45

Ludwig-Müller J, Prinsen E, Rolfe S, Scholes J (2009) Metabolism and plant hormone action during the clubroot disease. J Plant Growth Regul 28:229–244

Ludwig-Müller J, Jülke S, Geiß K, Richter F, Sola I, Rusak G, Mithöfer A, Keenan S, Bulman S (2015) A novel methyltransferase from the intracellular pathogen *Plasmodiophora brassicae* methylates salicylic acid. Mol Plant Pathol 16:349–364

Mancuso S, Azzarello E, Mugnai S, Briand X (2006) Marine bioactive substances (IPA extract) improve foliar ion uptake and water tolerance in potted *Vitis vinifera* plants. Adv Hortic Sci 20:156–161

Metting W, Zimmerman J, Crouch IJ, Van Staden J (1990) Agronomic uses of seaweed and microalgae. In: Akatsuka I (ed) Introduction to applied phycology. SPB, The Hague, pp 269–307

Mithen R, Magrath R (1992) A contribution to the life history of *Plasmodiophora brassicae*: secondary plasmodia development in root galls of *Arabidopsis thaliana*. Mycol Res 96:877–885

Moody SA, Piearce TG, Dighton J (1996) Fate of some fungal spores associated with wheat straw decomposition on passage through the guts of *Lumbricus terrestris* and *Aporrectodea longa*. Soil Biol Biochem 28:533–537

Muday GK, Murphy AS (2002) An emerging model of auxin transport regulation. Plant Cell 14:293–299

Murakami H, Tsushima S, Akimoto T, Shishido Y (2001) Reduction of spore density of *Plasmodiophora brassicae* in soil by decoy plants. J Gen Plant Pathol 67:85–88

Murakami H, Tsushima S, Kuroyanagi Y, Shishido Y (2002) Reduction in resting spore density of *Plasmodiophora brassicae* and clubroot severity by liming. Soil Sci Plant Nutr 48:685–691

Myers DF, Campbell RN (1985) Lime and the control of clubroot of crucifers: effects of pH, calcium, magnesium and their interactions. Phytopathology 75:670–673

Naiki T, Dixon GR (1987) The effects of chemicals on developmental stages of *Plasmodiophora brassicae* clubroot. Plant Pathol 36:316–327